ENGINEERING MANAGEMENT

WILEY SERIES IN ENGINEERING & TECHNOLOGY MANAGMENT

Series Editor: Dundar F. Kocaoglu, Portland State University

PROJECT MANAGEMENT IN MANUFACTURING AND HIGH TECHNOLOGY OPERATIONS
Adedeji B. Badiru, University of Oklahoma

MANAGERIAL DECISIONS UNDER UNCERTAINTY: AN INTRODUCTION TO THE ANALYSIS OF DECISION MAKING
Bruce F. Baird, University of Utah

INTEGRATING INNOVATION AND TECHNOLOGY MANAGEMENT
Johnson A. Edosomwan, IBM Corporation

CASES IN ENGINEERING ECONOMY
Ted Eschenbach, University of Missouri-Rolla

MANAGEMENT OF ADVANCED MANUFACTURING TECHNOLOGY: STRATEGY, ORGANIZATION, AND INNOVATION
Donald Gerwin, Carleton University, Ottawa
Harvey Kolodny, University of Toronto, Ontario

MANAGEMENT OF RESEARCH AND DEVELOPMENT ORGANIZATIONS: MANAGING THE UNMANAGEABLE
Ravinder K. Jain, U.S. Army Corps of Engineers
Harry C. Triandis, University of Illinois at Urbana-Champaign

STATISTICAL QUALITY CONTROL FOR MANUFACTURING MANAGERS
William S. Messina, IBM Corporation

KNOWLEDGE BASED RISK MANAGEMENT IN ENGINEERING: A CASE STUDY IN HUMAN-COMPUTER COOPERATIVE SYSTEMS
Kiyoshi Niwa, Hitachi, Ltd. and Portland State University

MANAGING TECHNOLOGY IN THE DECENTRALIZED FIRM
Albert H. Rubenstein, Northwestern University

MANAGEMENT OF INNOVATION AND CHANGE
Yassin Sankar, Dalhousi University

PROFESSIONAL LIABILITY OF ARCHITECTS AND ENGINEERS
Harrison Streeter, University of Illinois at Urbana-Champaign

ENGINEERING ECONOMY FOR ENGINEERING MANAGERS: WITH COMPUTER APPLICATIONS
Turan Gonen, California State University

FORECASTING AND MANAGEMENT OF TECHNOLOGY
Alan L. Porter, Jerry Banks, Georgia Institute of Technology
A. Thomas Roper, Thomas W. Mason, Rose-Hulman Institute of Technology
Frederick A. Rossini, George Mason University
Bradley J. Wiederholt, Institute for Software Innovation

ENGINEERING MANAGEMENT: MANAGING EFFECTIVELY IN TECHNOLOGY-BASED ORGANIZATIONS
Hans J. Thamhain, Bentley College

ENGINEERING MANAGEMENT

Managing Effectively in Technology-Based Organizations

HANS J. THAMHAIN

A WILEY-INTERSCIENCE PUBLICATION
JOHN WILEY & SONS, INC.
NEW YORK CHICHESTER BRISBANE TORONTO SINGAPORE

Figures 1.3, 2.4, 2.5, 4.5, 4.6, 5.1, 5.2, 5.4, 5.6, 5.7, 5.8, 5.9, 5.10, 5.11, 6.2, 6.4, 6.6, 9.1, 13.2, 13.3, A2.2, A2.3, and A2.4 are reprinted with permission from Hans J. Thamhain, *Engineering Program Management*, Wiley, 1986.

This text is printed on acid-free paper.

Library of Congress Cataloging in Publication Data:

Thamhain, Hans J. (Hans Jurgen), 1936-
Engineering management/Hans J. Thamhain.
p. cm. — (Wiley series in engineering and management technology)
Includes bibliographical references and index.
ISBN 0-471-82801-7
1. Engineering—Management. I. Title. II. Series.
TA190.T45 1992 92-2722
620'.0068—dc20

Printed in the United States of America

10 9 8 7 6 5 4 3 2 1

PREFACE

Technology has become an important element of global competitiveness. As leaders in business and government recognize the critical role of technology, R&D and engineering managers are under pressure to achieve marketable results focusing on quality, cost, and speed. This requires effective planning, organization, and integration of complicated multidisciplinary activities across functional lines and a great deal of people skills.

Time after time, managers have told me that the biggest challenge they encounter is not so much in understanding and applying technologies at the functional level, but in integrating and transferring technologies into marketable products and services. This requires skills for building cross-functional teams and leading them toward desired results. It involves motivation, power and resource sharing, communications both horizontally and vertically, and the ability to manage conflict effectively. To get results, R&D and engineering managers must relate socially as well as technically. They must understand the cultural and value system of the organization for which they work. The days of managers who get by with only technical expertise or pure administrative skills are gone.

This book is written for professionals and managers who must function effectively in technology-oriented work environments such as R&D, engineering, product management, manufacturing, and field services. It is also designed as a text for college courses in technology and engineering management. The book is an attempt to link today's engineering management practices with modern administrative techniques as well as contemporary concepts of organizations and behavior. Managers and professionals at all levels should find this information useful for better understanding the complex set of interrelated variables involving the organization, its environment, the tasks, and the people. Such insight can help in fine-tuning leadership style, resource allocation, and organizational development activities, and ultimately help in the continuous improvement of our organizations necessary to compete effectively in today's complex global markets.

This text is the result of 8 years of formal field research in the area of engineering and technology management. It also integrates the observations and experiences of my 20 years of R&D, engineering, project, and technical management with ITT, Westinghouse, General Electric, and GTE, prior to my current teaching, research, and consulting career.

I would like to express my appreciation to the many colleagues who were instrumental in encouraging and nurturing the development of this book. Special thanks go to the large number of professionals who contributed via field studies valuable data to the formal research integrated in this book. Further, the development of this book was supported by an *Institute Fellowship Grant* from Bentley College and other resources from the *Center for International Business Education and Research (CIBER)*, a joint-research venture of Bentley College and Tufts University Fletcher School of Law and Diplomacy, partially funded by a grant from the U.S. Department of Education. All of these supports are recognized with great appreciation.

HANS J. THAMHAIN

Waltham, Massachusetts
September 1992

CONTENTS

ENGINEERING MANAGEMENT

1

MANAGING IN ENGINEERING: A PERSPECTIVE

1.1 CHALLENGES OF MANAGING IN ENGINEERING

Managing engineering activities has always had its challenges: the technical complexities associated with risks and uncertainties, the special tools and techniques, which often require unique skills, the organizational dynamics. Certainly leading a variety of specialized people has always presented challenges to those who manage in engineering.

However, today's engineering environment is more challenging than ever before. With today's increased technical complexity and competitive pressures, the breed of managers that has evolved must confront new problems in managing complex tasks. Within their company they must be able to operate in a multidisciplinary environment, which requires dealing effectively with a variety of interfaces and support personnel over whom they often have little or no formal authority. Beyond the company, the engineering manager has to cope with the constant and rapid changes in technology, markets, regulations, and socioeconomic factors. Moreover, engineering management often relies on special organizational systems, such as the matrix, which is characterized by horizontal and vertical lines of communications and control, resource-sharing among various task teams, multiple reporting relationships to several bosses, and dual accountability.

To manage effectively in such a dynamic and often unstructured environment, engineering managers must understand the interaction of organizational, technical, and behavioral variables in order to build a productive engineering team.

All of these challenges are interrelated. However, an attempt is made in this chapter to break down the intricacies of this bewildering array of challenges by grouping them into eleven categories, as summarized in Table 1.1. Each of these categories is discussed below.

TABLE 1.1 Challenges in Engineering Management

Challenges	Impact Areas
Complex Tasks	Skill requirements, risks, work force, organizational design, innovation and creativity, multifunctional direction
End-date Driven Schedules	Risks, creativity, conflict, quality, doability, make–buy decisions
Limited Resources	Priorities, planning, conflict, power struggle, resource-sharing, creativity
Changing Technology	Forecasting, risks, flexible planning, market success, profits, opportunities
Uncertainty and Risks	Engineering success, budgets, schedules, profits, dynamic planning, dynamic leadership, priorities
Obsolescence	Training, hiring practices, management style, computer-aided decision support, CAD/CAM profits, organizational survival
Multifunctional Team-building	Leadership, motivation, decision-making, conflict, power struggle, commitments, organizational design
Matrix Leadership	Resource-sharing, dual accountability, power-sharing, conflict, planning, authority relations, control
Limited Rewards	Leadership and motivation, conflict priority shifts, turnover
Innovation and Creativity	Leadership, team-building, market success, profits, conflict, power plays
Resource Competition	Conflict, planning, leadership, top management control, priorities, budgets, commitment

CAD, Computer-aided design; CAM, Computer-aided manufacturing, CAE, Computer-aided engineering.

Complex Tasks. Engineering tasks are more complex and multidisciplinary than nontechnical activities. Consider, for instance, the development of electronic equipment or the construction of an office building, which requires the integration and coordination of an enormous number of different activities and subsystems, all involving people from various functional units with different skills, objectives, and desires. Leadership in an engineering organization requires specific knowledge of the products and services, the technologies involved, the product applications and the business environment, as well as the ability to facilitate multifunctional decision-making, innovation, and creativity.

End-Date Driven Schedules. Deadlines are realities in any business. They have a special meaning, however, to engineering managers, who often must find alternate solutions to their technical problems without compromising the quality, safety, economy, and technical performance of their creation. This often requires additional risk-taking and innovative tradeoffs among available alternatives in the areas of materials, components, technology, design, and product features.

Limited Resources. Almost any assignment is easy if we are given unlimited time and resources. However, engineering managers not only face tight budgets, but also limited availability of specially skilled people, shifting priorities, resource-sharing, limited authority and control over needed resources that are found in other departments, and multiple objectives. All of these factors impinge on the net resources available to engineering managers to perform their complex roles.

Changing Technology. Engineering managers have a great concern for properly utilizing existing technology and for future technological advances vis à vis designing, manufacturing, and servicing their products. Predicting technological changes and directions is critically important to many strategic business decisions, ranging from research activities to produce development and manufacturing methods. Coping with the inevitably changing technologies encompasses both challenges and opportunities, which can ultimately impact the company's success in the marketplace.

Uncertainties and Risks. Many of the challenges discussed in this section—from increasing technological complexity and competitive pressures to on-time, on-budget performance—contribute to the uncertainties and risks associated with managing engineering activities. In addition, the business environment itself is often unpredictable and uncertain. Examples include the various political, economic, and social changes that took place in the recent past and will continue to take place in the future; multiple business objectives; government regulations; and dependence on subcontractors and vendors. These challenges are, of course, not entirely new. But managers in the past operated in a more stable environment, where they could choose one set of primary objectives and live with the resulting performance. Today's engineering managers often find themselves in a dynamic, unsettled environment where long-range objectives must be constantly reassessed and incrementally adjusted. In fact, these realities frequently have led to the collapse of distinction between long- and short-range planning, between line and staff, and between administrative, support, and operating functions of a company. All of these factors involve risks and uncertainties, which require highly dynamic operating practices and a managerial style that facilitates a change-oriented mentality at all levels.

Obsolescence. The half-life of engineering knowledge seems to be somewhere between four and six years, depending on which assessment you read. That is frightening. It requires the equivalent of part-time graduate-level continuing education studies just to keep up with current technological advances. Technical expertise is only one part of the engineering manager's skill inventory; a similar obsolescence factor exists on the managerial side. Consider, for instance, the advances made over the last five years in administrative areas such as word-processing, electronic mail, electronic calendars, telephone communications practices, copying techniques, audiovisual aids, and management information systems; or, engineering support systems ranging from PC-based statistical programs to full-scale CAD/CAM (computer-aided design/computer-aided manufacturing) systems. Still other areas that have just begun to make an impact on engineering management are knowledge-based systems, expert systems, artificial intelligence, and various forms of computer-aided decision support. However, the biggest challenge for keeping abreast appears to be skill development in the area of managing people—the ability to lead, facilitate, and motivate people toward established goals. Leadership style that was effective not too long ago doesn't seem to work any more. Engineering managers must gain new knowledge and develop skills in team-centered leadership, team-building, group decision-making, conflict management, management by dynamic objectives, management of change, dealing with new work ethics, new personal needs and motivators, as well as dealing with a new mix of people in the work force and an array of growing government regulations.

Multifunctional Team-Building. Engineering activities are too multidisciplinary to be structured strictly along functional lines. They involve multifunctional task integration, broad organizational involvement, and effective teamwork in all areas. Building an effective engineering team involves a whole spectrum of management skills to identify, commit, and integrate various individuals and task groups from functional support organizations into a single management system capable of executing a complex engineering undertaking. Although this process has been recognized for centuries, it has become more intricate and requires more specialized management skills as bureaucratic hierarchies have declined and matrices replace the more traditional forms of functional organizations.

Leadership in Unstructured Environments. Engineering organizations are shared-power systems. Task leaders and managers must frequently step across functional lines and deal with people with whom they have little or no formal authority. They form multifunctional teams and depend on support from many other organizational units and sources outside the company. Traditional power bases, such as authority, reward, and punishment, are often shared with others and must be augmented by credibility, expertise, and even friendship. Engineering managers must earn their authority; that is, build an image of trust, expertise, and sound decision-making.

Limited Rewards. Traditional rewards, such as salary increases, bonuses, and promotions, are very scarce; furthermore, managers have limited control over these rewards. Managers must pay attention to their people's needs and create rewards in addition to the limited set of rewards based on money, status, and position. These so-called incremental rewards are based on recognition, accomplishments, freedom, work challenge, and professional development. The effective role performance of today's engineering manager is in part critically based on his or her ability to carefully observe and accommodate to others' needs and to foster a professionally stimulating and interesting work environment. All of this requires the integration of a variety of skills into one homogeneous management system.

Innovation and Creativity. The winning edge of an engineering organization is its innovative capacity. All organizations that compete in the same markets have access to the same technology, the same parts, materials, and methods. They have even access to the same people. What makes one organization more successful over another is often traced back to innovation and creativity. It is important in all areas of an organization; it seems to be especially crucial in engineering, which directly influences the features, quality, economy, serviceability and value of a product. Building an environment where people work innovatively and creatively is a prime responsibility of the engineering manager. It is also a major challenge.

Resource Competition. The complex organizational relations, the often divided responsibilities, the ambiguities of accountability and authority result in a highly dynamic control over internal resources. Combining this with the external pressures for efficiency and market performance often leads to intense competition over available resources. These resources include in the most direct way budgets, people, and facilities. In addition, these resources are also controlled via priorities, project selection, and schedules. To avoid power struggles, conflict, and mismanagement, strong, unified team-oriented direction and leadership is needed from upper management. It further requires skillful planning and resource negotiations on the part of the engineering manager and continuous refueling of resource commitments through the involvement with other resource managers and upper management.

1.2 THE ENGINEERING ORGANIZATION

With all the fuss about engineering management challenges, one might ask: what is different about these organizations? One of the distinguishing characteristics is that technology is a key element. Consequently, engineering is usually a critical link in the overall business strategy and long-range business plan of a company, institution, or government unit. The engineering capabilities critically influence new product strategies, company charters, and missions. A second distinguishing characteristic is the rapid rate of change to which engineering organizations are subjected. Changes in technology, market needs, competitive responses, and skill requirements are very intense and occur at a high rate in the engineering environment. As a result, product life cycles are often shorter than in other businesses, research and advanced development efforts are higher, and business risks are substantial. For example, in the 1990s, 60% of General Electric's annual sales came from products and services that did not exist five years earlier. Equally impressive, the company currently spends $2.5 billion for research and development (R&D) in support of their engineering businesses. By the way, this is slightly more than their total net earnings! The statistics become even more impressive with the increasing technology orientation of the business. Texas Instruments, Hewlett-Packard, Data General, and other high technology businesses derive over 75% of their revenues from products introduced within the last five years. Successful management within these engineering organizations requires the ability to cope with change and to deal effectively with the associated risks. The message sent by GE in its 1990 annual report is typical for the new managerial awareness of this decade: "GE people today understand the pace of change, the need for speed, the absolute necessity of moving more quickly in everything we do, from inventory turnover, to product development cycles, to a faster response to customer needs."

A third characteristic of engineering organizations is that they are densely populated with technical personnel. These people are well educated, highly individualistic, creative, but often not the greatest team players. To be effective, engineering managers must have the skills to unify the team behind the technical and business objectives and to lead and integrate their efforts toward the desired results.

Yet another distinguishing characteristic is that engineering organizations are by and large loosely structured. Driven by the multidisciplinary task requirements and project orientation of engineering work, traditional bureaucratic organizational hierarchies have declined and multifunctional work groups have evolved, together with new organizational systems such as the matrix. Today, task leaders and engineering managers must continuously step across functional lines and deal with personnel over whom they have little or no "formal" authority. They must deal with multiple accountabilities, multiple bosses, resource- and power-sharing, and conflicts. To be effective in such a dynamic environment, engineering managers must understand the interaction of organizational and behavioral variables. In short, to lead and manage effectively, they must be social architects who understand the cultures and value systems of their organizations.

1.3 SKILL REQUIREMENTS FOR ENGINEERING MANAGEMENT

Managing today's engineering function requires specific skills in leadership and in technical, interpersonal, and administrative areas. These requirements are not entirely new, but have evolved since the beginning of industrialization. However, these skill requirements became more complex and more crucial to effective role performance as bureaucratic hierarchies declined and multidisciplinary work groups evolved. Starting with the evolution of matrix organizations as formal management systems in the 1960s, managers have expressed increasing concern and interest in the identification and development of engineering management skills. As a result, many studies have been conducted that investigate technical management skills in a general context, contributing to the theoretical and practical understanding of effective engineering management performance.[1]

When asked about the specific skills necessary to manage effectively in an engineering environment, every manager has his or her own set of skills that seem to be particularly crucial for effective role performance. However, managers generally focus on three categories: leadership skills, technical skills, and administrative skills. Many of the components that make up each skill category overlap with the other areas, as shown in Table 1.2. The significance of defining and categorizing engineer-

[1]For an extensive bibliography on technical and engineering management skill development, see H. J. Thamhain "Developing Engineering Program Management Skills," Chapter 22 in D. F. Kocaoglu, editor, *Handbook on Management of R&D and Engineering,* New York, Wiley, 1992.

TABLE 1.2 Components of Engineering Management Skills

LEADERSHIP SKILL COMPONENTS

- Ability to manage in unstructured work environment
- Clarity of management direction
- Defining clear objectives
- Understanding of the organization
- Motivating people
- Managing conflict
- Understanding of professional needs
- Creating personnel involvement at all levels
- Communicating, in written and oral mode
- Assisting in problem-solving
- Aiding group decision-making
- Building multidisciplinary teams
- Credibility
- Visibility
- Gaining upper management support and commitment
- Action-orientation, self-starter
- Eliciting commitment
- Building priority image

TECHNICAL SKILL COMPONENTS

- Ability to manage technology
- Understanding of technology and trends
- Understanding of market and product applications
- Communicating with technical personnel
- Fostering innovative environment
- Unifying the technical team
- Aiding problem-solving
- Facilitating tradeoffs
- System perspective
- Technical credibility
- Integrating technical, business, and human objectives
- Understanding engineering tools and support methods

ADMINISTRATIVE SKILL COMPONENTS

- Planning and organizing multifunctional programs
- Attracting and holding quality people
- Estimating and negotiating resources
- Working with other organizations
- Measuring work status, progress, and performance
- Scheduling multidisciplinary activities
- Understanding of policies and operating procedures
- Delegating effectively
- Communicating effectively, orally, and in writing
- Managing changes

ing management skills is twofold: First, it offers some insight into the specific profile of engineering management skills needed for effective role performance, which might stimulate ideas for management training and development. Secondly, it establishes a typology for possible future management research, such as a more detailed investigation and analysis of these skills. It is interesting to note that although engineering managers described these skills in a broad variety of terms, when the components in Table 1.2 were shown to them, the managers rated 90% or higher of the skill components to be important to their ability to manage effectively.[2]

Leadership Skills

Effective engineering leadership involves a whole spectrum of interpersonal skills and abilities, as shown in Table 1.2: giving clear direction and guidance; integrating multidisciplinary efforts; planning and eliciting commitments; using communication skills; assisting in problem-solving; dealing effectively with managers and support personnel across functional lines, often with little or no formal authority; using information-processing skills, collecting and filtering relevant data valid for decision-making in a dynamic environment; and integrating individual demands, requirements, and limitations into decisions that benefit the overall engineering task. Effective engineering leadership further involves the program manager's ability to resolve intergroup conflicts and to promote team-building—important factors in overall program performance.

In engineering, quality leadership depends heavily on the manager's personal experience, credibility, and understanding of the interaction of organizational and behavioral elements. The engineering manager must be a social architect; that is, must understand how the organization works and how to work with the organization. Organizational skills are particularly important during the start-up of a new engineering program, when the manager forms the new team by integrating people from many different disciplines into an effective work group. It requires far more than simply constructing another organizational chart. At a minimum, it requires defining the reporting relationships, responsibilities, lines of control, and informational needs of each team member. Supporting skills are in the area of planning, communication, conflict resolution, and senior management support.

Many of the underlying leadership skills shown in Table 1.2 involve four primary components and can be further developed by thoroughly understanding them: people skills; management style; technological understanding; and organizational culture. To be effective, managers must consider all facets of the job. They must understand the people, the task, the tools and the organization. Further, their style must be conducive to the innovative, high-performance demands of the engineering organization. They must understand the dynamics of their organizations so they are able to diagnose potential problems and the need for change. The days of the manager who gets by with technical expertise alone or with purely administrative skills are gone.

[2]H. J. Thamhain, "Managing Technology: The People Factor," *Technical and Skill Training*, August/September 1990.

Interpersonal skills and leadership are developed through actual experience. However, formal MBA-type training, special seminars, and cross-functional training can help to improve the skills needed by managers in an engineering environment.

Technical Skills

The engineering manager rarely has all the technical expertise to direct the multidisciplinary activities single-handed. Nor is it necessary or desirable to do so. It is essential, however, that the engineering manager understand the technologies and their trends, the markets, and the business environment in order to participate effectively in the search for integrated solutions and technological innovations. Without this understanding, the consequences of local decisions on the total program, the potential growth ramifications, and the relationship to other business opportunities cannot be foreseen by the manager. Furthermore, technical expertise is necessary to communicate effectively with the project team and to assess risks and make tradeoffs between cost, schedule, and technical issues.

Administrative Skills

Administrative skills are essential. The engineering manager must be experienced in planning, staffing, budgeting, scheduling, and other control techniques. When dealing with technical personnel, the problem is seldom to make people understand and work with administrative tools such as budgeting and scheduling, but to impress upon them that costs and schedules are equally as important as elegant technical solutions.

While it is important that managers understand the company's operating procedures and the available tools, it is often necessary for the engineering manager to free himself or herself from administrative details. Particularly in larger departments, engineering organizational skills are normally developed through progressive professional growth and project-oriented assignments. Frequently, engineering work requires technical feasibility analyses, program definitions, and bid proposals. These are excellent opportunities for on-the-job training and testing of individuals who are thinking of advancing into engineering management. Not only does it provide an opportunity for hands-on skill development, but it also allows management to observe the candidate and judge his or her capacity for managing in an engineering environment.

As summarized in Table 1.2, the skills needed to effectively administer and manage an engineering organization seem to develop primarily through understanding and mastery of:

- The multifunctional work planning process
- Modular and incremental planning
- Written and oral communications
- Resource-planning
- Program measurements

- Potential problems and contingencies
- Motivational needs and self-actualizing behavior
- Administrative support activities.

Some Performance Correlates

Additional insight has been gained by investigating the association between managerial skill levels and managerial performance. Prior studies by Gemmill and Thamhain[3] clearly show significant correlations between managerial style and technical performance. What is interesting is the relatively strong level of correlation between the two factors.[4] Specifically, the Kendall's tau statistics shown in Table 1.3 indicate that the better engineering managers are perceived by their superiors in managerial skills, the better is the technical team's rating regarding work involvement; concern for quality; willingness to change; conflict minimization; on-time/on-budget performance; and overall engineering team performance.

Table 1.3 also shows that leadership and technical skills seem to have a particularly favorable effect on direct measures of engineering management performance, such as on-time, on-budget performance and overall performance, as perceived by upper management, and are also favorably associated with indicators of technical team performance, such as work involvement, concern for quality, willingness to change, and conflict minimization. Taken together, the data support the managerial statements that leadership and technical expertise are crucial for managing engineering organizations.

How Learnable Are These Skills?

Additional investigations into the learnability of engineering management skills reveals some good news.[5] Engineering managers feel that the skills needed to perform effectively in leadership positions are mostly learnable. The largest source for developing these skills is experiential; that is, by old-fashioned on-the-job training. Second to experiential learning, skills can be developed via professional activities such as seminars, readings, professional meetings, and special workshops. The third distinct category for managerial skill development consists of formal schooling. In addition, managers identified special sources, such as mentoring, job changes, and special organizational development activities, for building engineering management skills.

[3]G. R. Gemmill and H. J. Thamhain, "The Effectiveness of Different Power Styles of Project Managers in Gaining Project Support," *IEEE Transactions on Engineering Management,* May 1973; and H. Thamhain "Managing Engineers Effectively," *IEEE Transactions on Engineering Management,* August 1983.
[4]Kendall's tau rank-order correlation has been used to measure the association. Upper management was asked to score their engineering managers regarding three skill categories: innovative performance, on-time/on-budget performance, and overall managerial performance. Engineering managers were asked to score their technical teams on the remaining factors. Then all managers and their technical teams were rank-ordered, based on the perceived scores, and tau coefficients calculated, as shown in Table 1.3.
[5]See H. J. Thamhain, "Managing Technology: The People Factor," *Technical and Skill Training,* August/September 1990.

TABLE 1.3 Kendall's Tau Correlation Between Managerial Skill Level and Engineering Team Performance

	Involvement of Personnel	Concern for Quality	Willingness to Change	Conflict Minimization	Innovative Performance	On-Time On-Budget Performance	Overall Team Performance
Leadership skills	.55**	.35**	.30*	.35*	.25*	.20	.45**
Technical skills	.50**	.35**	.40**	.20	.50**	.35**	.50**
Administrative skills	.10	.10	.20	.10	.15	.35**	.30*

*p<.05
**p<.01.

The latter category represents less than 10% of the overall skills. Managers identified only a relatively small portion of skills as "not learnable." The portion of skills the manager "must be born with" averaged 8% and varied depending on the particular category. To summarize, engineering managers point at the enormous wealth of information and sources available for building and developing the skills needed to perform effectively in today's demanding engineering environment.

A Final Note

Solutions to engineering problems depend on the effective management of the technology via a team of task specialists from various organizations. Engineering managers thus must have skill in integrating specialists from many disciplines into working teams. In addition, the manager must live with constant change and be able to cope with the pressures of the evolving work environment. He or she needs to understand the interaction of organizational and behavioral elements in order to exert the influence required to build an environment conducive to the team's motivation. This will foster a climate of active participation and minimal detrimental conflict. The effective flow of communications is one of the major factors determining the quality of the organizational environment. Key decisions should be communicated properly to all project-related personnel. Engineering managers also must provide a high degree of leadership capability in unstructured environments. With increasing task complexity and organizational complexity, sophisticated engineering management skills are required. The engineering manager must have the capacity to deal seriously and carefully with:

- Selecting team personnel and integrating them into the work group
- Establishing a professionally stimulating environment
- Planning and organizing multifunctional programs
- Collecting and processing information
- Inspiring confidence
- Facilitating group decision-making
- Providing team visibility and upper-management involvement.

All of these skills will help the engineering manager to develop credibility among the peer group, team members, senior management, and the customer community. As a leader with clearly defined goals and operational tactics, the engineering manager can convert much of natural conflict that arises into useful energy. Above all, these skills will help the engineering manager to become a social architect who understands his organization, its culture and value system, its environment, and its technology.

1.4 EVOLUTION OF MODERN ENGINEERING MANAGEMENT

Engineering management is not a new idea; it has been around for thousands of years. As shown in Table 1.4, managerial principles already were being followed by the

TABLE 1.4 Evolution of Engineering Management Concepts

Time Frame	Typical Engineering Activities and Accomplishments	Management Concepts
5000 B.C.	Construction (Sumerians)	Recordkeeping, planning, and organizing
3000 B.C.	Pyramids, ship building, China wall	Planning, measuring, controlling of project activities
0	Roman roads, buildings, aqueducts, war machinery	Central control, communication networks, autocratic leadership
1800s	Industrialization	Functional processes
1900s	Brooklyn Bridge, transatlantic cable, *Titanic*	Scientific management, behavioral and quantitative approaches
1950s	Large-scale computers, *Sputnik*	Project management (formal)
1960s	St. Lawrence Seaway and Power	Matrix management
1970s	*Apollo*	Theory Z and contingency approaches to management
1980s	Space shuttle, personal computers, rail transportation	Team concepts, networking
1990s and beyond	Robotics, artificial intelligence, very large integrated circuits, genetic engineering	Continuous improvement of all functions via people involvement at all levels (Kaizen)

Sumerians in 5000 B.C. The Egyptian Pyramids, the Great Wall of China, and Roman roads and aqueducts are living proof that engineering management dates back a long way. Historic artifacts clearly show that basic management tools such as planning and control techniques, task definition, budgeting, and scheduling were already known in these early times. What has changed gradually is the complexity and interdependency of technical tasks, requiring more multidisciplinary support and managerial ability to integrate them. Together with this increased complexity emerged a body of organized knowledge on organizations and management, which helped to guide managerial actions and decisions.

Drivers Toward Formal Management Practices

Management formally emerged as an occupation during the Industrial Revolution of the 1700s, with its concentration of resources. The tremendous growth in factory capacity and organizational complexity required various layers of overseers and led to the emergence of the managerial and administrative function. This was also the time when engineering was formally recognized as a functional entity of an organization such as manufacturing the model shop or the design department. During this time, Adam Smith established some of the classical principles of management, specialization and division of labor. These basic principles, which were most visibly and successfully applied by industrialists such as Henry Ford in the late 1800s and early 1900s, form the pillars of modern engineering practices. In the years that followed, three major schools of management thought evolved: the classical approach to management, starting around 1900; the behavioral approaches to management, starting around 1925; and the quantitative approach to management, starting around 1925. These three schools provided the conceptual framework for the modern approach to management, which started around 1960.

The Classical Approach

The classical approach to management emerged at the beginning of the 1900s with three branches: Scientific management (its best known contributor, Frederick Taylor), administrative principles (Henri Fayol), and bureaucratic organization (Max Weber). This was the first time that principles and guidelines of management were formally established. As summarized in Figure 1.1, the classical approach to management emphasizes the rational and economic side of people and suggests ways to organize and manage tasks accordingly. While the classical approach established some of the fundamental guidelines to organization and management that remain valid to date and that are being practiced in every well-run organization, it ignored the human factor. Rooted in the famous Hawthorne Studies at the Western Electric Company, several alternative approaches to management emerged in the 1930s.

The Behavioral Approaches

Behavioral approaches to management focus on the intrinsic motivators of people. These theories assume that to be productive, people at work must have the desire to achieve certain goals, must find personal fulfillment in their work, and must have their social needs satisfied. Some of the best known theories that evolved in the behavioral school of management include Maslow's hierarchy of needs, McGregor's Theory X–Y, Herzberg's dual factor theory, and Argyris' writings on personality and organization.

The importance of the behavioral approaches is that they posit an organizational environment and employee personality that we find most likely in engineering. We have to work in an open, dynamic environment; deal with complex tasks, group dynamics, and multivalued group decisions; and work with people who can be characterized by McGregor's Theory Y, such as, they are willing to work, willing to accept responsibility, and capable of self-direction, self-control, and creativity. These findings suggest a managerial style that emphasizes human relations, individual responsibility identification, self-actuation, and team development; in fact, a style described by today's literature as *participative management.*

Quantitative Approaches

Yet another approach evolved in parallel with the behavioral school of thought. To facilitate managerial decision-making, quantitative techniques developed, which formed today's well-known disciplines of operations research, management science, and quantitative analysis. They form the basis for quantitative techniques such as linear programming, queuing, and simulation. Today, together with powerful, easily ac-

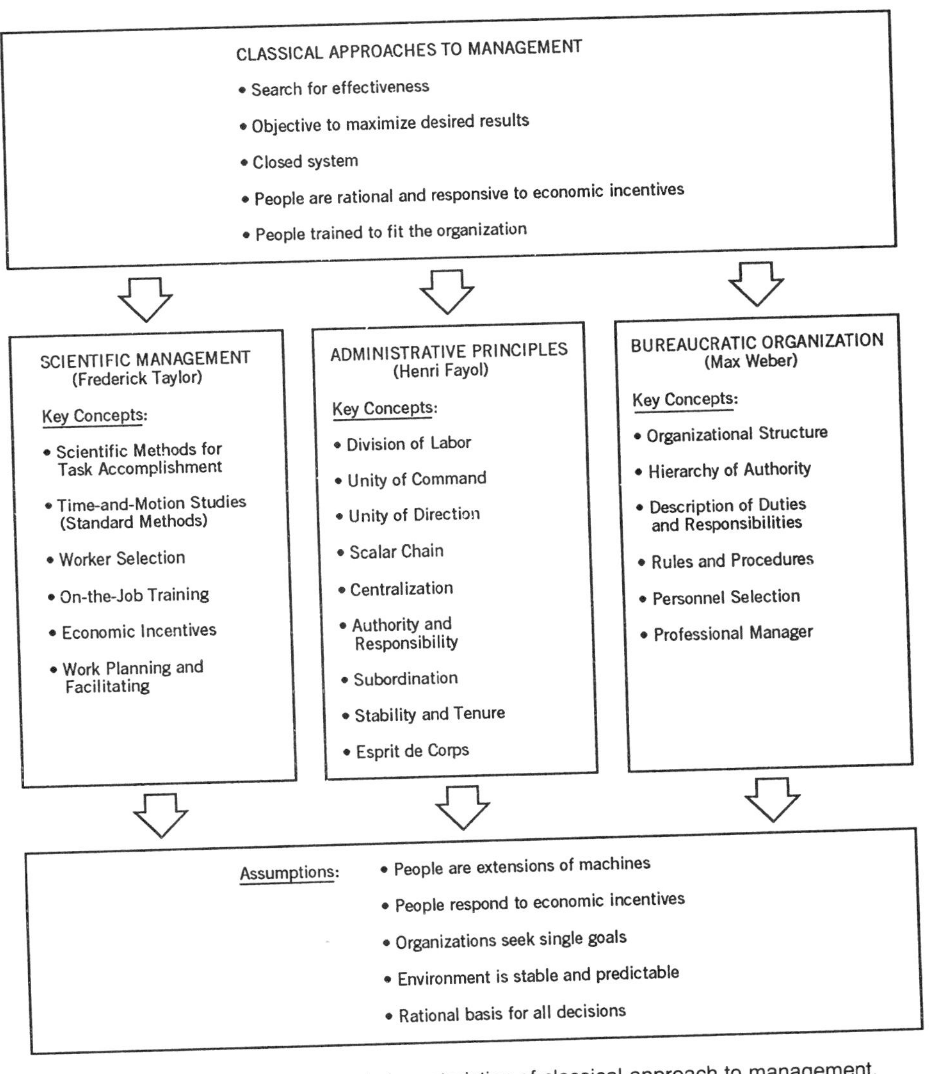

FIGURE 1.1 Components and characteristics of classical approach to management.

cessible computers, quantitative approaches play an important role in supporting managerial decision-making in all areas, including forecasting, financial analysis, production analysis, manufacturing resource planning, statistical process control, project evaluation and review techniques (PERT), and other decision support systems.

Modern Approaches to Management

Few practitioners could manage by using one particular management concept alone. Especially in engineering organizations, managers realize that the prescriptions of the classical, behavioral, and quantitative schools do not apply in every situation. Organizations are complex systems composed of psychological, sociological, economic, technical, and other elements, which are intricately interwoven, as symbolically shown in Figure 1.2. Recognizing these complexities, many practitioners and scholars have rejected attempts to neatly package conceptual components into one management theory. Such a package would be too simplistic and inflexible to respond to the realities of modern organizational systems.

In response to these realities, the modern approach to management evolved in the 1950s and gained acceptance in the 1960s. This approach builds from the base of contingency. No longer is there an attempt to find a single best way to manage. The modern approach views the total organization as a complex, open system whose various social, economic, and technical components interact with each other as well as with its external environment. This modern management approach leads to situational thinking, in which managers try to adapt their style to specific situations and constantly test the assumptions that underlie the various management guidelines provided (see Table 1.5).

One distinct difference between traditional and modern management theories is that the former are based primarily on a closed-system view, concentrating on the internal operation of the organization and trying to optimize a single value. The

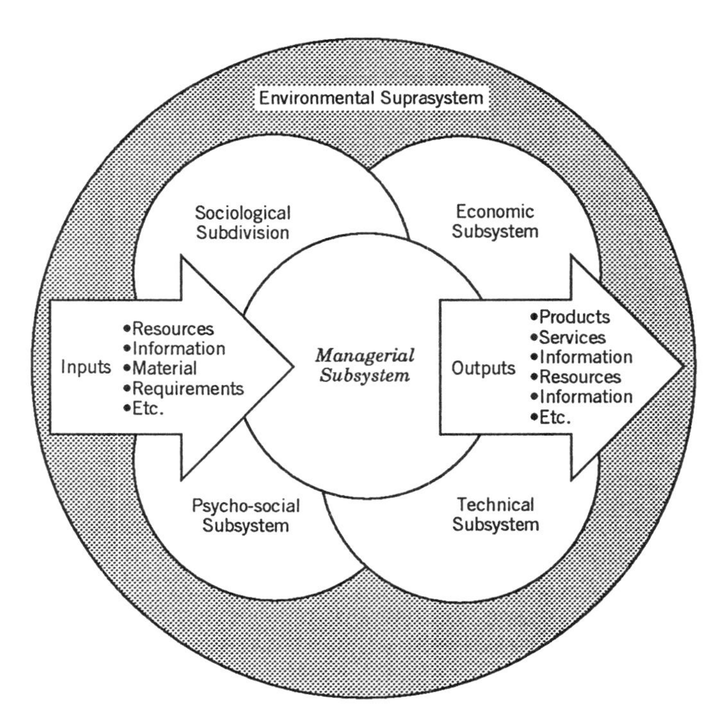

FIGURE 1.2 The various organizational subsystems as an integrated part of the environmental suprasystem.

TABLE 1.5 Characteristics of Two Contrasting Management Systems, Traditional versus Modern

Traditional Forms of Management (Stable Environment)	Modern Forms of Management Such as Engineering Management (Dynamic Environment)
Clear-cut tasks	Multidisciplinary tasks
Single clear-cut goals	Multivalued goals
Individual emphasis	Team focus
Authority by position	Earned authority
Strict division of labor	Multigroup dependencies
Conflict is undesirable	Conflict is unavoidable; maybe good
Stability-oriented	Change- and growth-oriented
Autocratic leadership	Participative, democratic leadership
People trained to fit organization	People shape the organization
Economic incentives	Economic and intrinsic rewards
Management controls	Commitment and self-actuating behavior

modern approach, on the other hand, takes an integrated systems view of the organization and its environment, which provides us with a macro paradigm for studying social organizations. It further accepts uncertainty and less deterministic approaches to problem-solving, as well as the complex and variable nature of human factors. As pointed out by Kast and Rosenzweig,[6] two of the pioneers who helped to formalize the modern approach to management, "One of the consequences of accepting systems concepts is the rejection of simplistic statements of universal principles for organization or management. Modern organization theory reflects a search for patterns and relationships, configurations among subsystems and a contingency view." It is this contingency view of management theory that has gained acceptance among today's managers, especially within engineering-oriented environments. Underlying this recent concept is the idea that for an organization to function effectively, the tasks, demands, and goals of the organization must be consistent with its structure, technology, external environment, and the needs of its people. Rather than searching for the one best way to organize under all conditions, managers tend more and more to examine the functioning of the organization relative to the needs of their members and the external requirements and pressures facing them.

Modern Engineering Practices

What do all these concepts really mean to engineering managers? Engineering executives were always more critical of the classical approaches to management than their counterparts in strictly administrative or production environments. Engineering

[6]F. E. Kast and J. E. Rosenzweig, *Contingency Views of Organization and Management,* Chicago: Science Research Associates, 1973.

always worked in a rather dynamic environment, associated with risks and uncertainties; an environment where people have to cross functional lines and solicit support from others over whom they have little or no authority. It is an environment that is usually driven by projects executed by task teams. These teams share resources, individual accountability, and power. This differs significantly from traditional management thinking, which stresses the importance of singular managerial accountability and control via scalar processes, the unity of command and chain of command principles. It is not surprising that some of the managerially more innovative engineering companies, both in the United States and abroad, were among the first to use modern management concepts and contingency views to organize and manage their businesses. As part of this innovative managerial thinking, many companies used the divisionalized structure to break down organizational complexities and to achieve improved accountability for business results and operational autonomy. Under pressure for business effectiveness during the 1960s, companies had opted for stronger functional organization rather than for contemporary systems, such as the matrix, which would have given them more flexibility in and adaptability to the fast-changing, complex business environment.

By going the functional route, managers could gain more economical use of their skilled personnel and achieve higher levels of training for their engineering, manufacturing, and marketing personnel. They attempted stronger technical leadership via centralized direction, and improved accountability for business results. In fact, by 1970, divisionalization was the prevalent organization structure among large companies.[7]

1.5 THE MATRIX APPROACH TO ENGINEERING

Gradually, another contemporary management concept emerged throughout the industrialized world. Many of its supporters claim that it is the first truly new approach to organization since product divisions were introduced in the 1920s. The matrix organization derives from the idea that managers can be accountable for business results although they have little or no direct control over the people and resources that support their activities. Operational responsibilities can, in effect, be divided along two axes, as shown in Figure 1.3. One axis contains all the responsibilities related to the management of the resources that are traditionally associated with the functional organization such as engineering or manufacturing, for example. The other axis contains the responsibilities associated with the management of the business. The business management axis, led by the program office, becomes an overlay to the functional organization. The program office is, in fact, contracting for specific services with the functional organization and managing the integration of the program.

This matrix approach to organization and management is not entirely new. Originally, companies that organized along matrix lines were typically in complex busi-

[7] A. R. Janger, *Matrix Organizations of Complex Businesses* (Research Report), New York: The Conference Board, 1979.

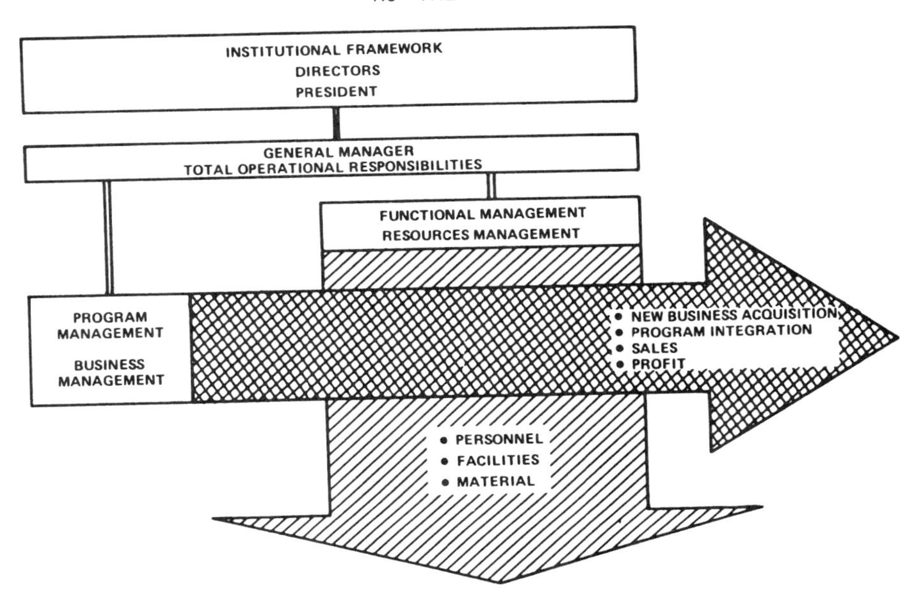

FIGURE 1.3 The two axes of matrix management.

ness or were engaged in engineering practices too multidisciplinary to be structured strictly on functional lines. Companies that are engaged in project work such as construction, research and development, aerospace, and defense have cross-functional coordination and dual management systems. However, its application to permanent organizational structures, rather than just to projects, is relatively recent and is still a major source of controversy and irritation among executives today.

For one thing, matrix management violates many of the fundamental ground rules of the traditional organization. While the traditional organization stresses the need for individual accountability, vested authority, and singular management control over personnel, matrix-organized businesses distribute managerial power among many managers, permit dual accountability and shared resources, count on personnel direction from various bosses, and have a management style that relies less on formal authority than on the manager's ability to deal across functional lines with personnel over whom he or she has little or no formal authority.

Despite the inevitable problems, intricacies, and understandable skepticism of executives, pressures were mounting as early as 1950 to find more suitable alternatives to strictly functional structures for managing engineering-oriented businesses and adapting to a more complex environment for the following reasons:

1. Technological undertakings had become too multidisciplinary to be structured strictly on functional lines.
2. Large program tasks interfered with the ongoing business of the functional organization.

3. Increasing work-force specialization necessitated more sharing of specialists throughout the organization.
4. The need for coordinating, integrating, and managing external activities increased beyond the capability of the traditional functional organization. Such activities include dealings with subcontractors, regulatory agencies, consumer groups, environmental concerns, customers, and user communities.
5. Associated with the increased engineering complexity, increased time and capital requirements often were tied to project specifications and mission requirements in a firm commitment with little flexibility.
6. Competitive pressures and increased task complexity gradually led to a specialized engineering function, organized along project lines with unique management systems and tools for planning, measuring, and control.

The major push toward further advances in matrix management came via project-oriented government business. With the success and publicity of the *Apollo* program, NASA, the Department of Defense, and other government agencies recognized the advantages of matrix management systems for their highly multidisciplinary technology undertakings. This resulted in pressures on companies doing business with the government to implement suitable project organizations and appropriate tools. Government contractors were the first to implement project management systems as an overlay to their engineering and manufacturing organizations. With this early lead, it is therefore not surprising that government contractors today often have the most advanced and mature engineering program organizations.

Program/Matrix Organizations and Divisionalized Structures

In responding to business realities, executives found that engineering project organizations are in fact compatible with divisionalized structures. A divisionalized company segment typically develops and produces products and services; a project-oriented company segment typically develops and produces a single product or service to particular customer specifications. Their organizational structures differ significantly. However, product companies often find similarities to program-oriented businesses that make the implementation of project organization in a divisionalized company compatible with the existing organizational structure:

- Both product and project organizations are overlays of business and functional resource organizations.
- Both product and project managers must cross functional lines to produce business results.
- Both product and project managers rely on decentralized decision-making.
- Both organizations rely on staff and support functions outside their organizational boundaries.
- Personnel in either organization often see themselves as having two bosses and having to live with dual accountability.

To the seasoned engineering executive, the problems of matrix organization are still acute today. Complex responsibility and personnel relationships, the inevitable dual accountability, and shared management powers make matrix organization a less robust and less organized management system than other functional counterparts. Organizational mortar is needed to hold this system together. The mortar is provided by clearly delineated organizational charters, policies, and directives, which define the organizational environment for all its members, including operating responsibilities, personnel relationships, reporting systems, personnel evaluation and reward system, and project controls, checks, and balances.

1.6 CAREER PERSPECTIVES

The emergence of technology as a dominant force in our lives makes engineering and its management a very desirable career choice for those who are willing to take on its challenges as well as enjoy its rewards.

A career in engineering management is not for everyone. It is also difficult to plan. Unlike pursuing a career in engineering, which has clearly prescribed milestones such as schooling, degrees, selection of employer, and "doing a good job," pursuing the managerial career seems to depend on a more subtle process. To be sure, it usually follows the same initial path into engineering. It is the advancement into *management* that seems to be difficult to achieve for some people, while it comes easily to others. As for any other job, we have to match career goals with individual capabilities and opportunities. Many engineers and scientists at the individual-contributor level are doubtful that a move into management would be a desirable career goal. Most dubious, of course, are those who have tried management and have returned to their individual-contributor positions. The undesirable aspects of engineering management often cited are:

- Long working hours
- Stress, anxiety, pressures
- Too much administrative work
- Loss of technical expertise
- Low job security
- Limited career mobility
- Too much politics
- Too much interpersonal conflict
- Limited opportunities for technological creativity.

There are, of course, many people in various firms and institutions who not only have successfully made the transition into engineering management, but also enjoy their work, the very rich variety of career challenges and opportunities, and the financial rewards that go with it. Those who enjoy their careers in engineering management or who aspire to such a career focus on the following rewards:

- Broader business responsibility
- Greater work challenges
- Freedom of action
- Chance to lead and motivate people
- Diversified career paths
- Opportunities for promotion
- Financial rewards
- High visibility and recognition
- Opportunity for creativity.

Whether or not a person likes or dislikes managerial jobs is not determined only by perception and attitude. Many researchers feel that intrinsic personal aptitudes, interests, and values play a major role in a person's career choices and desires.

Testing Managerial Potential

One good way to test managerial aptitudes and interests is via project management. Assignments to a particular task or project responsibility are made for a fixed time period—namely, the duration of the project. Thereafter, the individual returns to his or her previous position. This gives both management and the individual a chance to evaluate the individual's job performance on the prior assignment as well as his or her likes and dislikes regarding the new assignment. This process has a built-in fail-safe mechanism. In contrast to traditional management appointments, the assignment is not permanent. A reassignment of similar or higher-level responsibility depends on a mutually satisfactory performance assessment from the individual and his or her superior.

Career growth in engineering management can be effectively supported and enhanced by on-the-job training. A person interested in an engineering management career can develop the needed skills by taking an assignment as an assistant to a task manager, as a project engineer, or just by getting involved on a project team at any level. At more advanced stages, full project-management responsibilities are assigned. Furthermore, assignments can vary by project size, complexity, and duration to reflect the individual's experience level and career ambitions. Project assignments often provide excellent opportunities for the individual to gain a better understanding of the organization and its interfaces. The project activities cut across various functional lines, which requires dealing with a broad variety of personnel.

A Variety of Career Paths

Engineering management offers many career avenues. It provides opportunities for administrative as well as technical and business management positions. Tables 1.6–1.8 provide typical charters, job descriptions, and descriptions of skills and responsibilities of various managerial jobs in engineering. Career ladders in engineering

TABLE 1.6 Typical Engineering Management Positions and Their Responsibilities

Typical Managerial Responsibility	Primary Managerial Skill Requirements
Senior Engineer, Project Administrator, Project Coordinator, Task Manager, Project Engineer	
Coordinating and integrating subsystem tasks. Assisting in determining technical and personnel requirements, schedules, and budgets. Measuring and analyzing project performance regarding technical progress, schedules, and budgets	Technical expertise, planning, coordinating, analyzing, understanding the organization, assessing tradeoffs, managing task implementation, leading task specialists
Project Manager, Program Manager	
Same as above, but stronger role in project planning and controlling. Coordinating and negotiating requirements between sponsor and performing organizations. Bid proposal development and pricing. Establishing project organization and staffing. Overall leadership toward implementing project plan according to established schedule and budgets. Profit responsibility. New business development	Overall program leadership, team-building, resolving conflict, managing multidisciplinary tasks, planning and allocating resources, interfacing with customers/sponsors
Department Manager, Section Manager	
Establishes, develops, and deploys organizational resources according to business needs. Hires trains, and terminates personnel. Develops and maintains plant equipment. Directs functional personnel toward implementation of established program plans	Technical expertise and leadership, motivation, team-building, coordinating, communicating, administering, resolving conflict, developing personnel
Director of Programs, Vice-President of Program Development	
Responsible for managing multiprogram businesses via various project organizations, each led by a project manager. Focus is on business planning and development, profit performance, technology development, establishing policies and procedures, program management guidelines, personnel development, organizational development	Leadership, strategic planning, directing and managing program businesses, building organizations, selecting and developing key personnel, identifying and developing new business
Director of Engineering, Vice-President of Engineering	
Directs the activities and the development of the engineering organization according to established business plans. Focus is on directing overall organization toward desired business results. Customer liaison. Profit performance. New business development. Organizational development	Business leadership and management, building organizations, developing personnel, developing new business, strategic planning, communicating, social architecture

TABLE 1.7 Typical Charter of a Functional Engineering Manager

CHARTER: ENGINEERING MANAGER (ELECTRICAL DESIGN DEPARTMENT)

Position Title: Manager of Electrical Design

Authority: The functional manager of the electric design department has the delegated authority from the director of engineering to establish, develop, and deploy organizational resources according to the business needs, both long- and short-range. This authority includes the hiring, training, maintaining, and terminating of personnel; the development and maintenance of the physical plant, facilities, and equipment; and the direction of the functional personnel with regard to the execution and implementation of the various disciplines according to established plans, filling corporate engineering needs or specific customer project requirements.

Responsibilities: The functional manager is accountable to the director of engineering for all work within his or her functional areas in support of all business. He or she establishes organizational objectives and develops the resources needed for effective performance regarding the cost, schedule, and technical requirements of programs and activities executed within his or her department. Further seeks out and develops new methods and technology to prepare the company for future business opportunities.

management usually parallel those in other functions but additionally include experience in project management, with many crossovers. For example, an engineering professional might start a career as a designer, then take an assignment as a project engineer, later work as a project manager, and then return to a functional position such as product manager.

In addition, there are great opportunities to move across organizational lines from engineering into manufacturing, marketing, or field engineering. We see these moves more often today than in the past. They are more likely to occur in technology-oriented companies than in firms with more traditional businesses. Clearly, someone who has successfully managed a variety of different functions is often better positioned for vertical advancement toward general management than a person who stayed with one and the same function.

Taken together, engineering management can be a great career path for those who are willing to take the challenges and turn them into opportunities and rewards. It is also a career with tough prerequisites and skill requirements. Planning and preparing for such a career is a serious challenge, which is further discussed in Chapter 14, "Career Development in Engineering."

1.7 SOME NEW DIRECTIONS

Engineering is not deaf to the industrial nations' cry for increased productivity. Almost every issue of any management journal addresses some facet of the chal-

TABLE 1.8 Job Description of an Engineering Manager

JOB DESCRIPTION, ENGINEERING MANAGER, CUSTOM EQUIPMENT DESIGN

Function: Acts as project leader to originate, plan, and coordinate and/or supervise complete engineering projects related to the design and operation of petroleum and chemical facilities. Accepts complete responsibility for cost, time, and quality control on assigned projects.

Scope: Manager operates under the administrative supervision of the director of engineering. Receives little technical guidance on the technical aspects of the activity. The position is concerned with organizing, leading, and conducting engineering design projects for the parent company, its subsidiaries, and its affiliates relating to new construction and revamping of facilities, plants, and other locations, within the money and schedule allowed, and at a minimum cost consistent with recognized safety practices. The size or complexity of the design and development may require a task team approach; it may involve the use and technical direction of company or outside engineering, drafting, construction, and other needed personnel.

Design activities may include feasibility studies, economic justification, design, engineering, installation, inspection, startup, and initial operational check and repairs or any part thereof, as requested by sponsor. Task leader can either use own judgment regarding problems or consult with available specialists or consultants on certain phases of project.

Contacts are maintained with all departments and levels associated with the engineering activities and with contractors, engineering firms, governmental agencies, equipment manufacturers, etc.

Primary Duties:

1. Calls to attention of management opportunities for profitable engineering projects.
2. Plans the engineering scope of a project in time and money and secures necessary approval to proceed.
3. Discusses, with sponsors and others, the nature of the problem and design whatever facilities are necessary, sometimes with several alternative solutions.
4. Directs the technical efforts of others assigned to the project.
5. Estimates the investment and operating costs of such facilities and calculates the economic attractiveness.
6. Presents the results, conclusions, and appraisals to management for approval and for decision to proceed with facilities construction.
7. Instructs plant personnel on the proper use of such facilities.
8. Controls the expenditure of all project funds.
9. Reports status and progress of project regularly to project sponsor and to supervisors.
10. Keeps abreast of and advises on latest ideas, schemes, and equipment in manager's field of interest. Attends professional meeting, contacts manufacturers and suppliers, and consults with other interested parties both within and outside the company.
11. Maintains correct knowledge and adheres to engineering and company safety procedures, rules, and practices relating to a given project assignment.
12. Develops the equipment design function, its personnel and facilities to be at the appropriate state of the art, suitable for competitive operation in our type of business.

lenges faced by today's managers. However, unlike many other functions, engineering's overall productivity is difficult to quantify. Although attention to the operational level is important, many of the more senior executives believe that the engineering function must be designed, managed, and evaluated as an integrated part of the whole business or equivalent entity, such as a government agency. As these executives see it, there are four factors that are expected to put pressure on engineering organizations and their management to rethink the way they are currently operating. These factors may influence the leadership style of future managers, organizational structures, and most certainly will influence the skills required for effective managerial role performance. They are:

1. Emergence of new engineering support tools and techniques
2. Continued technological advances leading to easier-to-use materials and off-the-shelf system components
3. Continued organizational innovations toward better functional integration of engineering into the total organization, such as project management systems
4. Increased technical and managerial skill requirements and continuous managerial training and development. Let's briefly examine these trends and suggest ways to prepare for the future and take advantage of the underlying opportunities.

New Engineering Support Techniques

Driven by availability of powerful computers, from microcomputers to mainframes, many new tools and techniques emerged and have become widely available for supporting engineering activities on both the technical and administrative side. Most of these tools are not conceptually new; their wide-spread, low-cost availability at the individual-contributor level, as well as their user orientation and convenience, is new, however. These tools range from analytical support via spreadsheet and specialty software to simulations, CAD/CAM, material requisition planning (MRP), graphics, project tracking, and expert systems.

Engineering professionals who use these tools effectively typically cite several advantages. The professionals can do their jobs faster, more accurately, use fewer support people, yet unify a project team by using standard documentation and tracking systems; they are able to cope with higher complexities and become more adaptable to changes.

On the negative side, all of these tools require an initial investment of capital and time to install and learn the systems. They also require senior management commitment for continuous maintenance, support, and often large-scale integration with other functions and systems such as are needed for CAD/CAM or MRP. Further, the economic justification for any of these engineering tools is usually difficult. The addition of a new tool does not automatically guarantee improved operating efficiency. Problems can range from using the new tool as a toy to rejection or misuse leading to erroneous results, suboptimal decisions, or loss of creativity. To develop a new engineering support capability, management must truly understand its business, the task, and the technologies involved. It requires management to take an integrated systems approach to analyzing support requirements on a long-term basis in order to determine the engineering tools and support systems that will be most appropriate in the long run.

New Technological Advances

The availability of easier-to-use, more effective materials, and off-the-shelf components that perform more complex subsystem functions, such as integrated circuits and off-the-shelf software, has pushed engineering design and manufacturing more toward systems integration. In addition, the same technological advances lead to increased use of microprocessor electronics in previously mechanically engineered equipment such as automobiles, manufacturing systems, and household appliances. From an engineering design and manufacturing point of view, electronics not only replaces more and more mechanical components, but also introduces more advanced functions at lower cost, higher reliability, and with shorter design cycles. On the other hand, integrating electronic design into mechanical systems requires new skills, instrumentation, and facilities throughout the engineering organization, from R&D to manufacturing and field services. It also means a heavier dependence on a large number of outside suppliers for parts and service.

These trends are expected to continue. They affect many aspects of engineering management—the type of facilities and people needed in the future, product and service strategies, policies, and operating procedures, and, perhaps most importantly, the design of our engineering organizations.

New Organizational Designs

With an increase in functional complexity and a more competitive environment, pressures exist for better multidisciplinary integration, faster response times, and adaptability in a complex, uncertain business environment. In response to these realities, new organizational designs evolved, largely from the matrix, with the objective of achieving better integration. Examples are task teams, concurrent engineering, integral design/build, and various forms of organizational alliances. Much of today's organizational development (OD) is aimed at achieving better collaboration, teamwork, and integration among participants.

Ever since the evolution of formal project management systems in the 1960s, executives around the globe have taken a hard look around and are evaluating whether or not their organizations could benefit from the new structures, tools, and techniques available. Businesses with technology ventures first implemented formal project management systems; today, virtually every engineering function and many organizations in other segments of our society have taken advantage of the same management methodology and system that was originally developed for high-technology undertakings.

Currently, project management is most fully developed in organizations that are internally complex, in those with multinational involvement, or in those in the government and aerospace markets. On the other hand, there are many companies that have tried project management in various forms of depth and complexity but have abandoned the effort. Even those who regard project management as viable agree that there are many problems. However, organizational alternatives are limited. Many executives share the concept of *third generation management,*[8] that the business environment of the future will be even more dynamic and unpredictable, requiring organizations that are adaptable to rapidly changing conditions and responsive to specific problems to be solved. To be successful, organizations of the future will consist of temporary systems, organized around problems, to be solved by people of diverse professional skills. An increasing number of businesses will continue to replace rigid bureaucracy by more flexible temporary management systems. These temporary organizations will, however, continue to operate within a functional framework to provide the stability and efficiency needed for their ongoing businesses. The project management approach provides a powerful set of tools to engineering managers for tailoring organizational structures, procedures, and management systems to fit specific needs, preparing the organization for future survival and growth.

Contributions to the refinement of contemporary management systems, for example, project management and matrix organizations, come from many sources. It is delightful to see an increased emphasis on organization and management research focused on the engineering environment. The knowledge gained from both eclectic research and management practice will help to facilitate a more thorough understanding of the complex situations at hand. It will enable managers to design and fine-tune their engineering organizations to be most responsive to the environmental and technological challenges of the future.

Increased Skill Requirements

What all these changes to new organizational systems, new tools, techniques, and technological advances add up to is a more complex operational environment, which requires sophisticated management skills in leadership, administration, and technical areas. Skill requirements for today's engineering managers were discussed earlier in this chapter; however, a few comments about trends and professional development may be in order. A rather safe prediction is that skill requirements will be increasing even further.

Continuing advances in technology and engineering-support techniques, pressures for more effective operations, the influx of better-trained engineering personnel, and declining management hierarchies all provide pressures toward better-skilled managers. They further point toward additional leadership qualifications for senior technical staff below the managerial level, who have to coordinate tasks, build teams, and direct efforts across organizational lines through other people. Traditionally, all these hu-

[8]For a specific perspective on technology and engineering management, see P. A. Roussel, K. N. Saad, and T. J. Erickson, *Third Generation R&D Management,* Boston, Harvard Business School Press, 1991.

man interfaces were taken care of by line managers. However, with declining bureaucratic hierarchies and increased multidisciplinary complexities, human skills will become crucial to effective role performance of engineering personnel at any level. Therefore, professional development at all levels of the engineering function will be an ongoing activity in organizations of the future. Fundamentally, four types of training and development activities are seen to be useful: technical, administrative, interpersonal, and strategic activities. A combination of experiential skills development, classroom learning, professional information-sharing, simulated scenarios, and other methods can be used to develop the operating skills needed in engineering organizations of the future.

To be sure, these future organizations are not expected to become managerial training labs. Nor will they interrupt their operations to facilitate professional development, or wait until everyone is optimally trained. These practicing managers and their senior personnel must continue to function on a day-to-day basis while concurrently developing and updating the skills needed to function effectively in a probabilistic environment. What we predict is that managerial development will become a formal activity in *all* engineering firms, with clearly prescribed policies and long-range top management commitment similar to the practices that we find already in today's more progressive, and often larger, companies. What we further predict is that management development efforts will become more of an ongoing effort that is shared between the employee and the company, especially in terms of scope, initiative, and time.

BIBLIOGRAPHY

Adler, Paul S. and Aaron Shenhar, "Adapting Your Technology Base: The Organizational Challenge," *Sloan Management Review,* (Fall 1990).

Ait-El-Hadj, Smail, *Technoshifts,* Cambridge, MA: Productivity Press, 1991.

Aryee, Samuel, "Combating Obsolescence: Predictors of Technical Updating Among Engineers," *Journal of Engineering & Technology Management,* Volume 8, Number 2 (Aug 1991), pp. 103–109.

Ashton, W. Bradford, Bruce R. Kinzey, and Marvin E. Gunn, Jr., "A Structured Approach for Monitoring Science and Technology Developments," *International Journal of Technology Management,* Volume 6, Number 1 (1991), pp. 91–111.

Babcock, Daniel L., *Managing Engineering and Technology,* Englewood Cliffs, NJ: Prentice-Hall, 1991.

Badawy, Michael K., "Technology and Strategic Advantage," *Internaitonal Journal of Technology Management,* Volume 6, Number 2 (1991), pp. 205–215.

Bartlett, Christopher A. and Sumantra Ghoshal, "Matrix Management: Not a Structure, a Frame of Mind," *Harvard Business Review,* (July/Aug 1990), pp. 138–145.

Barton, Ron and Richard Bobst, "How to Manage the Risks of Technology," *Journal of Business Strategy,* (Nov/Dec 1988).

Bergen, S. A., *R&D Management: Managing Projects & New Products,* Cambridge, MA: Basil Blackwell, 1990.

Betz, Frederick, *Managing Technology,* Englewood Cliffs, NJ: Prentice-Hall, 1987.

Blanchard, Frederick L., *Engineering Project Management,* New York: Marcel Dekker, 1990.

Boxerman, Stuart B. and Ronald E. Gribbins, "Technology Management in the '90s," *Healthcare Executive,* Volume 6, Number 1 (Jan/Feb 1991), pp. 21–23.

Chapman, C. B., D. F. Cooper, and M. J. Page, *Management for Engineers,* New York: Wiley, 1987.

Cleland, David I., "The Age of Project Management," *Project Management Journal,* (Mar 1991).

DeSio, Robert W., "Management of Technology: A Prototype Graduate Program," *International Journal of Technology Management,* Volume 6, Numbers 1 & 2 (1991), pp. 51–58.

Drucker, Peter, "The Coming of the New Corporation," *Harvard Business Review,* (Jan/Feb 1988, pp. 45–53.

Erickson, Tamara J., Philip A. Russel, and Kamal Saad, "Third Generation R&D Management," *Prism,* Second Quarter (1991), pp. 5–17.

Fleming, Samuel C., "Using Technology for Competitive Advantage," *Research-Technology Management,* Volume 34, Number 5 (Sept/Oct 1991), pp. 38–41.

Gemmill, Gary R. and H. J. Thamhain "The Effectiveness of Different Power Styles of Project Managers in Gaining Project Support," *IEEE Transactions on Engineering Management,* Volume 34, Number 5 (Sep/Oct 1991), pp. 38–41.

Janger, A. R., *Matrix Organizations of Complex Businesses* (Research Report), New York: The Conference Board (1979).

Japan Human Relations Association (ed.), *Kaizen Teian-1,* Cambridge, MA: Productivity Press, 1992.

Kast, F. E. and J. E. Rosenzweig, *Contingency Views of Organization and Management,* Chicago: Science Research Associates (1973).

Katz, Ralph and Thomas J. Allen, "Project Performance and the Locus of Influence in the R&D Matrix," *Academy of Management Journal,* (Mar 1985).

Klimstra, Paul D. and Ann T. Raphael, "Integrating R&D and Business Strategy," *Research-Technology Management,* Volume 35, Number 1 (Jan/Feb 1992), pp. 22–28.

Michiyuki, Uenohara, "A Management View of Japanese Corporate R&D," *Research-Technology Management,* Volume 34, Number 6 (Nov/Dec 1991), pp. 17–23.

Nayak, P. Rauganath, "Managing Rapid Technological Developments," *Prism,* Second Quarter (1991), pp. 19–40.

Noori, Hamid, *Managing the Dynamics of New Technology,* Englewood Cliffs, NJ: Prentice-Hall, 1990.

Pinto, Jeffrey K. and Dennis P. Slevin, "Critical Success Factors in R&D Projects," *Research-Technology Management,* (Jan/Feb 1989).

Ramanathan, K., "Management of Technology: Issue of Management Skill and Effectiveness," *International Journal of Technology Management,* Volume 5, Number 4 (1990), pp. 409–422.

Riggs, Henry E., *Managing High-Technology Companies,* London: Lifetime Learning Publication (Wadsworth), 1983.

Rubenstein, Albert H., *Managing Technology in the Decentralized Firm,* New York: Wiley, 1989.

Roussel, Philip A., Karmal Saad, and Tamara J. Erickson, "The Evolution of Third Generation R&D," *Planning Review,* (Mar/Apr 1991).

Roussel, Philip A., Kamal Saad, and Tamara J. Erickson, *Third Generation R&D: Managing the Link to Corporate Strategy,* Boston: Harvard Business School Press, 1991.

Subramanian, D. K., " Managing Technology—The Japanese Approach," *Journal of Engineering and Technology Management,* Volume 6, Number 3 (May 1990), pp. 221–236.

Swierczek, Fredric William, "The Management of Technology: Human Resource and Organizational Issues," *International Journal of Technology Management,* Volume 6, Number 1 (1991), pp. 1–14.

Szakonyi, Robert, "101 Tips for Managing R&D More Effectively," *Research-Technology Management,* (Nov/Dec 1990).

Thamhain, Hans J., "Developing Engineering Program Management Skills", Chapter 22 in D. F. Kocaoglu, (ed.), *Handbook on Management of R&D and Engineering,* New York: Wiley, 1992.

Thamhain, Hans J., "From Engineer to Manager," *Training & Development,* Volume 45, Number 9 (Sep 1991), pp. 66–70.

Thamhain, Hans J., " Managing Technology: The People Factor," *Technical & Skills Training,* (Aug/Sept 1990).

Thamhain, Hans J., "Innovative Performance in Research, Development, and Engineering," *Engineering Management Journal,* (Mar 1990).

Thamhain, Hans J., "Managing Engineers Effectively," *IEEE Transactions on Engineering Management,* (Nov 1983).

Vessey, Joseph T., "Speed-to-Market Distinguishes the New Competitors," *Research-Technology Management,* Volume 34, Number 6 (Nov/Dec 1991), pp. 33–38.

Weimer, William A., "Education for Technology Management," *Research-Technology Management,* Volume 34, Number 3 (May/June 1991), pp. 40–45.

2
ORGANIZING THE ENGINEERING FUNCTION

2.1 ORGANIZATIONAL INTERDEPENDENCE

For many executives, engineering organizations are a bewildering array of management systems and techniques with one thing in common, an unconventional superior–subordinate relationship. In contrast to traditional businesses, engineering organizations are indeed more dynamic and intricate. They often focus on teamwork and the ability to coordinate work across functional boundaries. Consider, for instance, the management of a modern engineering development through the various phases of research, design, development, fabrication, testing, and installation. All work factors are interrelated and operate under the limited control of the engineering manager. Integration of the program requires commitment and coordination among all departments. Moreover, it involves subcontractors, vendors, government agencies, environmental considerations, licensing, and frequent interaction with the various sponsors or customers. The organization must have the ability to adapt to the changing work environment. Rigid rules, standards, procedures, and management processes are often ineffective. Managers must function in these engineering organizations with less formality and less line authority than their colleagues in more traditional environments. Communications do not always follow preestablished channels, and work directions do not necessarily follow the chain of command. Organizational interdependence is recognized at most levels and provides the catalyst for cooperation and teamwork.

During the 1960s and 1970s there was a strong push toward the introduction and development of new organizational structures. Management came to realize that engineering organizations must be dynamic and innovative in nature in order to cope with their demanding environments. These environmental factors evolved as a result of the increasing competitiveness of the market, changes in technology, and a requirement for better control of resources for multiproduct firms. As early as 1975, Grinnell and Apple[1] reported five general factors that indicated that a traditional structure might not be adequate for managing project-oriented engineering activities:

1. Management is satisfied with its technical skills, but projects are not meeting time, cost, and performance requirements.
2. There is a high commitment to getting engineering work done, but there are great fluctuations in how well performance specifications are met.
3. Highly talented engineering specialists feel exploited and misused.
4. Particular technical groups or individuals constantly blame each other for failure to meet specifications or delivery dates.
5. Engineering projects are on time, on budget, and to specifications, but groups and individuals are not satisfied with the achievement.

Unfortunately, many companies do not realize the necessity for organizational change until it is too late. For management it is often difficult to decide whether to seek solutions internally or externally. A typical example is in the area of new product management. Costs are continually rising, while the product life expectancy may be decreasing. Should emphasis be placed on lowering costs or on developing new products?

If we assume that an organizational system is composed of both human and technical resources, then we must analyze the sociotechnical subsystem whenever organizational changes are being considered. The social system is represented by the organization's personnel and their group behavior. The technical system includes the technology, materials, and machines necessary to perform the required tasks.

Behaviorists countered that there is no one best structure to meet the challenges of tomorrow's organizations. The structure used, however, must be one that optimizes company performance by achieving a balance between the social and the technical requirements. According to Sadler:[2]

> Since the relative influence of these (sociotechnical) factors changes from situation to situation, there can be no such thing as an ideal structure making for effectiveness in organizations of all kinds, or even appropriate to a single type of organization at different stages in its development. There are often real and important conflicts between the type of organizational structure called for if the tasks are to be achieved with minimum cost, and the structure that will be required if human beings are to have their needs satisfied. Considerable management judgement is called for when decisions are made as to the allocation of work activities to individuals and groups. High standardization of performance, high manpower utilization and other economic advantages

[1]S. K. Grinnell, and H. P. Apple, "When Two Bosses Are Better Than One," *Machine Design,* January 1975, pp. 84–87.
[2]Philip Sadler, "Designing an Organizational Structure," *Management International Review,* 1971, Vol. 11, No. 6.

associated with a high level of specialization and routinization of work have to be balanced against the possible effects of extreme specialization in lowering employee attitudes and motivation.

Even the simplest type of organizational change can induce major conflicts. The creation of a new position, the need for better planning, the lengthening or shortening of the span of control, the need for additional technology (knowledge), and centralization or decentralization can result in major changes in the sociotechnical subsystem. Argyris has defined five conditions that form the basis for implementing organizational change:[3]

1. Continuous and open access between individuals and groups
2. Free, reliable communication
3. Independence
4. Trust, risk-taking, and helping each other
5. Conflict is identified and managed effectively such that the destructive win–lose stances, with their accompanying polarization of views, are minimized.

Unfortunately, these conditions are difficult to create. There is a tendency toward conformity, mistrust, and lack of risk-taking among peers that results in focusing upon individual survival, requiring the seeking out of the scarce rewards, identifying one's self with a successful venture (be a hero), and being careful to avoid being blamed for or identified with a failure, thereby becoming a "bum." All these adaptive behaviors tend to induce low interpersonal competence and can lead the organization, over the long run, to become rigid, sticky, and less innovative, resulting in less than effective decisions, with even less internal commitment to decisions on the part of those involved.

Today, organizational structuring is a compromise between the traditional (classical) and the behavioral schools of thought; management must consider the needs of the individuals as well as the needs of the company. (For a detailed discussion of the classical concept of organization and management, see Chapter 1.) The organization is structured to manage both people and the work; but in the end, it's the people who perform the work.

2.2 WHY ORGANIZATIONAL STRUCTURES ARE CHANGING

For more than two centuries, the traditional management structure survived. However, starting with business developments in 1960s such as the rapid rate of change in technology and position in the marketplace, as well as increased stockholder demands, strains were created on the traditional management structure portrayed in Figure 2.1. Fifty years ago, companies could survive with only one or perhaps two

[3]Chris Argyris, "Today's Problems with Tomorrow's Organizations," *The Journal of Management Studies,* February 1967, pp. 31–55.

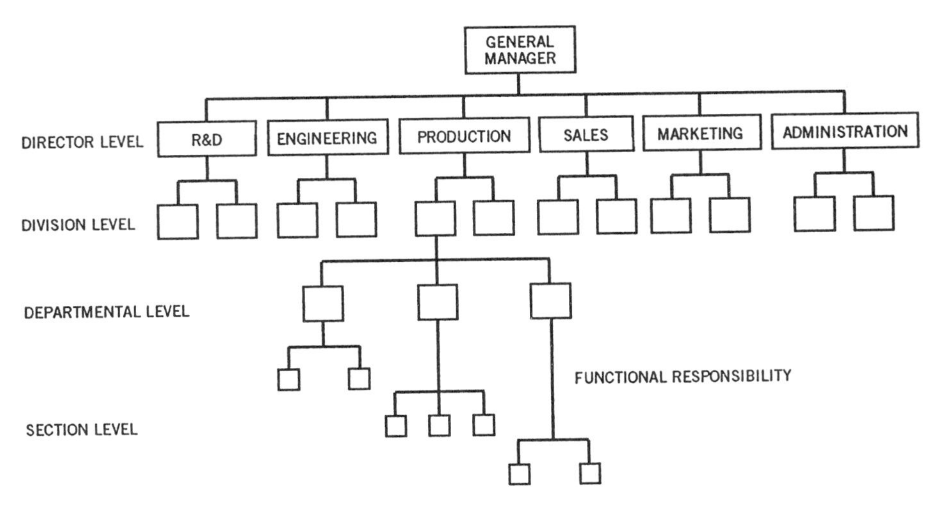

FIGURE 2.1 The traditional management structure of a functional organization.

product lines. The classical management organization was found to be satisfactory for control, and conflicts were at a minimum. (Many authors refer to classical organizations as *pure functional organizations.* Both names refer to a clear hierarchical structure of organizational units, as shown in Figure 2.1.)

However, as time elapsed, companies found that survival depended upon multiple product lines (i.e., on diversification) and on vigorous integration of technology into the existing organizations. As organizations grew and matured, managers found that company activities were not being integrated effectively and that new conflicts were arising in the well-established formal and informal channels. Managers began searching for more innovative organization forms that would alleviate the integration and conflict problems.

Before a valid comparison can be made with more contemporary organizational forms, the advantages and disadvantages of the traditional structure must be shown. Table 2.1 lists the advantages of the traditional organization. As seen in Figure 2.1, the general manager has beneath him all of the functional entities necessary to perform research and development (R&D) or manufacture a product. All activities are performed within the functional groups and are headed by a department head or, in some cases, a division head. Each department maintains a strong concentration of technical expertise. Since all of the project must flow through the functional departments, each project can benefit from the most advanced technology, thus making this organizational form well suited for mass production. Functional managers can hire a variety of specialists and provide them with easily definable paths for career progression.

The functional managers maintain absolute control over the budget. They establish their own budgets, upon approval from above, and specify requirements for additional personnel. Because the functional manager has manpower flexibility and a broad base from which to work, most projects are normally completed within cost.

In the traditional organization, both the formal and informal organizations are well established, and levels of authority and responsibility are clearly defined. Since each person reports to only one individual, communication channels are well structured. If the traditional structure has this many advantages, then why are we looking for other structures?

The traditional structure has many limitations and disadvantages for executing engineering work. As shown in Table 2.1, the majority of these are related to the fact that there is no strong central authority or individual responsible for project-oriented

TABLE 2.1 Advantages and Disadvantages of the Classical Organization

Advantages of the Classical/Traditional Organization

- Provides easier budgeting and cost control
- Provides better technical control
- Specialists can be grouped to share knowledge and responsibility
- Personnel can be used on many different projects
- All projects will benefit from the most advanced technology (better utilization of scarce personnel)
- Provides flexibility in the use of manpower
- Provides broad manpower base to work with
- Provides continuity in the functional disciplines; policies, procedures, and lines of responsibility are more easily defined and understandable
- Readily admits mass production activities within established specifications
- Provides good control over personnel since each employee has one and only one person to whom he or she reports
- Provides vertical and well-established communication channels
- Allows quick reaction capability, but reaction may be dependent upon the priorities of the functional managers

Disadvantages of the Traditional/Classical Organization

- No one individual is directly responsible for the total project (i.e., no formal authority; committee solutions)
- The project-oriented emphasis necessary to accomplish the project tasks is not provided
- Coordination becomes complex and additional lead time is required for approval of decisions
- Decisions normally favor the strongest functional groups
- There is no customer focal point
- Response to customer needs is slow
- There is difficulty in pinpointing responsibility; this is the result of little or no direct project reporting, very little project-oriented planning, and no project authority
- Motivation and innovation are decreased
- Ideas tend to be functionally oriented with little regard for ongoing projects

activities. As a result, integration of activities that cross functional lines becomes a difficult chore and top-level executives must get involved in the daily routines. Conflicts occur as each functional group struggles for power. The strongest functional group dominates the decision-making process. Functional managers tend to favor what is best for their functional group, rather than what is best for the project. Many times, ideas will remain functionally oriented with very little regard for ongoing projects. In addition, the decision-making process is slow and tedious.

Because there exists no central control point, all communications must be channeled through upper-level management. Upper-level management then acts as a clearinghouse for all problems and tries to resolve them by downward direction through the vertical chain of command to the functional managers. The response to any problem or customer need therefore becomes a slow and aggravating process because the information must be filtered through several layers of management. If problem-solving and coordination are required to cross functional lines, additional lead time is required for the approval of decisions. All tradeoff analysis must be accomplished through committees chaired by upper-level management.

Projects have a tendency to fall behind schedule in the classical organizational structure. Completing projects and tasks on time with a high degree of quality and efficient use of available resources is all but impossible within such a system. Large lead times are required. Functional managers attend first to those tasks that provide better benefits to themselves and their subordinates. Priorities may be dictated by requirements of the informal as well as the formal departmental structure.

2.3 FUNDAMENTALS OF ORGANIZING

As companies grew in size and technological complexity, more and more emphasis was placed upon the integration of multiple ongoing programs. Organizational pitfalls soon appeared, especially in the integration of the flow of work. As management discovered that the critical point in any program is the interface between functional units, the new concept of "interface management" developed. Let us discuss the fundamental concepts of organizational design and then examine the specific choices that organizations have in structuring their engineering functions.

Organizing is one of the principal functions of management. By definition, it involves the division of people and other resources into functional units that can perform the given tasks at hand most effectively. It is also one of the most challenging tasks of a manager, especially in the highly dynamic, multidisciplinary environment of engineering. This challenge is further reflected by the evolution of organization theory away from the earlier, simplistic views of closed systems toward recognition of the rich complexities of modern organizational systems. System concepts and contingency views provide the basis for understanding, designing, and managing today's organizations. Yet the very central issues remain unchanged, regardless of the type of organizational system, its complexity, or environment. These issues involve the division of work among organizational units and people, groups, departments, and divisions. They involve the definition of responsibilities and the process of coordination toward organizational objectives. In fact, these fundamental concepts of organizing date back to the classical theories influenced by the industrial engineering ideas of Frederick Taylor at the beginning of this century.[4] In this section we will state these organizational principles to develop a fundamental understanding of organizations and to lay the groundwork for designing today's complex and intricate engineering organizations.

The Elements of Organizing

The purpose and objective of organizing is to divide the work into manageable units. Regardless of the permanency of these functional units, organizing involves several basic elements, which must be carefully considered during the design of any organizational system.

1. Definition of Work. Management must be able to foresee the type, scope, and magnitude of work to be performed in the future.

2. Division of Work. The process of subdividing the work into doable and manageable units. The application of classical principles of management leads to a functional division; contemporary concepts lead project-oriented or task-oriented division.[5]

3. Division of Labor and Specialization. The subdivided work must be allocated to individuals and groups specialized in performing the function or task. It is via specialization that people develop particular skills and experience, and improve their performance effectiveness.

4. Authority and Responsibility. Each individual and each organizational unit must have clearly delineated authority defining the type and scope of work to be performed and how to work together, which includes the right to give orders and expend resources. Traditionally, organizational charts, charters, job descriptions, management directives, policies, and procedures are used to define authority and responsibility relations. In addition, task rosters, responsibility matrices, and policy directives are used to describe these relations for contemporary structures such as matrix organizations.

5. Unity of Command. One of the fundamentals of organization, originally delineated by Henri Fayol, is that an employee should receive orders from one superior only.[6] In spite of the dual accountability and power-sharing prevalent in contemporary organizations such as the matrix, each employee should be clear regarding what specific items he or she is responsible for and to whom. To establish a workable unity of command in contemporary organizations, management must employ some imag-

[4]See Frederick Taylor, *Shop Management,* New York: Harper & Brothers, 1903, and *The Principles of Scientific Management,* New York: Harper & Brothers, 1911.

[5]Yet another way to subdivide the work is via separation of different businesses, products, or markets. This method led to the introduction of the business division in the early 1920s and became the forerunner of strategic business units, divisionalization and conglomerated business organizations.

[6]Henri Fayol, *Industrial and General Administration,* Paris: DUAOD, 1925, and *Industrial and General Management,* London: Pitman & Sons, 1949.

ination and sophistication that goes beyond the traditional organization chart, including responsibility matrices and task descriptions that clearly define the various components of each task and function according to the various power and accountability axes of the organization.

6. Unity of Direction. Each set of activities must be clearly directed toward specific objectives, unified by one agreed-on plan and one leader. In a conventional, functional organization, this unity of direction is usually provided by the next-level supervisor through the unity of command channels. However, in a contemporary organization, task direction and leadership channels are more intricate. They often follow separate hierarchies, one for the functional and one for the project axis of the organization. It is the project axis of the organization that is usually responsible for the task integration and unity of direction, which is provided by the project manager via task leaders to individual team members. The management tool used for articulating integration responsibility and unified management direction is the policy directive or project charter.

7. Centralization. This is another one of Fayol's principles of management fundamental to an organization. It refers to the concentration of authority at the top level of an organization or any of its subsystems. The appropriate degree of centralization varies with the particular concern. In comparison with contemporary settings, this centralization is usually considerably stronger in traditional organizations.

8. Chain of Command. A chain of command is an unbroken line of authority, which links all persons to successively higher levels of authority. Every organization should be structured to provide a clear chain of command for every command axis of the organization. The process of following the chain of command for conveying orders and other forms of communication from higher to lower levels of authority is called the *scalar principle.*

The scalar principle ensures that all persons know to whom they report and from whom they should expect directions. A typical violation of the scalar principle occurs when, in an established chain of command, a higher-level manager bypasses a lower-level manager to give orders directly to the subordinates of that lower-level manager.

Although the chain of command and scalar principle were originally defined for, and work best for, the classic functional organization with its clearly structured hierarchy of authority, they apply also to contemporary organizations such as the matrix, as long as the basic chain of command can be defined for each axis of the organization. It should be observed, however, that today's engineering organizations, with their dynamic, often changing lines of responsibility and mostly task-centered team organizations, rely less on the vertical chain of command and the scalar principles in the classical sense, but more on decentralized group decision-making and communication networks, which are integrated through a network of task leaders. By comparison to the traditional organization, engineering organizations have a "permanent" chain of command and scalar principle established only for the vertical axis, or functional component, of their organization. The other overlying axes, such as the task, project, or committee axes, are highly temporary and their power structure is situational, rather than defined by the organization and established permanently.

9. Span of Control. Span of control is defined as the number of subordinates reporting directly to a manager. As with most other principles of organization, its definition dates back to the beginning of the 1920s, when the classical theory of organization and management was formulated. In today's organizations the span of control provides a conceptual reference point for structuring superior–subordinate and leader–follower relationships for the various organizational components within a functional unit, project team, or committee. The actual number of people in an ideal span of control cannot be calculated; certain factors help define the limits, however. These factors include:

a. Complexity of task or function supervised
b. Need for personal involvement (support) by leader
c. Similarity of tasks or functions supervised
d. Physical proximity of subordinate functions
e. Group/team characteristics of subordinate personnel
f. Degree of personnel administration versus task leadership only
g. Integration requirements among subfunctions
h. Overhead functions, responsibility such as planning, organizing, reporting, and staffing.

In engineering, we find a whole spectrum of leadership positions, each requiring a different span of control for effective operations. On one side of this spectrum there is the *functional manager,* who is responsible for the quality of the work and the effective organization and management of the resources. He or she is usually responsible, to some degree, for all factors in the above list. Therefore the span of control for managers in the functional layer of the engineering organization is relatively small, typically between 5 and 10 people. On the other side there is the *task manager,* who works in the project layer of the engineering organization. This supervisor is mostly concerned with the technical aspect of the task or project, its planning, developing, and building, but is less concerned with overhead functions such as resource development, organizational planning, and personnel maintenance.

Therefore, it is not surprising that the span of control for task leaders is usually broader, and might number up to 15 members within a well-developed task team.

Recent trends in American industry, driven by cost-cutting efforts, needs for better communications, and team-focused management, have led to the reduction of supervisory positions, often resulting in shorter chains of command and somewhat larger spans of control. It is a trend that is expected to continue for many more years to come.

The Role of Organizational Charts

There are few documents in an organization that are treated with more emotion and respect and carry more authority than an organizational chart. Yet, the same chart has been severely criticized by practitioners and scholars for its inflexibility and irrelevant presentation of real-world situations. It does not show the "informal organization" with its particular interfaces, culture, and value system. It does not really show who is in command of a particular program or mission and who are the real power brokers behind the position titles of the chart. Still another set of limitations derives from the organizational chart's inability to communicate the horizontal flow of information, directions, and transfer of resources, which eliminates the organizational charting of project activities.

With all of these limitations and shortcomings, why do companies spend so much time and energy drawing up organizational charts? This question seems especially relevant for engineering organizations, which require their operations to follow highly flexible, dynamic lines of authority and responsibility. They are often organized along networks of task teams, which cannot be charted very meaningfully using the stratified format of an organizational chart. In spite of all these limitations, the organizational chart serves several very important and useful functions, as summarized in Table 2.2. First, it provides a functional breakdown of the organization, dissolving its enormous complexity by showing its basic components, its lines of formal communication and command. In fact, the chart shows all the elements of the functional layer discussed in the previous section, such as division of work, authority and responsibility levels, chain of command, and span of control. Second, the organizational chart serves as a communications tool for showing who is who in the corporation. It is often used to familiarize new employees, customers, and business partners with the basic structure of the company and its key people. Third, the organizational chart fosters a "sense of belonging" for the people in each functional unit. People are proud to be part of the "Advanced Engineering Department" or the "Radar Systems Unit." The notion of a home office is legitimized by the organizational chart. Fourth, the organizational chart provides stability. It resembles a blue-

TABLE 2.2 The Role of Organizational Charts

Advantages	Limitations
• Shows functional subsystems	• Does not show informal organization
• Shows division of work level of authority and responsibility	• Overemphasizes status and position
• Shows level of management	• Cannot show power-sharing, dual accountability, and resource-sharing
• Functionally indicates reporting relations	• Does not show how the work flows
• Indicates chain of command	• Does not show how the work is performed and integrated
• Gives titles and positions	• Too static and inflexible
• Functionally shows span of control	
• Indicates vertical channels of communications and control	

print of the institutional framework of an organization, an area of slow change. As such, the organizational chart provides a stable framework for the various temporary organizational subsystems that are created and resolved on an ad hoc basis, a process typical for project-oriented activities.

Taken together, the organizational chart serves a very useful purpose as a management tool, defining and communicating the functional layer of an organization. Its limitations are basically in two areas. It shows only the functional structure of the organization and it does not explain the workings of the organization, in the same fashion that a circuit diagram does not explain the operation of an engineering system. To the competent engineering manager, the organizational chart is a very powerful tool that helps the manager in organizing and structuring the work and its environment. It helps in planning and executing the work assignment by aiding decisions on how to

- Divide the work into doable tasks or functions
- Budget and develop the necessary resources
- Recruit or assign the right people
- Build channels for communication, direction, and control
- Build support networks
- Integrate the work toward the final objectives.

For engineering managers, this type of organizational decision-making often results in an integrated management system built around the matrix structure. Obviously, none of the components of such a system are necessarily unique; they include elements from established organizational theories. The basic responsibility of every manager is to design and direct his or her work unit as productively as possible as part of the total organizational system. This requires the manager to be a social architect who understands not only his or her own work unit, but also the organizational interfaces and the culture, structure, and the subsystems of the organization as a whole.

Managerial Impact

As the increasing complexity and sophistication of modern engineering practices causes organizations to become more heterogeneous and change-oriented, managers will continue to experiment with contemporary organizational systems such as the matrix, with centralization versus decentralization, with conglomerate structures. Many of the concepts discussed in this chapter originated from the classical school of management. They form the pillars for understanding modern organizational theories and for fine-tuning today's management practices.

Organizational design should be kept simple. Often, traditional functional structures suffice where engineering tasks are routine and predictable. When activities are unique, or require group inputs and innovative solutions, more flexible organizations are needed to direct power and knowledge toward more creative decision-making.

Such situations most likely require team-centered approaches, which usually function with a matrix structure.

Centralized and decentralized structures should be balanced appropriately for effective organizational performance. In all of these choices it should be clear that modern organizations are not designed by polarizing their characteristics. Real-world organizations simply cannot be described as "centralized" or "decentralized," "closed" or "open." They have characteristics that are somewhere between these extremes. Accordingly, an organization leans more toward centralization if the following conditions are present:

- Close coordination of activities is required
- Resources can be pooled and shared
- The operation is capital-intensive
- Top-level strategy and direction are needed
- Advanced technology and skill developments are necessary
- Confidentiality of top-level decisions must be assured
- Clear channels of supply and distribution exist.

An organization leans towards decentralization under the following conditions:

- Operations and profit accountability can be separated
- Organization is large and in diversified businesses
- Resource-sharing is not critical
- The operation is labor-intensive
- The operation is project- or multiproduct-oriented
- Entrepreneurial drive, initiative, and commitment are needed from operating-level management.
- Fast reaction time in relation to the external business environment is necessary.

Early organizations were designed rather simplistically, following clear hierarchical structures of classical models. However, as the external environment became more turbulent, and internal operations became more dynamic and complex, the hierarchical model needed modification. The practical approaches to organizational design that evolved reflected the systems concepts and contingency views discussed in the previous chapter. They assist managers in the understanding of the interrelationships among the various subsystems of an organization, as well as the relationship between the organization and its environment. They emphasize the multivariate nature of organizations and direct managerial actions according to specific situations at hand.

Modern organization theory provides a general model or paradigm for the investigation of organizational designs. The concepts discussed here offer some guidelines for the structure of engineering organizations. Many of the conclusions are tentative and should be tested further in different situations. However, practicing managers in today's organization cannot wait until the ultimate body of knowledge emerges, but must function on a day-to-day basis. Currently, organization and management practices seem to confirm the contingency view. At the same time, the experiences gained in management practice have become an important input to the organization theory.

2.4 ORGANIZATIONAL CHOICES IN ENGINEERING

Fundamentally, organizations have two choices for structuring their operations: (1) functional or (2) projectized. The functional organization is the traditional structure, which is typically organized as shown in Figure 2.1. Resources are centralized and managed top-down through clearly established command and control channels. However, as a company engages in project activities, which require integration across functional lines, it establishes, formally or informally, horizontal channels of communication, command, and control. Therefore, as soon as a functionally organized company engages in such power- and resource-sharing, it operates in a *matrix* mode.

This transformation from a pure functional to a matrix structure, and eventually to a projectized organization, can also be expressed differently. As task leaders in a functional organization gain some power over resources in other departments, and as horizontal communication and command and control channels begin to form, the functional organization transforms into a *weak matrix.* As these horizontal channels become stronger, so does the matrix. Ultimately, task managers and project managers control all the resources needed to execute the project, which is known as a *projectized operation.*

Although there are situations that call for a purely functional or a purely projectized structure, most engineering organizations are too complex and dynamic to operate only at either end of the organizational spectrum. The matrix provides an elegant option for tailoring the organizational framework to specific organizational needs. It furthermore provides the flexibility for adjusting its organizational structure to changing requirements. The two basic forms of organization will be discussed below, with additional emphasis on matrix management.

The Functional Organization and Matrix

The engineering department can be organized functionally, that is, its company has a single engineering department that does all of the company's engineering work. There is only one of each specialty group, headed by a department manager or unit manager. Each resource is shared by all engineering activities that require it. When the engineering function has to perform several missions, tasks, or programs simultaneously, it automatically operates as a matrix, as is described below and shown symbolically in Figure 2.2.

An engineering function matrix is essentially an overlay of temporary project organizations on the functional engineering department. The engineering department

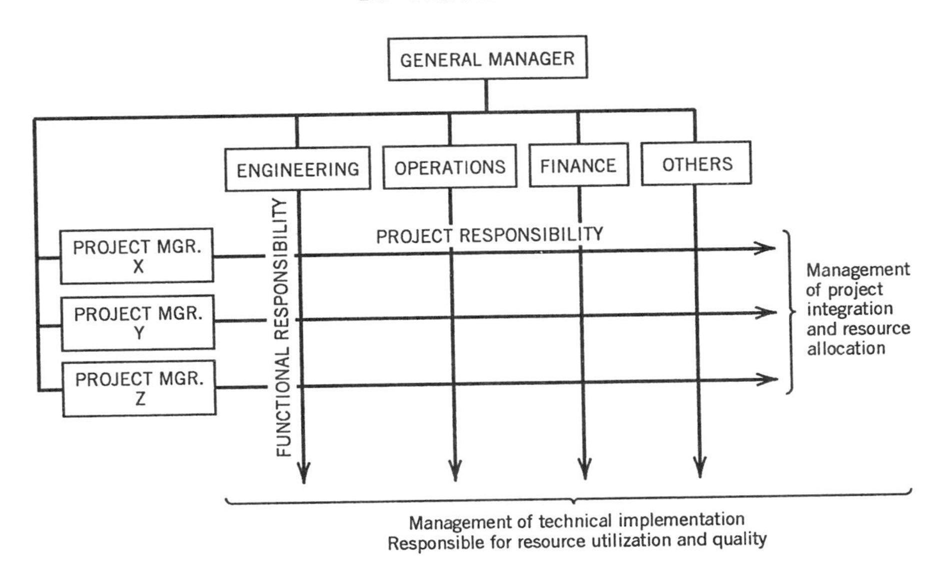

FIGURE 2.2 Pure matrix structure.

retains traditional functional characteristics. A clear chain of command exists for all organizational components within the engineering function. Department and unit managers direct the functional engineering groups. Task managers or project engineers perform the engineering-internal coordination and interface with external operating groups along established matrix lines.

The task managers essentially have two bosses; they are accountable (1) to the project manager, who may report to engineering or another function for the implementation of requirements, and (2) to the functional engineering manager for the quality of the work, the availability of necessary resources, and the use of these resources in accomplishing the desired results.

The advantages of this matrix overlay, summarized in Table 2.3, are a more rapid reaction time, better control, and more efficient integration of multidisciplinary activities. It also provides a high concentration of specialized resources and ensures their economical use. The additional integrator points, via engineering task/project managers, provide an effective interface with engineering-external organizations, while at the same time reducing the span of control needed by the program office.

On the other hand, the additional organizational overlay is likely to increase cost overhead and requires a more sophisticated management style in order to operate smoothly and efficiently within an environment that is characterized by extensive sharing of resources, power, and responsibility; the disadvantages are summarized in Table 2.3.

Furthermore, the operating environment of today's companies often includes additional dimensions, created by working across geographic, cultural, technological, and industrial boundaries, typical for companies conducting multinational business. Such a more complex typology is often reflected by additional axes of the matrix.

TABLE 2.3 Characteristics of the Matrix Organization

Advantages of a Pure Matrix Organizational Form

- The project manager maintains maximum project control over all resources, including cost and personnel (through the line managers)
- Policies and procedures can be set up independently for each project, provided that they do not contradict company policies and procedures
- The project manager has the authority to commit company resources, provided that scheduling does not cause conflicts with other projects
- Rapid responses are possible to changing conflict resolution and project needs
- The functional organizations exist primarily as support for the project
- Each person has a "home" after project completion. People are more susceptible to motivation and end-item identification. Each person can be shown a career path
- Because key people can be shared, program cost is minimized. People can work on a variety of problems (i.e., there is better people control)
- A strong technical base can be developed and much more time can be devoted to complex problem-solving. Knowledge is available to all projects on an equal basis
- Conflicts are minimal, and those requiring hierarchical referral are more easily resolved
- There is better balance between time, cost, and performance

Disadvantages of a Pure Matrix Organizational Form

- Company-wide, the organizational structure is not cost-effective because more people than necessary are required, primarily administrative
- Each project organization operates independently. Care must be taken that duplication of efforts does not occur
- More effort and time are needed initially to define policies and procedures
- Functional managers may be biased according to their own set of priorities
- Although rapid response time is possible for individual problem resolution, matrix response time is slow, especially on fast-moving projects
- Balance of power between functional and project organizations must be watched
- Balance of time, cost, and performance must be monitored

These regional, product, or technology axes overlay the functional organization in addition to other project axes. In describing these multidimensional matrices, executives usually reach for tools such as charters, directives, and policies other than conventional organizational charts, because the latter focuses on the primary command channels but often are not clear on the issues of dual accountability and multidimensional controls.

The Projectized (Aggregated) Organization

In the projectized organization, the resources of the engineering department, such as the people, facilities, and office space, are partitioned into project units. Each unit has a manager or leader, who is responsible for the proper management of that project. He or she also controls all the resources within that project unit without sharing them with other activities. This is shown in Figure 2.3 for a product structure and in Figure 2.4 for a project-oriented business organization.

The projectized or aggregated structure essentially organizes the engineering function into project groups. In its purest form, an engineering project manager is assigned for each project, directing a complete line organization with full authority over the personnel, facilities, and equipment, with the charter to manage one specific project from start to finish.[7] As summarized in Table 2.4, the projectized form represents the strongest form of project authority; it encourages performance, schedule, and cost tradeoffs; usually represents the best interfaces to engineering-external organizations; and has the best reaction time. However, by comparison with the matrix, the projectized organization usually requires considerable start-up and phase-out efforts and offers less opportunity to share production elements, to use economics of scale, and to balance workloads.

Because of these limitations, companies seldom projectize their engineering departments unless these projects are large enough to fully use the dedicated resources. What is more common is a partially projectized organization. That is, the project manager may fully control some resources that are particularly critical to the project and/or can be fully used over the project's life cycle, while other resources remain under the control of functional managers, who allocate them to specific projects as needed, based on negotiated agreements with the project managers.

2.5 REAL-WORLD HYBRIDS

For real-world engineering organizations, the choice among functional, divisionalized, projectized, or matrix organizations is not simple.

While from the top down a company that is in a number of unrelated businesses, such as a conglomerate, might find a divisional structure an obvious choice, the

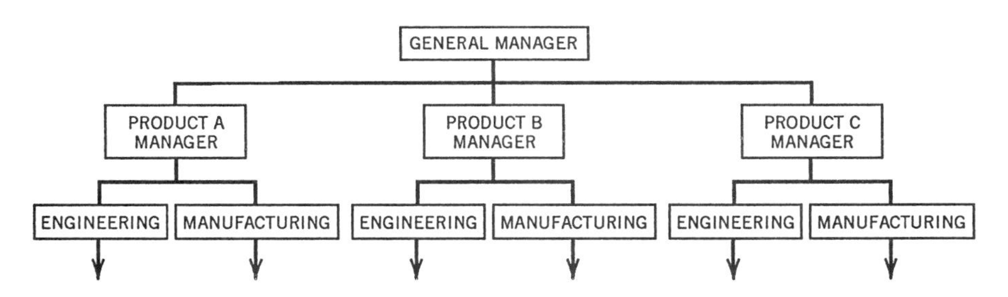

FIGURE 2.3 Pure product structure.

[7]For a complete set of procedural documents applicable to engineering project-orientated organizations see Harold Kerzner and Hans Thamhain, *Project Management Operating Guidelines: Directives, Procedures and Forms,* New York: Van Nostrand Reinhold, 1985.

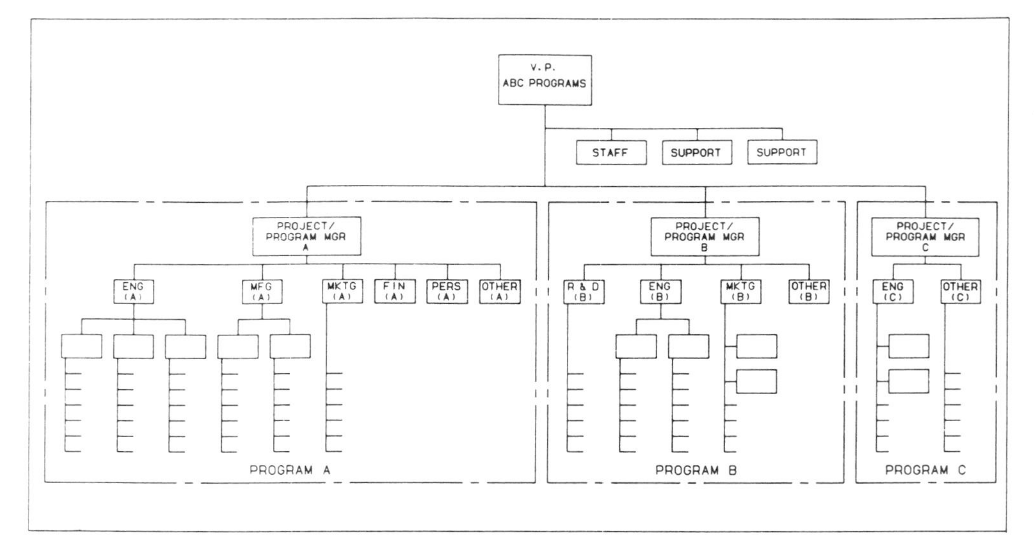

Advantages
- Strong control by single project authority
- Very rapid reaction time
- Encourages performance schedule and cost tradeoffs
- Personnel loyal only to project
- Interfaces well with out-of-company contacts

Disadvantages
- Inefficient use of production elements
- Does not develop technology for perpetuation
- Does not prepare for future business
- Less opportunity for technical interchange among projects
- Lack of career continuity for project personnel
- Difficulty in balancing work loads as projects phase in and out

FIGURE 2.4 Projectized (aggregated) structure typical of project-oriented business.

engineering departments within each division might operate as matrix, with a bewildering array of temporary organizational hybrids overlying the basic matrix structure. In a similar fashion, for a company that is essentially in one business, with products following a single engineering, production, and distribution process, the appropriate choice is most likely a functional organization such as is shown in Figure 2.1. However, at the same time, its engineering department may operate along matrix lines. Yet another example is a company in a strictly project-oriented business, such as we find in the construction, defense, and aerospace industries. Here the engineering department, and sometimes the whole company, is organized along projectized lines, as shown Figure 2.3.

Yet, the choice is even more difficult for executives whose companies are involved in a mixture of functional and multidisciplinary activities, or are multinational businesses. Their operations often involve long- and short-range engineering projects for outside customers as well as internal product and technology development, in addition to production and field support services. For these engineering executives, the choice is not simple. In fact, they cannot pick and choose from neatly defined organizational structures, but must design their own hybrid structures that are most conducive to their operational needs.

As these companies handle a mixture of large, small, short-range and long-range projects for internal as well as external clients through their engineering functions, they must have the ability to adapt continuously to changing multiproject requirements. Companies that fall into this category have often opted for hybrid structures that overlie the fundamental engineering organization with various organizational subsystems. These overlays may projectize certain engineering resources or establish miniature matrices within engineering, leading to so-called multidimensional matrices that are quite common for complex businesses. They also accommodate the widest

TABLE 2.4 Characteristics of the Projectized (Aggregated) Organization

Advantages of the Projectized Organizational Form

- This form provides complete line authority over the project (i.e., strong control through a single project authority)
- The project participants work directly for the project manager; unprofitable product lines are more easily identified and can be eliminated
- There are strong communications channels
- Expertise can be maintained on a given project without sharing of key personnel
- Very rapid reaction time is provided
- Personnel demonstrate loyalty to the project; better morale with product identification
- A focal point can be developed for out-of-company customer relations
- There is flexibility in determining time (schedule), cost, and performance tradeoffs
- Interface management becomes easier as a unit size is decreased
- Upper-level management maintains more free time for executive decision-making

Disdavantages of the Projectized Organizational Form

- Cost of maintaining this form in a mulitproduct company can be prohibitive due to duplication of effort, facilities, and personnel; inefficient usage
- There exists a tendency to retain personnel on a project long after they are needed. Upper-level management must balance work loads as projects start up and are phased out
- Technology suffers because, without strong functional groups, outlook of the future to improve company's capabilities for new programs is hampered (i.e., no perpetuation of technology)
- Interdisciplinary control of functional specialists is difficult and requires top-level coordination
- There is a lack of opportunities for technical interchange between projects
- There is a lack of career continuity and opportunities for project personnel

span of engineering activities with great flexibility while retaining most of the conventional functional stability and resource effectiveness.

These real-world project organizations are complex, both on paper and in practice. They must be carefully designed to accommodate the needed intrafunctional integration while maintaining effective use of engineering resources.

In many cases the executives who designed these engineering organizations are unable to classify their creations simplistically, but speak of overlays among various organizational forms. A typical complex project organization is illustrated in Figure 2.5 and in Table 2.5, showing the structure of GTE's Communications Systems division, a high-technology business operated along project lines. Although the overall structure is basically a matrix, we find many overlays of individual project, staff project, and projectized organizations. An individual project organization is a one-person program office chartered with the coordination of a single project across functional lines. It is simple, efficient, and quick to establish and to disband. It often exists as an overlay to other major organizational structures, such as the matrix or projectized organization. A staff project organization is a minimatrix, similar in characteristics to the individual project organization, but somewhat larger in resources. As the name implies, the staff project organization consists of a project manager with a small number of staff people reporting directly to him or her. Specifically, the right-hand side of Figure 2.5 represents the functional organizations, with directors heading up each of the disciplines. These executives are the resource managers who control the engineering, research, and operations facilities, the materials, and the services; above all, they manage the personnel who do the work, including hiring, work assignments, and professional development.

Overlying this functional organization is the project organization. Its program offices are chartered with the integration and management of the individual programs. Program managers report to a business area manager, who has the ultimate responsibility for developing and running the business sector according to established plans. However, depending on the size and nature of their programs, these managers have different charters, powers, and authorities, which are most visibly expressed by job titles. These titles range from *project administrator* for smaller programs to *project engineer, project manager, program manager,* and *executive program director.* The larger the programs, the more support staff is directly associated with the program office. Conversely, for very small jobs one project manager may handle several contracts. The amount of resources directly under the control of a program office varies with the type and size of the program.

For example, referring to Figure 2.5, Program I is fairly large and long-term. While all resources are managed by functional managers, the facilities and personnel from engineering, operations, and finance are assigned to the program office for the duration of the program. Other functional resources such as marketing, contracts, and personnel are shared with the other programs and made available only when needed. Figure 2.5 shows the dedicated resources as zoned-in areas, while shared resources are indicated by bullets connected with dashed lines to the program office. In contrast to Program I, Program II has a much weaker charter. Except for a dedicated financial specialist, all resources are shared with other programs.

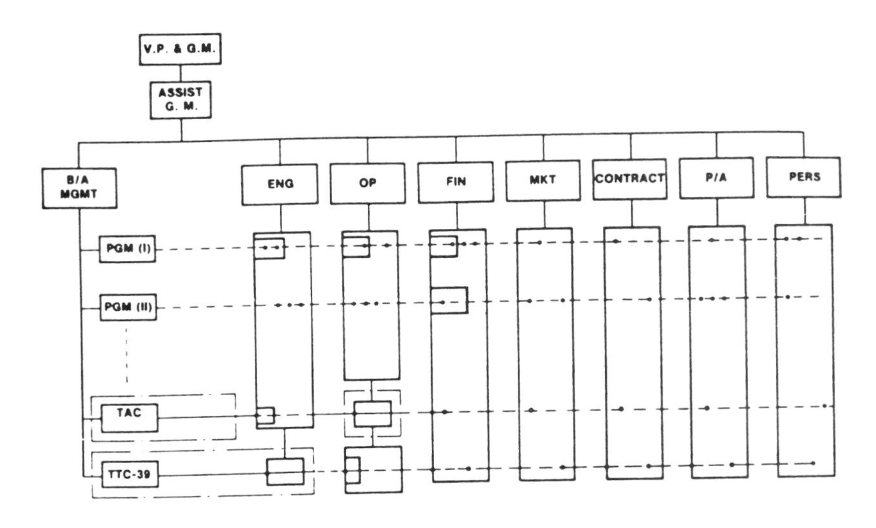

CHARACTERISTICS :

- Program organization is an overlay to functional organization
- Program manager is entirely responsible for project success
- Program manager reports to one of five business area managers
- Responsibility for functions/resources remains with functional managers
- Program responsibility within each functional area is assigned to a task manager who reports to both the functional manager, regarding the quality of the work, and to the program manager regarding implementation of the requirements, schedules, and budgets.

LEGEND :

B/A	Business Area
GM	General Manager
DP	Operations
P/A	Product Assurance
PERS	Personnel
PGM	Program or Program Office
TAC, TTC	Special Engineering Programs
VP	Vice President

PGM — Program office directs activities in functional area without personnel reporting to the program manager. Also, personnel may share their time and effort with other programs.

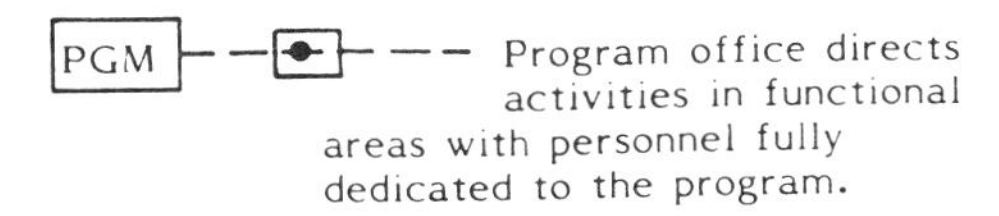

Program office directs activities in functional areas with personnel fully dedicated to the program.

FIGURE 2.5 Hybrid organization structured along matrix lines; business areas overlay the functional organization. B/A, business area; GM, general manager, OP, operations; P/A, product assurance; PERS, personnel, PGM, program or program office; TAC, TTC special engineering programs; VP, vice-president.

TABLE 2.5 Characteristics and Individual Responsibilities of Hybrid Organization

CHARACTERISTICS OF HYBRID ORGANIZATION

- Program organization is an overlay to functional organization
- Program manager is entirely responsible for project success
- Program manager reports to one of five business area managers
- Responsibility for functions and resources remains with functional managers
- Program responsibility within each functional area is assigned to a task manager, who reports to both functional manager, regarding the quality of the work, and to the program manager regarding implementation of the requirements, schedules, and budgets.

Individual	Responsible for What?	Responsible to Whom?
Business area manager	Business planning; program acquisition; program organization; establishing business objectives, policies, schedules and budgets	General management
Program manager	Proposal effort, program plan, resource negotiations, budget and schedule control, program integration, reporting, customer liaison	Business area manager
Task manager	Performance: specifications, budget, schedule. Technical direction and leadership	Functional and program manager
Functional manager	Technical direction and leadership; resource/personnel development; overhead control; quality of work	Functional manager
Engineer, scientist, technician	Task implementation, project support, proposal support, skill development	Functional manager and/or task manager

The most powerful charter has the TTC-39 (tactical switch) program.[8] (see Figure 2.5). The executive program manager is responsible for the supervision *and* the development of all program personnel and most facilities within engineering, operations, and finance. Although other services are shared, for all practical purposes the tactical switch can be considered a projectized organization.

[8]TTC-39 is a large, multi-year engineering development and small quantity production of special communications equipment (a defense contract).

In addition, many individual project organizations exist as an overlay to the principal organization to execute smaller short-term assignments, specialty tasks, or subsets of an ongoing program. Examples are intermix organizations,[9] which are often spontaneously formed to troubleshoot, advise on, or correct a critical situation, or to respond to a customer request for a bid proposal. These are the dynamics and complexities that typically exist in multilayered real-world engineering organizations.

The management tools available to define the actual operation of such hybrid organizations are charters, management directives, policies, and procedures, which will be discussed in the next section.

2.6 DESCRIBING AND COMMUNICATING THE ENGINEERING ORGANIZATION

To work effectively, people must understand where they fit into the corporate structure and what their responsibilities are. Especially with the internally complex workings of a modern engineering department, it is important to define the management process, the responsibilities, and the reporting relations for all organizational components within engineering, and at its organizational interfaces.

Too often the engineering manager, under pressure to "get started" on a new activity or program, rushes into organizing the engineering team without establishing the proper organizational framework. While initially the prime focus is on staffing in such a situation, managers will find that they cannot effectively attract and hold quality people until certain organizational pillars are in place. At a minimum, the basic tasks and reporting relations should be defined before the recruiting effort starts. Establishing the organizational foundation is necessary to communicate the project requirements, responsibilities, and relationships to new or prospective team members, but it is also needed to manage the anxiety that usually develops during the team formation.

Make Functional Ties Work For You

Strong ties of team members to other functional or project organizations may exist as part of the cultural network. These ties are desirable and often necessary for the successful management of multidisciplinary engineering efforts. An additional organizational dimension is introduced when engineering activities are directed via a project office, which may be chartered within the engineering department or external to it. To build the proper organizational framework, the engineering manager must understand the global organizational structure, culture, and value system. When

[9]Intermix organizations are created by splitting off resources from a functional department, such as individuals, teams, or complete operational groups. These resources are temporarily transferred to a new project organization to perform a specific, usually short-range task, such as making a bid proposal, merger, or feasibility assessment. The unique feature of an intermix organization is that it can be created "instantaneously" at the expense of other ongoing company activities.

working with such a program or project office, the two organizational axes are defined in principle as follows: The program office gives operational directions to the engineering personnel and is normally responsible for the budget and schedule. The functional/engineering organization provides technical guidance and is usually responsible for the quality of the work and for personnel administration. Both the engineering program manager and the functional engineering manager must understand this process and perform accordingly or severe jurisdictional conflicts can develop.

Structure Your Organization

The key to successfully building an organization for accommodating a new engineering program lies in clearly defined and communicated responsibility and reporting relationships. The tools come, in fact, from conventional management practices. They provide the basis for defining and communicating the organizational network and for directing engineering activities effectively. The principal tools are:

1. The Policy Directive. This top-level document describes the overall philosophy and principles of managing engineering activities within a particular organizational unit. A sample is shown in Table 2.6.

2. Procedure. This operational guideline describes the specific method of executing engineering programs or any of their subactivities, such as cost estimating or planning.

3. Charter of the Engineering Manager. The charter clearly describes the business mission and scope, broad responsibilities, authority, the organizational structure, interfaces, and the reporting relationship of the engineering organization. The charter should be revised if the scope or mission of the engineering organization changes, such as occurs with the addition of a major new product development or a new contract program. Sample charters of engineering managers and head engineers are provided in Table 2.7 and Table 2.8, respectively.

4. Organizational Chart. Regardless of the specific organizational structure and terminology used, a simple organizational chart will define the major reporting and authority relationships. These relationships are usually further clarified in a policy directive.

5. Responsibility Matrix. This chart defines the interdisciplinary task responsibilities: who is responsible for what. The responsibility matrix not only covers activities within the engineering organization, but also functional support units, subcontractors, and committees. Samples of the responsibility matrix are shown in Chapter 5.

6. Job Description. Similar to a charter, the job description defines more specifically the authority, responsibilities, and principal duties associated with a particular position or class of positions. Job descriptions should be developed for all key engineering personnel, such as managers, project leaders, senior engineers, and so on. The job descriptions are usually generic and hence portable among similar jobs. Job descriptions are modular building blocks which form the framework for staffing. The job description includes (1) the reporting relationship, (2) responsibilities, (3) duties, and (4) typical qualifications. A sample job description for an R&D Project Manager is shown in Table 2.9.

Examples of these management tools are shown in Tables 2.6 through 2.8. However, it should be emphasized that position titles, responsibilities, and matrix structures vary considerably in the industry. These examples primarily provide a framework to managers for developing their own tools.

TABLE 2.6 Policy Directive: Engineering

Company Logo
ABC DIVISION

Policy Directive No. 00.000.00
Engineering Management — Effective Date: May 1, 1988

1. Purpose

To provide the basis for the management of engineering activities within the ABC Division. Specifically, the policy establishes the relationship between program and functional operations and defines the responsibilities, authorities, and accountabilities necessary to ensure engineering quality and operational performance consistent with established business plans and company policies.

2. Organizations Affected

All Engineering Departments of the ABC Division.

3. Definitions

3.1 *Unit Manager.* Individual responsible for managing an engineering group. He or she reports to a department manager.

3.2 *Department Manager.* Individual responsible for managing an engineering department, usually consisting of various engineering groups, each headed by a unit manager. The department manager reports to the director of engineering.

3.3 *Lead Engineer or Task Manager.* Individual jointly appointed by the program manager and the functional superior. The lead engineer or task manager is responsible for the cross-functional integration of the assigned task.

3.4 *Task Authorization.* A one-page document summarizing the program requirements, including tasks, results, schedules, budgets, and key personnel.

3.5 *Annual Engineering Plan.* Document summarizing the specific goals and objectives of engineering as well as specific programs to be funded for the next year.

3.6 *Department Budget.* Document establishing the funding level for each engineering department based on activities and other organizational needs.

(continued)

TABLE 2.6 Policy Directive: Engineering *(continued)*

4. Applicable Documents

4.1 Supervisor's Handbook.

4.2 Program Management Guidelines.

4.3 Procedures on Engineering Management, No. 00.11.00.

5. Policy

5.1 All engineering departments are organized along functional lines with each individual reporting to one functional superior. The chain of command is established as: unit manager, department manager, director of engineering. The director of engineering is appointed by the division general managers.

5.2 Specific engineering departments are chartered by the director of engineering, who is also responsible for making the managerial appointments.

5.3 The engineering organization, through its various departments, is chartered to accomplish the following functions:

1. Execute engineering programs according to company-external or -internal requests.
2. Support the business divisions in the identification pursuit, and acquisition of new contract business.
3. Provide the necessary program direction and leadership within engineering for planning, organizing, developing, and integrating technical efforts. This includes the establishment of program objectives, requirements, schedules, and budgets as defined by contract and customer requirements.
4. Maintain close liaison with customer, vendor, and educational resources, keeping abreast of state-of-the-art and technological trends.
5. Plan, acquire, and direct outside and company-funded research, which enhances the company's competitive position.
6. Plan, utilize, and develop all engineering resources to meet current and future requirements for quality, cost-effectiveness, and state-of-the-art technology.

5.4 The activity level is defined by department budgets, which are based on company-internal or external requests for specific services or programs.

5.5 Resource requirements for new engineering programs can be introduced at any point of the engineering organization. However, these requests must be reported via the established chain of command to the director of engineering, who coordinates and assigns the specific activities or programs to organizational units.

5.6 The basis for resource planning is the annual engineering plan, which is to be coordinated with the division's annual business plan.

5.7 The program manager is responsible for defining a preliminary program plan as a basis for negotiating work-force resources with functional managers.

5.8 Each unit manager appoints a lead engineer or task manager for each program to be executed within his or her organization.

5.9 No engineering program will be started without proper program definition, which includes at the minimum (1) the specific tasks to be performed, (2) the overall performance specifications, (3) a definition of the overall task configuration, (4) the deliverable items or results, (5) a master schedule, and (6) the program budget. Five percent of the total program budget estimate is a reasonable outlay for this program definition phase.

5.10 Responsibility for *defining* an engineering program rests with the program manager, who must coordinate the requirements and capabilities among the sponsor and resource organizations, including engineering.

5.11 The principal instrument for summarizing contractual requirements of (1) technical requirements, (2) deliverable items, (3) schedules, (4) budgets, and (5) responsible personnel is the task authorization.

5.12 The lead engineer or task manager is responsible for the cross-functional integration of the assigned engineering task. He or she has dual accountability: (1) to his/her functional superior for the most cost-effective technical implementation, including technical excellence and quality of workmanship, and (2) to the program manager for the most efficient implementation of the assigned task, according to the established specifications, schedules, and budgets.

5.13 Functional management of their designated task leaders will comply with the directions of the program manager regarding the contractual requirements, including performance, schedules, budgets, and changes, as well as reporting requirements and customer presentations.

5.14 The responsibility for (1) quality of workmanship and (2) method of task implementation rests explicitly with the engineering organization.

5.15 Conflicts over administrative or technical items between the functional and program organizations must be dealt with in good faith and on a timely basis. If no resolution can be found, the problems must be reported to the next managerial level (if necessary up to the division general manager) that is responsible for resolving the conflict.

2.7 ORGANIZING A NEW ENGINEERING DEPARTMENT

Guidelines

The following 10-point list provides some guidelines for organizing a new engineering department or engineering task team.

Mission and Task Definition

1. *Mission Statement.* Define the overall mission for the new organizational unit, including global business objectives, technical objectives, responsibilities, and timing. Write the charter, such as that shown in Table 2.6, agreed upon by senior management.

TABLE 2.7 Typical Charter of the Engineering Manager

POSITION TITLE: MANAGER OF DESIGN ENGINEERING

Authority. The manager of design engineering has the delegated authority from the director of engineering to establish, develop, and deploy organizational resources according to established budgets and business needs, both long- and short-range. This authority includes the hiring, training, maintaining, and terminating of personnel; the development and maintenance of the physical plant, facilities, and equipment; and the direction of the engineering personnel regarding the execution and implementation of specific engineering programs.

Responsibility. The manager of design engineering is accountable to the director of engineering for providing the necessary direction for planning, organizing, and executing the engineering efforts in his or her cognizant area. This includes (1) establishment of objectives, policies, technical requirements, schedules, and budgets as required for the effective execution of engineering programs; (2) assurance of engineering design quality and workmanship; and (3) cost-effective implementation of established engineering plans. He or she further seeks out and develops new methods and technology to prepare the company for future business opportunities.

---

2. *Work Definition.* Define the type of work to be performed, both functions and projects. Use a functional organization chart or work breakdown structure as a planning model.
3. *Organizational Scope and Budget.* Define the functional capabilities that should be included in the new department versus those that can be contracted from other support departments. The global organization chart and budgets for personnel, all facilities, and organizational development provide the basic tools. The organizational scope must be agreed upon by senior management before proceeding to the next step.

Organization Design

4. *Organizational Structure.* Define the principal functional units of the new department or project and its reporting structure. The organizational chart and mission statements for each principal engineering unit are the basic tools.
5. *Reporting Relations and Control.* Define the specific reporting relations, responsibilities, and controls that make the new engineering department or project operational. These relations can be very intricate, especially in a matrix environment, and should be delineated in some basic policy statements, management directives, and organizational charters.
6. *Manpower and Facility Planning.* The specific staffing levels, facility, and equipment requirements should be delineated and summarized in an organiza-

TABLE 2.8 Typical Charter of Lead Engineer, Task Leader, or Task Manager (Matrix Organization)

POSITION TITLE: LEAD ENGINEER, PROCESSOR DEVELOPMENT

Authority. Appointed jointly by the director of engineering and the program manager, the lead engineer has the authority to direct the implementation of the processor development within the functional organization according to the established plans and requirements.

Responsibility. Reporting to both the director of engineering and the program manager, the lead engineer is responsible: (1) to the functional/resource manager(s) for the most cost-effective technical implementation of the assigned program task, including technical excellence and quality of workmanship; and (2) to the program manager for the most efficient implementation of the assigned program task according to the program plan, including cost, schedule, and specified performance.
Specifically, the lead engineer:

1. Represents the program manager within the functional organization.
2. Develops and implements detailed work package plans from the overall program plan.
3. Acts for the functional manager in directing personnel and other functional resources toward work package implementation.
4. Plans the functional resource requirements for his or her work package over the life cycle of the program and advises functional management accordingly.
5. Directs subfunctional task integration according to the work package plan.
6. Reports to program office and functional manager on the project status.

tion plan in a step-by-step fashion, showing: (1) the specific tasks necessary for implementing the organization, (2) the responsible individuals, (3) the timing, and (4) the budget.

Staffing

7. *Job Description.* Prepare job descriptions for all positions reporting directly to the head of the new department or project.
8. *Personnel Requisition, Hiring of Direct Reports.* Advertise your new positions, within and without the company. Interview and staff those positions that directly report to you.
9. *Complete Staffing.* Each manager should be permitted to hire his or her own staff. Therefore, staffing is performed in organizational layers, repeating steps 7 through 9 until the organization is fully staffed, a process that is often planned over longer periods of time to give the new organization the flexibility to grow gradually with its operational capabilty.

TABLE 2.9 Job Description for R&D Project Manager

JOB DESCRIPTION NO. 071285: R&D PROJECT MANAGER

The project manager is the key individual charged with the responsibility for the total project and is delegated the authority by management. He is also assigned a supporting staff, which may include individuals directly responsible to him or her (i.e., administratively and technically) as well as others technically responsible to the project manager though administratively located in other parts of the organization. The responsibilities and authorities of the project manager can be summarized as follows:

1. Establish and effectively operate an integration effort that is based on rapid feedback from each element of project activity.
2. Implement an appropriate philosophy of how project work will be done, avoiding cumbersome formality without sacrificing traceability and project intercommunications.
3. Plan the work and break it down into understandable elements with well-defined technical, schedular, and financial aspects. Maintain an up-to-date plan that integrates and reflects the actual work status of these work elements.
4. Analyze the total project work against allocated resources and establish an optimum project organization to achieve the project objectives. Reorganize this project organization as required to meet the changing needs of the project.
5. Assign responsibility for discrete elements of project work (contract and in-house) to members of his or her project organization and redelegate sufficient of his or her own authority to these members to enable them to contribute effectively to the successful completion of the project.
6. Serve as the focal point of responsibility for all actions relating to accomplishment of the total project, including both in-house and contract effort.
7. Assure, by direct participation and review of the work of others, that each contract effort in support of the project is well planned and fully defined during the preparation of the technical package, thus enabling the solicitation through requests for proposals (RFPs) of technically responsive offers from responsible offerors.
8. Assure by direct participation in negotiations, that the successful offeror and the company have a common understanding of the goals and requirements of each contract before contract execution.
9. Maintain a continuous surveillance and evaluation of all aspects of the work being conducted by each contractor, including technical, schedular, and financial status.
10. Identify, devise, and execute effective solutions to management and/or technical problems that arise during the course of the work.
11. Ensure the timely detection and correction of oversights in any aspect of each contract to minimize cost overruns, schedule delays, and technical failures.
12. Make final decisions within the scope of work of each contract.
13. Give guidance and technical direction for all elements of the project work, or concur in contractor actions in accordance with provisions of each contract.
14. Continually evaluate the quality of the work performed by each contractor and verify that the requirements are properly carried out by the contractor.
15. Ensure that configuration management requirements are adhered to by each contractor in accordance with the requirements of the contract.
16. Help to indoctrinate and regularly monitor contractor performance to ensure full reporting of all new technology evolved under each project contract.
17. Participate in contract closeout or in termination proceedings to ensure that the best interests of the company are safeguarded.

Organization Development

10. *Team-Building.* Building the personnel into a homogeneous team, unified and focused on the organizational goals, is a continuous process. It requires the whole spectrum of management skills and an understanding of the evolving culture and value system of the new engineering organization.

Eliciting Commitment

Establishing firm commitment from engineering personnel is important in any situation. It is absolutely critical, however, in situations where managers have to step across functional lines and have to deal with personnel over whom they have little or no control. In such a matrix environment, engineering management personnel must sell each assignment primarily on the basis of professional interest, perceived career potential, and satisfaction of higher-order needs, such as the need for recognition, visibility, praise, and accomplishment. Furthermore, the perception of how the task manager may influence the rewards administered by other functional organizations may influence the enthusiasm of personnel toward the new assignment. Therefore, it is critical for engineering task leaders to identify and satisfy the needs of their personnel, involve them during project start-up, and clearly discuss new assignments, including requirements and career implications. The initial job review is an important element in the process to involve people, assess their needs, and elicit their commitment. This interview should clearly address:

- Job requirements
- Project objectives, scope, and challenges
- Specific job content in terms of techniques, skills, experience, disciplines, and degree of difficulty
- Reporting relations and project organization
- Personnel transfer and assignment policy
- Risks and challenges
- Final project outcome and deliverables
- Specific measures for final success
- Reward system.

Such a dialogue will provide both parties with an insight into how compatible the job requirements are with personal goals on and capabilities. There are many additional benefits to a properly conducted interview that occurs prior to the assignment. It sets the tone of how the engineering task or project will be managed, recognizes individual needs and wants, stimulates thinking, creates involvement and visibility at many levels, and establishes lines of communication for the future. If properly done, interviewing for engineering assignments is a highly pervasive process that provides a unifying force toward integrating and building a committed engineering project team.

EXERCISE 2.1: MANAGING IN A MATRIX ENVIRONMENT

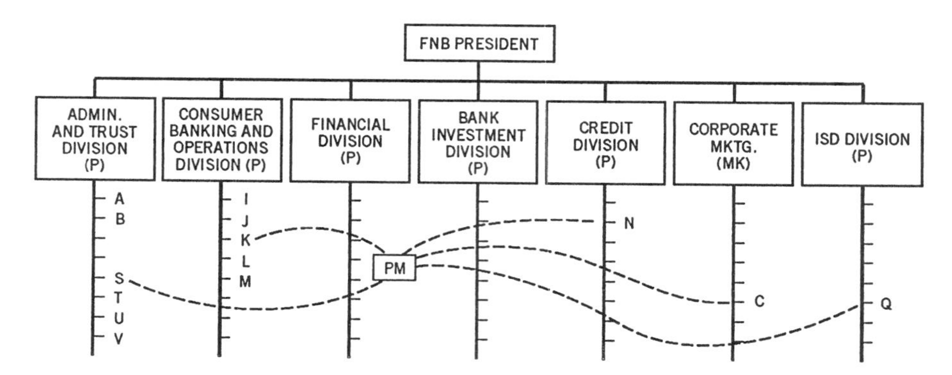

FIGURE 2.6 Organizational chart for Exercise 2.1. Dashed lines indicate cross-functional communication, command and control channels.

Preparation (3 Minutes)

1 Organize into three groups representing the Project Manager (PM), Marketing Manager (MK), and Marketing Specialist (C)
2 Read and analyze the scenario

Scenario

Project X, the acquisition and implementation of a new MIS, is being considered by the company for execution (see Figure 2.6). A project leader (PM) has been assigned and is currently trying to get the "right" people committed to her project.

Group Discussion

Each group listed below should discuss the scenario from its own vantage point, with focus on four questions. Record the essence of your answers on flip charts.

Project Manager Group (PM)

1 What are the challenges of getting resources (e.g., getting the "right" person from marketing)?
2 What are the "drivers" for getting the right person from marketing? What actions could you take?
3 What would minimize *interruptions* of the project (such as losing resource "C" of the marketing department, or budget cuts)?
4 Would there be an advantage (or disadvantage) in projectizing X?

Marketing Manager Group (MK)

1 Why would you provide resources to PM?
2 What situations would make you reluctant to provide resources to PM?
3 Why would you be reluctant to interrupt project X, even if you need C for an emergency?
4 How is your management performance being evaluated?

Marketing Specialist Group (C)

1 Why would you want to join the Project X team?
2 Why would you be reluctant to join the Project X team?
3 Why would you resist pressures to interrupt Project X and help out on Project Y?
4 Would you like to join PM on a projectized basis? Why or why not?

Intergroup Discussion

Share your findings and conclusions with other groups. Draw some overall conclusions.

BIBLIOGRAPHY

Akiyama, Kaneo, *Function Analysis,* Cambridge, MA: Productivity Press, 1991.

Allen, Thomas J., "Project Management Risk, Strategy and Structure in the Nuclear Industry: A Historical Perspective," *Project Management Journal,* Volume 20, Issue 4 (Dec 1989), pp. 21–30.

Anderson, Cindy Carpenter, and Mary M. K. Fleming, "Management Control in an Engineering Matrix Organization: A Project Engineer's Perspective," *Industrial Management,* Volume 32, Number 2 (March/April 1990), pp. 8–13.

Argyris, Chris, "Today's Problems with Tomorrow's Organizations," *Journal of Management Studies* (Feb 1967), pp. 31–55.

Babcock, Daniel L., *Managing Engineering and Technology,* Englewood Cliffs, NJ: Prentice-Hall, 1991.

Bacharach, Samuel B., "Organizational Theories: Some Criteria for Evaluation," *Academy of Management Review,* Volume 14, Issue 4 (Oct 1989), pp. 496–515.

Balinski, M. L. and G. Demange, "An Axiomatic Approach to Proportionality Between Matrices," *Mathematics of Operations Research,* Volume 14, Issue 4 (Nov 1989), pp. 700–719.

Bartlett, Christopher A., and Sumantra Ghosal, "Matrix Management: Not a Structure, a Frame of Mind," *Harvard Business Review,* Volume 68, Number 4 (July/Aug 1990), pp. 138–145.

Bergen, S. A., *R&D Management: Managing Projects and New Products,* Cambridge, MA: Basil Blackwell, 1990.

Boissoneau, Robert, "New Approaches to Managing People at Work," *Health Care Supervisor,* Volume 7, Issue 4 (July 1989), pp. 67–76.

Burns, J. N., "The Impact of Information Technology on Organizational Structure," *Information and Management* (1989), pp. 1–10.

Chambers, George J., "The Individual in a Matrix Organization," *Project Management Journal,* Volume 30, Issue 4 (Dec 1989), pp. 21–30.

Cleland, David I., "The Culture Ambience of the Matrix Organization," *Management Review,* (Nov 1981).

Cleland, David I., "Matrix Management: A Kaleidoscope of Organization Systems," *Management Review* (Dec 1981).

Cleland, David I., and William R. King (eds.) *Project Management Handbook,* New York: Van Nostrand Reinhold, 1988.

Cleland, David I., and Karen M. Bursie, *Strategic Technology Management,* New York: AMACOM, 1992.

Connolly, Terry, *Scientists, Engineers, and Organizations,* Monterey, CA: Brooks/Cole, 1983.

Dangot-Simpkin, Gilda, "Making Matrix Management a Success," *Supervisory Management,* Volume 36, Number 11 (Nov 1991), pp. 1–2.

Davis, S. M. and P. R. Lawrence, *Matrix,* Reading, MA: Addison-Wesley 1977.

Englehoff, W. G., *Organizing the Multinational Enterprise,* Cambridge, MA: Ballinger, 1988.

Fayol, Henri, *General and Industrial Management* (translated by Constance Storrs), London: Isaac Pitman & Sons, 1949.

Fetters, Drew B., and John Tuman, Jr., "Using Project Management to Create a Customer-Focused Nuclear Engineering Organization," *Project Management Journal,* Volume 21, Number 3 (Sept 1990), pp. 25–29.

Flores, Benito, and Clay D. Whybark, "The Strategic Management of Manufacturing Technology," *Engineering Costs & Production Economics* (Netherlands), Volume 17, Issue 1–4 (Aug 1989), pp. 293–302.

Forrester, Jay W., "A New Corporate Design," *Industrial Management Review,* (Fall 1965).

Galbraith, Jay W., "Matrix Organization Designs—How to Combine Functional and Project Forms," *Business Horizons* (Feb 1971).

Grinnell, S. K. and H. P. Apple, "When Two Bosses Are Better Than One," *Machine Design* (Jan 1975), pp. 84–87.

Heller, Robert, "Matrix Management Muddles," *Management Today,* (June 1991), p. 24.

Johnson, Eric, "The Priority of Engineering," *Chemical Engineering,* Volume 96, Issue 8 (Aug 1989), pp. 59–61.

Kerzner, Harold, *Project Management for Executives,* New York: Van Nostrand Reinhold, 1982.

Kerzner, Harold and Hans J. Thamhain, *Project Management Operating Guidelines: Directives, Procedures and Forms,* New York: Van Nostrand Reinhold, 1985.

Khan, Emdad H., "Organization and Management of Information Systems Functions: Comparative Study of Selected Organizations in Bahrain," *Information & Management,* Volume 21, Number 2 (Sept 1991), pp. 73–85.

Killian, William P., "Project Management—Future Organizational Concepts," *Marquette Business Review,* Number 2 (1971).

Killian, William P., "Matrix Organizations," *Business Horizons* (Summer 1964).

Kolondny, Harvey F., "Evolution to a Matrix Organization," *Academy of Management Review,* Volume 4 (1979).

Koontz, Harold and Heinz Weihrich, *Management,* New York: McGraw-Hill, 1988).

Kouskoulas, V. and G. Abutouq, "An Influence Matrix for Project Expediting," *European Journal of Operational Research* (Netherlands), Volume 43, Issue 3 (Dec 18, 1989), pp. 284–291.

Levin, David, "Today's Matrix Switches Blur Product Boundaries," *Network World,* Volume 7, Number 9 (Feb 1990), pp. 1, 43–44, 53.

McClenahen, John S., "This Matrix Was Made for You," *Industry Week,* Volume 240, Number 17 (Sept 1991), pp. 91–92.

McCollum, James K. and J. Daniel Sherman, "The Effects of Matrix Organization Size and Number of Project Assignments on Performance," *IEEE Transactions on Engineering Management,* Volume 38, Number 1 (Feb 1991), pp. 75–78.

Metzger, Robert O., "The Product/Market Matrix and Risk," *Financial Managers' Statement,* Volume 12, Number 3 (May/June 1990), pp. 36–39.

Middleton, C. J., "How to Set up a Project Organization," *Harvard Business Review* (Mar–Apr 1967).

Mintzberg, Henry, "Organization Design: Fashion or Fit?", *Harvard Business Review* (Jan–Feb 1981).

Pasmore, W. A., *Designing Effective Organizations,* New York: John Wiley & Sons, 1988.

Perry, James B., "Utility Engineering Work Management Systems (WMS)," *Project Management Journal,* Volume 21, Number 3 (Sept 1990), pp. 41–48.

Rader, David, "Team-Based Global Manufacturing," *Production & Inventory Management,* Volume 11, Number 7 (July 1991), pp. 28–29.

Plant, Roger, "Practical Ideas for Managing Change and Making It Stick," *Industrial & Commercial Training* (UK), Volume 21, Issue 5 (Sept/Oct 1989), pp. 15–17.

Riggs, Henry E., *Managing High-Technology Companies,* Belmont, CA: Lifetime Learning Publications (Wadsworth), 1983.

Roussel, Philip A., Kamal N. Saad, and Tamara J. Erickson, *Third Generation R&D,* Boston: Harvard Business School Press, 1991.

Sadler, Philip, "Designing an Organizational Structure," *Management International Review,* Volume 11, Number 6 (1971).

Schlesinger, Phyllis F., Vijay Sathe, Leonard A. Schlesinger, and Kotter, John P., *Organization,* Homewood, IL: Irwin 1992.

Souder, William E. and Suheil Nassar, "Managing R&D Consortia for Success," *Research-Technology Management,* Volume 33, Number 5 (Sept/Oct 1990), pp. 44–50.

Taylor, Frederick, *Shop Management,* New York: Harper & Brothers, 1903.

Thamhain, Hans J., "Production within a Matrix Environment," in David I. Cleland (ed.), *Matrix Management Systems Handbook,* New York: Van Nostrand Reinhold, Chapter 28 (1983).

Thamhain, Hans J., *Engineering Program Management,* New York: Wiley, 1986.

Van de Vliert, Evert and Boris Kabanoff, "Toward Theory-Based Measures of Conflict Management," *Academy of Management Journal,* Volume 33, Number 1, (Mar 1990), pp. 199–209.

Wallace, Don, "Get It Done!—Project Management Your Most Valuable Tool," *Success,* Volume 37, Number 2 (Mar 1990) pp. 46–47.

Warner, Alan, "Where Business Schools Fail to Meet Business Needs," *Personal Management (UK),* Volume 22, Number 7 (July 1990), pp. 52–56.

Wilson, Laurie J., "Corporate Issues Management: An International View," *Public Relations Review,* Volume 16, Number 1 (Spring 1990), pp. 40–51.

3
PLANNING THE ENGINEERING FUNCTION

3.1 PLANNING: THE CORNERSTONE TO EFFECTIVE ENGINEERING MANAGEMENT

Few engineering managers would argue the importance of proper planning for their engineering activities and organizational development. The challenge is to apply the available tools and techniques effectively; that is, to manage the effort by leading the multifunctional personnel toward the agreed-upon objectives within the given time and resource constraints. Even the most experienced practitioners often find it difficult to control engineering activities according to plan, in spite of sufficient detail and apparent personnel involvement and commitment.

In essence, effective engineering planning is related to properly defining the objectives, the work, the budgets, the schedules, and their monitoring processes. Equally important, it is related to the ability to keep personnel involved and interested in the work, to obtain and renew commitment from the team as well as from upper management, and to resolve some of the enormous complexities on the technical, human, and organizational side. Specifically what we find is that the usefulness of our engineering project depends on the following criteria. The plans must have:

- Detail regarding definition of the work and timing of resources and responsibilities
- Commitment from team management
- Trackability and measurability
- Cross-functional involvement
- Up-to-date content
- Simple structure, format, and ease of use.

These challenges are further enhanced by the realities of today's engineering business environment with its technological complexities, shorter product life cycles, resource limitations, and business uncertainties, to name just a few factors. Table 3.1 summarizes ten specific challenges that were pointed out during a survey of engineering managers, which shows the intricate nature of planning in today's technological environment. These challenges put more pressure on engineering managers for integrated planning and decision-making, which requires a systems approach to engineering management. To help managers better meet these challenges, let us develop some insight into the nature and purpose of managerial planning before addressing the specifics of the process.

The importance of managerial planning has been formally recognized since Henri Fayol wrote about it in 1916.[1] Fundamentally, the purpose of planning is to help people in the organization to achieve desired results with a given set of resources, capabilities, environmental factors, and changing conditions. While these principal objectives remain valid today, the nature, scope, and perceived benefits have changed in recent years with the new realities of the business environment. There are five interrelated forces that have affected the purpose and practice of planning, regardless of its type or level, as discussed below.

1. Limited Resources. Resources of all types are available only in limited quantities. This is the result of competitive pressures toward more effective operations as

TABLE 3.1 New Business Realities Put More Pressure on Engineering Managers for Integrated Planning and Decision-Making

Rank	Realities and Challenges
1	End-date driven schedules
2	Resource limitations
3	Effective communication among task groups
4	Commitment from team members
5	Establishing measurable milestones
6	Coping with changes
7	Project plan agreement with team
8	Commitment from management
9	Dealing with conflict
10	Managing vendors and subcontractors

Data were collected from a sample of over 400 engineering managers and engineering project leaders and are presented rank-ordered by their importance (1 = most important). For a detailed discussion see "Criteria for Controlling Projects to Plan," by H. Thamhain and D. Wilemon, *Project Management Journal,* June 1986.

[1]Henri Fayol considered planning the most important management function in his widely cited work *Administration Industrielle et Generale,* published in 1916 and translated into English in 1947.

well as of shortages, particularly in natural resources and skilled personnel. We have to learn to do more with less. Effective planning, which links the strategic and project levels, can potentially help to set priorities, define the most desirable objectives, select the best approaches, and manage project implementation.

2. *Task Complexity.* The ever-increasing complexity of engineering undertakings regarding their tasks, multifunctional integration, and skill requirements has put strong pressures on quality planning, to enable professionals to make a larger number of better decisions and to carry out these decisions more effectively and economically. Task complexity seems to further increase as companies put greater emphasis on systems rather than on components.

3. *Information Explosion.* Related to increasing task complexity is the trend toward increasing access to information. The new source of power, according to John Naisbitt,[2] is no longer money in the hands of a few but information in the hands of many. Hence, information becomes a strategic resource which, properly processed, can be transformed into knowledge and eventually into products, services, and money. Engineering organizations are information-processing organizations and their professional workers, such as engineers, analysts, accountants, programmers, scientists, and certainly managers, are information workers, as first pointed out by Naisbitt. All this adds up to additional pressures for more effective planning of information needs and their integrated processing and consolidation. This will help clarify the multifunctional decision-making responsibilities and provide the proper information and know-how when and where it is needed.

4. *Uncertainties and Risks.* The accelerated rate of change in our environment, labeled *transience* by Alvin Toffler,[3] is a constant source of uncertainty and risk. The ability to predict the future with accuracy diminishes as this rate of change increases. Planning becomes more demanding and sophisticated. Managers need planning systems that are dynamic, maybe even intelligent, to guide the team toward the desired results in spite of the fact that one cannot predict the future with accuracy.

5. *Power- and Resource-Sharing.* Engineering organizations are, in one form or another, matrix systems with sharing of resources and managerial powers. Decision-making is no longer functionally centralized and stratified at the top, but dispersed throughout the organization. This shift of power regarding resources and strategic decisions has changed the role of senior management. Management becomes a facilitator to decision-making through participative management and planning, with involvement and commitment from all key team members.

6. *Market Forces.* Organizations in all segments of industry, commerce, and government experience stronger competition to their products and services, with more interrelated markets, global operations, faster product obsolescence, and increased time-to-market pressures. Meeting a market entry target date can be crucial to new product success. In fact, in a survey of 200 new-product managers, who ranked the importance of 15 product-success factors, the ability to meet the market

[2] *Megatrends,* New York, Warner Books, 1982.
[3] *Future Shock,* New York, Bantam Books, 1970.

window was ranked second only to product quality.[4] All this requires more sophisticated planning to access the realities and feasibility of engineering development and to control it according to established plans.

3.2 THE PLANNING PROCESS

With the highly complex nature of engineering-oriented businesses, a new breed of managers has evolved, who plan more dynamically than their counterparts in traditional organizations. While conventional planning processes for traditional businesses stress the systematic development of the plan in a predominantly sequential mode, engineering organizations often use a more dynamic process, especially for plan implementation. The overall plan, regardless of its type, defines and aims at a global set of objectives, while leaving room and flexibility for short-range adjustments and refinements to adapt to environmental changes and contingencies. This process is an incremental short-range optimization of long-range plans, which will be discussed later on.

Fundamentally, the planning process can be discussed in terms of system input, output, and transformation, as shown in Figure 3.1. If done properly, planning of any type generates a roadmap for all downstream management functions such as staffing, organizing, leading, and controlling. Furthermore, it provides a link between the present and the future. From a managerial point of view, planning is a set of activities that consists of identifying, developing, and processing various environmental data so that they become useful as inputs for the planning process and eventually can be transformed into a specific plan that is useful as a managerial roadmap to reach specific results.

Using the system approach, planning is a task-oriented subset of the overall organizational framework characterized by the activities of:

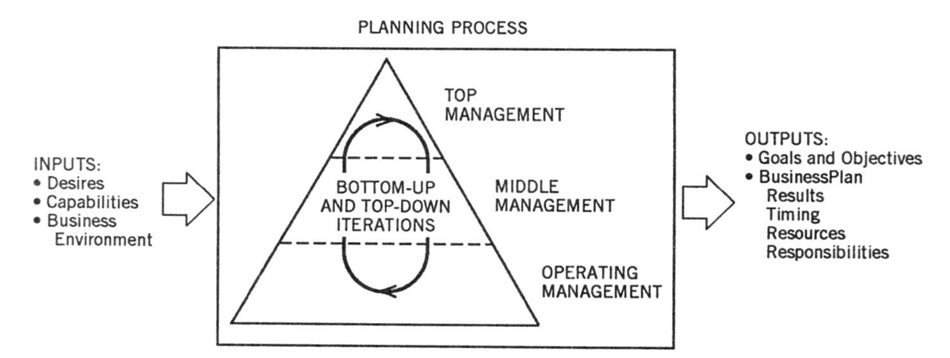

FIGURE 3.1 The input–output model for managerial planning processes.

[4]See Hans J. Thamhain, "Managing Technologically Innovative Team Efforts Toward New Product Success," *Journal of Product Innovation Management,* Vol. 7, No. 1, March 1990.

- Identifying organizational needs
- Accumulating information
- Relating this information to needs and premises
- Establishing objectives
- Forecasting future conditions
- Developing alternatives
- Ranking and selecting of plan alternatives that will achieve the best balance of ultimate objectives
- Establishing standards for measuring adherence to plan implementation
- Establishing policies.

Therefore, planning itself is a project. Depending on the type and nature of the plan, the organization setting, and the managerial leadership style, it is conducted in a variety of modes:

1. Top-down planning
2. Bottom-up planning
3. Outside-in planning
4. Inside-out planning
5. Participative planning
6. Directive planning
7. Incremental optimization

Most managerial planning combines several modes to optimize effectiveness in a specific situation. Let us briefly examine the characteristics of each mode.

Planning Modes

1. Top-Down Planning. Where should the planning process start? Top-down planning refers to a process in which upper management sets broad goals and objectives and directs lower levels of management in their implementation. The advantages of such a method are that the principal objectives are clear and a framework for the plan is available before any serious plan development starts. This can lead to a quicker, less confusing plan development, highly focused on the principal goals. The process is limited, however, to planning situations where upper management knows exactly what they want and are pretty sure of the plan's basic feasibility. Another disadvantage of top-down planning is the limited involvement of the lower-level people, which impedes communications, commitment, and their personal desire to succeed.

Few companies use top-down planning in a pure mode, but rather as a modified final phase in a series of planning stages. Planning situations that could follow a top-down process are: (1) a corporate directive to improve manufacturing productivity, (2) an organizational merger decision, (3) a corporate policy, or (4) a new product development.

2. Bottom-Up Planning. The impetus for the bottom-up plan comes from the operating level of the organization, often initially as a suggestion, or even as a complaint. The process may start at the bottom, based on some top-down suggestions or desires. In its pure form, bottom-up planning involves many people and many iterative cycles among individual contributors and their managers before a concept crystallizes into something that is acceptable to all people involved and is worthy of submission to the next-highest level of management. Eventually, a go decision is made at the managerial level that has the resources and authority to direct such an effort or redirect it. At that point the process reverts to being top-down.

The advantages of bottom-up planning are broad organizational involvement, which can (1) be persuasive in getting peoples' commitment to the final plan's development and its implementation, (2) test the basic feasibility of the plan during the process, (3) validate the basic assumptions underlying the plan, (4) create visibility and desire for the planned activities at all levels, and (5) foster an organizational environment conducive to cooperation among various functional groups. The limitations of the bottom-up approach are related to its complexity, which requires a sophisticated, team-centered management style. It takes more time and organizational effort, runs the risk of collusion and suboptimal solutions, and is inappropriate for confidential, sensitive issues.

3. Outside-In Planning. When management analyzes the external environment, such as the market, the economy, or a bid opportunity, and then positions itself via planning to exploit the opportunity, it is utilizing outside-in planning. Outside-in planning is used to find a unique niche for a particular capability or activity, such as finding a new customer for a special system. In essence, outside-in planning requires the organization to move with the environment rather than trying to shape it.

4. Inside-Out Planning. By definition, the inside-out approach to planning utilizes an organization's intrinsic strength to create a new market need or to try to position itself for the future. Product plannings of Sony's Walkman or Apple's Macintosh are examples of a mostly inside-out approach. But as we can see, there is a great deal of overlap between the two approaches of inside-out and outside-in planning. Most real-world planning includes both approaches. Regardless of the category label, successful planning requires a careful assessment of the environment and unique matching of the existing and potential needs with organizational capabilities.

5. Participative Planning. The involvement of people at various levels of the organization can enhance the quality of the plan. This involvement might lead to a better multifunctional understanding of the situation and a broader integration of expertise throughout the organization. It might stimulate interest for a new plan and help to facilitate agreement and commitment to the final plan's implementation. On the negative side, people may polarize over certain issues, interests may get in the way of optimal solutions, confidentiality is often difficult to maintain, and the planning process usually takes more time.

6. Directive Planning. In its pure form, directive planning is similar to top-down planning: both have clear goal directions coming from the top. However, in contrast to top-down planning, which allows for interaction and negotiations with top management, directive planning is basically a command for action. While directive planning is usually faster and more focused than any other method, it must be assumed, that upper management has most of the knowledge necessary for defining the broad parameters of the plan and for leading the plan development to its conclusion. Another major limitation of the directive approach is that it does not foster an environment conducive to change and the commitment needed for plan implementation. Because of its various strengths and weaknesses, the directive approach is not a mutually exclusive alternative to other methods, but effectively overlaps with them in many practical situations.

7. Incremental Optimization. This mode of planning looks toward plan implementation by breaking the total plan into smaller increments with somewhat flexible subobjectives. These subobjectives can be modified according to the changing business environment, and then they become the inputs for the next plan increment. Thus incremental short-range optimization of a plan is like shooting at a moving target while getting closer to it in the process. Not necessarily a simple process, it enables an organization to reach a desired goal in spite of a constantly changing target position. The process is discussed somewhat more technically in the next section.

Incremental Short-Range Optimization of Long-Range Plans

The previous sections provided an overview of the planning process and its basic modes. Traditional approaches to planning and subsequent plan implementation suggest the breakdown of the plan into short-range increments, each having specific objectives and resource requirements.

To be optimal, such a breakdown of the total plan into detailed steps assumes, however, that our forecast of the organizational environment is, at least in principle, valid for all time periods of the total plan. Hence, such an implementation procedure is not optimal if unpredicted changes occur in the environment. The traditional approach, furthermore, has no provisions for adjusting to contingencies, even if they were part of the original forecast. In essence, the traditional process of plan implementation will lead to optimal results only if: (1) the internal and external organizational environment do not change from their forecasted composition and interrelationship during plan implementation and (2) it was possible initially to predict the variables of the internal and external environment for the total planning period with sufficient accuracy.

Specific investigations into actual planning practices indicate that some firms recognize the highly dynamic nature of the organizational environment, with its continuously changing variables.[5] Accepting the notion of a changing environment

[5]The process of incremental short-range adjustments was formally researched and described in *Incremental Short-Range Optimization of Long-Range Corporate Plans,* Ph.D. dissertation by Hans J. Thamhain, Syracuse University, July 1974 (University Microfilms, Ann Arbor, Michigan, 1974).

and its inevitable deviation from their long-range forecast, these firms argue that the achievement and optimization of the long-range goals will depend in no small measure on the ability to adjust to prevailing environmental conditions and hence will depend on the ability to optimize the long-range plan in short-range increments.[6]

In this context, the "optimum increment" is defined as the specific short-range objective with its associated resource commitment. The incremental short-range optimization of long-range plans is a process for adjusting the long-range plan to prevailing conditions of the environment in which the organization operates, such that it either

1. Meets long-term objectives with a minimum of total resource expenditures, or
2. Maximizes long-term objectives for a given set of resources.

The process is symbolically shown in Figure 3.2.

Practical examples of the continuously changing environmental construct are economic variables originating from the consumer's as well as from the supplier's

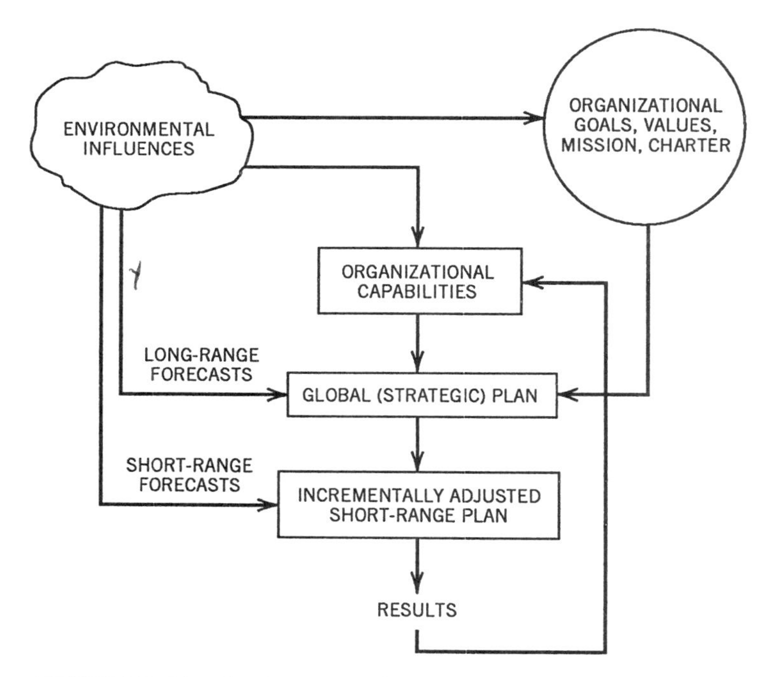

FIGURE 3.2 The planning process in a complex, changing environment.

[6]Short-range optimization of long-range plans has been practiced in industry for over 20 years. See S. P. Ehrhardt, "A Five-Year Plan, a Year at a Time," *Administrative Management,* January 1973. However, the more systematic and proceduralized approach to short-term plan adjustments to changing conditions gained increasing acceptance during the 1980s.

side. Competitor's strategy and labor conflicts are other examples of variables that are difficult to estimate with accuracy in short-term increments over a longer period of time, but they may be predictable rather precisely on a short-range basis.

Yet another need for systematic optimization occurs because all forecasts (e.g., of technology, sales of goods and/or services) are made before the market had a chance to react to this organizational output. A revision of these forecasts after the market reaction has been ascertained may be essential to optimize the results of any overall long-range plan.

In a dynamic society, the way in which organizations can adapt to changing requirements and constraints is through dynamic planning, a vehicle for reaching a desirable position while reacting and adapting to different internal and environmental forces.

For those engineering managers who already use short-range incremental optimization of their plans, the long-range technique offers an extension to the strategic planning process. The incremental optimization technique is described here in terms of its inputs, process, and output parameters, similar to the model in Figure 3.1.

1. Inputs from strategic planning subsystem:
   - Long-range goals
   - Outline for implementation
   - Time frame for implementation
   - Available total resources
   - Long-range forecasts and intelligence
2. Inputs from the external environment:
   - Economic conditions
   - Social change
   - Technological change
   - Competitor's reactions
3. Inputs from the internal (organizational) environment:
   - Current technical and organizational capabilities
   - Future technical and organizational capabilities
   - Reactions to competitors
   - Regression on past experiences.

The process of incremental short-range optimization of long-range plans can be described as follows:

1. Take long-range objectives, constraints, and resource commitments from strategic plan.
2. Derive detailed intelligence concerning the current environment, including short-range forecasts of the technological, economic, and market situation.
3. Evaluate past experience.

4. Determine deviations of current environment from previous forecasts. Update existing (older) forecasts.
5. Partition or repartition the not-yet executed part of the long-range plan into increments that will lead to an optimal utilization of available resources for the overall plan.
6. Develop alternative partitioning of the overall plan based on contingency estimates. Follow the same logic as discussed under 5 above.
7. Make final recommendations.

The output from incremental short-range optimization efforts consists of a set of short-range objectives and their associated resources and constraints, which are intended to optimize the overall strategic plan. Specific outputs include:

1. Specific short-range goals
2. Some broad concept for implementation
3. Risks and possible alternatives
4. Resources and constraints
5. Time schedules for implementation
6. Organizational responsibilities
7. Interface definitions
8. Specific deliverables (results).

3.3 TYPES OF PLANNING SYSTEMS

There are, of course, many types and forms of plans, all of them often overlapping and interrelated. But in general, these types follow a hierarchy, as shown in Figure 3.3. That is, the strategic plan provides the principal umbrella for delineating and unifying all other plans, which eventually translate into operational plans and project plans. Following this hierarchy helps in breaking down the complexity of and bringing some order into the discussion of the bewildering array of plans.

Six types of formal planning systems are of concern and interest to engineering managers: (1) strategic business planning; (2) administrative planning; (3) technology planning; (4) product planning; (5) operational planning; and (6) project planning. All of these planning systems must be integrated into a unified set of plans, consistent with the organization's strategic objectives and capabilities. To achieve integration there must be a great deal of overlap among these plans, their objectives, and their organizational involvement. This is shown conceptually in Figure 3.4. The procedures, tools, and techniques are similar for all types of planning, although the principal philosophy and focus may be quite different. Another difference among the five planning types becomes apparent when the level of planning effort is compared to the organizational level at which the planning occurs for each type. As shown in Figure 3.5, the operational level of management spends most of its planning efforts

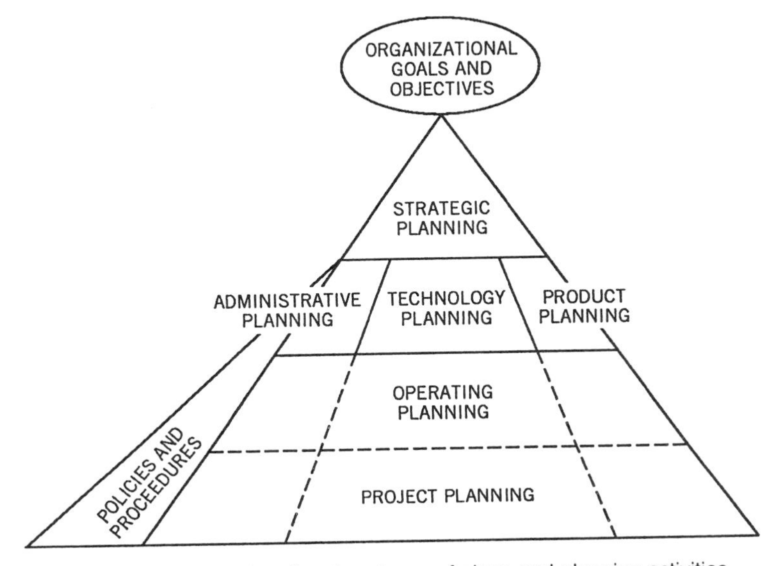

FIGURE 3.3 Hierarchy of various types of plans and planning activities.

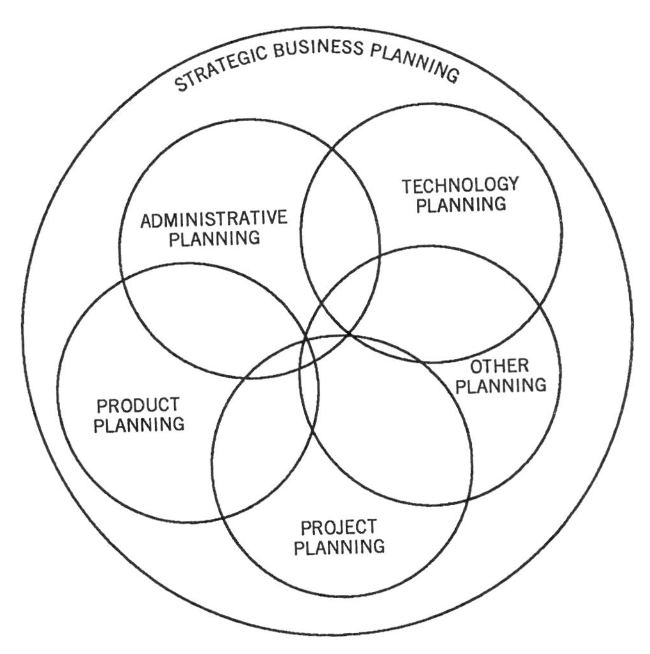

FIGURE 3.4 The overlapping nature of various types of planning within an integrated strategic business planning system.

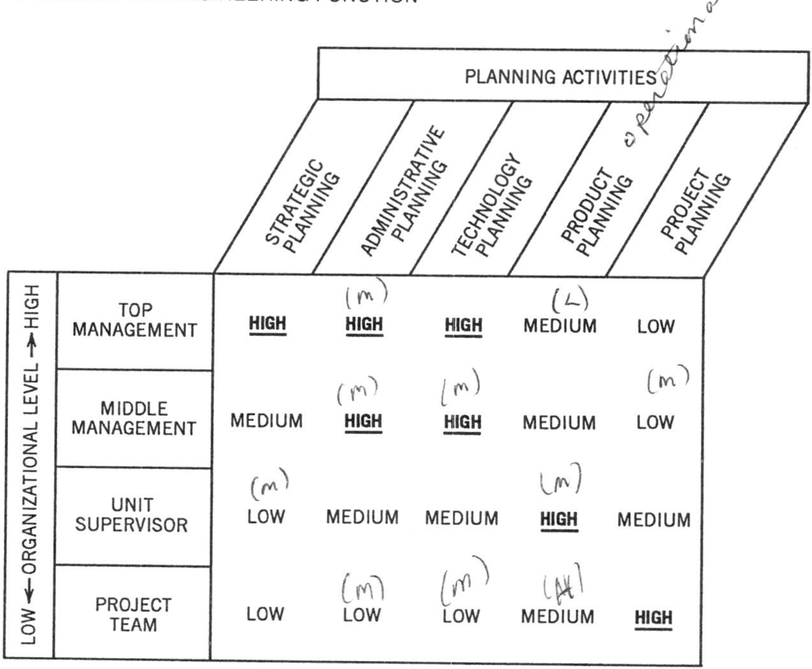

FIGURE 3.5 The relative level of effort of planning for each of the five planning types, relative to the organizational level.

on project and product planning, while higher organizational levels are mostly concerned with strategic and administrative planning.

The Format of a Plan

A business plan, regardless of its type, is a communication vehicle. It must clearly articulate *what* is to be done and *how*. Eventually, a plan must be "sold" to all of the constituents that can affect the plan and its outcome. Potentially, up to six categories of constituents can be involved:

1. *The Sponsor.* The sponsor could be an external customer, an internal department, the owner of a company, or a venture capital group.
2. *The Implementor.* The implementor could range from a senior executive to a project team. These are individuals accountable for overall results.
3. *The Direct Contributors.* These are the individual contributors and managers who are directly responsible for executing operational components of the plan.
4. *The Supporting Contributors.* These are the individuals or functions that indirectly contribute to the planned activities, such as support personnel, subcontractors, and vendors.

5. *The People of the Organizational Environment.* To be workable, many plans need the endorsement of people from the social, political, and economic environment in which the organization operates.
6. *The Regulators.* If the plan affects activities that are regulated, either company internally or externally, these regulators become part of the constituents to whom the plan must be sold.

The following topics should be considered for any type of plan:

- Executive summary
- Objectives and goals
- Mission statement
- Environmental analysis
- Strategy and action plan
- Schedule
- Resource requirements, budgets
- Organization and delegation
- Applicable policies and procedures.

Six Planning Systems

Six specific planning systems are briefly discussed next.

Strategic Business Planning. Strategic planning is the ongoing process used by an organization to position itself uniquely in today's complex, ever-changing world. It is an attempt to match internal capabilities with external opportunities in the most advantageous way, consistent with long-range organizational goals. Drafting a strategic business plan of a company, a division, or a profit center is one of the most important responsibilities of senior management. It usually coordinates and synthesizes many different forms of planning and effectively blends the strategic planning, implementation, and control process.

Effective strategic planning involves more than following a prescribed process. It requires serious and intense organizational involvement to assure reality, doability, and commitment. Engineering companies were among the first of many diversified organizations to use a "bottom-up" approach to strategic planning. Each functional unit submits its plan to the profit center manager, who coordinates and fine-tunes the plan and submits it to his or her boss for further coordination. This process helps to make the engineering plan realistic and responsive to the operational capabilities of the company.

Using the systems concept, strategic planning can be described in terms of inputs, outputs, and process, as illustrated in Figure 3.1. While the process of strategic planning may vary considerably with the type of long-range strategic planning that is undertaken, the kind of business the organization handles, and the general process adopted by the responsible individuals, the basic steps, as shown in Table 3.2, remain the same for most strategic planning.

The output from the strategic planning system consists, in essence, of broad long-range objectives that are in accord with the organizational desires and capabilities, plus guidelines for the plan implementation specifying timing, milestones, resources, and responsibilities, as summarized in Table 3.2. The long-range objectives provide the framework and guidance for organizational policy and decision-making in the development and management of resources at the various levels of the organization.

Administrative Planning. Administrative planning is primarily the responsibility of the functional axis of the engineering organization. It includes manpower planning and its broader set of resource planning; inventory planning; market planning; and other forms of functional planning necessary to prepare the organization for the effective and efficient support of future business activities. Normally, these planning activities lead to capabilities that support many activities and projects throughout several functional units over longer periods of time.

Technology Planning. Technology planning is concerned with providing the technical capability for performing the activities needed for effective organizational performance. It involves the knowledge, skills, techniques, processes, equipment, and facilities used in the transformation of inputs into outputs. As such, the plan provides specific guidelines as to how a particular technical capability is to be developed, given schedule and resource limitations. The plan must reflect current and future technological realities as well as organizational desires. Technology planning is strategizing new product ventures within their portfolio of existing offerings. The concept has several implications for management. First, it makes managers aware of the interdependencies among the various operations such as research and development, engineering, production, marketing, field services, finance, and legal services. Second, it provides a time-phased picture of the portfolio of existing products and services. Third, it portrays the timing of the product management effort from its inception to maturity and beyond. This can provide the framework for a new-product development procedure that describes the various phases the development passes through—research and development, engineering, production, etc.—including a definition of the principal activities, key personnel, interfaces, documentation, reviews, and sign-offs. Fourth, the PLC concept is used to stimulate market strategy and overall product strategy in combination with other management tools such as the growth-share matrix,[7] PIMS model,[8] Price–performance curve, the experience

[7]For a detailed description of the growth-share matrix, see Bruce D. Henderson, *The Growth Share Matrix of the Production Portfolio,* formerly researched and described in *Incremental Short-Range Optimization of Long-Range Corporate Plans,* Ph.D. dissertation by Hans J. Thamhain, Syracuse University, July 1974 (University Microfilms, Ann Arbor, Michigan, 1974).

[8]For more detail on the PIMS model, see Sidney Shoeffler, Robert Buzzell, and Donald Heany, "Impact of Strategic Planning on Profit Performance, *Harvard Business Review,* March–April 1974.

curves,[9] and the Porter curve. The growth-share matrix, developed by the Boston Consulting Group (BCG), maps the product life cycle into four quadrants with various degrees of growth potential, market share, and cash flow.[7] The PIMS (profit impact on market strategies) model, developed by the Strategic Planning Institute of Cambridge, Massachusetts, evaluates a data pool of company experiences to determine specific correlations between business strategy measures and profit performance.[8] The price–performance curve maps the relative value of a product in relation to other competitors. The experience curve is a concept for tracking the cost or price of goods and services over time. Plotted against a log-log scale, cost trends become obvious and can be used in long-range strategy formulation.[9] The Porter curve, based on field research by Michael Porter, a professor at Harvard University, shows the generic correlation between market share and new product profitability. His findings show that companies with either very high or very low market share are most likely to be profitable, while companies in the middle of the market are likely to experience lower profitability.

Operational Planning. Operational plans are developed at the operational level of the organization. As shown in Table 3.2, they take the outputs from a higher level plan, such as a strategic technology plan, and translate them into action items that must be accomplished on time with available resources. There is a great deal of overlap joint responsibility of functional managers at all levels. The final planning is normally driven by the strategic business objectives of an organization and the technical trends that are forecast. A great deal of similarities exists between operational planning and project planning, especially in areas of programs generated within the company, such as technology development or capability development. However, managers often distinguish between two basic types of operational plans: single-use plans and standing-use plans. Single-use plans are designed to achieve specific, one-time objectives, such as capability development, a new material requisition planning (MRP) system implementation, a production plan, a staffing plan, or development of a new product. Essentially, single-use planning is for a project plan that terminates with the accomplishment of the project. Standing-use plans are developed for repeat use in standardized, predictable situations, which are expected to recur. Typical examples are policies, procedures, operating guidelines, and standards.

Product Planning. Product planning is oriented toward the creation of a new product service, the enhancement of an existing product, or the search for a new product or customer. In today's complex world, most engineering companies consider the total product life cycle in their product planning process. These companies in essence practice the contingency approach to management and derive their strategies from the situation at hand rather than from fixed rules and objectives. In this context, the product becomes a variable rather than a fixed physical entity. The product is defined as a complex set of technical, economic, legal, ethical, and personal

[9]For details see *Perspectives on Experience,* written and published by the Boston Consulting Group, Boston, 1972.

TABLE 3.2 Comparison of Three Planning Processes: Strategic, Operational, and Project Planning

Strategic Planning	Operations Planning	Project/Program Planning
	Inputs to Planning System	
FROM EXTERNAL ENVIRONMENT • Economic conditions • Social change • Technological forecast • Competitive assessment • Government regulations FROM ORGANIZATIONAL ENVIRONMENT • Organizational desires • Capabilities, resources • Limitations, risk	FROM THE STRATEGIC PLANNING SYSTEM • Long-range goals • Specific short-range objectives • Concept for implementation • Available resources FROM THE EXTERNAL ENVIRONMENT • Economic conditions • Competitive behavior • Social/political climate • Technological status FROM THE ORGANIZATIONAL ENVIRONMENT • Current technical and organizational capabilities • Future technical and organizational capabilities • Past experiences extrapolated into the future	FROM THE PROJECT SPONSOR • Goals, objectives, requirements • Results and deliverables • Timing • Budgets FROM THE PERFORMING ORGANIZATION • Technical capabilities • Resources • Risks and alternatives
	Planning Process	
1. Establish long-term objectives for the firm 2. Prepare long-range forecasts (technological, economic, market) 3. Establish specific long-range performance objectives for the firm	1. Analyze strategic requirements 2. Derive detailed intelligence on current and future technological, economical, and market situation 3. Assess internal capabilities including resources 4. Break strategic goals into smaller operational objectives	1. Gather information 2. Analyze requirements 3. Define project 4. Define reviews, reports, controls

4. Develop and evaluate alternatives 5. Decide upon the course of action to be undertaken 6. Obtain approval of the plan	5. Define results and deliverables 6. Define management responsibilities and controls 7. Formulate actions and procedures for achieving defined objectives incrementally 8. Define evaluation and feedback system for managerial tracking and control	
	Outputs from Planning System	
• Specific long-range goals • Implementation concept • Implementation time frame • Resource estimate • Risk and contingencies assessment • Contingency plans • Specific organizational responsibilities	• Specific operational goals and objectives • Deliverable items and results • Specifications and statements of work • Concept for implementation • Time schedules • Budgets • Individual responsibilities • Management controls	• Specific operational goals and objectives • Deliverable items and results • Specifications and statements of work • Concept for implementation • Time schedules • Budgets • Individual responsibilities • Management controls

parameters, all interrelated and dependent on both the company and its environment. This means we must consider not only the physical product development, but also its production, distribution, marketing, technical field support and service, updates, and enhancements. Considering all of these factors in a rapidly changing environment over the anticipated product life cycle makes product planning a highly flexible and dynamic subset of the strategic planning system.

The product life cycle (PLC) concept, as summarized in Figure 3.6, gives managers an important tool for planning and strategizing product-oriented activities, considering all phases of their life cycles.

Project Planning. Within the organizational hierarchy of plans, shown in Figure 3.3, project planning is at the implementation level. That is, the output of the project plan is specific task statements, schedules, and budgets, which translate higher-level objectives into final operating results. A summary of the input–output process is shown in Table 3.2. Like operating plans, project plans also fall into two classes. One consists of projects that originate via the internal planning process, with objectives trickling down from the top as part of the strategic planning process, as shown in

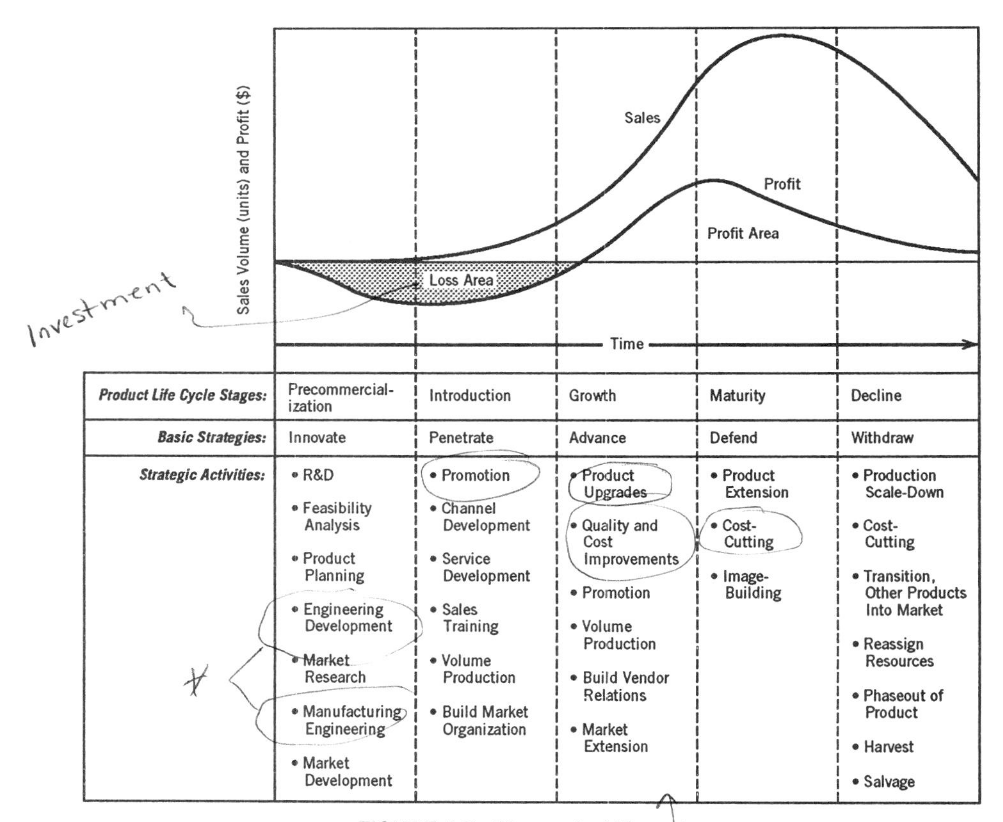

FIGURE 3.6 The product life cycle.

Figure 3.3. The second class consists of projects that originate externally from a customer request and are turned into a contract via bid proposals and negotiations.

The particular tools and techniques available for project planning, tracking, and control are discussed in detail in the next two chapters. Managing engineering projects can be a highly complex job, requiring sophisticated skills for effective integration of multidisciplinary activities under demanding cost, schedule, and technical requirements.

Plans That Sell

Plans are also sales vehicles, whose purpose is to obtain funding or raise venture capital, or just to obtain approval for an action or procedure. In almost every case, the plan must be approved, or "bought," by a higher authority. For a plan to sell, it must clearly present a "can-do" image and must articulate the benefits to the sponsor. A winning plan must address the following seven issues:

1. *Purpose:* A clear statement of the goals, objectives, and intended results, including their relationship to other established plans or missions
2. *Understanding:* Demonstrated understanding of the existing challenges and problems to be solved
3. *Feasibility:* Demonstrated feasibility of the technical approach, the market acceptance or its equivalent, the financing, and the profit or cost objectives
4. *Results:* Clearly stated results and deliverables
5. *Alternatives:* Options and alternatives to the primary plan, including tradeoff analysis
6. *Risks:* Clear statement of risks in the major plan areas, such as technical, financial, human resources, and marketing, including impact assessment
7. *Benefits:* Specific benefits to the plan sponsor. Quantifiable data wherever possible.

3.4 PLANNING PRACTICES TODAY

Effective planning is important for any business or program; it is essential, however, for the successful management of an engineering organization in today's complex world. An increasing number of executives, management consultants, and management studies support this proposition for the full spectrum of business practices, from strategic to operational and project activities. Managers are willing to allocate a reasonable amount of scarce resources to formal planning in the hope that the overhead for this effort will return dividends in the form of a better business position, aimed at future profitability and growth. They also hope that proper operational and project planning will make plan implementation simpler and more efficient. Few engineering managers go into a planning activity lightly, however. Many of them worry that planning itself can stifle innovation and creativity downstream in the organization. Planning also takes time and uses scarce resources, but the benefits are not always obvious.

Responding to these interests and concerns, this section examines planning practices, focusing on engineering-oriented companies. We also discuss the potential benefits and limitations of these practices and conclude with a set of recommendations for effective engineering planning.

The most fundamental purpose of any plan is to provide a roadmap for implementing a particular objective; that is, to show how to get from here to there. To be meaningful, the plan must clearly define the starting point (situation), the end point (results), and the method of getting there (tasks). The plan must further define individual responsibilities, resources, and timing. Furthermore, a workable plan must have checkpoints and measurable milestones for tracking and control. These ingredients must be present for the implementation of any plan, whether it involves driving a car from Boston to Philadelphia for a 6:00 p.m. meeting, developing a computer system, building a space vehicle, or developing a new engineering capability such as CAD/CAM. Engineering managers certainly understand these facts. Few managers would go into an engineering program without a clearly defined and detailed plan.

However, what today's engineering leaders often find frustrating is the inability to track and control their efforts in spite of a detailed plan and commitment to specific results. A closer look at this dilemma reveals that the ability to control engineering activities depends not only on the ability to define the work, timing, budgets, and responsibilities, but also on the ability to resolve the enormous complexities of the technical, organizational, and human issues that are involved in today's engineering undertakings of any size.

To be effective, the planning process must be conducive to resolving these larger organizational and technological issues. Quality planning involves more than the generation of paperwork. It requires the participation of the entire engineering team, including support departments, subcontractors, and top management.

Why bother to plan? The benefits of properly conducted planning activities are to be found in many areas. Most important, they lead to more realistic engineering project and activity plans, with people who are involved and interested in the work and committed to its end objectives. Properly conducted planning also makes everyone's job easier and more effective, because the planning activities themselves foster an environment conducive to teamwork and goal achievement. Specifically, properly conducted planning has the following benefits, which form the pillars for effectively organizing and managing engineering programs:

- Establishing a comprehensive roadmap
- Providing insight and perspective
- Providing a basis for setting objectives
- Defining tasks and responsibilities
- Providing a basis for tracking and controlling
- Providing a basis for decision-making

- Building engineering teams
- Minimizing paperwork
- Minimizing confusion and conflict
- Building confidence
- Helping to cope with changes
- Leading to computer-aided decision support
- Enhancing communications throughout the organization.

The benefits of a quality planning process are briefly discussed regarding their specific content, and some suggestions of *how* to obtain the best results are given.

Establishing a Comprehensive Roadmap. One of the key objectives of any engineering plan is to establish a detailed roadmap for the intended activity, one which leads people who will implement the plan to the desired results. This roadmap must define the overall scope of the work and its major activities as well as their specific results, timing, resources, and responsible personnel. Such a detailed and comprehensive roadmap is the prerequisite for making proper personnel assignments, obtaining commitment, and establishing measurability and trackability of the planned activities.

Providing Insight and Perspective. Dwight Eisenhower once proclaimed, "the outcome of a planning activity is useless but its process is indispensable." Although engineering managers might not agree with the first part of Eisenhower's statement, the second part is indicative of one of the major side benefits of planning. Planning generates involvement among the participants and provides additional insight and perspective of the complexities of a given project and its methods of execution. All this involvement can be very persuasive. It leads to a better understanding of activities and personnel requirements. It might stimulate personal interest for planned activities and provide the basis for commitment.

Providing a Basis for Setting Objectives. Planning translates broad strategic goals into an operational plan with specific and detailed goals, objectives, and outcomes. For example, a strategic objective to develop a new technological capability or a new product is broken down into specific operational objectives as part of the detailed planning process. These operational objectives evolve with the details of the plan and its feasibility.

Defining Tasks and Responsibilities. This is yet another side benefit of planning. Through the involvement and perspective generated by planning, it becomes clear how best to partition the overall project into work packages and who might be best qualified to lead each of these tasks.

Providing a Basis for Tracking and Controlling. Only a plan with sufficient detail has a basis for measurability. Measurable milestones can be defined, which permits performance-measuring and tracking of the engineering program throughout its life cycle. Measurability is a crucial prerequisite for managerial direction and control.

Providing a Basis for Decision-Making. Proper planning establishes the benchmarks for problem identification. Problems often surface and can be dealt with during the planning phase. The resulting roadmap with measurable milestones is the basis for managerial decision-making, selective redirection, and control of the engineering activities.

Building Engineering Teams. Planning is persuasive. It not only gets people involved, but also nurtures team spirit. People become personally and professionally interested in a particular task or program during the planning process. They identify with the program, its objectives, and its mission, developing a sense of pride and ownership in the program. This process stimulates team-building, a very important ingredient in the success formula of multidisciplinary engineering undertakings.

Minimizing Paperwork. Contrary to some beliefs, good planning practices actually minimize paperwork. The most effective way to matriculate and document an engineering plan is to use established management tools, such as work breakdown structures, budgets, task rosters, and schedules. Paperwork is usually minimal, even for the most complex program plans. However, engineering managers often learn the hard way that paperwork can become extensive during a poorly managed program when unexpected problems occur, progress and status cannot be measured, and milestones are not being met. At that point, extensive reports, detailed analyses, explanations, and meetings are often requested by upper management or by the activity sponsor. This not only results in unproductive paperwork, at least from an operational manager's perspective, but it also comes at a time when all the energy is needed to rectify a problem, rather than to explain it in a report.

Minimizing Confusion and Conflict. Good planning practices lead to a clear understanding of the work to be done. They further document the execution plan, which cross-validates the agreements worked out with team members and helps to clear up any potential misunderstandings or confusions before the plan gets executed. In addition an agreed-upon, well-documented plan minimizes conflict over the life cycle of an engineering program. This point is made in Figure 3.7, which shows that a well-managed engineering program, although starting off initially with a high level of conflict, becomes reasonably smooth during its execution. Programs that have performance problems have an opposite conflict profile. In explaining this finding, engineering manages point out that planning activities provoke a great deal of discussion and debate over the type of work, its need for resources, time requirements, and who is the best person for the job. For well-run engineering programs, these issues are therefore settled upfront, often generating considerable conflict—beneficial conflict, which brings out new information and ideas and leads to an agreement on how to accomplish the final results. This type of conflict is mostly technical in nature and has few interpersonal aspects. On the contrary, engineering

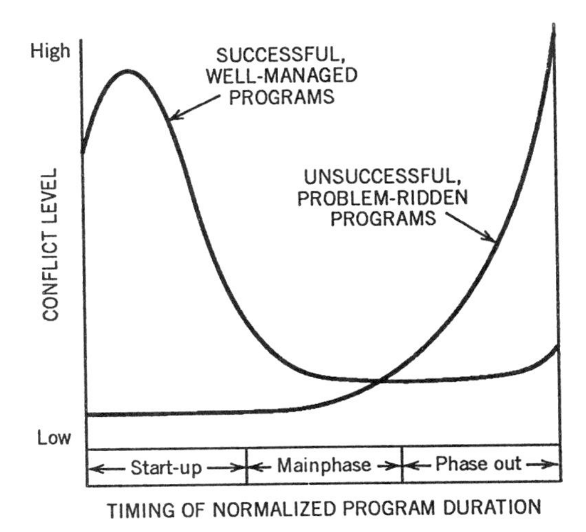

FIGURE 3.7 Conflict levels of successful and unsuccessful engineering programs over their life cycles. Programs were classified as successful if completed as planned within 25% of their budgeted time and money. Programs were defined as unsuccessful if they overran their budgets and/or schedules by more than 60%. Over 50% of the "problem" programs were eventually canceled.

programs that ran into technical problems, overran their budgets, and missed milestones were found to be low on conflict during the program-definition phase. Planning was minimal and so was the level of involvement, debate, disagreement, and conflict. However, as these programs progressed, things did not work out. The various subsystems did not integrate, confusion and conflict over what had been agreed upon developed, and often finger-pointing and witch hunts started to create an unpleasant work environment. Such an environment is retroactive and defensive; it is laced with interpersonal conflict and unproductive activities to defend individual territories and to blame problems on someone else.

Building Confidence. A well-documented plan inspires confidence and builds a "can-do" image with upper management, the customer community, and the engineering team. The image of an experienced and seasoned engineering manager often helps to enhance the manager's credibility and earned power base, which is needed to cross functional lines and to direct multidisciplinary support efforts, in which the manager has little formal authority.

Helping to Cope with Changes. A well-documented plan is a sound basis for identifying and assessing changes from the original plan. These changes may be unavoidable, having been caused by internal or external contingencies. Their impact should be assessed and the original plan should be revised. Similarly, a documented plan makes a "what-if" analysis much easier, which often enables the team to work around a problem or to support a complex decision.

Leading to Computer-Aided Decision Support. The use of standard management techniques for project-oriented planning is consistent with the data format, structure, and methods needed for computer-aided decision support systems, such as the many software packages available to support project planning, tracking, and control activities on minicomputers or microcomputers.

Enhancing Communications. Planning enhances communications throughout the organization by developing the vehicles and the channels for transmitting information. Standard management tools such as schedules, budgets, and task descriptions are very effective for establishing and communicating objectives. The planning process, through its involvement of personnel throughout the organization, helps to define the proper channels for communicating the data. Equally important, during the process of planning many of the potential risks, uncertainties, and problems are identified early on as a part of the enhanced team communication activity. Potential problems become visible throughout the organization, which often includes the customer and the subcontractors. Many engineering managers utilize phased approaches for their planning activities, with clearly defined interfaces and sign-offs to enhance communications by early involvement of all key personnel.[10]

To sum up, planning the engineering function for its organization development and the execution of its specific programs is one of the primary responsibilities of the engineering manager. Most of the manager's organizational performance can be directly related to the quality and effort of planning conducted during the early stages of a new program. To work properly, the process of planning must involve both the performing and the sponsoring organizations. This involvement creates new insights into the intricacies of the plan and its management objectives. It also leads to visibility of the mission at various organizational levels and to management involvement and support. It is involvement at all organizational levels that stimulates interest in the plan and the desire for success and fosters a pervasive reach for excellence, which unifies the management team. It leads to commitment to establish and reach the desired objectives and leads to a self-actuating management system in which people want to work toward the established objectives.

3.5 RECOMMENDATIONS FOR EFFECTIVE PLANNING

With all the pressures for more accurate planning in a more complex world with limited resources, planning practices must help people in an organization to prepare for the future despite the fact that the future often seems unpredictable. The greatest risk in establishing plans in today's environment is to fail to acknowledge these

[10]Phased approaches to program planning break the total engineering program into natural phases, such as requirements analysis, design, prototyping, and testing and integration. When each phase has clearly defined inputs, outputs, and interface responsibilities, each phase can be managed as a separate project. Continuous involvement of other related subprograms should be maintained during the entire life cycle via the review and sign-off process.

dynamics and uncertainties, and therefore to design plans that maintain the status quo—that is, people design plans that are safe and doable, but they make the organization ultra-stable. These plans are doable in a procedural sense. They produce apparently proper intermediate results; but the final output is useless, a classical example of an *activity trap.* The "activity trap" is a situation in which every task is being done correctly but no one is evaluating whether or not *the correct task is being done.* An example is a product development that is executed precisely according to a previously agreed-upon plan, while the market in the meanwhile has shifted, resulting in an inferior or useless product.

In today's environment, planning can be defined as an ongoing, dynamic process of systematically making managerial decisions associated with risks in a constantly changing environment. This process involves a systematic assessment of current capabilities, future environments, and organizational objectives. It also involves a systematic effort to organize the resources for carrying out these decisions, including the means of tracking performance progress via measurable milestones and managerial control toward a set of moving targets.

To be effective, planning practices and their resulting plans should have the following characteristics, regardless of their specific application to strategic, operational, or project situations:

1. Realistically Assess Desires and Opportunities Compared with Capabilities and Priorities. Effective planning attempts to quantify and analytically define most of the key variables affecting plans. The process itself is pervasive, leading to a better definition of goals and objectives and their feasibility and impact on other organizational factors.

2. Involve the Key Personnel with Cross-Functional Activities Early in the Planning Process. This will lead to more realistic inputs to the plan, a better integration of multifunctional components, and a better chance for cooperation, commitment, and self-actuating behavior during the implementation phase.

3. Keep the Planning Process Simple. Minimize paperwork and meetings. Utilize administrative support personnel for data-gathering and analysis, standard forms, and electronic data processing whenever possible. Use modular and phased approaches.

4. Use a Modular Approach. The modular approach to planning breaks the overall task of establishing and executing a plan into major task modules or work packages, such as (1) strategic goal definition, (2) competitive assessment, (3) technology assessment, (4) resource requirements, and (5) the implementation plan, to give a few examples. Each module has its detailed task definition for establishing the plan element, has a time schedule, resources, and a responsible individual named.

5. Use Phased Approach. As the plan can be broken into modules, the complex plan also can be broken down by subdividing it into specific phases, such as (1) goals definition, (2) feasibility analysis, (3) produce development, and (4) production. Each phase can be treated as a separate plan, with separate funding, schedules, responsibilities, and deliverables. Much care should be taken, however, to unify the various phases into one integrated plan.

6. Be Willing To Iterate. Especially during the initial plan-formation stage, it is important that the planning team be willing to iterate several times through the stages of goal definition, customer or sponsor requirements, team capabilities, and final deliverables. In this iterative process, which is schematically shown in Figure 3.8, the realities of the sponsor requirements are assessed against the operational capabilities, after which both may have to be modified in order to obtain a workable plan. For most applications, iteration is far more useful than one-way sequential plan development with go/no-go decision points.

7. Assure a Nonthreatening Work Environment. If people do not feel comfortable within the organization regarding the acceptability of well-intentioned failures, they will not accept work challenges, but rather formulate safe, predictable, and doable plans. Such behavior usually does not lead to sophisticated plans and high organizational performance in a highly dynamic and competitive environment. To solve problems in today's type IV organizations, which are complex, dynamic, and organic in nature, we must engage in type IV problem-solving methods, which include risk-taking, innovative untried approaches, and double-loop learning.[11]

8. Keep Planning Flexible. In our dynamic, changing environment, we are shooting at moving targets. The planning process must acknowledge these dynamics and must have flexibility to respond to these seemingly unpredictable changes in our environment. Regular reviews of the plan and its execution vis-à vis environmental

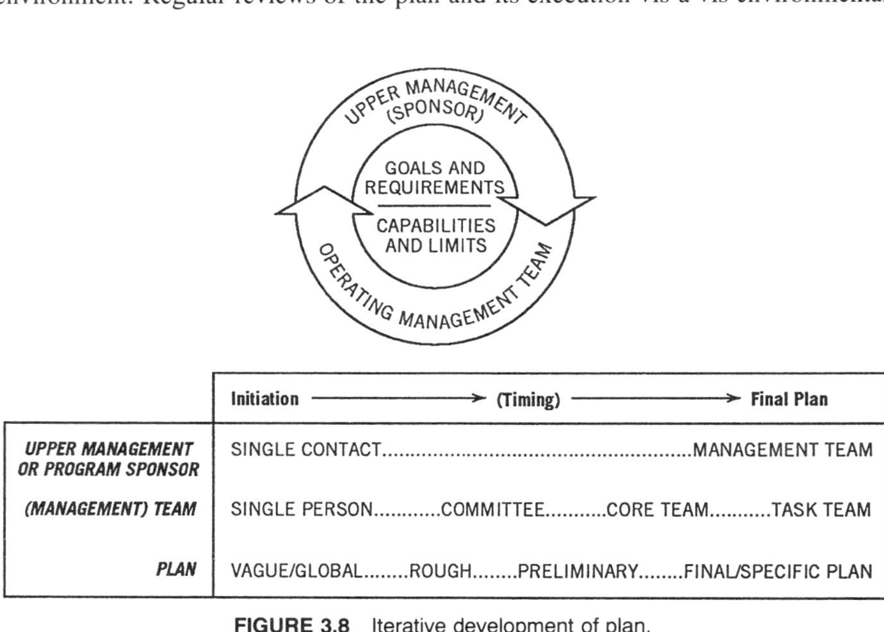

	Initiation ——→ (Timing) ——→ Final Plan
UPPER MANAGEMENT OR PROGRAM SPONSOR	SINGLE CONTACT..........MANAGEMENT TEAM
(MANAGEMENT) TEAM	SINGLE PERSON..........COMMITTEE..........CORE TEAM..........TASK TEAM
PLAN	VAGUE/GLOBAL........ROUGH........PRELIMINARY........FINAL/SPECIFIC PLAN

FIGURE 3.8 Iterative development of plan.

[11]Rensis Likert defined four systems of organization and management: (I) autocrative, (II) benevolent autocratic, (III) consultative, and (IV) participative, group-oriented (*The Human Organization,* New York: McGraw-Hill, 1967).

assumptions and any contingencies minimizes the risk of implementing a plan that in the end produces useless or even detrimental results. Even past success is no guarantee for future success, as the evaluation of comparative performance records such as the Fortune 500 listing quickly shows. Unless an organization reexamines its strategic position, its goals, and its approaches regularly, its efforts may soon be obsolete and inappropriate.

9. Don't Substitute Procedures for Managerial Leadership and Drive. No plan is a substitute for managerial judgment, courage, or even intuition. A plan is a roadmap developed by the management team for their guidance. The plan may be useless in the hands of an unprepared or unwilling manager. Proper leadership further includes fostering a team-oriented work environment, where motivated people with a pervasive desire for excellence make a commitment for reaching desirable, dynamically qualified results.

The next chapter will discuss the specific tools available for planning of the engineering function, with a focus on project-oriented activities.

BIBLIOGRAPHY

Adler, Paul S., William D. McDonald, and Fred McDonald, "Strategic Management of Technical Functions," *Sloan Management Review,* Volume 33, Issue 2 (Winter 1992), pp. 19–37.

Ashmore, G. Michael, "Bringing Information Technology to Life," *Journal of Business Strategy,* Volume 9, Issue 3 (May/June 1988), pp. 48–50.

Ashmore, G. Michael, "Putting Information Technology to Work," *Journal of Business Strategy,* Volume 9, Issue 5 (Sept/Oct 1988), pp. 52–54.

Ayal, Igal and Joel Raban, "Developing Hi-Tech Industrial Products for World Market," *IEEE Transactions on Engineering Management,* (August 1990), pp. 177–183.

Barton, Laurence, "The Use of Scenario-based Planning for Management Executives," *Industrial Management,* Volume 33, Number 6 (Nov/Dec 1991), pp. 8–11.

Batson, Robert G., "Critical Path Acceleration and Simulation in Aircraft Technology Planning," *IEEE Transactions on Engineering Management,* Volume EM-34, Issue 4 (Nov 1987), pp. 244–251.

Bergen, S. A., *R&D Management: Managing Projects & New Products,* Cambridge, MA: Basil Blackwell, 1990.

Boston Consulting Group, *Perspectives on Experience,* Boston: The Boston Consulting Group, 1972.

Brownlie, Douglas T. "The Strategic Management of Technology: A New Wave of Market-Led Pragmatism or a Return to Product Orientation?" *European Journal of Marketing* (UK), Volume 21, Issue 9 (1987), pp. 45–65.

Collins, James C. and Jerry I. Porras, "Organizational Vision and Visionary Organizations," *California Management Review,* Volume 34, Issue 1 (Fall 1991), pp. 30–52.

Contractor, Farok J. and V. K. Narayanan, "Technology Development in the Multinational

Firm: A Framework for Planning and Strategy," *R&D Management* (UK), Volume 20, Issue 4 (Oct 1990), pp. 305–322.

Crow, Michael M., "Assessing Government Influence on Industrial R&D," *Research-Technology Management,* Volume 31, Issue 5 (Sept/Oct 1988), pp. 47-52.

Currid, Cheryl, "IS Managers Can Expect a Year of Planning and Much Change," *InfoWorld,* Volume 14, Issue 2 (Jan 13, 1992), p. 44.

Dean, Burton V., "Multiple Objective Selection Methodology for Strategic Industry Selection Analysis," *IEEE Transaction on Engineering Management,* (Feb 1991), pp. 53–62.

Drigani, Fulvio, *Computerized Project Management,* New York: Marcel Dekker, 1989.

Dyson, Esther, "Re-Creating DEC," *Forbes,* Volume 149, Issue 7 (Mar 30, 1992), pp. 124.

Ehrhardt, S. P., "A Five-Year Plan, a Year at a Time," *Administrative Management,* (Jan 1973).

Engel, Alan K., "How to Profit from Japanese Technology," *Directors & Boards,* Volume 12, Issue 4 (Summer 1988), pp. 14–17.

Fayol, Henri, *Administration Industrielle et Generale,* Paris: Dunod, 1925.

Fayol, Henri, *Industrial and General Management,* London: Pitman & Sons, 1949.

Gilmartin, Patricia A., "Defense Planners Refocus R&D to Meet Changing Military Threats," *Aviation Week & Space Technology,* Volume 136, Issue 11 (Mar 16, 1992), p. 37.

Gilreath, Art, "Participative Long-Range Planning," *Industrial Management,* Volume 31, Number 6 (Nov/Dec 1989), pp. 13–15.

Golden, Brian R., "SBU Strategy and Performance: The Moderating Effects of the Corporate-SBU Relationship," *Strategic Management Journal,* Volume 13, Issue 2 (Feb 1992), pp. 145–158.

Goldsmith, N. "Linking IT Planning to Business Strategy," *Long Range Planning,* Volume 13, Issue 2 (Feb 1992), pp. 119– 134.

Hammerly, Harry, "Matching Global Strategies with National Responses," *Journal of Business Strategy,* Volume 13, Issue 2 (Mar/Apr 1992), pp. 8–12.

Harrison, E. Frank, "Strategic Control at the CEO Level," *Long Range Planning,* Volume 24, Issue 6 (Dec 1991), pp. 78–87.

Henderson, Bruce D., *The Growth Share Matrix of the Production Portfolio,* Boston: Boston Consulting Group, 1971.

Henderson, John C. and N. Venkatraman, "Understanding Strategic Alignment," *Business Quarterly* (Canada), Volume 55, Issue 3 (Winter 1991), pp. 72–78.

Hinterhuber, Hans H. and Wolfgang Popp, "Are You a Strategist or Just a Manager?" *Harvard Business Review,* Volume 35, Issue 1 (Jan/Feb 1992), pp. 22–28.

House, Charles H. and Raymond L. Price, "The Return Map: Tracking Product Teams," *Harvard Business Review,* (Jan/Feb 1990), pp. 92–100.

Humphreys, Kenneth, K. and Paul Wellman, *Basic Cost Engineering,* New York: Marcel Dekker, 1987.

Kerzner, Harold, "Organizing and Staffing the Project Office and Team," in *Project Management for Executives,* New York: Van Nostrand Reinhold, 1982, Chapter 4.

Kleindorfer, Paul R. and Fariborz Y. Partovi, "Integrating Manufacturing Strategy and Technology Choice," *European Journal of Operational Research* (Netherlands), Volume 47, Issue 2 (July 25, 1990), pp. 214–224.

Klimstra, Paul D. and Ann T. Raphael, "Integrating R&D and Business Strategy," *Research-Technology Management,* Volume 35, Issue 1 (Jan/Feb 1992), pp. 22–28.

Koenrer, Elaine, "Technology Planning at General Motors," *Long Range Planning* (UK), Volume 22, Issue 2 (Apr 1989), pp. 9–19.

Likert, Rensis, *The Human Organization,* New York: McGraw-Hill, 1967.

Martin, Dean and Kathleen Miller, "Project Planning as a Primary Management Function," *Project Management Quarterly,* (March 1982).

Mitchell, Graham, "Options for the Strategic Management of Technology," *International Journal of Technology Management,* Volume 3, Number 3 (1988), pp. 253–262.

Monday, Gregory M., "Business Plan: Outlining the Future," *Business Age,* (Sept 1989), pp. 42–45.

Naisbitt, John, *Megatrends,* New York: Warner Books, 1982.

Naisbitt, John and P. Aburdene, *Megatrends 2000,* New York: William Morrow, 1990.

Newman, William H., "Focused Joint Ventures," *Academy of Management Executive,* Volume 6, Issue 1 (Feb 1992), pp. 67–75.

Obradovitch, Michael M. and S. E. Stephanou, *Project Management: Risks and Productivity,* Bend, OR: Daniel Spencer, 1990.

Peevey, Michael R., "Transforming Organizational Structure," *Executive Speeches,* Volume 6, Issue 6 (Jan 1992), pp. 6–9.

Powell, Thomas C., "Organizational Alignment as Competitive Advantage," *Strategic Management Journal,* Volume 13, Issue 2 (Feb 1992), pp. 119–134.

Randolph, W. Alan and Barry Z. Posner, *Effective Project Planning and Management,* Englewood Cliffs, NJ: Prentice-Hall, 1988.

Shina, Sammy G., *Concurrent Engineering and Design for Manufacture of Electronics Products,* New York: Van Nostrand Reinhold, 1991.

Shoeffler, Sidney, Robert Buzzell, and Donald Heany, "Impact of Strategic Planning on Profit Performance, *Harvard Business Review,* (Mar/Apr 1974).

Silverman, Melvin, *The Art of Managing Technical Projects,* Englewood Cliffs, NJ: Prentice-Hall, 1987.

Singer, Joseph F., "Strategic Management Planning Process for Acquisitions Decision-Making in Consulting Engineering Firms," *IEEE Transactions on Engineering Management,* (May 1988), pp. 114–117.

Stalk, George, Philip Evans, and Lawrence E. Shulman, "Competing on Capabilities: The New Rules of Corporate Strategy," *Harvard Business Review,* Volume 70, Issue 2 (Mar/Apr 1992), pp. 57–69.

Steger, W. A., "How to Plan for Management in New Systems," *Harvard Business Review,* (Sept/Oct 1962).

Szakonyi, Robert, "Critical Issues in Long-Range Planning," *Research-Technology Management,* Volume 32, Issue 3 (May/June 1989), pp. 28–32.

Thamhain, Hans J., "Managing Technologically Innovative Team Efforts Toward New Product Success," *Journal of Product Innovation Management,* Volume 7, Number 1 (Mar 1990), pp. 5–18.

Titus, George J., "Five-Step Procedure for Auditing R&D Planning and Budgeting Processes," *IEEE Transactions on Engineering Management,* (May 1991), pp. 171–177.

Toffler, Alvin, *Future Shock,* New York: Bantam Books, 1970.

Varadarajan, P. Rajan, Terry Clark, and William M. Pride, "Controlling the Uncontrollable: Managing Your Market Environment," *Sloan Management Review,* Volume 33, Issue 2 (Winter 1992), pp. 39–47.

Villareal, Louis, "Strategic Planning, Company Beliefs and Applied IE," *Industrial Management,* Volume 33, Number 6 (Nov/Dec 1991).

Wiersema, Margarethe F. and Karen A. Bantel, "Top Management Team Demography and Corporate Strategic Change," *Academy of Management Journal,* Volume 35, Issue 1 (Mar 1992), pp. 91–121.

Yamanouchi, Teruo, "Why is Globalization of R&D Necessary Today?" *Journal of Science Policy and Research Management,* Volume 2, Number 2 (1987), pp. 123–130.

4
ENGINEERING PROJECT PLANNING

4.1 INTRODUCTION

For many engineering managers involved in day-to-day operations, engineering management is synonymous with project management.[1] Their organizations evolve around task teams rather than functional units to accommodate multidisciplinary work integration. Control and information must flow both vertically and horizontally. They have to deal with several bosses, share power and resources, and deal with support departments over which they have little or no direct control.

In such an open and dynamic environment, planning is crucial, not only at a strategic level, but also at the operational level, which is often project-oriented. In addition, many engineering departments execute large numbers of specific projects or programs in parallel, such as new product development and external contracts.

Engineering managers today have a set of powerful tools available that have proven capability to plan and control multidisciplinary activities effectively. Most of these tools were originally developed within the aerospace and construction industries. They are, however, equally effective and find increasing applications in other areas, ranging from new product development to engineering training programs and R&D concept studies. Like strategic planning, project planning establishes a comprehensive roadmap for the work to be done and fosters a work environment of involved, interested, and committed personnel, who focus on the objectives and direct the project toward its goals.

[1]The terms *project* and *program* are often used interchangeably. However, if both terms are used within the same organization or write-up, *program* refers to a bigger entity, while *project* refers to a small unit of work, or to a subsystem that is to be integrated with a program. For simplicity, this chapter uses the term *project,* most of the time.

The Challenges of Project Planning

With all these benefits, why is there often so much reluctance to do proper project planning and sometimes an outright rejection of the planning process? The reluctance centers around four problems: start-up delays, stifled creativity, rigid controls, and lack of funding. First, formal planning often repels the action-oriented manager, who wants to start work and see some business activities. Second, the planning system is perceived as stifling creativity. The comment of one engineering project manager may be typical of many situations: "My support personnel feels that we spend too much time planning a project up front; it creates a very rigid environment that stifles innovation. The only purpose seems to be establishing a basis for controls against outdated measures and for punishment rather than help in case of a contingency." This comment is echoed by many project managers. It also illustrates the third problem, the misuse of planning to establish unrealistic controls and penalties for deviations from the program plan, rather than helping to find solutions. Whether these fears are real or fantasy does not change the situation. It is *perceived coercion* that leads to the rejection of the planning system. The fourth problem relates to the lack of funding. Proper project planning requires the involvement and participation of personnel from all parts of the performing organization, an effort that takes time and resources. Unless this activity is planned and agreed upon with the customer, it is difficult to get such a program definition phase funded as an afterthought to an already established contract.

Solving Project-Planning Problems

Few companies have introduced project-planning procedures with ease. Most have experienced problems ranging from skepticism to sabotage of the planning system. Realistically, however, program managers have not much of a choice, and have to introduce project planning, especially for the larger, more complex programs. Every project manager who believes in planning has his or her own success story. It is interesting to note, however, that many use incremental approaches to project-planning in order to solve their problems. The incremental approach recognizes the pressures for initial operating efficiency by getting the work started, while developing the project plan incrementally. For example, at the time a particular work package is executed, all parameters must be clearly defined. It is, however, not necessary, and often impractical, to establish these details at the outset of the program. What is required is a process that provides the discipline for incremental detailed development of the measurable parameters as the program progresses.

Developing and implementing such a project plan incrementally is a multifaceted challenge to management. The problem is seldom to understand the techniques, such as budgeting and scheduling, but to involve the project team early in the planning process, to get their inputs, support, and commitment, and to establish an environment that unifies and pervades. Such an environment will help build a self-activating system, where people work toward realistic, agreed-upon objectives and are determined to meet them. Furthermore, project personnel must have the feeling that the project plan facilitates communication, is flexible and adaptive to the changing environment, and provides an early warning system through which project personnel obtain assistance rather than punishment in case of a contingency.

4.2 PLANNING FOR ENGINEERING PERFORMANCE

By definition, a project plan is a roadmap that defines (1) the task to be done, (2) the timing, (3) the resources, and (4) the responsible personnel. To be effective, however, several other conditions must be met. Involvement and participation of all project team members and their managers is usually necessary for the development of a realistic plan, one which is operationally feasible and will eventually obtain the acceptance and commitment of all key project personnel. A quality project plan must have measurable milestones, each of which has specific deliverable items or results defined. This will make it possible to measure project status and performance and provide the crucial inputs for controlling the project to achieve the desired results.

Effective project planning requires particular skills far beyond writing a document with schedules and budgets. It requires communication and information-processing skills to define the actual resource requirements and administrative support necessary. It also requires the ability to negotiate the necessary resources and get commitments from key personnel in various support organizations, although one may have little or no formal authority over them. Taken together, effective planning requires skills in the areas of:

- Information processing
- Communication
- Resource negotiations
- Securing commitments
- Incremental and modular planning
- Assuring measurable milestones
- Facilitating top-management involvement
- Motivation and leadership.

In addition, the project manager must ensure that the plan remains a viable document. Changes in project scope and depth are inevitable. The plan should reflect necessary changes through formal revisions and should be the guiding document throughout the life cycle of the project. Nothing outlives its usefulness faster than an obsolete or irrelevant plan.

Finally, project managers need an awareness that planning can certainly be overdone. If not controlled, planning can become an end to itself and a poor substitute for innovative work. Individuals retreat to the utopia of no responsibility where innovative actions cannot be taken "because it is not in the plan." It is the responsibility of the project manager to build enough flexibility into the plan and police it against such misuse.

In addition to having specific skills and disciplines, engineering project managers must be able to recognize potentially harmful organizational barriers, which can impair successful planning and plan implementation. Project managers must be able to diagnose the type of tools and the level of involvement needed in each particular case. The following list describes specific challenges and potential problems. The discussion should help engineering project managers to understand the complex relationships among organizational variables and project management effectiveness.

1. Management tools for project planning and control are no substitute for innovative thinking and sound management practices. Network techniques, such as program evaluation and review techniques (PERT) or critical path method (CPM), task matrices, work breakdown structures, budgets, and even computer-aided project tracking systems are often misused to fabricate desirable results and to justify poor project performance. Support departments are annoyed if data are taken out of context. The worst situation occurs when project management personnel try to manage the tools rather than the project. The remark of one executive is typical of the frustrations experienced: "Too often the program managers spend a lot of time to determine where and why certain schedules or budget variances occurred so they can reschedule the job properly and generate all the backup reports. Little is done to catch these problems before they get messy, rectify them, and get the project back on its original track."

2. Not every management tool is universally appropriate for all projects. This is particularly true for the more sophisticated computer-aided tools, such as network (PERT), variance analysis, line of balance (LOB), or performance measurement system (PMS) techniques. Management tools are seldom too simple, but often too complex. Indicators of inappropriate tools are (1) program management personnel spends disproportionate amounts of time in preparing and evaluating program controls, (2) only a few people can interpret the generated data, (3) the information generated is too late to be relevant to ongoing management activities, and (4) the tools do not really contribute to the effectiveness of managing and controlling the program activities.

3. Project planning is the responsibility of the project leader. Planning is time-consuming. It is tempting, therefore, especially for larger engineering departments, to delegate planning responsibilities to staff personnel, or to create a special project planning group. This practice, although convenient, may have unintended consequences as the planning department becomes institutionalized and uses its own techniques, jargon, and standards.[2] Project planning performed by such staff groups is often perceived as isolated from the real-world situation and noncommittal to the project office.

Planning of the project is clearly the responsibility of the project leader. Only he or she can participate effectively in the search for integrated solutions; assess risks; make tradeoffs among cost, schedule, and technical issues; and ultimately elicit the

[2]A more detailed discussion of these staff planning problems is provided by Louis A. Allen in *Managerial Planning,* New York: McGraw-Hill, 1982.

commitment of project personnel to agreed-upon objectives. This does not mean, however, that the project leader cannot get assistance from competent staff personnel. Proper assistance is particularly important for larger programs, where the program manager does not have the time nor all the special administrative skills to plan the program single-handed. However, the final responsibility for integrating the overall program plan, the evaluation of tradeoffs and risks, and the commitment of support personnel to the plan are the ultimate responsibility of the program manager or project leader.

4. Senior management must provide global direction and leadership. Senior level management direction, involvement, interest, and support are usually necessary to create proper visibility for the project. It is this visibility that stimulates involvement and interest among project team members. It builds the desire for success, a pervasive reach for excellence that unifies the team and leads to commitment, the cornerstone of project planning.

Engineering projects and programs vary widely in scope, complexity, and duration. Thus plans and control systems for the execution of these projects must vary in depth and content. To achieve some standard topology in their planning, most project managers use a phased approach for organizing and executing projects. Accordingly, their project plans are partitioned into those specific phases, as will be discussed next.

4.3 THE PHASED APPROACH TO PROJECT PLANNING

Most engineering projects and programs are too complex to be managed effectively in a single step. The enormous intricacies among the technical, organizational, and human components make disciplined planning, tracking, and controlling of a project very difficult, unless specific work packages can be separated and managed as independent projects. In fact, the breakdown of a project into its natural work packages has been practiced ever since the formal introduction of the work breakdown structure in the 1950s. However, its application to time-phased partitioning of a program, with clearly defined inputs, outputs, and interface criteria, is a relatively new development.

What many engineering managers did was to break up the complexities of their projects and make them more manageable via phased approaches. This appears to offer many organizations a way out of the problems that are associated with the complexities of modern technical projects, including obscured measurability and bewildering arrays of interdependencies.

The Basic Concept

The principal concept of phased project planning is simple and straightforward. It relies on the program manager's ability to break up the overall project into discrete phases, with measurable outputs and multifunctional involvement.

Engineering managers use the phased approach for organizing projects in a disciplined and simplified manner. The phases may follow functional lines, such as system design, engineering development, and prototyping. The phases may follow business cycles, such as requirements analysis, bid proposal, prototyping, production, and so on, or they may follow any other logic. Although the type of phases chosen may vary with different project situations, project managers should make an effort to define natural phases that divide the overall project into logical sets of activities with specific outputs.

For purpose of illustration, we have divided the overall project and its planning activities into six phases. This provides a basic topology for project planning and management that is applicable to a wide range of technical projects and programs.

1. Conceptual project planning
2. Defining the project
3. Defining project integration and control
4. Project organization and start-up
5. Project execution
6. Project phaseout.

The primary features of the phase management system are: (1) it breaks down the complexity of the planning and execution effort of a program and (2) it provides an integrated system for interdisciplinary reviews and integration. Before a project can progress from one phase to the next, specific requirements regarding deliverables and sign-offs must be met. This establishes a discipline, checkpoints, and interdisciplinary involvement, which ultimately facilitate the orchestration and integration of a multifunctional project.

As for other planning activities, each step should involve the key personnel affected by the plan and should lead to a basic consensus. Several iterations of activities, reviews, and decisions might be necessary during the first three steps. The phased project planning process can be useful regardless of the project type, complexity, and duration, but is specifically designed for the more complex, multidisciplinary programs of longer duration. The principal stages of the phased planning process are summarized in Table 4.1 and briefly discussed below.

Phase 1: Conceptual Project Planning

Conceptual planning is the first formal activity phase of a new project under consideration. The objective of the conceptual planning phase is to define the scope of the project and its principal feasibility, quickly and inexpensively. Brainstorming sessions involving all key players from the engineering and business community of the company have been found effective in assessing the basic ability of the project at the conceptual stage.

The first half of this phase involves defining the global goals and objectives of the project as well as its target parameters such as (1) functional specifications and

TABLE 4.1 Project Organization and Management Using the Phased Approach

Phase	Key Activities	Key Results	Responsible Person(s)	Relative Effort
1. Conceptual planning	Defining requirements Program definition Feasibility Management approach	Global goals Target budget Key milestones Key concepts Major deliverables Key responsibilities	Engineering manager Program manager	2%
2. Project definition	Identify discrete program phases Detailed program planning Review, reporting, and control system definition	Program phases Responsibilites Program objectives Work definition Schedules Budgets Task rosters Deliverables	Program manager Task leaders	8%
3. Defining project integration and control	Interface definition Program review definition Sign-off process Performance measurements, tracking and control definition	Phased results Sign-off personnel Integration tasks Key review dates Target integration dates Check points	Program managers Task leaders	2%
4. Project organization and start-up	Kickoff (commencement) Team organization	Program plan Resources	Program manager Task leaders	5%
5. Project execution	Project management Coordination, integration Review, reporting Controlling	Deliverables Project reports Change proposals Technical results Documents	Program manager Task leaders	78%
6. Phaseout	Resource transfer Follow-on business	Program organization dissolved Contract closure	Program manager	5%

features, (2) major deliverables items, (3) key milestones, (4) budget limitations, and (5) business objectives regarding market, technology, and financial targets.

The second half, often conducted simultaneously, is to determine project feasibility from both technical and business points of view—that is, to determine whether the requirements set forth by the project sponsor and articulated in the target parameters can realistically be met. The question is: do we have the capability to perform?

Often, the conceptual planning phase also addresses the principal management approach to the project, including organizational issues and make–buy alternatives.

Accordingly, the requirements analysis, like the conceptual analysis, requires inputs from various organizational components: (1) the customer, (2) the company's planning system, (3) the company functions and (4) the company's socioeconomic environment, as is graphically shown in Figure 4.1.

No engineering project can be organized without first establishing the basic feasibility of and investigating the desirability of the effort. The starting point for Phase 1 is often the emergence of a new contract opportunity, the need for a new product development, or the redirection of an ongoing project.

The objective of the preliminary analysis effort is twofold: (1) establish a basis for management decision-making regarding whether to proceed and how to proceed with the project; (2) define the scope and baseline of the projects, hence generating the inputs for the project definition phase, which will include detailed plans for start-up or a bid proposal.

The requirements analysis phase is a project by itself, which usually has specific deadlines, resource limitations, and results. Like any other task, the preliminary analysis varies in scope and deliverables depending on the specific objectives and the nature of the upcoming project. However, the basic results required from the analysis are often common among many projects. Typically, the efforts leading to these results include the following tasks:

1. Determine user or sponsor requirements regarding baseline configuration, specifications, functional description, features, budget, schedule, and deadlines.
2. Assess technical feasibility.
3. Determine principal resource requirements, including work force and key personnel, facilities, special resources, and funding (budgetary estimate).
4. Assess risk areas, including (a) technical needs, (b) scope of schedule and budget, (c) staffing, (d) market contract award, product marketing, and so on, and (e) financial risks (cash flow, return on investment, cost overruns, and so on).
5. Assess program feasibility (requirements versus capabilities) re: technical needs, timing, costs, resources.

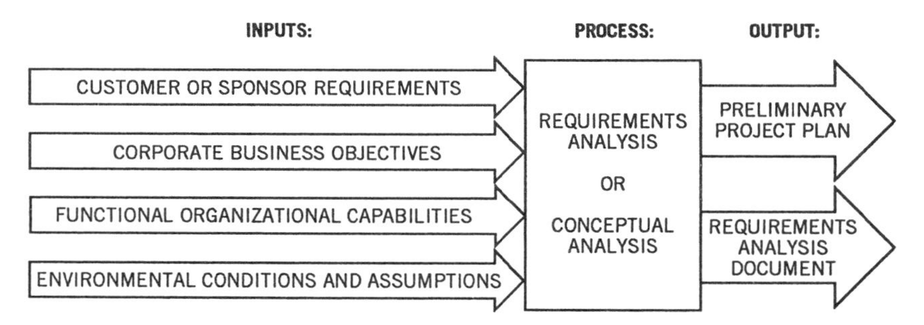

FIGURE 4.1 Inputs and outputs of the requirements analysis process.

6. Analyze alternatives regarding (a) different technical approaches of features, (b) trade-off analysis of concepts, products, vendors, (c) make–buy decisions, (d) teaming, and (e) timing of effort.
7. Assess potential business success, regarding (a) competitive assessment, (b) contract award, (c) product acceptance, (d) pricing, (e) financial performance.
8. Establish project management approach.

The outcome of this preliminary project analysis is usually a concise report, which summarizes the project effort regarding technical, resource, and timing requirements. The report then compares the requirements with the organization's capabilities and assesses overall project feasibility. Finally it comments on the potential for business success and the project's desirability.

The report provides the basis for a management decision on whether and how to proceed with the project. In competitive bidding situations, the bid decision is usually supported by such an analysis.

Performing a requirements analysis is hard work. One person alone can seldom do an adequate job. It requires the involvement of key personnel from the sponsor and from all organizations that are expected to support the project later on. The engineering manager often acts as a facilitator who brings together the right people from the sponsor, functional organizations, subcontractors, and top management. Moreover, the process of determining and matching the sponsor requirements with the capabilities of the performing organization is highly interactive. It usually takes several rounds of meetings before a basic agreement on the project baseline is reached that is both doable and affordable.

In parallel to the requirements analysis, the communications network with the customer and sponsor community must be developed. This is a complex task, which involves arrangements for meetings among the proper functional groups of both the customer and of the performing organizations. These customer contact meetings provide the communication vehicle for working out the project details and matching and fine-tuning the customer requirements with the contractor's capabilities and available resources. The meetings further lead to involvement and commitment of all parties engaged in the project.

Management responsibility for conceptual project planning is usually shared between the engineering manager and project manager.

Phase 2: Defining the Project

The second phase involves the partitioning of the overall project into subprojects or activity phases, which flow naturally and can be executed to minimize multifunctional involvements. As an example, a typical engineering development might involve the following activity phases: (1) requirements analysis, (2) project definition, (3) project organization and start-up, (4) design and development, (5) prototyping, (6) testing and integration, (7) pilot production, (8) volume production, (9) market development, and (10) product enhancements and logistics.

Many of these activity phases are overlapping and will be executed concurrently, but in principle the phase concept provides a method for breaking up the enormous complexity of today's engineering projects and establishes a framework for project planning. In fact, the phases can often be transcribed, one for one, into the major subsystems of the project such as the work breakdown structure, master schedule, the budget, and the key responsibilities.

The responsibility for breaking the project into phases usually rests with the project manager. Some companies have procedural guidelines to assist the project leader in this task. This is a relatively small but important effort, which often is made part of the conceptual planning phase.

Developing the Program Plan. The project definition phase is often the first organized effort in determining what needs to be done, how the task can be accomplished most effectively, and who is best qualified to do the job. Project definition is a complex, well-orchestrated "impedance-matching effort" among customer and sponsor requirements, company capabilities, and available resources. Table 4.2 summarizes the principal steps of the requirements analysis, which eventually lead to the project plan. This project plan also forms the cornerstone for the bid proposal, which is required if the sponsor is an external customer. Regardless of the type of sponsor involved, the experienced engineering manager will always require a detailed project plan that clearly shows how the project is to be executed and what is to be delivered at the end, even for company-internal projects. This is a good practice, which clarifies many potential problem areas and leads to a clear contractual agreement between the sponsor and all performing organizations. A detailed outline for a typical engineering project management plan is provided in Table 4.3.

One of the most crucial inputs to project definition comes from the requirements analysis, which is made either before or during the time of project definition. The requirements analysis consists of a series of meetings, analyses, reports, and other forms of communication. It involves all supporting organizations, potential subcontractors, customer personnel, and top management. The process, discussed under Phase 1, leads to the preliminary project plan, which presents the project requirements, capabilities, and resources together in an organized fashion, most likely for the first time, including risks and alternatives.

For larger projects it is often beneficial to break the total project plan into modules or subplans. This reduces complexity and increases the clarity of the plan. It also facilitates multifunctional group involvement and plan development in different time phases. It further eases plan updates and changes. Table 4.4 presents examples of individual subplans that might be required in support of a total engineering project plan. However, management must be sure that planning activities do not proliferate, since this might lead to additional and unnecessary paperwork, extra overhead expenditures, and start-up delays. Management must have a good practical sense of how much modularity and detail is beneficial, so that it does not become an administrative burden on the team.

TABLE 4.2 Sequence of Activities in Project-Definition Phase

Step Effort (and Who is Responsible)	Key Interfaces	Description of Effort	Output and Results
1. Start (management)	Top management Functional directors Project managers	Kickoff by management. Project manager starts with the basic requirements and global objectives available at the outset.	Basic charter Global objectives Initial scope
2. Management commitment and seed money (project manager)	Top management Functional directors	Obtain management endorsement for project formation and seed money. (In dealing with company-external project opportunities, this is commonly referred to as a *preliminary bid decision.*)	Initial commitment Initial bid decision Seed money
3. Functional disciplines (project manager)	Functional managers	Define the key disciplines involved in the project and build an executive project steering committee representing these disciplines.	Project formation/acquisition plan
4. Data file (project manager)	Project originators Customer/sponsor Marketing	Develop project data file. Collect as much information as practical on established customer requirements, analyses, preliminary budgets, estimates, schedules, specifications, and documented communications.	Project data file: • Requirements • Budgets • Schedules • Applications
5. Key project personnel (project manager)	Functional managers	Form a team of key project personnel to (1) analyze the requirements, (2) negotiate with the customer, (3) plan, and (4) organize the project. (In dealing with an outside customer this team most often becomes the bid proposal team. Later on the proposal team assumes project responsibility and forms the management network for this particular project or program.)	Core team Proposal team

(continued)

TABLE 4.2 Sequence of Activities in Project-Definition Phase *(continued)*

Step Effort (and Who is Responsible)	Key Interfaces	Description of Effort	Output and Results
6. Requirements analysis (project manager and key personnel)	Customer/sponsor Functional support organizations Subcontractors	Analyze all available data relevant to the customer requirements, internal capabilities, and available resources. This analysis must involve all supporting functional organizations, potential subcontractors, the customer's personnel, and top management. It will lead to a preliminary project plan.	Project plan: • Scope and global project objectives • Work breakdown structure • System specifications • System diagrams • Functional block diagrams • Cost estimates • Statement of work • Project organization and task matrix • Skill and work force requirements • Milestone schedule • Hardware and software tree • Tradeoff analysis • Make–buy plan • Bill of material • Purchasing specifications and requests for quotations • Risk analysis • Start-up plan • List of deliverable items

7. Cost estimate (functional managers)	Customer/sponsor Functional support organizations Subcontractors	Generate a parametric estimate, top-down, first. Develop detailed bottom-up estimates based on work breakdown structure as cost model. Cost estimating is usually an iterative process which is cycled through the fine-tuning of customer requirements.	(Project plan as in Step 6).
8. Customer contacts (project manager and key personnel)	Customer/sponsor Functional support organizations Subcontractors	Develop the communication network with the customer community. Arrange meetings among the proper functional groups of both the customer and the performing organizations. Work out the project details, matching and fine-tuning the customer requirements with the contractor capabilities and available resources.	(Project plan as in Step 6).
9. Iterate (project manager and key personnel)	Customer/sponsor Functional support organizations Subcontractors	Perform more detailed requirement analyses and develop the specifics of the project plans. This is a team effort, which requires close cooperation of all personnel from both the customer and the performing organizations. In essence, it is an iterative process of steps 6 and 9.	(Project plan as in Step 6).
10. Project plan (project manager)	Customer/sponsor Functional support organizations Subcontractors	Integrate the results from the requirements analysis into the project plan. Use modular approach.	(Project plan as in Step 6).
11. Proposal (project/proposal manager)	Top management Functional support organizations Subcontractors Marketing	If required, develop a project proposal for the customer or sponsors, to document how the project will be managed and how the requirements will be implemented against the established specifications, work statements, schedules, and budgets.	Technical proposal Management proposal Cost proposal

TABLE 4.3 Outline of a Typical Engineering Project Management Plan

Section 1. Executive Summary. A top-level overview of the project, including:

Global project objectives
Brief description of tasks to be performed
Resource requirements
Key milestone schedule
Project organization
Management team
Management approach
Technical approach
Risks

Tools: Work breakdown structure, master schedule, budget, project organization chart

Section 2. Project Scope. This section *briefly* describes the overall project and its operational interfaces, indicating:

What needs to be done
How the project fits into the larger system or mission
Financial and scheduling boundaries
Major results and deliverables

The scope section should cross-reference the work breakdown structure, statement of work, specifications, master schedule, and project budget.

Tools: System block diagram, functional flow diagram, hardware and software tree, work breakdown structure

Section 3. Project Schedules. This section shows a simple project master schedule, referencing the key milestones and integration points. It lists all subordinate schedules and their reference documents, such as:

Development schedule
Production schedule
Test schedule
Integration schedule
Time phasing

Tools: Project master schedule, milestone chart

Section 4. Project Organization. This section describes the organizational structure, authority, and reporting relationships, including:

Work force allocation for project
Company organization in relationship to project
Project organization and rationale for its structure
Key personnel
Task responsibilities

Supporting organizations
Subcontractors
Work force loading
Job descriptions
How to build the new project organizations
Key interfaces to customer or sponsor organization
Related policies and procedures

Tools: Organization chart, project charter, task matrix, work force loading chart, job description, policies, and procedures

Section 5. Staffing and Personnel Development. This section describes the specifics for recruiting, training, and maintaining the necessary project personnel over the entire project life cycle:

Work force loading over project life cycle
Listing of skill requirements versus task descriptions
Key personnel requirements
Inventory of available talents
Analysis of organization and task environment
Methods of recruiting or assignment
Training and development plan
Methods of performance appraisal and rewards
Professional growth potential
Related policies

Tools: Job descriptions, task matrix, work force loading chart, policies, and procedures

Section 6. Project Management. This section describes *how* the project will be managed. It refers to and integrates other operational sections of the project plan. Topics include:

Project management structure and philosophy
Support organizations
Subcontractor management
Advisory groups and committees
Cost accounting system
Work planning, budgeting, and authorizing
Management controls: task authorizations, schedule monitoring, budget monitoring, technical performance measurement, integrated cost, schedule, and performance measurements and control, management reviews, technical reviews, approval system
Related policies and procedure
Project management forms
Project management directives and guidelines

Tools: Project organization chart, task matrix, work breakdown schedule dictionary, management guides, operating instructions, policies, and procedures

(continued)

TABLE 4.3 Outline of Typical Engineering Project Management Plan *(continued)*

Section 7. System Engineering Management (SEM). This section describes the plan for *producing and integrating* the various engineering specialties into the total system most cost-effectively. Typically this section includes:

System engineering objectives
Life cycle cost (LCC) considerations
Tradeoff studies
Make–buy plan
Design-to-cost (DTC) program
Configuration management
Logistics support
Design engineering program
Producibility program
System integration and testing
Reliability program
Quality program
Standardization
Human factor engineering
Safety program

Tools: Tradeoff studies, make–buy analysis, parametric cost estimates, operational scenarios, engineering plans

Section 8. Technical Approach. Starting with the system block diagram and the system design/integration framework set in Section 7, System Engineering Management, this section describes *how* the various subsystems and system components are to be developed, tested, and produced, including:

Engineering process
Design philosophy
Relationship among interdisciplinary support groups
Monitoring and controlling design and development activities
Configuration management and document control
Change management

A section on each system component containing (1) reference to work breakdown structure, specifications, and statement of work, (2) technical approach, (3) tradeoffs and options, (4) risks, and (5) criteria for design release.

Section 8 is often just a summary of the technical volume which describes the implementation of each subsystem in detail.

Tools: Work breakdown structure, specifications, statement of work, system engineering management plan, system block diagram, functional flow diagram, hardware and software tree, policies, and procedures.

Section 9. Facility Support. This section describes the type of engineering and other facilities needed to complete the project. Facilities may include:

Plant
Office
Machinery
Roads
Transportation
Test beds
Laboratories
Design support equipment such as CAD/CAM

For each new construction or development required a detailed plan should be prepared.

Tools: Organization chart, department charters, facility description, plant and office layout

Section 10. Resource Requirements. This section summarizes the project resources, time phased, over the project cycle:

Budgets
Personnel
Management
Support departments
Special equipment

Tools: Budgets, schedules, task matrix, work-force loading chart

Section 11. Risk Analysis and Contingency Plans. This section analyzes major risk areas and presents contingency plans:

Technical risks
Innovations
Market risks
Financial risks
Subcontractor risks
Contractual risks and penalties
Terminations
Critical personnel
Long-lead items
Technical exposures

Section 12. List of Deliverable Items. This section summarizes all major items deliverable under the project agreement, including:

Equipment
Software

(continued)

TABLE 4.3 Outline of Typical Engineering Project Management Plan *(continued)*

Section 12. (continued)
Documentation
Tooling
Training
Spare parts

This is a short list of only major items, referring to more detailed contractual documents. The list of deliverables should be fully consistent with the work breakdown structure and the statement of work and should show for each item (1) a target date and (2) a budget or cost account.

The typical outputs of the project definition phase are the following:

- Work breakdown structure
- Task matrix
- Statement of work
- Cost estimates and project budget
- Milestones and master schedule
- Make–buy plan
- Purchasing specifications and requests for quotations
- Work packages
- Project team organization.

Phase 3: Defining Project Integration and Control

One of the most difficult tasks of the project leader is to achieve proper integration of the various subsystems. This challenge exists, regardless of whether the leader is using or not using a phased approach to program management. However, if the project is partitioned into specific phases, the potential problems usually surface during the planning stage.

As the project moves through its life cycle, it passes from one phase to the next. In each phase, specific tasks must be completed and integrated with the next phase. At the transition points, such as bringing the project from engineering into manufacturing, certain deliverable items must be available and procedures for the project's transition must be followed. In essence, each phase consists of a number of activities, inputs, and outputs, which must be defined and orchestrated:

1. Project tasks
2. Project deliverables prepared to specifications
3. Project logistics and management control
4. Documentation

TABLE 4.4 Types of Program Subplans

Type of Subplan and Description
Budget: How much money is allocated to each activity?
Configuration management: How are technical changes made?
Cost control: How are expenditures against established program performance and schedule controlled?
Documentation: How is program documentation recorded and tracked?
Facilities: What facility and plant resources are available?
Human factors: How is the human interface engineered?
Logistics support: How will training, maintenance, and repair be handled?
Management: How is the program organized and managed?
Manufacturing: What are the time-phased manufacturing events?
Procurement: What is the program acquisition process and capture plan?
Producibility: How is the equipment manufactured to given requirement?
Product support: How are marketing, distribution, sales, and service handled?
Quality assurance: How are specifications and requirements met?
Research and development: What are the advanced technical activities?
Scheduling control: How do program activities compare to time?
System engineering management: How is the multidisciplinary engineering design and development effort managed?
System integration: How are the various components integrated and tested as a total system?
Test: How are the various systems tested against specifications?
Tooling: What are the time-phased tooling requirements?
Transportation: How will goods and service be delivered?
Vendor selection: How are qualified vendors identified and selected?

5. Sign-offs (criteria and responsible individuals for final acceptance of deliverable items for each phase).

Some operational guidelines are provided to assist managers in achieving effective project integration. The management tool is the integration plan, a procedural document that links the various subsystems and phases of the project and facilitates its integration. The integration plan includes definitions of the following categories of information, which can be summarized for each phase on a form sheet such as is shown in Figure 4.2.

1. Task Summary. Task leaders for each project phase define their specific work package, including the tasks, timing, responsibilities, and resource requirements.

2. Project Deliverables. All specific results or deliverable items should be defined in a listing that is referenced to requirements documents, specifications, and schedules. In general, deliverables may include both technical and logistical items. In an effort to keep the process simple, we have separated the logistical from the managerial items such as reviews, sign-offs, and documentation. In this discussion we

INTEGRATION PLAN PHASE NO. 4: DESIGN AND DEVELOPMENT

Program Name: ______________________ **Review Date:** ____________

Program Manager: ______________________________

Phase/Workpackage Manager: ______________________________

Sponsor/Customer: ______________________________

1. TASK SUMMARY

2. PROJECT DELIVERABLES

1. ______________________ 6. ______________________
2. ______________________ 7. ______________________
3. ______________________ 8. ______________________
4. ______________________ 9. ______________________
5. ______________________ 10. ______________________

3. LOGISTICS AND MANAGEMENT CONTROLS

TASKS	DONE		DUE DATE	RESPONSIBLE INDIVIDUAL
1. ________	Y	N	________	________
2. ________	Y	N	________	________
3. ________	Y	N	________	________
4. ________	Y	N	________	________
5. ________	Y	N	________	________
6. ________	Y	N	________	________

4. DOCUMENTATION

TASKS	DONE		DUE DATE	RESPONSIBLE INDIVIDUAL
1. ________	Y	N	________	________
2. ________	Y	N	________	________
3. ________	Y	N	________	________
4. ________	Y	N	________	________
5. ________	Y	N	________	________
6. ________	Y	N	________	________

5. SIGN-OFF:

______________________ ______________________

______________________ ______________________

______________________ ______________________

FIGURE 4.2 Sample layout of phase integration plan, Phase 4, design and development.

confine the project deliverables to the narrow set of "physical" outputs, such as prototypes, mockups, coding sheets, manuals, and structures.

3. Logistics and Management Controls. All project control activities and their specific outputs should be defined for each phase. Examples of these items include feasibility assessments, design reviews, produceability plans, tooling requirements, and cost reviews. The summary form in Figure 4.2 provides the space and format for a listing of items to be considered for the design phase of a typical engineering program. Equally important, an agreement should be reached among the interfacing parties on the specific project and design reviews and their timing relative to the master schedule.

4. Documentation. The principal documents should be defined for each phase's input and output. These documents may include specifications, drawings, release schedules, and financial documents. It is the responsibility of the leader of the following phase to assure that the required documentation is properly defined by receding project group.

5. Sign-off. The key responsible individuals of both interfacing groups, plus the manager(s) with overall program responsibility, should be identified for each program or project phase. This leads to involvement and commitment of both groups early in each project phase and helps to build interdisciplinary teams and accountabilities.

Phase 4: Project Organization and Start-Up

The primary objective for Phase 4 is to define the work and operating environment of the program or project in enough detail so that it can be communicated and agreed upon by all project team members. One key milestone toward the end of the start-up phase is the project kickoff (or commencement), which ratifies the assignments and activates the project budget. Many of the start-up activities overlap with the project definition phase and sometimes continue into the main project execution phase (Phase 5). However, during project organization and start-up efforts typically include the following steps:

1. Finalize the project's baseline definition (requirements, specifications, block diagrams, hardware and software tree).
2. Write the project charter and obtain management endorsement (for example, see Tables 2.7 and 2.8 for similar position charters).
3. Define all activities that will be directed by the project office. Then define, consistent with the project charter, which functions should report directly to the project office and which ones should be assigned on a task basis.
4. Develop the project organization plan, including delineating the structure, key responsibilities, reporting relationships, and controls for the project organization.
5. Finalize the project plan, including work breakdown structure, task matrix, master schedule, statement of work, and project budgets.

6. Define specific work-force requirements and develop a staffing plan in co-operation with resource and task managers. Obtain commitment to the plan.
7. Write job descriptions for all direct reports to the project manager. Define candidates and negotiate assignments.
8. Organize project team via task managers and direct reports. Establish a project team roster.
9. Complete open items left over from the project definition phase. Update and revalidate.
10. Define work packages together with key project personnel and negotiate the assignments with the corresponding task managers.
11. Establish a project control and reporting system.
12. Organize and conduct kickoff meeting.

One of the key issues in setting up a project organization is the span of control at the project office and at the various integrator points. The span of control is basically determined by the number of subsystems or tasks to be integrated by one lead individual. As a guideline, ten subsystems may be a comfortable upper limit. For larger projects, organizational layering is used to reduce the width of the organization to a manageable level. The price of such layering is an additional overhead expense and, often even more serious, longer lines of communications and command, which extend through several organizational layers.

For those large projects where layering is unavoidable, some project administrative personnel can help in making communications more fluid and the management of the project more efficient. As a rough guideline, the administrative overhead of a project *should not exceed 10%* of the project budget. This guideline makes it difficult to justify any dedicated administrative support to smaller projects. Hence a project with an annual budget of $2 million can barely afford a single administrator (assuming a $80,000 salary plus 1.5 overhead factor). For larger projects the following administrative support personnel are commonly appointed and should be considered, depending on actual needs and budget constraints:

- Project manager
- Deputy project manager
- Administrative manager or assistant
- Secretary
- Manager of plans and plan implementation
- Technical director
- Manager of configuration management
- Contracts manager
- Subcontracts manager
- Financial analyst
- Change management administrator.

Many of these support functions can also be filled by part-time appointments, who are shared among several projects. The 10% rule is a good checkpoint, at least for the initial assignment. At a later point the real need for a specific support function may emerge, and a management decision based on the cost–benefit ratio must be made.

The start-up phase is also the time to establish all subplans, such as quality plans and test plans, needed for the management of the total project. Often it is impossible, however, to develop detailed subplans this early in the project life cycle because of a lack of technical details or work force. In this case it is important to define at least the types of plans needed and their timing.

The kickoff meeting is scheduled at the end of the project planning and organization activities. The kickoff signals the formal project start-up. If planned properly, the meeting provides an excellent focus on the end results expected from the start-up phase. Kickoffs can also be made incrementally, that is, the project manager can kick off the requirements analysis, the work-force planning, or any other project activity. There is an advantage to starting the main project phase as one integrated effort. All interrelated activities can be planned and managed in a natural flow, at agreed-upon integrated timing, which is difficult to achieve if activities are kicked off at different times.

The kickoff of the main-phase effort is the official confirmation of the project plan to all team members and signals the release of task funding. Team members expect that all basic plans are final, and everyone is ready to move according to these plans. Therefore a well-organized kickoff meeting will set the stage for the subsequent project effort. It will further help to unify the work group and build the team spirit necessary for high-quality committed teamwork. An agenda and some procedural guidelines for a typical project kickoff meeting are shown in Table 4.5

Phase 5: Main Phase—Project Execution

A detailed project plan, agreed to by all key personnel, is the ideal starting condition for the execution of a project. However, such an ideal situation does not always exist. Often some plan details must still be worked out after the main phase has been kicked off. The reason could be an oversight, timing, or changing conditions. These are the realities of project management, but, if handled properly before the particular task is started, the impact on overall project performance may be minimal.

The process of managing the project during the main phase consists of implementing the project plan according to preestablished objectives. For the project management personnel this means providing leadership for:

- Directing the multidisciplinary activities
- Providing the necessary resources
- Directing the cost-effective use of resources
- Facilitating interdisciplinary communication
- Sensing potential problems and searching for solutions
- Assisting in problem-solving

TABLE 4.5 Checklist for a Kickoff Meeting for a Typical Project

AGENDA

1. Message from senior management (*5 minutes*)
2. Project overview (baseline, application, business impact) (*10 minutes*)
3. Key milestones (*5 minutes*)
4. Project organization and management (*10 minutes*)
5. Major criteria for success (*5 minutes*)
6. Adjourn (*1 minute*)

HANDOUT MATERIAL

1. Project summary plan
   - Baseline configuration
   - Project objectives
   - Project scope
   - Management approach
   - End-item specifications
   - Target schedules
   - Resource requirements
   - Work force/staffing plan
   - Risk areas
2. Global customer requirements
3. Work breakdown structure
4. Statement of work
5. Task matrix
6. Major milestone schedule
7. Budgets
8. Project personnel roster

- Facilitating a stimulating work environment
- Measuring technical progress against established schedules and budgets
- Reporting project status to top management and the sponsor
- Reacting to changing conditions and requirements by replanning and redirecting task efforts.

The fundamental challenge for the management team is to provide the direction and leadership for the execution and integration of the multidisciplinary tasks against established but often changing requirements. Moreover, the baseline to be implemented often calls for innovative solutions to complex engineering problems, which may exceed the established schedule and resource limitations. In such an environment, keeping on top of technical activities is crucial. Technical problems must be detected and dealt with early in their development, *before* they lead to major impacts on the schedule and, eventually, on the project budget. Measurability of technical progress is one of the most crucial requirements for managing an engineering project. Regularly scheduled review meetings at all levels of the project organization seem to be one of the best management tools to keep abreast of technical progress and to communicate project status and potential problems to key personnel. Moreover these meetings can provide an excellent forum for addressing potential or actual problems early in their development and structuring action plans for their resolution.

Another facet of managing the project is maintaining senior management involvement, interest, and support. Top management support is often an absolute necessity for dealing effectively with interface groups and assuring continued resource commitment. Furthermore, continued senior management involvement and support is helpful and often necessary for dealing effectively with the customer or sponsor community. Effective reporting, project visibility, and on-target performance are usually the ingredients that stimulate continued senior management support and commitment.

Phase 6: Project Phaseout

By definition, projects have a finite duration. The final project phase is often characterized by a diminishing need for personnel and other resources. At the same time, pressures on personnel are great for closing on open items, getting customer acceptance, integrating the activities toward the end product, and meeting all contractual requirements. The challenge for the project manager is to sustain the team effort in a work environment that is often high on anxiety, uncertainty, and conflict.[3] The major driving forces toward successful completion of the project are a clearly established project plan, the desire of the team members to succeed, senior management support, and resource commitment.

One of the most severe restraining forces is the uncertainty of project personnel regarding their next assignment. Such uncertainty will affect project performance in several ways. First, people are not motivated to finish in a hurry. Adverse actions may range from foot-dragging to sabotage. Second, people can be expected to spend considerable time and effort in searching for a new job opportunity. Even worse, they may find a new assignment and leave the project prematurely, during a period when their efforts are critically needed. Third, the conflict that is generated over these anxieties and uncertainties is a significant barrier to effective team performance.[4] Project managers must understand existing performance drivers and barriers and find ways to minimize or neutralize barriers while enhancing drivers. For example, if

[3]In a research paper by H. J. Thamhain and D. L. Wilemon, "Conflict Management in Project Life Cycles" (*Sloan Management Reviews,* Spring 1975), the authors discuss the specific sources of conflict during project phaseout. These sources are (with most intense sources of conflict listed first): schedules, personality, resources, priorities, cost, and technical issues.

[4]For a detailed discussion, see H. S. Dugan, H. J. Thamhain, and D. L. Wilemon, "Managing Change in Project Management," *Proceedings of the 19th Annual Symposium of the Project Management Institute,* October 1979.

identified in time, negative attitudes and low motivation of personnel can possibly be dealt with through rational problem-solving and senior management involvement. Earlier team-building effort may pay off during this final phase. A truly dedicated and intrinsically motivated team that finds the project work stimulating and the final accomplishments professionally rewarding is more likely to withstand the pressures of the final project phase and to produce the needed results.

During the final phase, the project manager has very specific responsibilities for closing out the project, which should be defined in a separate project plan. Such a project closeout plan should delineate the responsible individuals, resources, and timing for all closeout activities, as with any other project plan. Below is a sample listing of typical closeout activities in six areas:

1. Documentaion
   - Complete and file all engineering baseline documentation.
   - Update configuration management system.
   - File project plan.
   - Summarize results in project bulletin.
   - Document all final tests.
   - Document vendor parts lists.
   - Generate operation and maintenance manuals.
   - Establish microfilm file.
   - Prepare final project report.
2. Contract administration
   - Obtain customer sign-off and acceptance for all contractual items.
   - Compile and submit all final contract documents.
   - Close out work orders and subcontracts.
3. Financial management
   - Close out all work orders and accounts.
   - Prepare and review financial summary report.
   - Audit financial report.
   - Collect all receivables from customer.
   - Record all financial transactions.
4. Program management
   - Coordinate project delivery and closeout.
   - Reassign or terminate all personnel directly reporting to project office.
   - Prepare and submit performance appraisals through the established channels for all personnel directly reporting to project office.
   - Close all charge numbers.
   - Sell off or return to resource departments any remaining assets.
   - Conduct and document postcompletion evaluation.

- Deliver all permanent records, documentation, and files to permanent quarters.
- Close project office and other facilities chartered to project.

5. Marketing
   - Prepare public relations announcement of project completion (e.g., press release).
   - Document potential new business, such as extensions, follow-on, spares, training, or new customers.
   - Discuss potential for new business opportunities with customer. Transfer all new business activities into functional marketing department.
   - Generate promotional documents for future reference, such as system/product bulletin, project summary report, and picture file.
6. Final management review
   - All reviews are to be held by senior functional managers and key project personnel.
   - Review project performance against management objectives laid out in project and business plan. Take final corrective actions; record lessons learned.
   - Review new business potential, such as follow-on contracts, extensions, spares, training, or new customer. Assign action items.
   - Review financial summary statement and accounts receivable for project. Assign action items. The final financial summary is to be signed off by profit center manager.
   - Review project management process and major problem areas throughout the project life cycle. Document lessons learned and recommendations. Issue policy directives. Develop or update procedures selectively.

4.4 INTEGRATED PROJECT PLANNING

Planning the project or program for its multifunctional integration throughout its life cycle is one of the most important and challenging responsibilities of the project manager.

Before we discuss the specific management tools of project planning and control in the next chapter, an integrated approach to project planning is presented here, which should help to unify the various concepts and provide an overall perspective.

Much of the planning techniques discussed in this book rely on the phased approach, which utilizes the chronological flow of activities, as shown in Figure 4.2, as a framework for project planning and controlling. It offers a disciplined, systematic process, regardless of project size and complexity. As the planning activities progress and more details become available, the project execution phase (Phase 5) can be expanded to reflect the actual activities of design, prototyping, etc., as shown in Figure 4.4.

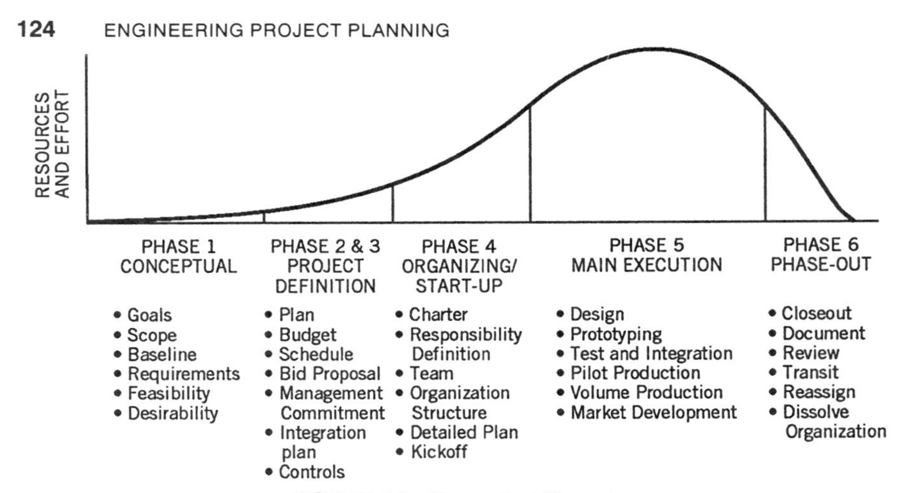

FIGURE 4.3 The project life cycle.

As the project moves through its life cycle, the focus of managerial activities shifts from planning to controlling, as illustrated in Figure 4.4. Many of the activities are interrelated and cannot be confined to only one particular project phase. This interrelationship is especially strong for planning activities, which are part of an ongoing process throughout the project life cycle. Plans are managerial tools. They are seldom final and should not be rigged. The purpose of the plan is to provide the basis for organizing the project, defining resource requirements, setting up controls, and eventually guiding the activities. As the various elements of the plan are integrated and actual operations begin, modifications of the original program plan may become necessary. Continuous reviews and updates of all components of the program plan are needed throughout the project life cycle if the plan is to remain a useful reference and guidance document.

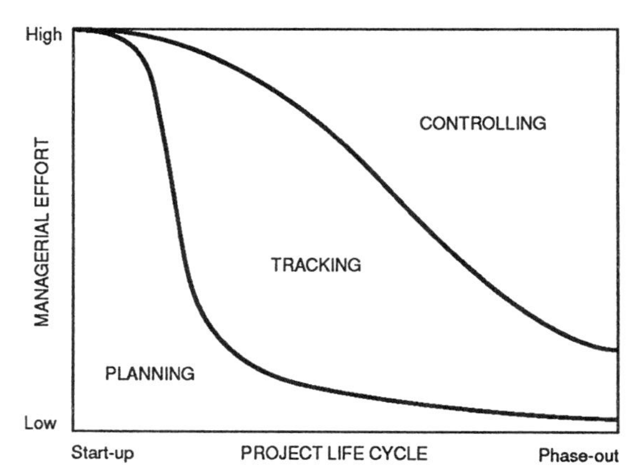

FIGURE 4.4 Managerial efforts shift over the life cycle of a project, from planning to tracking and controlling.

Benefits of the Phased Approach

To many engineering managers, the phased approach seems relatively new. However, it has been used extensively for planning and implementing construction projects and high-technology product and systems development. By and large, companies that have used the phased approach for managing projects have found that it offers many benefits, as summarized in Table 4.6. Among the most obvious benefits cited are simplicity of planning, procedural modularity, directions for assigning lead personnel, logic in defining measurable milestones and reviews, and the possibility of incremental funding and program direction.

In addition, many of the more senior managers point out that the method promotes a total-program perspective and interfunctional involvement, as well as better visibility of cross-functional activities and overall objectives; stimulates interest in work; and, perhaps most important, includes constant reinforcement of the personal commitment of key contributors and managers, A detailed description of the activities recommended for setting up and executing a project is provided next, using the framework of the five "generic" project phases. It should be emphasized that the recommended activities apply to all projects, regardless of size and complexity. What varies with the project size is the depth and detail of the plan, but the principal elements of the project plan should be in place for any project.

Implementing the Concept

The plan for an effective integrated project management system must include and integrate 11 specific areas:

1. *Objective.* Specific result, goal, target, or quota to be achieved within certain time and resource limitations

TABLE 4.6 Benefits of the Phased Approach

- Modular, flexible, practical, and participative planning.
- Reduced complexity of the total program.
- Overall program and integrated systems perspective.
- Creative ideas about implementation.
- Proper identification of key personnel.
- Involvement and visibility across functional lines.
- Identification and development of critical interfaces.
- Definition of specific results, deliverables, and performance measurements.
- Definition and scheduling of reviews and management decisions.
- Incremental funding.

2. *Program.* The strategy to be followed and major actions to be taken in order to achieve or exceed objectives
3. *Deliverables.* The principal items to be delivered and results to be produced, including hardware, software, services, and documentation
4. *Performance.* Global performance requirements and specifications
5. *Schedule.* A plan showing when individual or group activities or accomplishments will be started and/or completed
6. *Budget.* Planned expenditures and other resources required to achieve or exceed objectives
7. *Forecast.* A projection of what will happen by a certain time, including contingencies
8. *Organization.* Design of the number and kinds of positions, along with corresponding duties and responsibilities, required to achieve or exceed objectives
9. *Policy.* A general guide for decision-making and individual actions
10. *Procedure.* A detailed method for carrying out a policy
11. *Standard.* A level of individual or group performance defined as adequate or acceptable.

The first step in total program planning is understanding the project objectives. Such objectives may be to develop expertise in a given area, to become competitive, to modify an existing facility for later use, or simply to keep key personnel employed.

Objectives are generally interrelated, both implicitly and explicitly. Many times it is not possible to satisfy all objectives. At this point, management must prioritize the objectives as to which are strategic and which are not.

Another important issue is upper-level management involvement. Many project managers view the first critical step in planning as obtaining the support of top management. Once top management expresses an interest in the project, the functional managers are more likely to respond favorably to the project team's request for support, since the new project is now part of the functional support mission.

Executives are also responsible for selecting the project leader. One of the critical skills is for project planning. Unfortunately, not all technical specialists are good planners. As Rogers points out:[5]

> The technical planners, whether they are engineers or systems analysts, must be experts at designing the system, but seldom do they recognize the need to "put on another hat" when system design specifications are completed and design the project control or implementation plan. If this is not done, setting a project completion target date or a set of management checkpoint milestones is done by guesswork at best. Management will set the checkpoint milestones, and the technical planners will hope they can meet the schedule.

[5]Lloyd A. Rogers, "Guidelines for Project Management Teams," *Industrial Engineering,* December 12, 1974.

Senior managers must not set unrealistic or arbitrary milestones and then force line managers to fulfill them. Both project and line managers should try to agree on doable milestones. However, in case of an impasse, the executive should yield, or allow for a compromise on project performance. It must be realized that project parameters (technical, timing, or budget) cannot be forced on a team but must be negotiated, agreed upon, and committed to. A manager who tries to force these parameters on the project team will find hostility, increased conflict, and waning enthusiasm and commitment to previously agree upon parts of the project plan. In the end, the best performance that can be achieved is the plan that was originally agreed upon by the program organization. In many cases the results are much lower. As an example, a company executive took the 6-month completion milestone and changed it to 3 months. The project and line managers rescheduled all of the other projects to reach this milestone. The executive then did the same thing on three other projects, and again the project and line managers came to his rescue. The executive began to believe that the line people did not know how to estimate and that they probably loaded up every schedule with "fat." So the executive changed the milestones on all of the other projects to what his "gut feeling" told him was realistic. The reader can imagine the chaos that followed.

Executives should interface with project and line personnel during the planning stage in order to define the requirements and establish reasonable deadlines and budgets. Executives must realize that creating an unreasonable deadline may require the resetting of priorities for other projects and hence upsetting established business plans and commitments.

It is a challenge for the manager, who often works with a skeleton team at the program initiation stage, to define the program in sufficient detail so that additional specialists can be assigned for further program definition and planning. For this reason it is quite common that project planning remains highly iterative, as shown in Figure 4.5. The project gets defined in more depth and detail as it progresses through the program feasibility and definition phases, and sometimes such incremental refinement of the plan continues into the actual execution phase.

Integrating Technical Performance, Schedule, and Cost Variables

Overall project performance is a function of three major interdependent variables: technical performance, schedules, and costs. To develop an integrated project management system, these three variables need to be considered simultaneously in all three project management functions: planning, tracking, and controlling. Moreover, the information developed and communicated must be suitable for all members of the organization: project team members, project manager, customer or sponsor, functional management, support organizations, and top management. The interrelationship of all these components of the project management system, shown in Figure 4.6, was formally researched and presented by Hopeman and Wilemon in 1973 as a result of a NASA-sponsored study into the *Apollo* program.[6] Today this concept is an

[6]Richard J. Hopeman and David L. Wilemon, "Project Management/Systems Management Concepts and Applications," Syracuse, N.Y., NASA/Syracuse University, 1973.

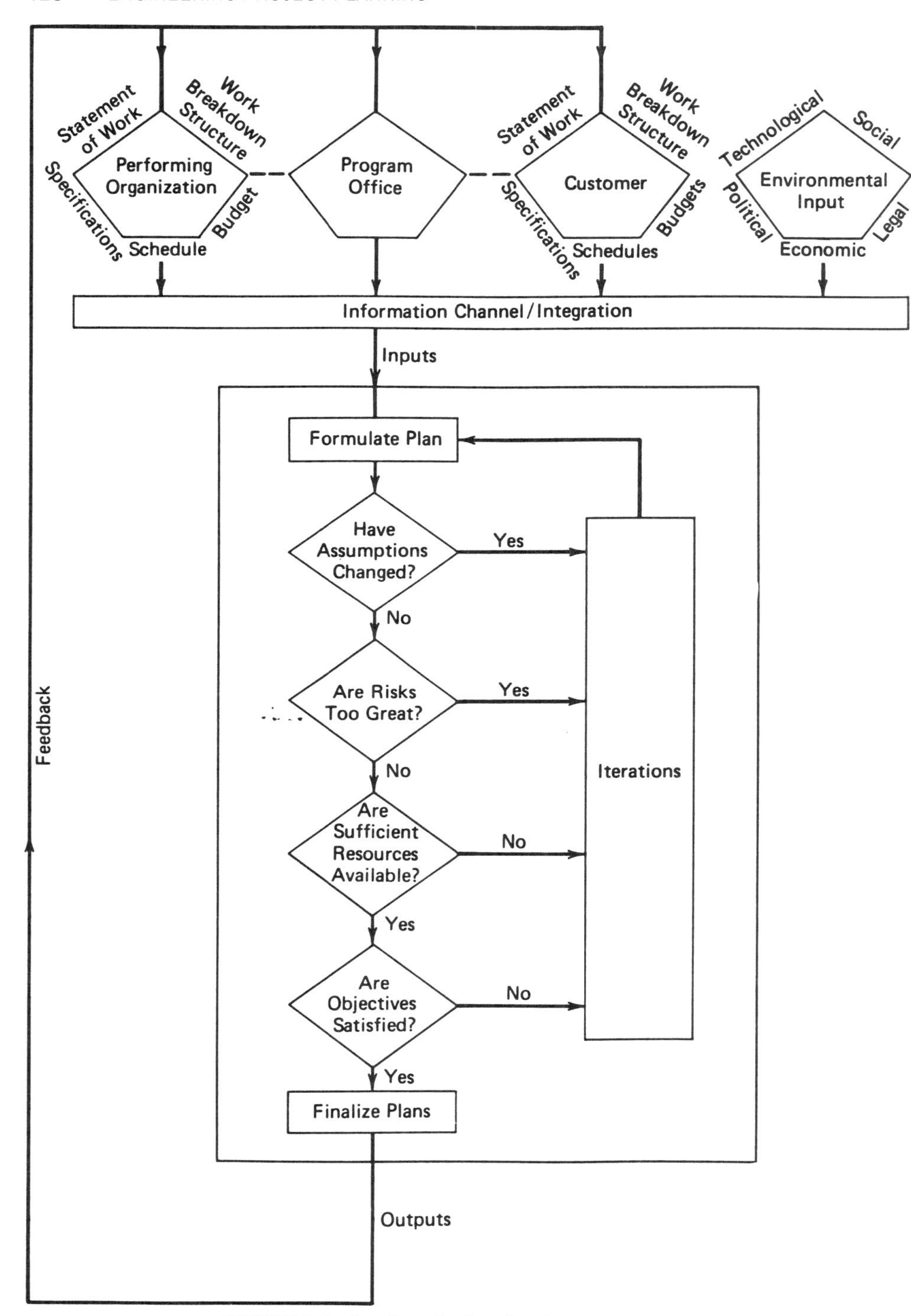

FIGURE 4.5 Iterations for the planning process.

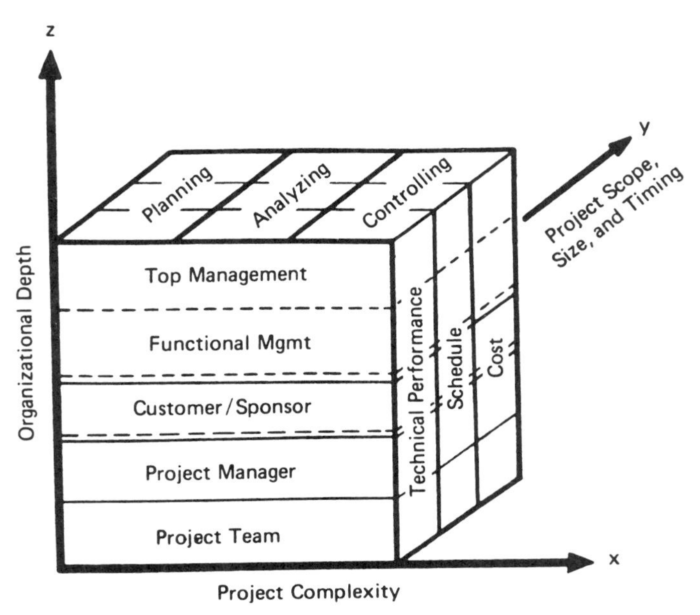

FIGURE 4.6 Integration of performances, schedule, and cost variables with planning analysis, and control sequence within the organizational framework.

important cornerstone of all project and program planning and control systems. To summarize:

1. Planning requires an integrated consideration of technical performance, schedules, and costs, since these variables are interdependent. Changes in one variable usually affect all three. Tradeoffs among technical performance, schedules, and costs are common and should be formally considered during the planning phase.

2. Tracking the program requires the ability to simultaneously measure and analyze the technical performance against budget and schedule (a topic that is discussed in detail in the next chapters). Such integrated tracking is important for determining total project performance, which includes quality, cost, timing, and technical parameters. Proper tracking is also important for evaluating alternative courses of action prior to redirecting a program because of customer-requested changes or contingencies.

3. Controlling the project requires an integrated consideration of technical performance, schedule, and cost variables, because management control directed toward one part or one variable of the project can potentially affect all other variables. A typical example is the attempt to expedite the schedule by changing from a custom-designed subsystem to an off-the-shelf component, which clearly affects the technical specifications as well as cost and timing.

4.5 RECOMMENDATIONS FOR EFFECTIVE PROJECT PLANNING

A number of specific recommendations are stated below, to help engineering managers and project leaders to develop realistic, agreed-on plans that are doable and represent the best resource utilization for accomplishing the established project objectives.

1. *Detailed Project Planning.* Develop a detailed project plan involving all key personnel, defining the specific work to be performed, the timing, the resources, and their responsibilities.
2. *Break the Overall Program into Natural Phases and Subsystems.* Use a work breakdown structure (WBS) as a planning tool (described in the next chapter).
3. *Requirements.* Define the program objectives and requirements in terms of specifications, schedules, resources and deliverable items for the total project and its subsystems.
4. *Deliverables and Results.* Define the specific deliverable items to be produced at the end of each project phase.
5. *Measurable Milestones.* Define measurable milestones and checkpoints throughout the project. Measurability can be enhanced by defining specific results, deliverables, and technical performance measures against schedule and budget.
6. *Management Controls.* Define the specific procedures and controls for managing the project according to plan, such as reviews, quality plans, producibility plans, releases, and signoffs.
7. *Phase Overlap.* Permit project phases to overlap in time.
8. *Key Personnel.* Define key personnel in the project life cycle and involve them in the project planning and definition phase.
9. *Business Objectives.* Communicate overall project and business objectives to all team personnel.
10. *Commitment.* Obtain commitment from all key personnel regarding the project plan, its measures, and results. Commitment can be enhanced and maintained by involving the team members early in the project planning. It is through this involvement that the team members gain a detailed understanding of the work to be performed and develop professional interests in the project, and desires to succeed.
11. *Intra-program Involvement.* Assure that the interfacing project teams, such as engineering and manufacturing, work together not only during the task transfer, but during the total life of the project. Involvement is enhanced by clearly defining the results and deliverables agreed upon by both parties for each interphase point. In addition, a simple signoff procedure for these deliverables is useful in establishing clear checkpoints and enhancing involvement and cooperation of the interphasing team members.

12. *Project Tracking.* Define and implement a proper project tracking system, which captures and processes project performance data and conveniently summarizes it for reviews and management actions.
13. *Measurability.* Assure accurate measurements of project performance data, especially technical progress against schedule and budget.
14. *Regular Reviews.* Projects should be reviewed regularly, both at the subsystem and total-project level.
15. *Signing On.* The process of signing on (recruiting) project personnel during the initial phases of the project seems to be very important to proper understanding of the project objectives, the specific tasks, and personal commitment. The sign-on process, so well described in Tracy Kidder's book, *The Soul of a New Machine* (New York: Avon, 1981), is greatly facilitated by sitting down with each team member and discussing specific assignments, overall project objectives, professional interests, and support needs.
16. *Communication.* Good communication is essential for effective project planning. It is the responsibility of the project manager to provide appropriate communication tools, techniques, and systems, which include status meetings, reviews, and reporting systems. To minimize chances of any mix-up, there should be only *one* master plan, which is continuously updated.
17. *Leadership.* Assure proper direction and leadership throughout the project planning cycle. This includes project definition, organization, task coordination.
18. *Phase Transition.* Establish cross-functional task teams to plan and direct the transition of project work across functional interfaces, such as from engineering to manufacturing.
19. *Visibility.* Make project effort visible throughout the organization by involving senior management and making personnel aware of the business impact. It is easier to obtain resource commitments and personal support for projects that are viewed as being "important."
20. *5% Rule.* As a ground rule, set aside a minimum of 5% of the total project effort for up-front planning and project definition.

4.6 A FINAL NOTE

Project planning means charting a course within a forecasted environment. The quality of the final plan as a management tool for project implementation will largely depend on the reality level at which the plan is perceived by the key decision-makers of the project team. Early involvement of key personnel from the customer, the performing organization, and upper management, together with the leadership of an experienced, capable program and project manager, will help foster the right environment, conducive to early team-building, organizational visibility and support, individual interest, and a desire to participate. This is a work environment that leads to an understanding of requirements and options. It is conducive to realistic estimating of resources and timing, and ultimately to an agreed-on plan and commitment to its implementation.

Once it has been agreed upon by the customer or sponsor and the key personnel of the performing organization, the project plan becomes a contract among all parties involved. It also becomes a roadmap to guide them through all phases of the project implementation.

It should be realized, however, that, except for very simple undertakings, project plans must remain flexible to accommodate changes, contingencies, or unexpected difficulties that surface during the implementation.

EXERCISE 4.1: HIGH-TECH PROJECT "A"

Team A Assignment: Project Definition

Learning Objective

Apply knowledge of project planning, with a focus on high technology. The exercise can be completed individually or in a group. It should help to build skills for contemporary management problem-solving of:

- Defining project requirements
- Using standard project plan documentation
- Establishing planning processes
- Defining management controls
- Assessing challenges of project planning and start-ups
- Validating project plans
- Establishing checkpoints
- Establishing performance measurements
- Soliciting commitment.

Situation

You are a project team at Hewlett-Packard. You have accepted the responsibility for evaluating the feasibility of and eventually developing a new DeskJet Printer. Your team consists of four key individuals who agreed to take on and lead the following tasks: Rob is the project leader. He will also organize and coordinate the feasibility analysis, which includes 3 major tasks. Al is in charge of the functional design of the printer, a task which includes the system design, structural design, reliability planning, and producibility planning. Beth is responsible for all detailed design and development of the printer including: electrical, mechanical, human factors, prototyping, and testing. Chris is in charge of production setup. It includes preparing for design to production transitioning, volume production setup, pilot run, and quality control. Some of the agreed-on milestone dates are: Project kickoff (1/3), feasibility assessment completed (3/10), functional design completed (6/15), transitioning to production (9/15), pilot production (12/30). Budgets are allocated to each task or project leader as follows: Al ($200,000), Beth ($500,000), Chris ($200,000), and Rob ($100,000).

Your Assignment

1. Document your project plan by using standard forms/formats.
2. What problems and challenges do you anticipate in executing your project plan?
3. What management controls would you set up as project manager to ensure that the project will run according to plan?
4. You are now facing a new situation! In a major reorganization, funding for your project has been canceled. Further, the project has been transferred to another division for reevaluation. However, your group receives a new project assignment, which has been completely planned and documented and is ready for immediate execution (For the exercise, exchange your project plan with the plan of "team B")
5. Evaluate the new project plan. (1) Are you clear on your new assignment? (2) Do you have all the information to kick off the project? (3) Validate the new project plan.

In addressing the 3 items under #5 above, prepare some view graphs or flip charts useful for presenting to upper management regarding your assessment of the project, including: (1) missing information, (2) clarity of the plan, (3) how to proceed, (4) checks and balances, and (5) support needs.

You will be asked to make a commitment to the project plan. Or, if you have any problems with the plan, explain what you need.

EXERCISE 4.2: HIGH-TECH PROJECT "B"

Team B Assignment: Project Definition

Learning Objective

Apply knowledge of project planning, with a focus on high technology. The exercise can be completed individually or in a group. It should help in building skills for contemporary management problem-solving of:

- Defining project requirements
- Using standard project plan documentation
- Establishing planning processes

- Defining management controls
- Assessing challenges of project planning and start-ups
- Validating project plans
- Establishing checkpoints
- Establishing performance measurements
- Soliciting commitment.

Situation

You are an R&D project team at Hewlett–Packard. You have accepted the responsibility for a software project of developing a new PC operating system. Your team consists of four key individuals, who agreed to take on and lead the following tasks: Sue is the project leader. She will also be responsible for developing the project plan, including task definition, scheduling, cost modeling and estimating, budget planning, and management approvals. Art is in charge of the software architecture. This includes a requirements analysis, software specs, system diagrams, protocols, and timing and controls. Bob is responsible for all development activities for the new software, up to a prototype demonstration. It includes design, coding, testing, and debugging of 6 major software subsystems, rapid prototyping of all software, including integration and testing, and documentation. Carl is in charge of providing liaison to the product development and marketing groups. Some of the agreed-on milestone dates are: Project kickoff (1/3), budget proposal completed (2/10), system architecture defined (5/1), software documentation completed (11/15), project completed (12/30). Budgets are allocated to each task/project leader as follows: Art ($21 million), Bob ($5 million), Carl ($.1 million), and Sue ($.9 million).

Your Assignment

1. Document your project plan by using standard forms and formats.
2. What problems and challenges do you anticipate in executing your project plan?
3. What management controls would you set up as project manager to ensure that the project will run according to plan?
4. You are now facing a new situation! In a major reorganization, funding for your project has been canceled. Further, the project has been transferred to another division for reevaluation. However, your group has received a new project assignment, which has been completely planned and documented and is ready for immediate execution (for the exercise, exchange your project plan with the plan of "team A.").
5. Evaluate the new project plan. (1) Are you clear on your new assignment? (2) Do you have all the information to kick off the project? (3) Validate the new project plan.

In addressing the 3 items under #5 above, prepare some view graphs or flip charts useful for presenting to upper management regarding your assessment of the project, including; (1) missing information, (2) clarity of the plan, (3) how to proceed, (4) checks and balances, (5) support needs.

You will be asked to make a commitment to the project plan. Or, if you have any problems with the plan, explain what you need.

BIBLIOGRAPHY

Adams, John R. and S. E. Brandt, "Organizational Life Cycle Implementations for Major Projects," *Project Management Quarterly,* 1978.

Allen, Louis A., *Managerial Planning,* New York: McGraw-Hill, 1982.

Archibald, Russell D., *Managing High-Technology Programs and Projects,* New York: Wiley, 1976.

Babcock, Daniel L., *Managing Engineering and Technology,* Englewood Cliffs, NJ: Prentice-Hall, 1991.

Clark, John R., "Practical Specifications for Project Scheduling," *Cost Engineering,* Volume 32, Number 6 (June 1990), pp. 17–21.

Cleland, David I. and Karen M. Bursic, *Strategic Technology Management,* New York: AMACOM, 1992.

Cleland, David I. and William R. King, *Systems Analysis and Project Management,* New York: McGraw-Hill, 1983.

Cleland, David I., *Matrix Systems Management Handbook,* New York: Van Nostrand Reinhold, 1983.

Cleland, David I., *Project Management: Strategic Design and Implementation,* New York: McGraw-Hill, 1990.

Cleland, David I. and William R. King (eds.), *Project Management Handbook,* New York: Van Nostrand Reinhold, 1988.

Derks, Richard P., "CIM Planning's Hidden Dangers," *Information Strategy: The Executive's Journal,* Volume 5, Number 4, (Summer 1989), pp. 31–34.

Dugan, H. Sloane, Hans J. Thamhain and David L. Wilemon, "Managing Change in Project Management," *Proceedings of the 19th Annual Symposium of the Project Management Institute,* October 1979.

Dunnington, Judith I., "Successful HRIS Implementation Planning," *Personnel Journal,* Volume 69, Number 2 (Feb 1990), pp. 78–84.

Easton, James L. and Robert L. Day, "Planning for Project Management," *The Implementation of Project Management: The Professional's Handbook,* Project Management Institute, Reading, MA: Addison-Wesley, 1982, Chapter 3.

Fereig, Sami M., Nabil H. Qaddumi, and Amro El-Akkad, "Computer Applications in the Kuwaiti Construction Industry," *Computers in Industry,* Volume 13, Number 2 (Nov 1989), pp. 135–140.

Finnigan, Orville E., and James R. Wardin, "An Integrated Approach to Substation-Betterment Design," *Transmission & Distribution,* Volume 42, Number 3, (Mar 1990), pp. 32–36.

Gupta, Yash P. and T. S. Raghunathan, "Impact of Information Systems (IS) Steering Com mittees on IS Planning," *Decision Sciences,* Volume 20, Number 4 (Fall 1989), pp. 777–793.

Harris, Michael J., "A Planning Framework for Systems Management," *Journal of Information Systems Management,* Volume 8, Number 1 (Winter 1991), pp. 8–16.

Humphreys, Kenneth, K. and Paul Wellman, *Basic Cost Engineering,* New York: Marcel Dekker, 1987.

Hopeman, Richard J. and David L. Wilemon, "Project Management/Systems Management Concepts and Application," Syracuse, NY: NASA/Syracuse University, 1973.

Izuchukwu, John I., "Project Management: Shortening the Critical Path," *Mechanical Engineering,* Volume 112, Number 2 (Feb 1990), pp. 59–60.

Jackman, Hal, "State-of-the-Art Project Management Methodologies: A Survey," *Optimum* (Canada), Volume 20, Number 4 (1989/1990), pp. 24–47.

Julian, John C. and Remo J. Silvestrini, "Variations in Project Planning Intensity," *AACE Transactions,* (1990), pp. H.2.1–H.2.7.

Kartum, Nabil A. and Raymond E. Levitt, "An Artificial Intelligence Approach to Project Planning Under Uncertainty," *Project Management Journal,* Volume 22, Number 2 (June 1991), pp. 7–11.

Kernaghan, John A. and Robert A. Cooke, "Teamwork in Planning Innovative Projects: Improving Group Performance by Rational and Interpersonal Interventions in Group Process," *IEEE Transactions on Engineering Management,* Volume 37, Number 2 (May 1990), pp. 109–116.

Kerzner, Harold and Hans Thamhain, *Project Management for Small and Medium Size Businesses,* New York: Van Nostrand Reinhold, 1984.

Kerzner, Harold and Hans Thamhain, *Project Management Operating Guidelines,* New York: Van Nostrand Reinhold, 1986.

Kerzner, Harold, *Project Management for Executives,* New York: Van Nostrand Reinhold, 1982.

Kezsbom, Deborah S., "Match Strategies to Structure with a Project Management Requirements Analysis," *Industrial Engineering,* Volume 23, Number 4 (Apr 1991), pp. 56–58.

Kirby and East, *A Guide to Computerized Scheduling,* New York: Van Nostrand Reinhold, 1990.

Laszcz, John F., "Take a Close Look at EMS and Its Role in Today's Manufacturing Environment," *Industrial Engineering,* Volume 21, Number 12, (Dec 1989) pp. 22–24.

Laufer, Alexander, "Project Planning: Timing Issues and Path of Progress," *Project Management Journal,* Volume 22, Number 2 (June 1991), pp. 39–45.

Lenss, Maris G., "The Use of CPM Through Knowledge-Based Information Systems," *AACE Transactions,* (1988), pp. G.4.1–G.4.5.

Liberatore, Matthew and George J. Titus, "The R&D Planning-Business Strategy Connection," *Journal of the Society of Research Administrators,* Volume 20, Number 4 (Spring 1989), pp. 17–26.

Love, Sydney F., *Achieving Problem Free Project Management,* New York: Wiley, 1989.

Lowery, Gwen, *Managing Projects with Microsoft for Windows,* New York: Van Nostrand Reinhold, 1992.

McCusker, Tom, "Project Planning Made Easy," *Datamation,* Volume 35, Number 20 (Oct 15, 1989), pp. 49–50.

Martin, Dean and Kathleen Miller, "Project Planning as a Primary Management Function," *Project Management Quarterly* (Mar 1982).

Michael, Stanford B., and Linn C. Stuchenbruck, "Project Planning," in *The Implementation of Project Management: The Professional's Handbook,* Project Management Institute, Reading MA: Addison-Wesley, 1982, Chapter 7.

Miller, Michael J., "Symantec Takes Automation Route with On Target Project Management," *Infoworld,* Volume 13, Number 16 (Apr 22, 1991), p. 1105.

Mills, Nancy L., "The Development of a University Sports Complex: A Project Management Application," *Computers & Industrial Engineering,* Volume 17, Number 1–4 (1989), pp. 149–153.

Moder, Joseph J. and Cecil R. Phillips, *Project Management with CPM and PERT,* New York: Van Nostrand-Reinhold, 1970.

Murphy, John A. "Software Improves Work-Group Productivity," *Today's Office,* Volume 25, Number 6 (November 1990), pp. 42, 44.

Nicholas, John M., *Managing Business & Engineering Projects,* Englewood Cliffs, NJ: Prentice-Hall, 1990.

Obradovitch, Michael M. and S. E. Stephanou, *Project Management: Risks and Productivity,* Bend, OR: Daniel Spencer, 1990.

Pincus, Claudio, "An Approach to Plan Development and Team Formation," *Project Management Quarterly,* (Dec 1982).

Prentis, Eric L., "Master Project Planning: Scope, Time and Cost," *Project Management Journal,* Volume 20, Number 1 (Mar 1989), pp. 24–30.

Radding, Alan, "Management Software: Project Management," *Bank Management,* Volume 66, Number 5 (May 1990), pp. 62–67.

Randolph, W. Alan and Barry Z. Posner, *Effective Project Planning and Management,* Englewood Cliffs, NJ: Prentice-Hall, 1988.

Ricciuti, Mike, "Easy Ways to Manage Multiple IS Projects," *Datamation,* Volume 37, Number 23 (Nov 15, 1991), pp. 61–62.

Riggs, Henry E., *Managing High-Technology Companies,* Belmont, CA: Lifetime Learning Publication (Wadsworth), 1983.

Rogers, Lloyd A., "Guidelines for Project Management Teams," *Industrial Engineering,* (Dec 12, 1974).

Schei, Kenneth G., "Small Project Management," *Civil Engineering,* Volume 60, Number 1 (Jan 1990), pp. 42–44.

Shah, R. K. D., "Strategic Planning for Power Equipment Manufacture," *Long Range Planning* (UK), Volume 22, Number 5, (Oct 1989), pp. 98–111.

Shina, Sammy G., *Concurrent Engineering and Design for Manufacture of Electronics Products,* New York: Van Nostrand Reinhold, 1991.

Silverberg, Eric C., "Predicting Project Completion," *Research-Technology Management,* Volume 34, Number 3 (May/June 1991), pp. 46–49.

Silverman, Melvin, *The Art of Managing Technical Projects,* Englewood Cliffs, NJ: Prentice-Hall, 1987.

Smith, L. Murphy, "Using the Microcomputer for Project Management," *Journal of Accounting & EDP,* Volume 5, Number 2 (Summer 1989), pp. 30–37.

Tabucanon, M. T. and N. Dahanayaka, "Project Planning and Controlling Maintenance Overhaul of Power-Generating Units," *International Journal of Physical Distribution & Materials Management* (UK), Volume 19, Number 10 (1989), pp. 14–20.

Thamhain, Hans J., and David L. Wilemon, "Project Performance Measurement, The Keystone to Engineering Project Control," *Project Management Quarterly* (Jan 1982).

Thamhain, Hans J., *Engineering Program Management,* New York: Wiley, 1984.

Vacca, John R., "Project Management Techniques," *Systems 3X/400,* Volume 19, Number 9 (Sept 1991), pp. 56–63.

5
TOOLS AND TECHNIQUES FOR MANAGING ENGINEERING PROJECTS

5.1 ENGINEERING ACTIVITIES ARE PROJECT-ORIENTED

Engineering organizations perform project work. Engineering tasks have typical project characteristics that are single, nonrepeat undertakings that have specifically defined results, schedules, resources, and responsibilities. Moreover engineering work requires that work status and progress can be measured over the project's life cycle, against often challenging performance parameters. Therefore, it is not surprising that engineering companies have long been accustomed to working with project management tools such as work breakdown structures, schedules, statements of work, and task matrices.

Why Is Management Reluctant To Plan?

Formal planning and control systems are by no means universally used among engineering organizations. Especially in a smaller company, such as an engineering laboratory, design office, or small job shop employing 5, 10, or 20 people, management is concerned about additional paperwork, lack of flexibility and creativity, and, in the end, too much attention being paid to the plan at the expense of managing the work. This concern is valid. Many of the smaller engineering tasks can be effectively organized and controlled with a very short and simple planning document, provided that it captures the four key parameters: (1) work to be done. (2) schedules, (3) resources, and (4) individual responsibilities, in addition to agreement and commitment by the team.

However, we often find that engineering managers who are skeptical toward formal plans and controls also lack the basic understanding of available, modern tools and techniques. As a result, their work plans require more paperwork, although they are less efficient and useful for managing activities. The secret lies in using the tools appropriately, scaled up or down according to the job requirements. One would expect that a professional carpenter brings a basic set of tools to work regardless of whether he or she works on a big or small job. However, engineering managers often apply different criteria for their work. One of the objectives of this chapter is to show how to use the project management tools effectively regardless of the size of the job or of the performing organization.

Moving Toward More Formal Systems

For larger engineering undertakings, managers don't have much of a choice but to use formal project management techniques. It just gets too complicated to keep it all in one's head and to communicate it among a large number of people. According to some studies, the threshold between informal and formal planning and control techniques seems to lie at approximately 20–50 people working on a single program. However, as shown below, there are many other pressures that drive engineering managers to the use of formal project tools. Managers are required to use formal tools so that they can better respond to:

- Specific contract requirements, especially from government clients
- Better performance measurements of work in progress
- Shorter product development cycles
- Organizing multifunctional teams affectively
- Questions of individual accountability
- Changing requirements
- Negotiating resource requirements
- Potential conflict and confusion over plan
- Personnel changes
- Priority shifts
- Subcontractor support requirements
- Geographically dispersed work units
- Language and cultural barriers.

All of these reasons provide pressures on engineering managers to use a formal planning system with standard project management tools for defining, organizing, and directing their work. In addition, no less significantly, these systems lay the basis for individual accountability and managerial performance appraisal and rewards.

Using Project Management Tools Properly

Fundamentally, project management tools are communication devices. They have been designed to *define* and *communicate* the requirements to all parties involved, to *measure* performance, and finally to *direct* and *control* the effort toward the preestablished requirements. Effective project management tools have the following characteristics:

- They assure measurability of all parameters used
- They use standard formats for all tools throughout the organization
- They use a uniform number system for project elements, consistent with all tools (e.g., a project subsystem number should be the same in the work breakdown structure as it is in the statement of work or cost account).
- They do not clutter the managers tools with too much detail. They use modular concepts.
- They keep the manager's documents current.

Some suggestions are made below for using project management tools properly:

1. An agreed-upon program plan is absolutely essential. The plan defines the requirements in measurable steps and is the key to successful project performance. Measurability of technical status is crucial for engineering projects. If we cannot measure, we cannot control.

2. The use of standard formats for all tools is another recommendation. Nothing can be more confusing than looking at two different schedule or budget forms for the same project. It is the responsibility of the project manager to issue a standard set of forms if the organization does not have the planning process already proceduralized.

3. Number systems are often another area of unnecessary conflict and confusion. Many of the tools, such as schedules, work breakdown structures, and task matrices, label each task with a number. There is no need to label a particular task differently in the work breakdown structure than in the schedule for example. All tasks should be easily traceable throughout the project plan.

4. Use modular concepts to break down the complexity of your plans. For example, rather than showing all activities in a schedule in detail, partition the schedule into a master schedule, which provides just an overview, a subsystem A schedule, a subsystem B schedule, and so on. Or, alternately, use time phasing to break down the complexity, issuing a system design schedule, prototype design schedule, prototype fabrication schedule, and so on.

5. Keep your documents current. Nothing outlives its usefulness faster than an outdated document. Review your plans regularly and make revisions as needed. A document control system should be maintained by an assigned individual such as a secretary, who makes sure that agreed-upon changes get properly recorded and distributed.

The following specific tools and management techniques are commonly used for organizing, tracking, and controlling engineering projects. These tools can be classified into six principal categories based on their primary management focus: (1) Overall project; (2) work; (3) responsibilities; (4) resources; (5) timing; and (6) tracking.

Accordingly, the following project management tools are used to achieve those purposes:

1. *For Overall Project Definition*
   - Work Breakdown Structure
   - Work Breakdown Structure Dictionary
   - Task Authorization
2. *For Work Definition*
   - Statement of Work
   - Specification
   - Work Package
   - Task Authorization
3. *For Responsibility Definition*
   - Task Matrix
   - Task Roster
4. *For Resource Definition*
   - Resource Plan
   - Manpower Plan
   - Budget
5. *For Timing Definition*
   - Milestone Schedule
   - Bar Graph Schedule (Gantt Chart)
   - Network: Program Evaluation and Review Techniques (PERT) Critical Path Method (CPM) etc.
6. *For Tracking*
   - Program Evaluation and Review Techniques (PERT)[1]
   - Line of Balance (LOB)
   - Variance Analysis
   - Performance Measurement System (PMS).

All of these tools and their applications are described briefly in the remainder of this chapter.

5.2 THE WORK BREAKDOWN STRUCTURE

The work breakdown structure (WBS), sometimes called the *project breakdown structure* (PBS), is a family tree of project elements. It defines the various hardware,

[1]There are many other, similar program tracking techniques such as Critical Path Method (CPM) and Presidence Method (PM). Many of the available computer software packages actually combine various techniques and capabilities.

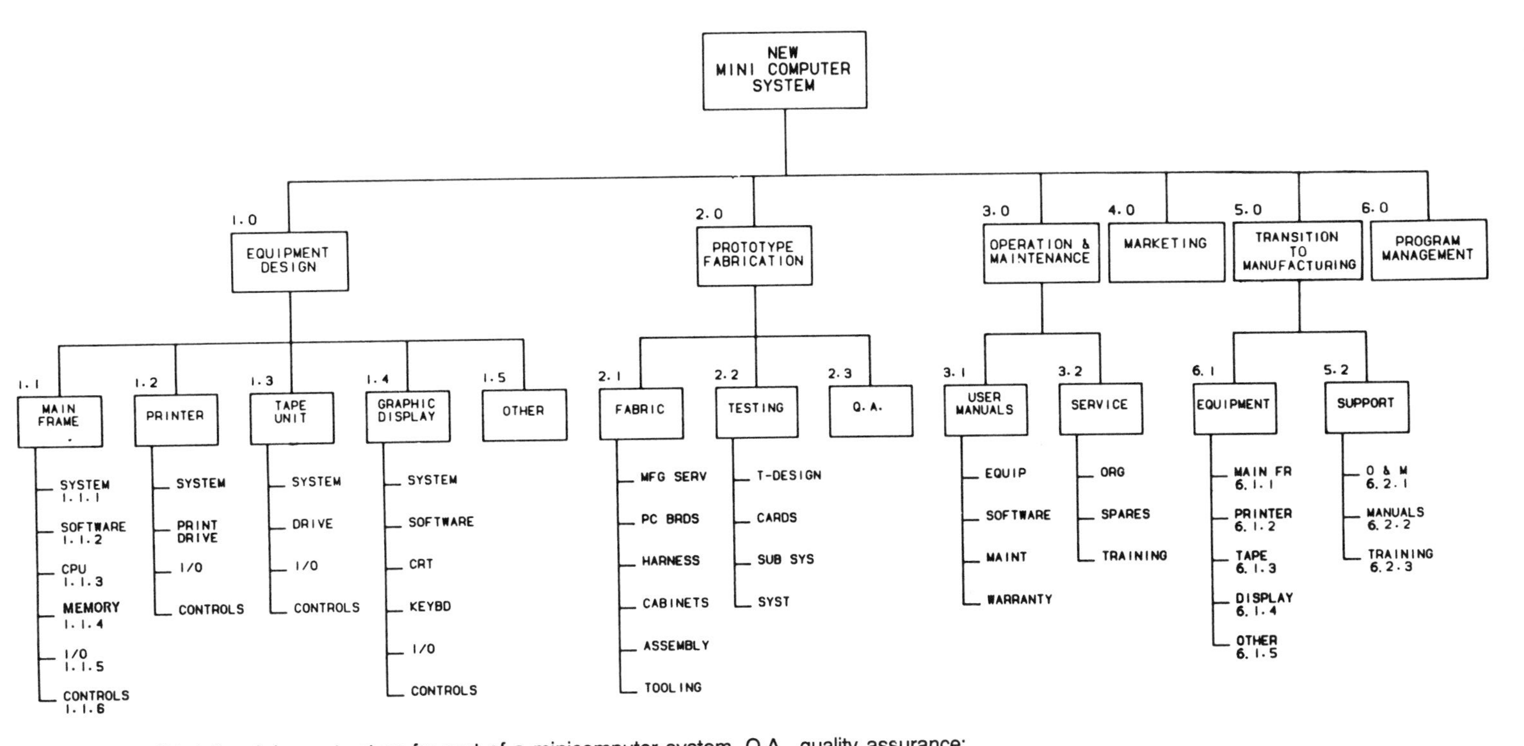

FIGURE 5.1 Work breakdown structure for part of a minicomputer system. Q.A., quality assurance: O&M, operation and maintenance: I/O, input/output.2p

software, and service systems and their subcomponents that make up the specified project or program. For example is shown in Figure 5.1 for part of an engineering development plan for a minicomputer prototype system. The work breakdown structure provides the framework for dividing the total program into manageable tasks. It must include all deliverable items, and functions and processes to conceive, design, fabricate, test, and deliver these items. The work breakdown structure provides the framework for all program planning, budgeting, and scheduling. It is the backbone of the program status analysis, reporting and control. Therefore it is important that all other program management tools, such as schedules, task descriptions, and budgets, correspond clearly to the work breakdown structures: that is, the work breakdown structure comes first, and everything else is derived from it.

The Level of Detail Depends on the Task

Normally three levels of breakdown are sufficient to define a project. Anything more detailed clutters the chart and destroys its flow and simplicity. As an example, the work breakdown structure in Figure 5.1 provides sufficient detail to the program manager for integrating the new computer systems development. The program office is not really concerned about the specifics of designing and integrating the central processing unit (CPU) into the computer mainframe, for example. However, someone else is. The CPU task manager must develop his or her own project plan, starting with a work breakdown structure for the CPU, which dovetails into the overall plan. Therefore, depending on the working level of the program, modulars of the work breakdown structure are subdivided into smaller components until a level of detail, scope, and complexity is sufficiently defined for its proper management.

Numbering the Tasks

In order to reference specific tasks consistently throughout all planning and control documents, a decimal system is commonly used. The system is illustrated by listing the Level I and II tasks from Figure 5.1:

1.0 Equipment design
    1.1 Mainframe
    1.2 Printer
    1.3 Tape unit
    1.4 Graphic display
    1.5 Other
2.0 Prototype fabrication
    2.1 Fabrication
    2.2 Testing
    2.3 Quality Assurance

3.0 Operation and maintenance
   3.1 User manuals
   3.2 Service
4.0 Marketing
5.0 Program management
   5.1 Equipment
   5.2 Support

Such a listing is called a *WBS index*. If in addition each element is described with a brief statement of work to be performed, it becomes a *WBS directory*. Lower levels can be included by breaking the decimal system down further. Alternatively, a modular approach to the WBS directory can be chosen by selecting specific elements and providing a detailed work breakdown just for those elements of interest in a separate list.

Grouping of Program Elements

This is a critical step. If the work breakdown structure is used properly, the various tasks are integrated and controlled along the hierarchical lines of the work breakdown structure. Therefore the work breakdown structure should be arranged to satisfy two criteria: (1) it should flow according to the integration of the program, and (2) the tasks should be grouped to minimize interfunctional involvement for each subset of activities. Accordingly the work breakdown structure shown in Figure 5.1 is appropriate if the minicomputer company has organizational units responsible for equipment design, fabrications, and operations and maintenance. A different structure might be advisable if the company is structured into product groups for mainframe, printers, and so on, and if each group has its own design, fabrication, and operations and maintenance departments. In this case, the Level 1 integration is probably most effectively performed at the product group level. One should emphasize however, that the work breakdown structure is not an organizational chart. Attempts to force congruence often result in a confused and suboptimal work breakdown structure.

Regardless of how carefully the work breakdown structure has been fine-tuned to accommodate the existing organization, specific responsibility and authority relationships must be defined with a task matrix and the statement of work, which are described in Sections 5.2 and 5.3. Finally, it should be noted that a specific work breakdown structure may be given by the customer, forcing the program office to organize, integrate, and manage the project along a preestablished structure.

The Work Breakdown Structure As Cost Model

Because the work breakdown structure breaks down the overall program into its subsystems, tasks, and deliverable items, it provides an effective and convenient model for estimating the program cost at various task levels. It is equally useful for top-down parametric estimating as well as for bottom-up estimating. As a cost model, the work breakdown structure further provides a management tool for defining cost drivers, budgetary estimating, economic feasibility studies, and designates cost studies.

In summary, *the work breakdown structure is the backbone of the program.* It has benefits for both the customer and the contractor because it:

- Defines program building blocks
- Breaks down program complexity
- Allows further expansion via modules
- Resembles a cost model of the program
- Provides a basis for all program planning and control: (1) specification, (2) responsibility, (3) budget, and (4) schedule.

5.3 THE TASK MATRIX AND TASK ROSTER

The task matrix is a simple but powerful tool to define responsibility relationships among the various program tasks and the performing organizations. The task matrix, also known as *responsibility matrix* or *linear responsibility chart*, evolved out of frustrations over conventional organization charts, which do not show relationships "within" and "between" organizational subsystems—that is, they do not show who is responsible for what at the task level. Equally important, the task matrix has a great amount of flexibility and does not restrain the system by emphasizing status and position, which is an awkward limitation of conventional organizational charts. The task matrix is derived from the work breakdown structure, therefore, all task descriptions should correspond with it. Figure 5.2 illustrates this management tool by defining the task responsibilities for the minicomputer development shown in the work breakdown structures of Figure 5.1.

The task descriptions on the left-hand side of the matrix are derived directly from the work breakdown structure and should correspond exactly to its reference numbers and structured hierarchy as shown in Figure 5.1. The right-hand side shows either the responsible organizations with their section heads or individual task managers. This structure should correspond to the actual performing organizations, which might be submerged within a functional, projectized, or matrix framework. The relationship between tasks and responsibilities is indicated by a symbol in the matrix. It is common to use P for "prime responsibility" and S for "supporting responsibility." However, quite elaborate schemes for identifying various degrees of responsibility and participation can be devised if needed.

During the program planning phase, the task matrix can be used effectively as a cost model for budgeting. Cost estimating data are collected by writing the working hours on dollar estimates or budget to perform at the crossing points of tasks and

TASK DESCRIPTION	WBS NUMBER	WBS LEVEL	R & D (RAY)	ENG SECTION A (ALLEN)	ENG SECTION B (BROWN)	ENG SECTION C (CHEN)	OPERATIONS SECTION X (DOE)	OPERATIONS SECTION Y (ED)	• • • • •
EQUIPMENT DESIGN	1	I		P					
MAIN FRAME	1.1	II		P					
SYSTEM	1.1.1	III	S	P	S				
SOFTWARE	1.1.2	III	S		P	S			
CPU	1.1.3	III	S	S	S	P			
MEMORY	1.1.4	III		S	S	P			
IN/OUTPUT	1.1.5	III		S	S	P			
CONTROLS	1.1.6	III			S	P			
PRINTER	1.2	II			P				
SYSTEM	1.2.1	III	S	P			S		
PRINT DRIVE	1.2.2	III		P		S	S	S	
IN/OUTPUT	1.2.3	III		S	S	P			
CONTROLS	1.2.4	III			P				
• • • •									
PROTOTYPE FABRICATION	2	I					P		
FABRICATION	2.1	II					P		
MFG SERVICES	2.1.1	III					P		
PC BOARDS	2.1.2	III					S	P	
HARNESS	2.1.3	III	S				S	P	
CABINETS	2.1.4	III					S	P	
• • • •									

P - PRIME RESPONSIBILITY

S - SUPPORTING RESPONSIBILITY

FIGURE 5.2 Task matrix based on work breakdown structure (WBS) of Figure 5.1 (minicomputer development).

adding up the rows and columns provides a convenient summary of the projected program cost, separated by task and performing organizations.

In summary, the task matrix provides a single framework for planning and negotiating program assignments throughout the program organization, including support departments, subcontractors, and the customer. The task matrix provides excellent visibility throughout all organizations regarding who is responsible for what. It further provides some information on the organizational interfaces involved for integrating particular subtasks through the organization.

An alternative to the task matrix is the *task roster*. It has a simple format, making it especially attractive for smaller projects. As shown in Figure 5.3, the task roster is simply a listing of the team members of a project, including their organizational affiliation, telephone number, and task responsibility, referenced against the WBS. Although the task roster is less sophisticated, it is often preferred over the task matrix because of its clarity and flexibility regarding change. The task roster is, furthermore, a team-building device. It summarizes in a simple format, usually on one sheet of paper, who is responsible for what. It recognizes the project contributors individually and as a team, regardless of their status and position within the company, and regardless of their full-time or part-time contribution to the project. More than any other management tool, we find that the task roster fosters a sense of belonging, pride of ownership, and commitment to the project. Because of these motivational benefits, the task roster is often used in large projects in addition to the task matrix.

TASK ROSTER

New Microcomputer Development Project

Project Leader: Jim Brown

Customer/Sponsor: Computer Product Group

Responsible Organization: Engineering

Responsible Individual	Organization/ Telephone	Task
Al	DE.103/x759	Equipment Design (1.0)
Beth	DE.255/x315	Printer Design (1.2)
Charlie	MF.007/x701	Test Sub System (2.2.3)
Don	DE.250/x333	Printer I/O (1.2.3)
Eric	MK.051/x653	Marketing (4.0)
Fran	MF.156/x618	Trans to MF (6.1.6)
George	QA.807/x161	ETC
Helen	ETC	ETC
Ike	ETC	ETC
John	ETC	ETC
Karrin	ETC	ETC

FIGURE 5.3 Task roster; example is consistent with WBS of Figure 5.1.

5.4 DEFINING THE PROJECT ORGANIZATION

People must understand where they fit in the organization. The traditional organizational chart is a common and usually sufficient device to define the command channels of conventional superior-subordinate relationships within project organizations. However, additional tools are needed to describe the specific authorities, responsibilities, and interfaces associated with each key position. Organization charters, task matrices, and job descriptions (as were shown in the tables of Chapter 2) are common devices for delineating the roles of key project personnel and for communicating them to all team members. In addition, companies may use the management directives to clarify definitions and organizational relations, especially for joint responsibility.

Policies and procedures are yet another set of management tools for defining the inner workings of a project organization. They present operating guidelines for running business in a proven, standard way. The depth of these documents can range from simply acknowledging project management to detailed chapter definitions and procedural guidelines, which carry the project team through the complete project life cycle. Unless there are extenuating circumstances, policies and procedures must be followed. To assure integration of these guidelines, they should be tested and fine-tuned before being formally issued. These documents must be written broadly enough to leave operating leeway to management and to accommodate the variety of projects that run through the organization.

Project managers who find themselves without adequate guidelines often develop and issue their own procedures at the project levei. This is an excellent way to communicate the established operating standards. Most likely, these project-level procedures will be reviewed for possible adoption at a higher organizational level.

5.5 WORK DEFINITION AND SPECIFICATION

The technical work and its organization within the construct of the overall program are most commonly described within the format of (1) the statement of work, (2) the specifications, and (3) the work package. A brief description of each of these management tools is provided in this section.

Statement of Work

The statement of work (SOW) describes the actual work to be performed. Together with the specifications, it forms the contractual basis for any program. Related to the task elements in the work breakdown structure, it describes precisely what is to be accomplished. The statement of work includes: (1) a definition of the task with reference to the corresponding work breakdown structure element, (2) a description of the task, (3) the results and deliverable items to be produced regarding the system, hardware, software, tests, documentation, training, and so on, (4) references to specifications, standards, directives, and other documents, and (5) all inputs required from and to other tasks. The statement of work is often a key document as part of a customer's request for proposal. Its development, particularly on large programs, is a project by itself, which requires a considerable amount of interface with all parties from the user, the customer community, and the performing organizations.

Specifications

Specifications (specs) are the descriptions of the program elements to be delivered. They form the baseline for developing, producing, and controlling the technical part of the program. Good specifications relate to the work breakdown structure. they describe the desired characteristics of the various subsystems in a modular fashion. Depending on the program, there might be (1) an overall system specification, as well as (2) hardware, (3) software, and (4) test specifications for the various program components to be delivered. Although specifications should be developed with focus on measurability in order to define and verify the end product, there should also be flexibility for the changes that inevitably evolve, particularly during a development program.

Work Package

A work package is a total program that relates to specific elements of the work breakdown structure. Each work package uniquely represents a specific unit of work, defined at a level at which the unit is managed. Regarding the work breakdown structure of Figure 5.1, easily definable work packages would be clearcut subsystems, such as 1.1 (mainframe) or 1.2 (printer) or, at a higher level, 2.0 (prototype fabrication). In providing the basis for proper management, the work package must define (1) the work to be performed, with reference to the corresponding statements of work and specifications, (2) the responsible organization or individual, (3) the resource requirements, and (4) the schedules. The task authorization is the proper tool for summarizing and communicating work package data to the project team.

5.6 TASK AUTHORIZATION

All work must be properly defined and authorized by the project manager. This also applies to subcontracted items. The task authorization is a convenient way to summarize the requirements, as well as the budget and schedule constraints, for a particular project subsystem, which is often a work package. As shown in Figure 5.4, the format of a task authorization form provides for summary and reference data in four major categories, (1) responsible individual organization, (2) schedule, (3) budget and (4) work statement. A one-sheet task authorization form is recommended, regardless of the task magnitude. The form is intended as a summary of key data, referencing pertinent documents for detailed specifications, statement of work, deliverable items, and quality standards.

TASK AUTHORIZATION

NO.

PROGRAM NAME

BUSINESS AREA

DATE

REVISION NO.

CUSTOMER ____

USER ____

PRO MGR.

INITIAL CONTRACT VAULE ____

TASK NAME:

WBS NUMBER

APPROVALS

PROGRAM OFFICE/DATE	LINE ORG/DATE	FINANCIAL CONT/DATE

CONTRACT NO:

PROG/PROJ MANAGER:

RESP. TASK MANAGER:

RESP. ORGANIZATION:

1. SCHEDULE

TASK DESCRIPTION	RESPONSIBLE INDIVIDUAL	COMPLETION DATE	COST ACCOUNT	$ BUDGET

2. BUDGET DOLLARS & MANHOURS

3. WORK STATEMENT (INCLUDING APPLICABLE REFERENCE DOCUMENTS)

FIGURE 5.4 Task authorization form.

The task authorization form is the written contract between the project manager and the performing organization, usually represented by a task manager. To be meaningful, the task authorization must have been worked out together with the key personnel who will perform the work. Moreover, an agreement on the task feasibility, its schedule, and budget must exist within the team to have some assurance of self-actualizing behavior to reach the established objectives.

5.7 PROJECT BUDGET AND COST ACCOUNTS

Project budgets are important management tools for defining resource requirements and profit expectations. Budgets are based on cost estimates. Usually the first cost estimate is needed to support a feasibility analysis long before the project baseline is fully defined. Such early cost estimates are referred to as *budgetary estimates*. Later on, during the project definition phase, the project requirements are known in enough detail to develop a more accurate estimate, often called a *bottom-up estimate*, which provides a critical input to management decisions on bidding, pricing, new product development, and strategic business plans.

Cost-Estimating Philosophy

In spite of the sophisticated approaches and computer-aided models available today, estimating the cost of an engineering development remains an art, at least to some degree. The driving forces toward a quality estimate are (1) knowledge of the project baseline, (2) historic cost availability, (3) competence of the estimator, and (4) willingness of the estimator to estimate realistically. In addition, it is important to involve the project team in the cost estimate. Without such an involvement, the commitment to the budget does not develop and design-to-cost concepts will not work.

Who Should Perform the Estimate?

Ultimately it is the project manager who is responsible to management for the cost estimate. However, many staff groups, such as cost accounting, can help to support the effort by developing the cost model and coordinating and integrating the many activities that come into play during a major cost estimating effort. The critical question is, who should make the estimate? And at what level of detail should it be made? The answer depends on the detail of the project concept available at the time of the estimate.

During the early stages of the project formation where little detail is known, the estimate should be performed on a global level by a few senior people who, by their own experience, can judge the effort involved. Breaking the overall project into its major subsystems, such as Level I work breakdown structure, may serve as a sufficient cost model.

Later on, a detailed cost estimate is needed as a basis for contract pricing, operating budgets, and cost control. The level of estimating detail depends on the level at which the project cost is to be controlled. As a ground rule, the estimate should be performed at one level below the cost accounts. The cost account becomes the level at which the cost is tracked and controlled.

The work breakdown structure is always a convenient cost model. To illustrate, reference is made to the work breakdown structure of Figure 5.1. If the minicomputer project should be controlled at Level II, cost accounts must be established for 1.1 (mainframe), 1.2 (printer), and so on. Therefore the estimates must be performed at Level III for each of the subsystem elements—1.1.1, 1.1.2, 1.1.3, and so on. For contract work, the cost account and cost estimating detail is often specified by the customer, while for company-internal projects, the judgment is left to management or the project leader. The best-qualified person to make the estimate may not be the specialist, intimately familiar with the project element, but someone who has a broader overview, who can judge multidisciplinary integration efforts as well as task redundancies, and, above all, is motivated to provide a realistic estimate as part of his or her performance.

Unit Measurements

Eventually, the question of unit measurement must be addressed. Budgets are usually established and controlled in dollars or some other local currency. This is convenient because it normalizes labor, material, travel, overhead, and other cost components. Furthermore, money becomes the ultimate measure of financial performance. However, when estimating the labor component of an engineering project, it is advisable to estimate the actual effort in worker hours, worker weeks, and worker months. Later on, a cost accountant can translate these figures into dollars by properly considering labor rates, overhead, and cost escalators. The cost-estimating form provides a convenient format for compiling labor time units associated with their job classification and the distribution of effort over time periods. For larger projects, the aid of a computer for number crunching and formatting becomes almost a necessity. In our relatively small minicomputer project of Figure 5.1, over 50 elements of cost must be calculated and summarized. Each element might require the efforts of several people, with different rates and overheads. Moreover, cost estimating often requires many iterations. Once the cost model has been set up on the computer, changes in the original estimate can be handled very quickly. An additional advantage of the computer is its ability to perform parametric comparisons, relative cost distributions, listings of cost drivers, and other checks and balances. As a rule of thumb, computers should be used to prepare cost estimates for projects in excess of 10 cost accounts. However, once established, many companies find computer-aided estimating beneficial and economical for projects of all sizes.

Estimating The Target Cost

Budgetary constraints are realities in today's world. The feasibility of a new project often hinges on its affordability. Therefore the process of conceptualizing a project based on requirements and then estimating its cost is often reversed. The project baseline may have to be designed to a given budget. The challenge here is to come up with a design that is acceptable to the sponsor without exceeding a preestablished budget. This design-to-cost approach usually leads to compromises in features and timing. An additional challenge exists if the project is a prototype development that must be designed to meet a specified unit production cost. Such design-to-unit-production-cost (DTUPC) efforts are major efforts that are quite common for new product developments or specialty production such as ships, airplanes, or telephone switching systems.

The Budgeting Process

The budgeting process is one that releases many emotions. Each group tries to negotiate for enough money to perform the work comfortably, maybe to even fund some other developmental or professional activities. While the project manager is under pressure from the sponsor to commit to a limited budget. Moreover, project personnel usually does not see any incentive for a low cost estimate, and this attitude can easily turn into a cost overrun during project execution.

It is up to the project manager to make the project desirable to the team and to convince them that getting the project started will depend on the team's ability to come up with an affordable budget. Second, the project manager should emphasize, and get upper management endorsement for the philosophy that employee performance is not measured against an arbitrary cost overrun or underrun, but is measured by the employee's (1) willingness and effort to design to established cost targets, (2) contribution to winning the contract, (3) innovative performance in coming up with cost-saving ideas, (4) ability to handle risks and contingencies, (5) support of overall team efforts, and (6) effort and success in supporting customer or sponsor relations. If the project manager fosters the right environment, which focuses on the work challenges, recognition for achievements, and professional rewards, but downplays penalties for cost overruns, the team may be able to cost-estimate more aggressively and make an effort to define a project baseline that is within an affordable budget range.

Techniques for aiding a team in designing and estimating a project concept while fulfilling a limited budget rely on the design-to-cost approach. Cost targets are established for all major efforts and then renegotiated, the task manager, depends on cost–performance tradeoffs. The key steps of a typical cost-estimating effort are stated below:

1. Define the target cost for the total project.
2. Develop the cost model, such as the work breakdown structure.
3. Establish the target budgets for labor, material, travel, and so on.
4. Allocate relative (percent) efforts to all project subsystems at work breakdown structure Level I. The total project must add up to 100%.
5. Establish worker-hour target budgets for each subsystem, based on the percentage allocation of the total budget.
6. Distribute all target cost information to all team members.
7. Ask functional managers, ultimately responsible for each subsystem, to estimate the work force needs and other costs together with their task teams. For the first round, the estimate should be performed at a high project level, such as work breakdown structure Level I. This should be a top-down estimate based on past experiences for parametric comparisons. After the initial estimates begin to converge toward the target costs, more detailed estimates are recommended, which eventually will be one level below the cost accounts.
8. The estimators provide four sets of data: (1) best estimate of effort, (2) cost drivers, (3) cost–performance tradeoff analysis, and (4) alternative baseline for given target budget.

9. Analyze cost–performance tradeoffs. Select acceptable approaches for negotiations with sponsor.
10. Recompute the relative (percent) efforts for all subsystems and tradeoffs.
11. Select the cost driver subsystems of the project and try to work out alternative solutions. Consider the total project and its concepts. The functional personnel have to work as a team together with the project office and sponsor or customer community.
12. Go back to step 6 and repeat the process until a solution satisfactory to all parties is worked out.

The Cost Account

The cost account is the lowest level at which program performance is measured and reported. It represents a specific task, which is identifiable on the work breakdown structure, it contains its relevant schedule, budget, task specifications, and responsible organization. Budgets for each cost account are time-phased over the program life cycle and become the focal mechanism for program budget planning, cost control, and reporting, especially for larger engineering programs.

A simple cost-account tracking mechanism is the time card, a document which is required for most employees in most organizations. The time card contains already prerecorded information on the employee and his or her organization. If project personnel are required to record on the weekly time card the number of hours worked on each task, which is classified by a cost account, then all cost information is available for data processing. Cost reports can be generated, which feature various summaries for convenient cost tracking and ultimate cost control. Examples are listed below:

- Actual total project cost versus planned cost, both in weekly increments and cumulatively
- Variance reports
- Expenditures, by task categories and project subsystems
- Listing of project personnel and their time charges.

If established within a proper organizational support system, cost accounts, together with their reporting systems, can be powerful tools for the project manager for measuring and controlling project cost. The cost account is also the lowest level at which actual cost budgets, schedules, and technical performance measures can be integrated.

5.8 SCHEDULES AND NETWORKS

Schedules are the cornerstones in any program-planning and control system. They present a time-phased picture of the activities to be performed and highlight the major milestones to be tracked throughout the program. Although schedules come in many forms and levels of detail, they should be related to a master program schedule and their activity structure should be consistent with that of other planning and control documents, particularly the work breakdown structure. For larger programs, a modular arrangement is a necessity to avoid cluttering; the program activities should be partitioned according to their functions or phases and then indentured to their operating levels. For example, as suggested in Figure 5.5. This provides a scheduling system that is workable for all project disciplines and at all levels.

Schedules are working tools for program planning, evaluation, and control. They are developed via many iterations with project team members and the sponsor. They should remain dynamic throughout the program life cycle. Every program has unique management requirements. The most comprehensive schedule is not necessarily the best choice for all programs. Selecting the right schedule is important. There are two principal schedule types that are most commonly used in program management. Gantt charts (bar graphs) and network techniques, as described later in this chapter.

The Milestone Chart

A good way to start any schedule's development is to define the key milestones for the work to be scheduled. Once agreed upon, the milestone chart becomes the skeleton for the master schedule and subsequent subsystem schedules.

A key milestone is defined as an important event in the project life cycle, such as the start of a new project phase, a status review, a test, or the first shipment of a deliverable. Ideally, the completion of a key milestone should be easily verifiable. In reality, however, most milestones, although crucial, are not easily verifiable. System design completed and first article test, final design review are examples of potential key milestones that must be defined in special detail to be measurable and useful for subsequent project control.

Prior to start-up key milestones should be defined for all major phases of the program. The type and number of these milestones must be carefully determined in order to make meaningful tracking of the program possible. If the milestones are

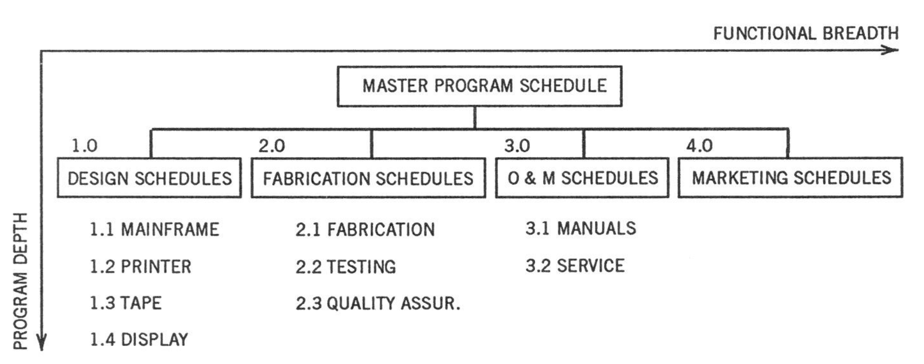

FIGURE 5.5 Modular array of program schedules for minicomputer development shown in work breakdown structure of Figure 5.1. O&M, operation and maintenance.

spread too far apart, continuity problems in program tracking and control can arise. On the other hand, too many milestones can result in unnecessary busywork, confusion, inappropriate controls, and increased overhead cost. As a guideline for multiyear programs, four key milestones per year seem to provide sufficient inputs for detailed program tracking without overburdening the system.

The program office typically has the responsibility for defining key milestones, in close cooperation with the customer and the performing organization.

Selecting the right *type* of milestone is critical. Every key milestone should represent a checkpoint for a large number of activities at the completion of a major program phase. Examples of milestones that encompass significant program segments with well-defined boundaries are listed below:

- Project kickoff
- Requirement analysis complete
- Preliminary design review
- Critical design review
- Prototype fabricated
- Integration and testing
- Value engineering review
- Start volume production
- Promotional program defined
- First shipment
- Customer acceptance test complete.

The Bar Graph

The most widely used management tool for project schedule planning and control is the bar graph. Its development dates back to work by Henry L. Gantt during World War I, which is the reason bar graphs are often referred to as Gantt charts. Figure 5.6 illustrates the basic features of the bar graph by showing a partial master schedule for the minicomputer development of the work breakdown structure of Figure 5.1.

The tasks on the left-hand side of the schedule should correspond directly to the work breakdown structure and its numbering system. In fact, there should be no tasks in the schedule which cannot be found in the work breakdown structure. If it becomes necessary to introduce a new task on the schedule, the work breakdown structure should be revised accordingly, which is part of the iterative nature of project planning. Often, however, the bar graph includes milestones such as project kickoff and critical design review, which are listed along with the tasks.

Bar graphs are simple to generate and easy to understand. They show the schedule start and finish of the tasks to be managed. Also, bar graphs can be modified to indicate project status and the most critical activity. While the hatched areas indicate the approximate project completion status as a given review date. According to Figure 5.6, task items 2, 5 and 8 are most critical. An activity is called most critical if it takes the longest time path through the project. In Figure 5.6 task item 2 is 100% complete, item 4 is 65% complete, item 7 is 50% complete, while task items 5, 6, and 8 have not been started.

Bar graphs can be further modified to show budget status by adding a column that lists planned and actual expenditures for each new task. Many variations of the original bar graph have been developed to provide more detailed information to project managers. One commonly used method is shown in Figure 5.7, which replaces the bars with lines and triangles to indicate original project status and revisions. To explain the features, let us take a look at task 2 in Figure 5.7. Using the code given in the lower left-hand corner of Figure 5.7 we see task 2 had been rescheduled three times, was finally started in February, and was completed by the end of July.

The problem with these additional features is that they take away from the clarity and simplicity of the original bar graph. However, in many cases the additional information gained from communicating the project status and subsequent control is worth the additional effort in generating and interpreting the data.

The major limitation of the bar graph schedule is its inability to show task interdependence and time–resource tradeoffs. Network techniques, which usually work together with modern data processing, have been developed, especially for larger projects. These are powerful but also expensive techniques, which help project managers to plan, track, and control their larger, more complex projects effectively.

Network Techniques

Several techniques evolved in the late 1950s for the planning and tracking of projects with large numbers of interdependent activities. Best known today are PERT and

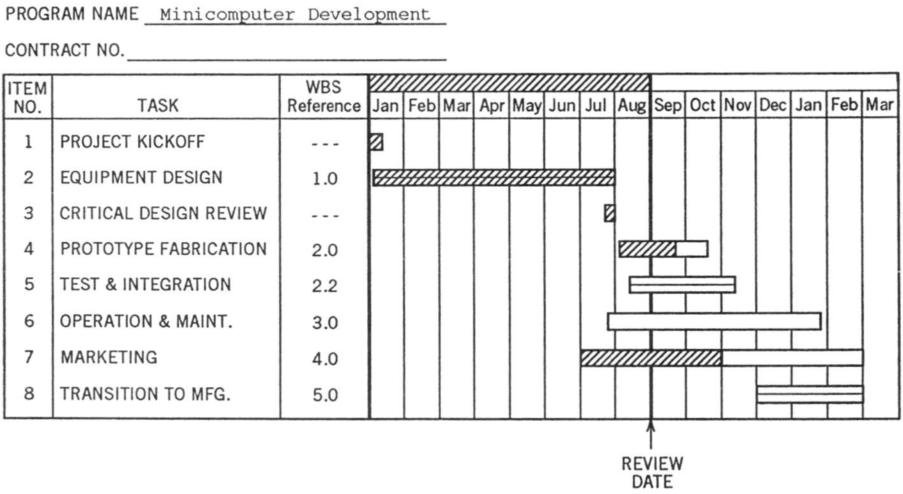

PROGRAM NAME Minicomputer Development

CONTRACT NO.

ITEM NO.	TASK	WBS Reference	Jan	Feb	Mar	Apr	May	Jun	Jul	Aug	Sep	Oct	Nov	Dec	Jan	Feb	Mar
1	PROJECT KICKOFF	---															
2	EQUIPMENT DESIGN	1.0															
3	CRITICAL DESIGN REVIEW	---															
4	PROTOTYPE FABRICATION	2.0															
5	TEST & INTEGRATION	2.2															
6	OPERATION & MAINT.	3.0															
7	MARKETING	4.0															
8	TRANSITION TO MFG.	5.0															

REVIEW DATE

FIGURE 5.6 Bar graph schedule for minicomputer development example. Hatched areas indicate approximate completion status at review (end of August). Horizontal line through bar indicates a critical activity. WBS, work breakdown structure.

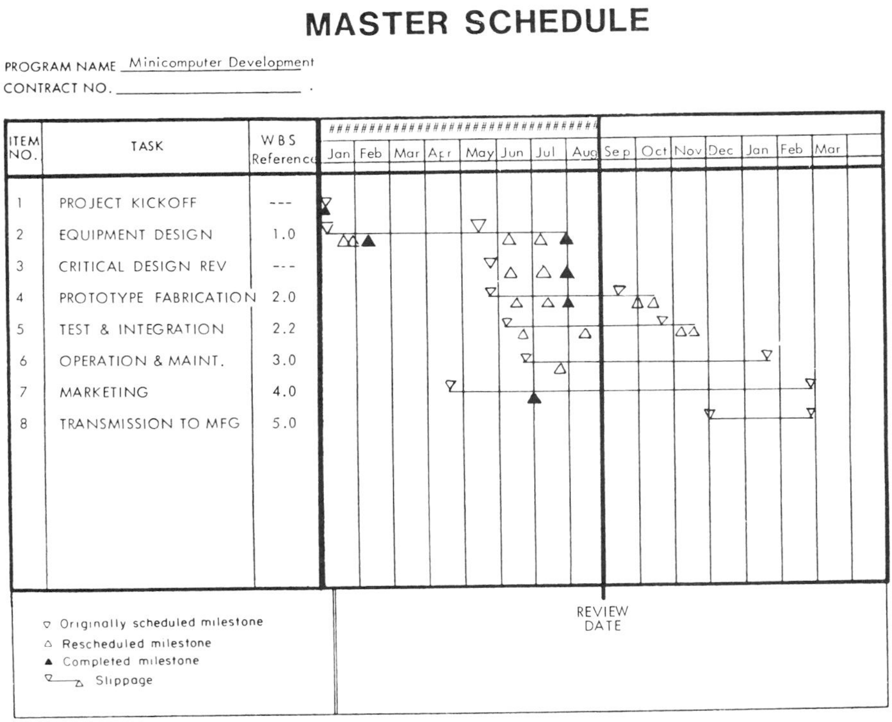

MASTER SCHEDULE

PROGRAM NAME Minicomputer Development
CONTRACT NO. ____________________

ITEM NO.	TASK	WBS Reference
1	PROJECT KICKOFF	---
2	EQUIPMENT DESIGN	1.0
3	CRITICAL DESIGN REV	---
4	PROTOTYPE FABRICATION	2.0
5	TEST & INTEGRATION	2.2
6	OPERATION & MAINT.	3.0
7	MARKETING	4.0
8	TRANSMISSION TO MFG	5.0

FIGURE 5.7 Line–bubble schedule for minicomputer development example (Figure 5.1).

CPM. PERT (program evaluation and review technique) was developed by the U.S. Navy in the late 1950s to aid in the management of the *Polaris* missile program; CPM (critical path method) was jointly developed by DuPont and Remington Rand, also in the 1950s. CPM originally had an additional feature in comparison to PERT in that it could track resource requirements. While originally each technique had its own unique features, today's commercially available computer-aided project tracking systems, as shown in Appendix 3, combine both features and are often referred to as PERT/CPM. The features of PERT/CPM are discussed with the aid of the network in Figure 5.8, which is based on our minicomputer development schedule shown in the bar graph of Figure 5.6.

It is clear that the network diagram is more powerful than the bar graph in describing the interdependent project activities. It provides a dynamic picture of the events and activities and their interrelationships. However, this is only the tip of its capability. Its main value lies in its ability to track the timing and cost[2] of all project activities. The system also assesses project duration as well as slippages and provides data for schedule tradeoffs.

[2]PERT/COST is an off shoot development from the earlier PERT and CPM techniques. It provides for tracking time, cost, and technical progress, similar to the techniques described under Variance Analysis.

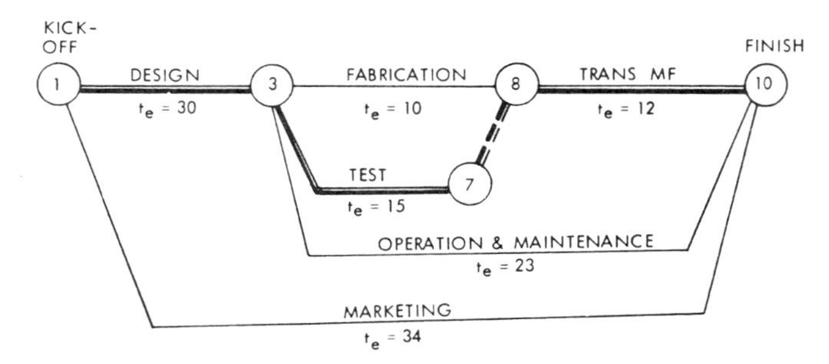

FIGURE 5.8 Network diagram for minicomputer development example. t_e, expected duration of activity in weeks. TRANS MF, transition to manufacturing. See Table 5.1.

PERT/CPM Example PERT/CPM is an excellent system for keeping track of all activities in a large project where there are thousands of interdependent tasks. Moreover, it is a powerful tool for a quick impact assessment of contingencies during the execution of a major project. To explain the principal quantitative methods and features of PERT/CPM we use the earlier example of the minicomputer development. (Figure 5.1 and the bar graph in Figure 5.6). The methodology for setting up the system is as follows:

1. Define the Work. The work breakdown structure, as shown in Figure 5.1, usually provides a good framework for delineating the work and deciding on the level or subsystem to be prepared for PERT/CPM. Let us assume that we want to prepare the network on approximately Level I of the work breakdown structure.

2. Define the Resources. On the level for which we want to perform the analysis, we have to list all the activities, develop the man-hours budget, and decide, based on available work force resources, on the duration of each activity.[3] The summary listing may look like Table 5.1.

Please note that the expected duration of an activity is not necessarily obtained by dividing total effort in man-weeks (MW) by available personnel P, since the applied work force is unlikely to be 100% over the duration of the task. The important figure for continuing the exercise is the expected duration for each task.

3. Establish the Skeleton Diagram. The activities from Table 5.1 must be organized manually into a skeleton diagram or network to reflect their existing interdependence, as shown in Figure 5.8. Note that the circles are *events*, the lines are *activities*. All events must be uniquely numbered, but do not have to be sequential. The skeleton diagram becomes the basis for setting up the PERT/CPM table.

Important Rule. Each activity must be uniquely defined by a *preceding* and a *succeeding event*. For example, the prototype fabrication activity is uniquely defined

[3]Although it is recognized that the work force consists of both male and female personnel, the expressions man-month (MM), man-week (MW), and man-hours (MH) are used in this text to be consistent with currently established terminology. For the time being, management uses these terms generically, applying them to both male and female resources.

by events 3 and 8. In order to define the test and integration stage uniquely, that is, as different from prototype fabrication 3–8, an additional event, 7, must be inserted, resulting in the dummy activity 7–8, shown as a dashed line on Figure 5.8.

4. *Define Schedule Limitations.* Let us assume that the project kickoff cannot take place before January 1 and the total project should be completed by December 31. Therefore, the *earliest start* (ES) = week 1, and the *latest finish* (LF) = week 52. Also, start and finish events must be specified.

5. *Generate Computer Inputs.* The computer must be instructed regarding (1) how the network is constructed and (2) the time duration for each activity. According to Figure 5.8 the computer input is as follows:

6. *Generate PERT/CPM Table.* The computer now generates a table (Table 5.1) based on the following calculations, which refer to a string of sequential activities n, with $n - 1$ representing the preceding activity and $n + 1$ the succeeding activity.

$$
\begin{aligned}
ES_n &= EF_{n-1} + 1 \\
EF_n &= ES_n + t_e - 1 \\
LS_n &= LF_n - t_e + 1 \\
LF_n &= LS_{n+1} - 1 \\
TS_n &= LF - EF = LS - ES \\
FS_n &= TS_n - TS_{MCP}
\end{aligned}
$$

where

ES = Earliest Start
EF = Earliest Finish
LS = Latest Start
TS = Total Slack
FS = Free Slack
CP = Critical Path is an activity behind schedule
MCP = Most Critical Path is the longest path
PE = Preceding event
P = Number of full-time personnel
MW = Work Effort in man-weeks
t = Expected duration (in weeks)

The most critical path (MCP) is the longest activity path through the network. TS_{MCP} is the total slack of activities on the most critical (longest) path in the network. (See #7 for further definition of terms.)

Although only the slack figures are of real significance, the completely calculated PERT/CPM table is shown in Table 5.1.

7. *Interpretation of Results.* Looking at the start and finish figures identifies the specific dates, here given in weeks, when work can begin and must be finished. In our small example, this information might not look like a big help to the project manager. However, when we deal with the thousands of activities, this type of bookkeeping is very helpful and leads to other types of automatic recordings, checks, and balances.

TABLE 5.1 Summary of PERT Calculations, for PERT/CPM Example

Activity (*n*)	WBS Reference	Total Work Effort (man-weeks)	Available Personnel (P)	Expected Duration (t_e) (weeks)	Preceding Event (PE)	Succeeding Event (SE)	Earliest Start (ES)	Earliest Finish (EF)	Latest Start (LS)	Latest Finish (LF)	Total Slack (TS)	Free Slack (FS)
Equipment design	1.0	2000	100	30	1	3	1	30	−4	25	−5	0
Prototype fabrication	2.0	500	60	10	3	8	31	40	31	40	0	+5
Test and integration	2.2	250	20	15	3	7	31	45	26	40	−5	0
Operation and maintenance	3.0	100	5	23	3	10	31	53	30	52	−1	+4
Marketing	4.0	100	5	34	1	10	1	34	19	52	+18	+23
Transition to manufacturing	6.0	50	6	12	8	10	46	57	41	52	−5	0
Dummy	—	0	0	0	7	8	45	45	40	40	−5	0

The more powerful indicators are the slack figures. *Total slack* indicates the number of time units that we can spare and still meet the schedule constraints. In our example, a TS of −5 for equipment design indicates that we will be 5 weeks behind schedule with this activity and any other activity in the same path, unless we shorten the expected duration of t_e. Therefore scanning the TS column quickly indicates which activities need help and where we have the luxury of slack. Any activity with *negative* total slack is termed a *critical activity*. The activities on the longest path, that is, those activities that are most negative or least positive, are called *most critical*. Therefore the longest activity path through a network is the *most critical path*.

Free slack indicates the amount of time we can spare without impacting the schedule outcome. For example, activity 3–10, operation and maintenance, will be late by one week, as indicated by TS = −1. However, activity 8–10, transition to manufacturing, will be late by five weeks (TS = −5). Therefore, activity 3–10 (TS = −1) really has another 4 weeks before an impact on the delivery date of the overall project is being felt.

Computer-Aided Technology Today

Computer software and time-sharing services for managing engineering projects are now offered by more than 100 companies. A detailed description of the features provided in project management software is given in the next chapter. These systems have the capacity to do in-depth analysis of your project data, far beyond the example given here to illustrate the basic methods and features. Project-management software available today aids in all facets of project planning, reporting, and control. This includes both numerical and graphic aids to scheduling, budgeting, and resource allocations.

In spite of all these sophisticated techniques, it is left to the project manager and his or her team to structure the framework of the project and define the key players and their relationships. In this context, the fundamental tools of modern project management, such as work breakdown structure, task matrix, bar graph, skeleton network, resource budget, and statement of work are indispensable. The computer is certainly a powerful tool to help sophisticated project managers with their administrative functions. However, it is not a substitute for innovative and creative managerial behavior.

5.9 EARNED-VALUE SYSTEMS

In an effort to measure actual project performance, several systems evolved in the 1960s that compare schedule and/or budget expenditures with technical progress, measured against an established baseline. One of the first such techniques was the line-of-balance (LOB) method, designed to measure the production status of repetitive items against budget and schedule. Other techniques that followed were originally developed for the management of government contracts. They include the cost/schedule control system criteria (C/SCSC) and the performance measurement system (PMS). Today these techniques find increasing acceptance and use among project managers of all industries.

These systems are all very similar in nature and are referred to in this text as *earned-value systems*. Their methodologies and calculations are straightforward. They consist of establishing a project baseline and then measuring technical progress against an established budget and schedule. However, *the crucial part is the ability to assess technical progress*. The following example illustrates the earned-value system.

How Does An Earned-Value System Work?

Let us assume that the spending profile of an engineering project was originally planned as shown in Figure 5.9. Let us also assume for this example that the budget includes only labor costs rather than material and other items.[4] The object is to perform a project status assessment in May, by the end of the scheduled milestone B. The project was originally scheduled for completion by the end of October. Terminology and definitions are listed in Table 5.2. All numerical data used in the example are summarized in Table 5.3.

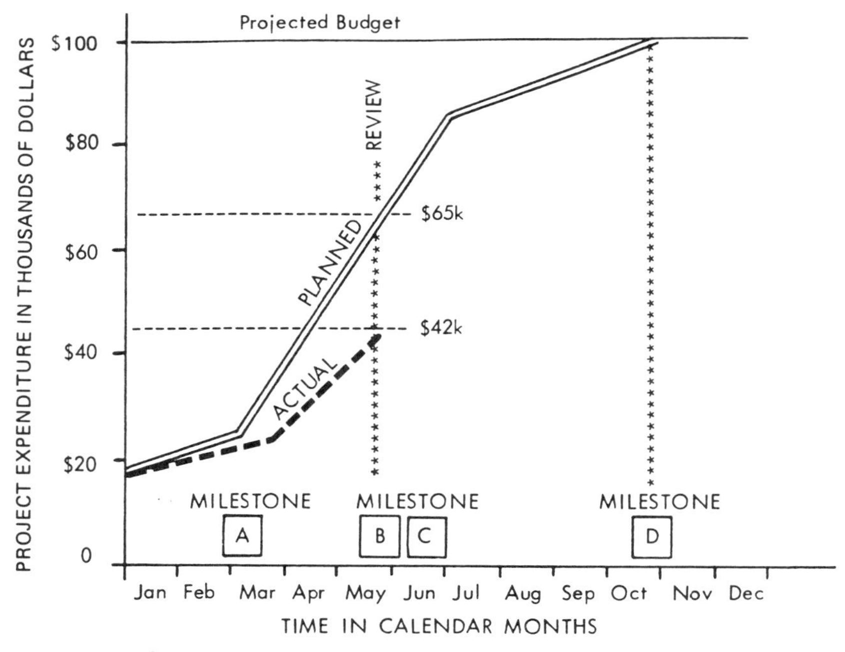

FIGURE 5.9 Spending profiles of sample engineering project.

[4]In actuality, project budgets contain labor, material, and other cost elements. The earned-value analysis must be performed separately for each element and then summed up in the same fashion, as shown in this example for the different milestone efforts. For instance, if milestone A is all labor and milestone B is all material, the sample calculations would again produce the correct results and analysis.

TABLE 5.2 Terminology and Formulas Used for Earned Value Calculations

Performance Parameter	Formula	Data Needed
BU Agreed-on budget	Negotiated Resource	• Original budget estimate
SPENT Money spent on project to date	SPENT = CAC − CTC = PCC × CAC	• Actual Cost Record • Actual estimate of cost at completion, cost to complete (CTC), and percentile of project completion (PCC).
CTC Expected cost to complete	CTC = CAC − SPENT $CTC = \frac{SPENT\ (1 - PCC)}{PCC}$	• Actual estimate of cost to complete, *or* • Revised/total project budget (CAC), expenditures to date or percent completion
CAC Expected cost at completion	CAC = SPENT + CTC $CAC = \frac{SPENT}{PCC}$	• Expenditures to date (SPENT), estimate of cost to complete (CTC) or percent of project completion (PCC)
VAR Variance (expected cost overrun)	VAR = CAC − BU	• Revised/total project budget estimate (CAC) and agreed-on or contracted budget (BU)
PCC Percent completed of project	$PCC = \frac{SPENT}{CAC} = \frac{EV}{BU}$	• Expenditures to date (SPENT), revised/total project budget (CAC)
EV Earned value	EV = PCC × BU	• Percent of project completion (PCC)—either direct estimate or calculation—and agreed-upon or contracted budget (BU)
PI Performance index	$PI = \frac{EV}{SPENT}$ $PI = \frac{BU}{CAC}$	• Calculated earned value and expenditure to date, or agreed-on budget and revised/total project budget (CAC)

Each of the milestones—A, B, C, and D—is measurably defined so that a judgment regarding its percentage of completion can be made at any point in the life cycle of the project. The budgets originally established for each set of tasks, associated with each milestone, are as follows: A = \$20,000, B = \$40,000, C = \$10,000, and D = \$30,000, leading to the "planned" spending profile shown in Figure 5.9, which adds up to a total project budget of \$100,000. The cost report, presented at review time in May, shows an actual expenditure of A = \$22,000, B = \$16,000, C = \$1,000, and D

TABLE 5.3 Summary Data for Earned-Value Calculations

Parameter	Cost and Performance Measures for Each Milestone				Total Project
	A	B	C	D	
Agreed-upon budget (BU)	$20,000	$40,000	$10,000	$30,000	$100,000
Actually spent by May (SPENT)	$22,000	$16,000	$ 1,000	$ 3,000	$ 42,000
Cost at completion (CAC)	$24,400	$53,300	$ 5,000	$30,000	$112,700
Cost to complete (CTC)	$ 2,400	$37,300	$ 4,000	$27,000	$ 70,700
Percentage completed (PCC)	90%	30%	20%	10%	
Earned value (EV)	$18,000	$12,000	$ 2,000	$ 3,000	$ 35,000
Performance index (PI)	.82	.75	2.00	1.00	.83
Variance (VAR)	$ 4,400	$13,300	−$ 5,000	0	$ 12,700

= $3,000 (Table 5.3). These expenditures add up to a total of $42,000 in May versus a planned expenditure of $65,000 by May, which is shown graphically in Figure 5.9. Although we spent less money than budgeted, obviously we cannot draw any conclusions on project performance, since we do not yet have data on the technical progress of the project, that is, how much work has been completed.

At this point let us assume that we can actually measure the status of all tasks leading to each of the four milestones. This status measure is the starting point for all earned-value calculations. There are two approaches to determine the percent completion of the project:

1. Estimate the relative completion of each milestone, measure in percent (called the *percent completion*). For example, if we estimate a 50% completion for a milestone against which we already have spent $80,000 in labor, we can extrapolate the cost at completion of this milestone to be $160,000.
2. Estimate the cost or effort to complete the milestone, measured in dollars or labor-hours. For example, if we estimate that it will require an additional $80,000 to complete the milestone and we already have spent $80,000, the milestone is 50% complete.

Using the simple example above, either approach seems to result in the status measure. After all, both approaches relate to the source formulas given in Table 5.2. However, from a practical point of view, approach number 2 is preferred because it leaves less room for misinterpretation. For many practitioners, percent completion is not straightforwardly related to cost to completion. For example, a project leader may state "We are 80% complete, but the remaining 20% will require 50% of the project budget." He is really saying that he is only 50% complete from the view of cost to completion. You don't run into these arguments if you *estimate* the cost to complete the effort and *calculate* the percent completion, which is approach number 2.

Cost at Completion and Percent Completion

Let us work through the example in Table 5.3, starting with a cost-to-complete estimate for each of the four milestone activities:

STEP 1. Estimate Cost to Complete (CTC)

For milestone A	$2,400
For milestone B	$37,300
For milestone C	$4,000
For milestone D	$27,000
For total project	$70,700

STEP 2. Calculate Cost at Completion (CAC):
SPENT + CTC = CAC

For milestone A	$22,000 + $2,400 = $24,400
For milestone B	$16,000 + $37,300 = $53,300
For milestone C	$1,000 + $4,000 = $5,000
For milestone D	$3,000 + $27,000 = $30,000
For total project	$42,000 + $70,700 = $112,700

STEP 3. Calculate Percent Completed (PCC):
SPENT/CAC = PCC

For milestone A	$22,000/$24,400 = .90 (90%)
For milestone B	$16,000/$53,300 = .30 (30%)
For milestone C	$1,000/$5,000 = .20 (20%)
For milestone D	$3,000/$30,000 = .10 (10%)
For total project	$42,000/$112,700 = .37 (37%)

Using the earned-value system in this fashion, the only estimated variable is the cost to complete. All other measures are calculated in a very mechanical way, according to the formulas defined in Table 5.2. The *earned value* (EV) is another important measure that can be calculated for each milestone and the total project. The formula is

$$EV = PCC \times BU$$

where

EV = earned value
PCC = percent completed of project
BU = agreed on budget.

It should be noted that the earned value has nothing to do with the actual expenditures, but depends only on the technical performance relative to the agreed-upon budget. According to this definition, the earned values are as follows for our milestone in Table 5.3: A = $180,000, B = $12,000, C = $2,000, and D = $33,000, adding up to $35,000 earned value for the total project. Now we have a basis for assessing project performance. While we spent $42,000 (SPENT), we earned only $35,000 (EV). The $35,000 is the amount that can be billed to the customer if we work on a contract.

Performance Index (PI)

A convenient performance index (PI) can be computed to express the ratio of earned versus spent resources:

$$PI = EV/SPENT.$$

At the May review, our performance index was $35,000/$42,000 = 0.83 indicating that we accomplished only 83% of what we planned to accomplish for the money spent.

Variance

Finally, we can compute the cost variance (VAR) for each milestone task and the total project. By definition, the variance is the difference between the agreed-on budget (BU) and the projected cost at completion (CAC).

$$VAR = CAC - BU.$$

Therefore, a positive variance indicates a projected budget overrun, while a negative variance indicates an underrun.[5] For our example in Table 5.3, we calculate a budget overrun projection of VAR = $112,700 − $100,000 = $12,700.

Performance Measurements

Project status reviews and subsequent controls focus on measuring actual performance versus expended resources and schedule. The earned-value system provides a framework for quantifying and summarizing project performance data, performing variance analysis, rebudgeting and rescheduling, and negotiating changes with the sponsor. Above all, the analysis shows the manager where the project is heading in terms of overall budget and schedule, providing a basis for renegotiations with the sponsor and redirection of the efforts at selected points in the project life cycle. The ability to summarize the performance data in terms of earned value, performance index, schedule slippage, cost variance, cost to complete and estimated completion cost is especially useful for multiproject management and the management of large programs, which need to be measured and controlled at a subsystem level. The

[5]Caution should be exercised in using this definition. Some project control systems define the variance, VAR = BU−CAC. In this case a negative variance indicates an overrun.

validity of the earned-value analysis hinges, however, on the ability to measure technical progress, which is related to the willingness of the team to define measurable milestones, a topic which is discussed in Chapter 6.

Variance Analysis

Variance analysis compares budget and earned-value data in a systematic way to define current and future financial project performance. The concept can also be extended to include schedule data, hence providing a more comprehensive project performance measure.

Sample Analysis. The variance analysis system is illustrated in Figure 5.10, the terms used are defined in Table 5.4. Let us assume that a project was originally planned for completion in October within a budget (PS) of $100,000. During the project status review in May, it is determined, by using the line-of-balance method, that the work actually completed (earned value, EV) is worth $50,000 based on the agreed-upon original budget (PB). On the other hand, the accounting department

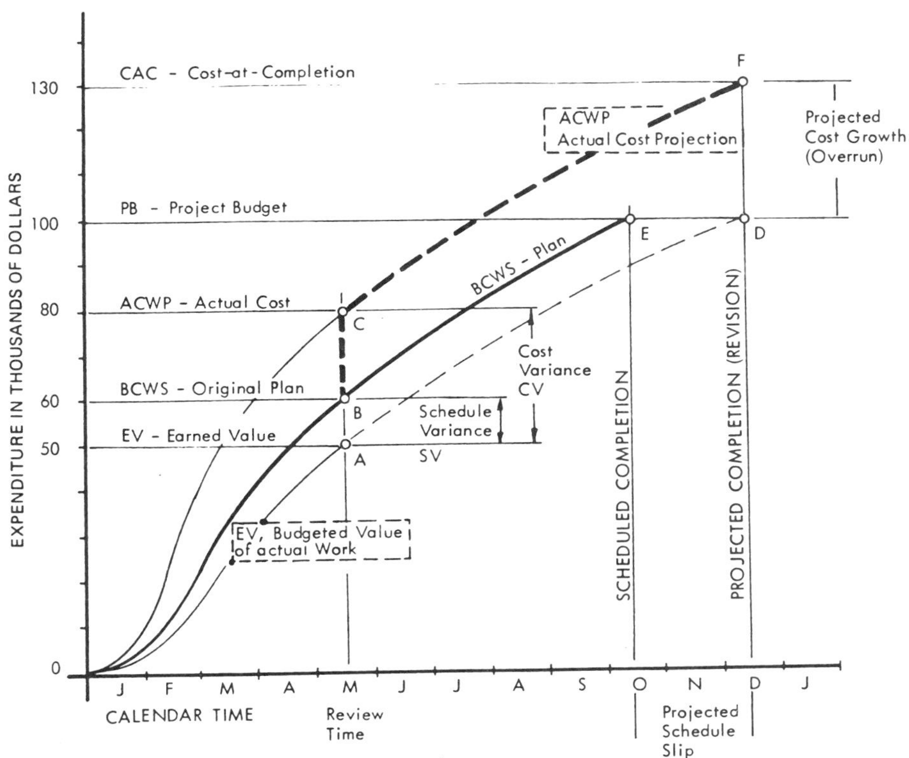

FIGURE 5.10 Graphic presentation of variance analysis and project performance measurement system.

verifies that project expenditures (SPENT) accumulated to $80,000 against a planned budgeted cost of work scheduled (BCWS) of $60,000 for May. This is all the information needed to calculate the variance data summarized in Table 5.5, using the simple formulas of Table 5.4. The only data that must be estimated graphically are the project schedule slippages. This is done by extending the expected-value curve beyond the review date *in parallel* to the planned expenditure (BCWS) curve until it intersects with the project budget (PB) line. The timing point of this intersection is the projected completion date. Therefore the slippage is the difference from the originally scheduled completion. Our example of Figure 5.10 projects a schedule slippage of 2 months.

While the variance and performance figures represent the actual project status up to the review date, the *projected* cost and schedule data underlie the assumption that progress from now on parallels the original project plan. That is, the projected EV and ACWP curves are assumed to parallel the planned expenditure (BCWS) curve, and therefore the planned expenditure curve moves from the original path B–E to B–C–F (see Figure 5.10). If there are any reasons to believe there are additional contingencies, the project plan should be revised accordingly.

A clear indication that revisions of plans are needed is a changing projection of completion cost (CAC) and date with each review as shown in Figure 5.11, where project A's cost at completion estimate grows with each review. The diverging review points A–A', B–B', and so on, indicate runaway cost and schedule, which means that the project will never be completed. The situation is more favorable for project B in Figure 5.11. Even though the projected completion cost still keeps increasing at

TABLE 5.4 Definitions for Variance Analysis and Performance Measures

Project budget (PB): Originally planned and agreed-upon budget for total project

Budgeted cost of work scheduled (BCWS): Funding profile as established in the project plan, distributed incrementally or cumulatively over the original project schedule

Earned value (EV): Budgeted cost of the work actually completed over time, that is, how much did was budgeted for the work completed so far, EV = BCWS × (PCC)

Actual cost of work performed (ACWP): Incurred cost of work actually completed and actual cost projection

Cost variance (CV), projected cost overrun: CV = ACWP − EV

Schedule variance (SV): A cost measure, comparing planned and actual work, SV = BCWS − BCWP

Projected completion date (PCD): Graphic projection of earned-value curve to intersection with original project budget

Projected schedule slippage (PSS): PSS = PCD − (originally planned completion date)

Performance index (PI): PI = EV/ACWP

Estimated cost at completion (CAC): Projected completion cost, CAC = PB + CV

Cost to complete (CTC): Funds needed to go from current status to project completion, CTC = CAC − ACWP

TABLE 5.5 Variance Analysis for Sample Project

Measurements/Projections	Original Plan (January)	First Review (May)	Second Review (July)	Third Review (October)
Cost at completion (CAC)	$100,000	$130,000		
Projected completion date (PCD)	October	December		
Budgeted cost of work scheduled (BCWS), also called "planned expenditure"	$ 0	$ 60,000	$75,000	$100,000
Actual cost of work performed (SPENT)	$ 0	$ 80,000		
Earned value (EV)	$ 0	$ 50,000		
Cost variance (CV)	$ 0	+$ 30,000		
Schedule variance (SV)	$ 0	+$ 10,000		
Performance index (PI)	1.0	0.62		
Projected cost overrun [equals cost variance (CV)]	$ 0	$ 30,000		
Projected schedule slippage (PSS)	$ 0	2 months		

review 3, the review points converge toward a final cost at completion, CAC_F of $155,000 and a 17-month completion date T_F. Plotting the projected performance for each review date can show trends of continuous problems and contingencies. It also provides a judgment of how accurately project performance was measured during the last status review. Continuously slipping target costs and schedules may indicate serious technical difficulties or project management problems that need careful and immediate attention by the project manager.

5.10 GOVERNMENT REPORTING REQUIREMENTS

Government contracts, especially those for the Department of Defense, the Department of Energy, and the National Aeronautics and Space Administration, require particular project planning and reporting formats for their larger programs. The requirements vary with contract type and size, but contracts over $1 million are expected to have some formal reporting requirements specified to the contractor. The PERT–COST requirements imposed during the 1960s have been eased in favor of less costly methods. Currently nearly all engineering projects contracted by the U.S. government require detailed scheduling, budgeting, and reporting, however.

Many specialized systems have been devised by various government agencies. Most widely used is the performance measurement system (PMS), also known as cost/schedule control system criteria (C/SCSC). Although these systems are based on conventional tools and techniques, such as work breakdown structure, task matrix, network techniques, activity cost accounting, earned-value concepts, and variance

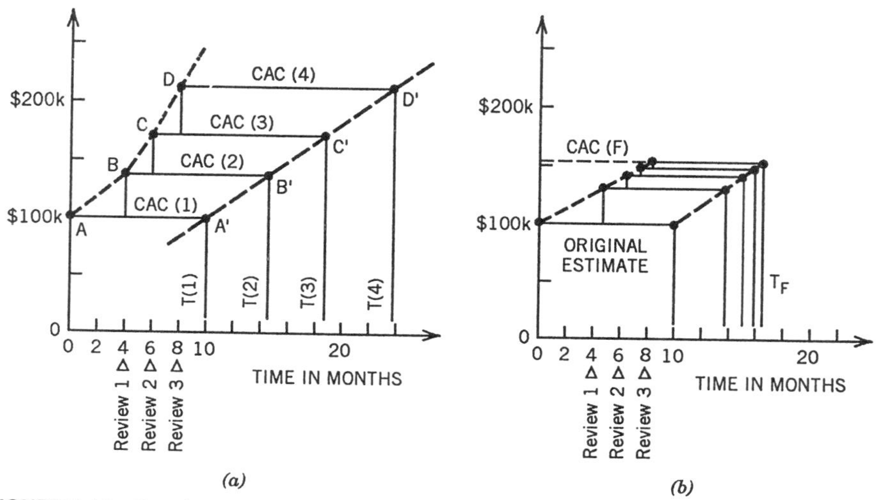

FIGURE 5.11 Trend analysis of projected performance. (*a*) Project A—runaway cost and schedule. (*b*) projected cost at completion, $155,000. CAC, cost at completion; T, time; T(F), completion date.

analysis, their understanding and implementation at the operational level often require considerable effort, cost and time. It is not unusual for a company with government contracts to maintain a special support department for establishing and administering these report requirements, as well as for training project personnel in the use of these techniques. Special publications are available from procurement offices of government agencies, which describe the specific requirements and techniques in detail.[6]

EXERCISE 5.1: PROJECT TRACKING AND REVIEW

The date is August 10. You just have been given full project responsibility for the XYZ project, a new product development project in its third month of activity. Your predecessor just quit his job to take on a new assignment elsewhere. He was kind enough however to brief you on the project and furnish you with a complete set of project status data as of August 6 (see Table 5.6). The former project manager told you that all was going fine so far. All activities are on schedule and the total cost is running below budget. To ensure proper communications, the former project manager discussed the assignment with each of the 16 task leaders individually, prior to project kick-off on June 1. The overall project plan had been worked out in detail by the XYZ Products Division, which is also funding the new development.

[6]Listings of specific procurement related publications are available from the *Government Printing Office*, Washington, D.C. 20548.

TABLE 5.6 XYZ Project, Complete Set of Project Status Data

Activity Report

Project: DEVELOP **Date: Aug 6, 19_ _ 12:00 AM**

Activity Number	Activity Description	Scheduled Start Date	Scheduled Finish Date	Total Cost
*1	Gen mktg. plans	Jun 3, 19_ _ 1:00 PM	Jun 24, 19_ _ 12:00 PM	$2492.30
*2	Assign responsibilities	Jun 24, 19_ _ 1:00 PM	Jul 1, 19_ _ 12:00 PM	$1131.92
3	Consolidate plans	Jun 24, 19_ _ 1:00 PM	Jul 1, 19_ _ 12:00 PM	$1546.34
4	Review product lines	Jun 3, 19_ _ 1:00 PM	Jun 24, 19_ _ 12:00 PM	$4326.92
*5	Hire prototype artist	Jul 1, 19_ _ 1:00 PM	Jul 17, 19_ _ 12:00 PM	$1304.30
*6	Design prototypes	Jul 17, 19_ _ 1:00 PM	Aug 28, 19_ _ 12:00 PM	$10500.00
7	Hire layout artist	Jul 1, 19_ _ 1:00 PM	Jul 8, 19_ _ 12:00 PM	$543.46
8	Hire new production crew	Jun 24, 19_ _ 1:00 PM	Jul 8, 19_ _ 12:00 PM	$1086.92
9	Train new production crew	Jun 10, 19_ _ 1:00 PM	Jul 8, 19_ _ 12:00 PM	$4200.00
*10	Review prototypes	Aug 21, 19_ _ 1:00 PM	Aug 28, 19_ _ 12:00 PM	$1442.30
*11	Final selection	Aug 28, 19_ _ 1:00 PM	Sep 25, 19_ _ 12:00 PM	$5769.23
*12	Prepare national ads	Sep 25, 19_ _ 1:00 PM	Oct 23, 19_ _ 12:00 PM	$13920.00
*13	Approve advertising	Oct 23, 19_ _ 1:00 PM	Oct 30, 19_ _ 12:00 PM	$574.61
*14	Produce advertising	Oct 30, 19_ _ 1:00 PM	Nov 20, 19_ _ 12:00 PM	$2287.50
15	Draft press releases	Sep 25, 19_ _ 1:00 PM	Oct 9, 19_ _ 12:00 PM	$1248.46
16	Approve press releases	Oct 9, 19_ _ 1:00 PM	Oct 16, 19_ _ 12:00 PM	$488.07
*17	Press ready	Nov 20, 19_ _ 1:00 PM	Nov 20, 19_ _ 12:00 PM	$0.00
				$52862.38

(continued)

TABLE 5.6 XYZ Project *(continued)*

Earned Value

Project: DEVELOP **Date: Aug 6, 19_ _ 12:00 AM**

Activity Number	Activity Description	Total Cost (F)	% Complete	Cost to Date (A)	Cost to Date (F)	Cost to Date Variance
*1	Gen mktg. plans	$2492.30	100%	$2492.30	$2492.30	$0.00
*2	Assign responsibilities	$452.76	100%	$1131.92	$452.76	$679.15
3	Consolidate plans	$3092.69	100%	$1546.34	$3092.69	–$1546.34
4	Review product lines	$4326.92	100%	$4326.92	$4326.92	$0.00
*5	Hire prototype artist	$1304.30	100%	$1304.30	$1304.30	$0.00
*6	Design prototypes	$10500.00	45%	$4725.00	$6650.00	–$1925.00
7	Hire layout artist	$543.46	100%	$543.46	$543.46	$0.00
8	Hire new production crew	$1086.92	100%	$1086.92	$1086.92	$0.00
9	Train new production crew	$4200.00	100%	$4200.00	$4200.00	$0.00
*10	Review prototypes	$1442.30	0%	$0.00	$0.00	$0.00
*11	Final selection	$5769.23	0%	$0.00	$0.00	$0.00
*12	Prepare national ads	$13920.00	0%	$0.00	$0.00	$0.00
*13	Approve advertising	$574.61	0%	$0.00	$0.00	$0.00
*14	Produce advertising	$2287.50	0%	$0.00	$0.00	$0.00
15	Draft press releases	$1248.46	0%	$0.00	$0.00	$0.00
16	Approve press releases	$488.07	0%	$0.00	$0.00	$0.00
*17	Press ready	$0.00	0%	$0.00	$0.00	$0.00
		$53729.57		$21357.16	$24149.35	–$2792.19

*Asterisk indicates activities on most critical path (MCP).

The next day you have lunch with two senior managers of your division, who are supporting your project via resource personnel since June. These managers voice some strong concern over the handling of the project so far. They are greatly worried about project integration, that is the proper interfacing of the activities, such as the prototype, with other project phases. There seems to be a general confusion over specific task responsibilities and deliverables and how the new project fits into the company's business plan. Back in your office, you begin to get nervous about your new responsibilities. Your boss stops in to inform you that a decision was made to formally review the XYZ project in 3 days. Senior managers from both divisions are expected to attend the 1-hour review session.

Your Assignment

A: Discuss in an essay:
1. The type of problems you suspect to exist in the current project, and what might have caused them.
2. The type of preparations you would make for the review meeting; how would you determine the true project status; and how you would run the project review.
3. What controls you would suggest as a project manager to ensure proper completion of the project on time and within budget.

B: Using the attached charts (the "complete set of project status data"), prepare standard-format project documentation that can be effectively communicated to all parties. If you have insufficient data, generate a list of data needed. Show the format of all documents you wish to use.

C: Perform an earned value analysis on the available project status data (August 6), include determining of budget data, percent completion, performance index, and earned value. Analyze (discuss) what your calculations mean regarding project performance. Can you draw any conclusions from the data regarding schedule performance?

BIBLIOGRAPHY

Akiyama, Kaneo, *Function Analysis,* Cambridge, MA: Productivity Press, 1991.

Allen, William F., Jr., "Project Control Engineering: Setting the Pace in the Nineties," *Cost Engineering,* Volume 32, Number 10 (Oct 1990), pp. 9–13.

Anderson, Brant W., "Tracking Projects-at-a-Glance: Applying the SAS? Graph Product as a Project Management Tool," *Computers & Industrial Engineering,* Volume 17, Number 1–4 (1989), pp. 154–158.

Anderson, Phillip and Michael L. Tushman, "Managing Through Cycles of Technological Change," *Research-Technology Management,* (May/June 1991), pp. 26–31.

Archibald, Russell D., *Managing High Technology Programs and Projects,* New York: Wiley, 1976.

Babcock, Daniel L., *Managing Engineering and Technology,* Englewood Cliffs, NJ: Prentice-Hall, 1991.

Becker, Robert H., Paula K. Martin, and Robert D. Wilbur, "Management by Accountability," *Research-Technology Management,* Volume 34, Number 1 (Jan/Feb 1991), pp. 41–43.

Beckert, Beverly A., "SDRC's New Concurrent Engineering Tools," *CAE,* Volume 10, Number 4 (Apr 1991), p. 22.

Bergen, S. A., *R&D Management: Managing Projects and New Products,* Cambridge, MA: Basil Blackwell, 1990.

Beyer, Mark W., "A Project Control to Meet Project Change," *AACE Transactions,* (1988), pp. P.5.1.–P.5.7.

Bjurstrom, Edward E. and Bonnie J. Smelser, "Commercializing Biotechnology Processes, *Chemical Engineering,* Volume 95, Number 1 (Jan 18, 1988), pp. 81–84.

Bogard, Tim, B. Hawisczak and T. Monro, "Concurrent Engineering Environments," *Printed Circuit Design,* Volume 8, Number 6 (June 1991), pp. 30–37.

Brazier, David and Mike Leonard, "Concurrent Engineering: Participating in Better Designs," *Mechanical Engineering,* Volume 112, Number 1 (Jan 1990), pp. 52¬53.

Bush, Dennis H., *New Critical Path Method,* Probus Publishing Company, 1991.

Carrabine, Laura, "Concurrent Engineering: Narrowing the Education Gap," *CAE,* Volume 10, Number 10 (Oct 1991), pp. 90–94.

Christian, Peter H., "Capital Tracking and Project Control," *Computers Industrial Engineering,* Volume 20, Number 1 (1991), pp. 71–75.

Clark, John R., "Practical Specifications for Project Scheduling," *Cost Engineering,* Volume 32, Number 6, (June 1990), pp. 17–21.

Clark, Douglas W., "Bugs are Good: A Problem-Oriented Approach to the Management of Design Engineering," *Research Technology Management,* (May/June 1990), pp. 23–27.

Cleland, David I. and William R. King (eds.), *Project Management Handbook,* New York: Van Nostrand-Reinhold, 1988.

Cleland, David I., *Project Management: Strategic Design and Implementation,* New York: McGraw-Hill, 1990.

Cleland, David I., "Product Design Teams: The Simultaneous Engineering Perspective," *Project Management Journal,* Volume 22, Number 4 (Dec 1991), pp. 5–10.

Cleland, David I., and Karen M. Bursie, *Strategic Technology Management,* New York: AMACOM, 1992.

Cleland, David I., *Matrix Systems Management Handbook,* New York: Van Nostrand-Reinhold, 1983.

Coughlan, Paul D. and Albert R. Wood, "Developing Manufacturable New Products," *Business Quarterly,* Volume 56, Number 1 (June 1991), pp. 49–53.

Coulter, Carleton, III, "Multiproject Management and Control," *Cost Engineering,* Volume 32, Number 10, (Oct 1990), pp. 9–13.

Creese, Robert C. and L. Ted Moore, "Cost Modeling for Concurrent Engineering," *Cost Engineering,* Volume 32, Number 6 (June 1990), pp. 23–27.

D'Abadie, Catherine A., M. Funk, and D. Nelson, "The Fast Decision Process: Enhancing Communication in Product Design," *IEEE Spectrum,* Volume 24, Number 6 (June 1991), pp. 23–26.

Danziger, Michael R. and Phillip S. Haynes, "Managing the Case Environment," *Journal of Systems Management,* Volume 40, Number 5, (May 1989), pp. 29–32.

Drigani, Fulvio, *Computerized Project Management,* New York: Marcel Dekker, 1989.

Fish, John G., "Cost Control in Design-Build," *AACE Transactions,* (1991), pp. M2(1)–M2(4).

Fukada, Ryuji, *Managing Engineering,* Cambridge, MA: Productivity Press, 1986.

Gates, Dermit H., III and Daniel A. Wickard, "Construction-CAE: Integrated Planning and Cost Control," *AACE Transactions,* (1989), pp. C.2.1–C.2.4.

Gervais, Bernard, "Planning for Technology Transfer," *International Journal of Technology Management* (Switzerland), Volume 3, Number 1–2 (1988), pp. 217–224.

Goodwin, Barry L., "The Development and Use of Progress Curves," *AACE Transactions,* (1990) pp. H.4.1–H.4.6.

Greene, Alice H., "Concurrent Engineering: Improving Time to Market," *Production & Inventory Management Review & APICS News,* Volume 10, Number 7 (July 1990), pp. 22, 25.

Hartley, John R., *Concurrent Engineering,* Cambridge, MA: Productivity Press, 1991.

Henderson, Thomas R. and Paul J. Buckholtz, "Comparative Cost Estimates for Project Control," *AACE Transactions,* (1989), pp. 17–26.

Henrick, Thomas R. and Timothy J. Greene, "Using the Nominal Group Technique to Elicit Roadblocks to and MRP II Implementation," *Computers & Industrial Engineering,* Volume 21, Number 1–4 (1991), pp. 335–338.

Hickman, Anita M., "Refining the Proccess of Project Control," *Production & Inventory Management,* Volume 12, Number 2 (Jan 1992), pp. 26, 29.

Hughes, David, "Growing Use of CAD/CAM Workstations Leading to Paperless Design," *Aviation Week & Space Technology,* Volume 135, Number 7 (Aug 1991), pp. 44–46.

Humphreys, Kenneth, K. and Paul Wellman, *Basic Cost Engineering,* New York: Marcel Dekker, 1987.

Ishikawa, Junichi, *Productivity Through Process Analysis,* Cambridge, MA: Productivity Press, 1991.

Izuchukwu, John I., "Project Management: Shortening the Critical Path," *Mechanical Engineering,* Volume 112, Number 2 (Feb 1990), pp. 59–60.

Izuchukwu, John, "The Design Team Approach to Class 'A' Engineering Is Working for Digital," *Industrial Engineering,* Volume 23, Number 1 (Jan 1991), pp. 37–39.

Jain, R. K. and H. C. Triandis, *Management of R&D Organizations,* 1990.

Jo, Hyeon H., Parsaei, and J. Wong, "Concurrent Engineering: The Manufacturing Philosophy for the '90s," *Computers and Industrial Engineering,* Volume 21, Issue 1–4 (1991), pp. 35–39.

Kernaghan, John A. and Robert A. Cooke, "Teamwork in Planning Innovative Projects: Improving Group Performance by Rational and Interpersonal Interventions in Group Process," *IEEE Transactions on Engineering Management,* Volume 37, Number 2 (May 1990), pp. 109–116.

Kerzner, Harold and Hans J. Thamhain, *Project Management Operating Guidelines,* New York: Van Nostrand Reinhold, 1986.

Kerzner, Harold, *Project Management for Executives,* New York: Van Nostrand Reinhold, 1982.

Kerzner, Harold and Hans J. Thamhain, *Project Management for Small and Medium Size Business,* New York: Van Nostrand Reinhold, 1984.

Kezsbom, Deborah S., Donald Schilling, and Katherine A. Edward, *Dynamic Project Management,* New York: Wiley, 1989.

Kirby and East, *A Guide to Computerized Scheduling,* New York: Van Nostrand Reinhold, 1990.

Kochan, Anna, "Simultaneous Engineering Puts the Team to Work," *Multinational Business* (UK), Number 1 (Spring 1991), pp. 41–48.

Krouse, John, R. Mills, B., Beckert, and P. Dvorak, "Successfully Aplying CAD/CAM," *Machine Design,* Volume 62, Number 15 (July 1990), pp. CC50–CC61.

Kunz, Gerald, "Project Controls: Management's Decision-Making Tool," *Cost Engineering,* Volume 30, Number 1 (Jan 1988), pp. 16–22.

Kuo, Way and J. P. Hsu, "Update: Simultaneous Engineering Design in Japan," *Industrial Engineering,* Volume 22, Number 10 (Oct 1990), pp. 23–26.

Laufer, Alexander, "A Micro View of the Project Planning Process," *Construction Management & Economics,* Volume 10, Number 1 (Jan 1992), pp. 31–43.

Lennark, Raymond H., "Grass Roots Project Control," *AACE Transactions,* (1990) pp. P.7.1–P.7.6.

Lennark, Raymond H., "Buyer Beware!—Procurement and Project Controls," *AACE Transactions,* (1991) pp. K1(1)–K1(4).

Liberatore, Matthew and George J. Titus, "The R&D Planning-Business Strategy Connection," *Journal of the Society of Research Administrators,* Volume 20, Number 4 (Spring 1989), pp. 17–26.

Lichtenstein, Chase W., "Bar-Net Schedule Eases Project Control," *Chemical Engineering,* Volume 95, Number 4 (Mar 28, 1988), pp. 53–56.

Love, Sydney F., *Achieving Problem Free Project Management,* New York: Wiley, 1989.

Lowery, Gwen, *Managing Projects with Microsoft for Windows,* New York: Van Nostrand-Reinhold, 1992.

McKim, Robert A., "Project Control-Back to Basics," *Cost Engineering,* Volume 32, Number 12 (Dec 1990), pp. 7–11.

McMullan, Leslie E., "Cost Control—The Tricks and Traps," *AACE Transactions,* (1991) pp. 05(1)–05(6).

Meyer, Marc H., "Locus of Organizational Control in the Development of Knowledge-Based Systems," *Journal of Engineering & Technology Management,* Volume 8, Number 2 (Aug 1991), pp. 121–140.

Michael, Stanford B. and Linn C. Stuckenbruck, "Project Planning," *The Implementation of Project Management: The Professional's Handbook,* Project Management Institute, Reading, MA: Addison-Wesley, 1982, Chap. 7.

Mills, Robert, B. Beckert, and L. Carrabine, "Concurrent Engineering: The Future of Product Development," *CAE,* Volume 10, Number 10 (Oct 1991), pp. 38–46.

Mills, Robert and Beverly A. Beckert, "Concurrent Engineering: Software Tools," *CAE,* Volume 10, Number 10 (Oct 1991), pp. 58–76.

Mills, Robert, "Linking Design and Manufacturing," *CAE,* Volume 10, Number 3 (Mar 1991), pp. 42–48.

Mills, Robert, "Concurrent Engineering: Hardware Architecture," *CAE,* Volume 10, Number 10 (Oct 1991), pp. 78–86.

Moder, Joseph J. and Cecil R. Phillips, *Project Management with CPM and PERT,* New York: Van Nostrand Reinhold, 1970.

Moore, John M., "Effective Use of Management Control Systems," *AACE Transactions,* (1990), pp. P.5.1–P.5.4.

Nicholas, John M., *Managing Business & Engineering Projects,* Englewood Cliffs, NJ: Prentice-Hall, 1990.

Obradovitch, Michael M. and S. E. Stephanou, *Project Management: Risks and Productivity,* Bend, OR: Daniel Spencer, 1990.

Olesen, Douglas E., "Six Keys to Commercialization." *The Journal of Business Strategy* (Nov/Dec 1990), p. 43–47.

Parker, Miles, "A New Manufacturing Mind-Set," *Manufacturing Systems,* Volume 9, Number 6 (June 1991), PP. 42–44.

Perry, Tekla, "Teamwork Plus Technology Cuts Development Time," *IEEE Spectrum,* Volume 27, Number 10 (Oct 1990), pp. 61–67.

Pincus, Claudio, "An Approach to Plan Development and Team Formation," *Project Management Quarterly,* (Dec 1982).

Port, Otis, "A Smarter Way to Manufacture," *Business Week (Industrial/Technology Edition),* Number 3157 (Apr 1990), pp. 110–117.

Postula, Frank D., "WBS Criteria for Effective Project Control," *AACE Transactions,* (1991) pp. I6(1)–I6(7).

Puttre, Michael, "Product Data Management," *Mechanical Engineering,* Volume 113, Number 10 (Oct 1991), pp. 81–83.

Randolph, W. Alan and Barry Z. Posner, *Effective Project Planning and Management,* Englewood Cliffs, NJ: Prentice-Hall, 1988.

Reddy, Ramana, R. Wood, and J. Cleetus, "The Darpa Initiative: Encouraging New Industrial Practices," *IEEE Spectrum,* Volume 28, Number 7 (July 1991), pp. 26–30.

Rigg, Michael, "Breakthrough Thinking-Improving Project Effectiveness," *Industrial Engineering,* Volume 23, Number 6 (June 1991), pp. 19–22.

Rosenbaum, Eugene S., and Frank D. Postula, "Computer-Aided Engineering Integrates Product Development," *AACE Transactions,* (1991), pp. Q16–Q18.

Rosenblatt, Alfred, "Success Stories in Instrumentation, Communications—Case History 3: Raytheon," *IEEE Spectrum,* Volume 28, Number 7 (July 1991), pp. 34–36.

Rosenblatt, Alfred, "Success Stories in Instrumentation, Communications—Case History 4: ITEK Optical Systems," *IEEE Spectrum,* Volume 28, Number 7 (July 1991), pp. 36–37.

Roussel, Philip A., Kamal N. Saad, and Tamara J. Erickson, *Third Generation R&D,* Boston: Harvard Business School Press, 1991.

Rowen, Robert B., "Software Project Management Under Incomplete and Ambiguous Specifications," *IEEE Transactions on Engineering Management,* Volume 37, Number 1 (Feb 1990), pp. 10–21.

Sasaki, Toru, "How the Japanese Accelerated New Car Development," *Long Range Planning,* (Jan 1991), pp. 15–25.

Schei, Kenneth G., "Small Project Management," *Civil Engineering,* Volume 60, Number 1 (Jan 1990), pp. 42–44.

Schopbach, Stephen C., "Key to Concurrent Engineering: Managing Product Data," *Automation,* Volume 38, Number 8 (Aug 1991), pp. 36–37.

Shina, Sammy G., "New Rules for World-Class Companies," *IEEE Spectrum,* Volume 28, Number 7 (July 1991), pp. 23–26.

Shina, Sammy G., *Concurrent Engineering and Design for Manufacture of Electronics Products,* New York: Van Nostrand Reinhold, 1991.

Shoor, Rita, "Information Management: Keeping Track of Designs," *CAE,* Volume 7, Number 6 (June 1988), pp. 80, 82.

Siegel, Bryan, "Organizing for a Successful CE Process," *Industrial Engineering,* Volume 23, Number 12 (Dec 1991), pp. 15–19.

Silverberg, Eric C., "Predicting Project Completion," *Research-Technology Management,* Volume 34, Number 3 (May/June 1991), pp. 46–49.

Silverman, Melvin, *The Art of Managing Technical Projects,* Englewood Cliffs, NJ: Prentice-Hall, 1987.

Smith, Bruce A., "Lockheed Advance Planning Prepares Engineers for ATF Product Development," *Aviation Week & Space Technology,* Volume 134, Number 17 (Apr 29, 1991), pp. 26–28.

Smock, Robert, "Fast-Track, Low-Cost Construction Starts with the Owner," *Power Engineering,* Volume 96, Number 2 (Feb 1992), pp. 19–23.

Speed, William S., "Cost Scheduling Control of Paper Machine Rebuilds," *AACE Transactions,* (1990), pp. I.1.1–I.1.3.

Szakonyi, Robert, "Establishing Discipline in the Selection, Planning, and Carrying Out of R&D Projects," *Technovation,* (Netherlands), Volume 10, Number 7 (Oct 1990), pp. 467–484.

Taylor, Allen G., Robb Ware, and William Duncan, "Project Management Software," *Computerworld,* Volume 25, Number 10 (Mar 11, 1991), pp. 55–64.

Thamhain, Hans J., *Engineering Program Management,* New York: Wiley, 1984.

Tullock, Vincent T., Robert H. Murphy and Paul F. McManus, "Project Controls Using Integrated Computer Modeling," *AACE Transactions,* (1990), pp. N1.1–N.1.5.

Turino, Jon, "Making It Work Calls for Input from Everyone," *IEEE Spectrum,* Volume 28, Number 7 (July 1991), pp. 30–32.

Vasilash, Gary S. and Robin P. Bergstrom, "Concurrent Engineering: Yes, This May Have A Familiar Ring," *Production,* Volume 5, Number 2 (May 1991), pp. 23–33.

Vasilash, Gary S., "How Your Team Can Fly as High as an SR-71," *Production,* Volume 104, Number 2 (Feb 1992), pp. 62–66.

Vesey, Joseph T., "Speed-to-Market Distinguishes the New Competitors," *Research-Technology Management,* Volume 34, Number 6 (Nov/Dec 1991), pp. 33–38.

Voelcker, John, "An Engineering Manager Who Thrives on Challenge," *IEEE Spectrum,* Volume 26, Number 10 (Oct 1989), pp. 39–42.

Wheeler, Roy, "Success Stories in Instrumentation, Communications—Case History 1: Hewlett–Packard," *IEEE Spectrum,* Volume 28, Number 7 (July 1991), pp. 32–33.

Whiskerchen, Michael J. and Bruce R. Pittman, "Dynamic Systems-Engineering Process: The

Application of Concurrent Engineering," *Engineering Management Journal,* Volume 1, Number 2 (June 1989), pp. 27–34.

Whiteley, Linda, "New Roles for Manufacturing in Concurrent Engineering," *Industrial Engineering,* Volume 23, Number 11 (Nov 1991), pp. 51–53.

Woodruff, David, "GM: All Charged Up Over the Electric Car," *Business Week,* Number 3236 (Oct 1991), pp. 106, 108.

Ziemke, M. Carl, "Warning: Don't Be Half-Hearted in Your Efforts to Employ Concurrent Engineering," *Industrial Engineering,* Volume 23, Number 2 (Feb 1991), pp. 45–49.

6
CONTROLLING AND MEASURING ENGINEERING WORK

6.1 CHARACTERISTICS OF EFFECTIVE CONTROLS

The various plans established for engineering organizations, are working tools for the engineering manager. However, plans by themselves turn into literature unless they are used to manage the organization to move toward the established objectives. Hence managerial control is the process of taking the necessary corrective actions for implementing a plan and reaching its objectives, in spite of changing, often unpredictable conditions.

Managerial controls have many dimensions and forms. Traditionally, budgets and performance appraisals were primarily used in addition to reports, procedures, and directives to control the implementation of managerial plans. These tools are still very important for today's engineering managers, however, they need to be augmented to control today's dynamic, complex, and sometimes unplannable ventures, such as the development of a financially successful new product. Implementing these plans requires much more than directives, budgets, and reports. It requires a highly dynamic control system, which uses all facets of management.

Every engineering manager has his or her own set of criteria for effective management controls. But in general, the characteristics have some common points, which are summarized below.

1. Realistic Plans. Plans are implemented by people. It is difficult, if not impossible, to control activities toward objectives if the objectives are perceived as unrealistic. The people who are ultimately held accountable for producing the results should be involved in the planning. This leads to a better understanding of the requirements, personal involvement on tasks, and possibly to a commitment to the plan and its objectives. It also leads to more realistic plans, which are agreed upon by implementors.

2. *Commitment.* Control of engineering activities depends a great deal on self-imposed control through the personal commitment of the individual contributors. Without this commitment, controls will focus on dealing with excuses for slipping performance. The same commitment to the plan, its resources, and priorities must also come from senior management.

3. *Competence.* The people implementing the plan must be professionally competent. The best plans and strongest commitments are useless unless the people have the capacity to perform. Attracting and holding quality people is a primary mission of engineering managers. Furthermore, the sign-on process of personnel to a new task or project can help considerably in matching personal skills, capabilities, and interests with the project requirements.

4. *Measurability.* If you can't measure, you can't control. Activities and milestones must have measurable results, which should be compared against the planned resources and schedules on an ongoing basis.

5. *Appropriate Control.* Management controls must be meaningful and appropriate for the type of mission. This includes both the management's actions and timing. For example, incentive bonuses and close supervision may be appropriate for controlling sales activities, but of much less value in R&D. Many controls work only on concert with other means, such as review meetings, management actions, and reports.

6. *Focus on Key Objectives.* Peter Drucker says, "Controls must follow strategy."[1] Managerial control should be related to the key objectives. Trivia should not be measured or controlled. It only masks the principal objectives and leads us into the "activity trap." That is, we get busy earning brownie points, maybe following procedures or writing memos, but not really controlling activities toward the desired key objectives.

7. *Simplicity and Adaptability.* Management controls must be simple. The purpose is to identify problems early on and take corrective actions. Good controls are quick reactions to potential problems. Further, good controls have the ability to monitor feedback and to adapt dynamically to a changing situation or emerging problem.

8. *Early Problem Detection.* Controls should focus on problem prevention and early problem detection. Direct management involvement and review meetings are more likely to have this focus than reports which often focus on problem documentation and justification.

9. *Controlling Authority.* Management controls should be in the hands of the people who are accountable for the results. As we move up the management hierarchy, management is accountable for more global results and should also exercise more global controls.

[1]Peter F. Drucker, *Innovation and Entrepreneurship,* New York: Harper & Row, 1985.

6.2 WHY ENGINEERING PROGRAMS FAIL

Few engineering managers would argue with the need for controlling their projects according to established plans. The challenge is to apply the available tools and techniques effectively, that is, to manage the effort by leading the multifunctional personnel toward the agreed-on objectives within the given time and resource constraints. Even the most experienced practitioners often find it difficult to control engineering activities in spite of apparent detail in the plan and personnel involvement or even commitment. Effective engineering management is a function of properly defining the work, budgets, and schedules and then monitoring progress. Equally important, control is related to the ability of keeping personnel involved and interested in the work, obtaining and refueling commitment from the team as well as from upper management, and to resolving some of the enormous complexities on the technical, human, and organizational side.

6.3 WHAT ARE THE CHALLENGES—A FIELD STUDY OF MANAGEMENT CONTROL

Many managers argue that engineering work *should* not be more difficult to control than any other work. Others claim that the technical complexities, the multidisciplinary nature of most engineering programs, and the multitude of superimposed regulations and restrictions make it indeed more challenging to control engineering work according to original plans. This argument is echoed especially by those managers involved in the research, development, and design phases of engineering work. A survey of 300 engineering managers indicated that, regardless of their functional responsibility, 70% of these managers feel their work is more difficult to control according to plan than other business activities, such as a promotional campaign or departmental reorganization.[2] Table 6.1 summarizes the specific challenges as seen by the engineering managers, while Table 6.2 lists reasons for overruns. This perception is further supported by statistics showing that almost every engineering program requires more time and resources than originally were estimated.

Responding to this interest, we conducted a field study to investigate the practices of engineering and project managers regarding their experiences in controlling engineering tasks (projects) according to plan.[3] Data were collected over a period of three years from a sample of over 400 engineering professionals, including 125 engineering managers. The results help to set the stage for the conceptual discussions and recommendations of this chapter.

[2]Data were collected from engineering managers during interviews. Table 6.1 summarizes their responses, rank-ordered by the frequency and significance of these problems on schedule and budget performance. For method and additional discussion see "Criteria for Controlling Projects According to Plan," *Project Management Journal,* June 1986.

[3]H. J. Thamhain and D. L. Wilemon, "Criteria for Controlling Projects According to Plan," *Project Management Journal,* June 1986, pp. 75–81.

TABLE 6.1 Challenges of Managing Projects According to Plan, and as Perceived by Engineering Managers

Rank Order	Challenge	Frequency
1	Coping with end-date-driven schedules	85%
2	Coping with resource limitations	83%
3	Communicating effectively among task groups	80%
4	Gaining commitment from team members	74%
5	Establishing measurable milestones	70%
6	Coping with changes	60%
7	Working out project plan agreement with team	57%
8	Gaining commitment from management	45%
9	Dealing with conflict	42%
10	Managing vendors and subcontractors	38%
11	Other challenges	35%

The Reasons for Poor Project Control

First, we analyzed the reasons for poor management control as they related to budget overruns and schedule slips. Second, we discussed the less tangible criteria for these control problems. We found that many of the reasons blamed for poor project performance, such as insufficient front-end planning and underestimating the complexities and scope, were really rooted in some less obvious organizational, managerial, and interpersonal problems. We studied the relationship between engineering management performance and management problems and summarized the criteria for effective project controls.

Figure 6.1 summarizes our investigation into 15 problem areas regarding their effects on poor project performance. We found that engineering managers (EMs) perceive these problem areas in a somewhat different order than their superiors (general managers).

Engineering managers most frequently blame the following reasons for poor project performance:

1. Customer and management changes
2. Technical complexities
3. Unrealistic project plans
4. Staffing problems
5. Inability to detect problems early.

Senior Management, however, most frequently cites the following:

1. Insufficent front-end planning
2. Unrealistic project plans

TABLE 6.2 Potential Problems (Subtle Reasons) Leading to Schedule Slips and Budget Overruns, and Their Impact on Performance

Rank by Frequency	Problem	Impact*
01	Difficulty of defining work in sufficient detail	H
02	Little involvement of support personnel during planning	H
03	Problems with organizing and building project team	H
04	No firm agreement to protect plan by functional support groups	H
05	No clear charter for key engineering personnel	L
06	Insufficiently defined project team organization	M
07	No clear responsibility definition within team	M
08	Rush into project kickoff	M
09	Project perceived as not important or exciting	M
10	No contingency provisions	H
11	Inability to measure true project performance	H
12	Poor communications with upper management	H
13	Poor communications with customer or sponsor	M
14	Poor understanding of organizational interfaces	M
15	Difficulty in working across functional lines	M
16	No ties between job performance and reward system	L
17	Poor leadership	H
18	Weak assistance and help from upper management	L
19	Engineering leader not involved with team	M
20	Ignorance of early warning signals and feedback	M
21	Poor ability to manage conflict	M
22	Credibility problems with task leaders	M
23	Difficulties in assessing risks	M
24	Insensitivity to organizational culture/value system	L
25	Insufficient formal procedural guidelines	L
26	Apathy or indifference by team or management	M
27	No mutual trust among team members	M
28	Too much unresolved/dysfunctional conflict	H
29	Power struggles	H
30	Too much reliance on cost accounting system	L

*Impact listed as perceived by engineering managers. H, High; M, Medium; L, Low.

3. Underestimation of project's scope
4. Customer and management changes
5. Insufficient contingency planning.

On balance, the data supported the findings of subsequent interviews: engineering managers are more concerned with external influences such as changes, complexities, staffing, and priorities, while senior managers focus on what should and can be done to avoid problems.

Rank by			Frequency of Occurrence				
General Managers	Engineering Managers	Reason or Problem	Rarely 1	Sometimes 2	Often 3	Most Likely 4	Always 5
1	10	Insufficient Front-End Planning					
2	3	Unrealistic Project Plan					
3	8	Project Scope Underestimated					
4	1	Customer/Management Changes					
5	13	Insufficient Contingency Planning					
6	12	Inability to Track Progress					
7	5	Inability to Detect Problems Early					
8	9	Insufficient Number of Checkpoints					
9	4	Staffing Problems					
10	2	Technical Complexities					
11	6	Priority Shifts					
12	10	No Commitment by Personnel to Plan					
13	7	Sinking Team Spirit					
14	14	Unqualified Project Personnel					

FIGURE 6.1 Directly observed reasons for schedule slips and budget overruns. Solid bar, engineering managers' ranking; twisted bar, general managers' ranking.

We statistically measured the differences between engineering managers and senior management's perceptions by using a Kruskal–Wallis analysis of variance by ranks. At the 90% confidence level, engineering managers disagreed with their superiors on the ranking by importance of all but six reasons. While both management groups agree and the basic reasons behind schedule slips and budget overruns, they attach different weights to them. The practical implication of this finding is that senior management expects proper activity planning, organizing, and tracking from engineering leaders. They further believe the "external" criteria such as customer changes and project complexities impact on project performance only if the engineering activity has not been defined properly and sound management practices are ignored. On the other side, senior management's view that some of the subtle problems, such as sinking team spirit, priority shifts, and staffing, are of lesser importance may point to a potential problem area. Upper management may be less sensitive to these struggles, get less involved, and provide less assistance in solving these problems than is needed.

Less Obvious Reasons for Poor Performance

Managers at all levels had long lists of the "real" reasons why the problems identified in Figure 6.1 occurred. They pointed out, for instance, that while insufficient front-end planning eventually got the project into trouble, the real culprits were much less obvious and visible. These subtle reasons, summarized in Table 6.2, have a common theme. They relate strongly to organizational, managerial, and human aspects. In fact, the most frequently mentioned reasons for poor project performance can be classified in five categories:

1. Problems with organizing project team
2. Weak project leadership
3. Communication problems
4. Conflict and confusion
5. Insufficient upper-management involvement.

Most of the problems in Table 6.2 relate to a manager's challenge to foster a work environment conducive to multidisciplinary teamwork, rich in professionally stimulating and interesting activities, involvement, and mutual trust. The ability to foster a high-performance project environment requires sophisticated skills in the leadership, technical, interpersonal, and administrative areas. To be effective, project managers must consider the task, the people, the tools, and the organization. The days of the manager who gets by with technical expertise or purely administrative skills alone are gone. Today, the engineering manager must relate socially as well as technically. He or she must understand the culture and value system of the organization. Research and experience show that effective project management is directly related to the level of proficiency at which these skills are mastered.[4] Furthermore, there is a strong correlation between the potential problems (Table 6.2) and the directly observed reasons for schedule slips and budget overruns (Figure 6.1). These correlations were measured by using Kendall's tau rank-order statistics at a strength of approximately $\tau = +.50$ and a confidence level of 99%. This indicates that the more potential problems a project leader or engineering manager experiences, the more actual performance problems are observed. The factors seem to be strongly related.

Management Practice and Project Performance

From the foregoing discussion, we saw that managers appeared very confident in citing actual and potential problems. However, we could relate these problems to poor management performance only with the help of additional statistical tests, which indeed found strongly negative correlations between (1) project performance and potential problems ($\tau = -.55$) and (2) project performance and actual problems ($\tau = -.40$). The presence of either sort of problem indeed resulted in lower performance.

[4]For a detailed discussion of the skill requirements of project managers and their impact on project performance, see H. J. Thamhain and D. L. Wilemon, "Skill Requirements of Project Managers," *Convention Record, IEEE Joint Engineering Management Conference,* Denver, Colorado, October 1978, and H. J. Thamhain, "Developing Engineering Management Skills," in *Management of R&D and Engineering,* New York: John Wiley and Sons, 1992.

Specifically, the stronger and more frequently an engineering manager experienced these problems, the lower he or she was judged regarding overall engineering management performance.

Further, an examination of specific problem areas showed the strongest negative association with performance for the following categories of problems, which we denoted *barriers*:

- Team organization problems and staffing problems
- Work not perceived as important, challenging, or having growth potential
- Little team and management involvement during planning
- Conflict, confusion, power struggles occurred
- Commitment by team and management lacking
- Poor project definition
- Difficulty in understanding and working across organizational interfaces
- Weak project leadership
- Measurability problems
- Changes, contingencies, and priority problems
- Communications, management involvement, and support were poor.

To be effective, engineering leaders must not only recognize the potential barriers to performance, but also must know where in the life cycle of the project they are most likely to occur. The effective engineering manager takes preventive actions early in the project life cycle and fosters a work environment that is conductive to active participation, interesting work, good communications, management involvement, and low conflict.

The remainder of this chapter will discuss the process of establishing measurability for engineering activities as well as the elements of engineering program control. Finally, a set of recommendations is being presented for managing and controlling engineering activities.

6.4 MEASURING TASK PERFORMANCE

The ability to measure the status of an engineering task or project is crucial for assessing engineering performance. It is also crucial for moving the project toward its established objectives. *If you can't measure, you can't control.* The ability to measure is often a determinant of a project's success or failure. This need for measurability is clear to most managers; it is the operational decision of *how* and *in what detail* to measure these parameters that causes problems to even the most experienced practitioner.

The fundamental problem is that project managers must take a snapshot simultaneously of all three measures—budget, schedule, and technical accomplishments—before the project status can be determined. In particular the technical progress, often measured as a percentage of the total project effort, is the most difficult to determine. It is also associated with the largest degree of uncertainty.

Measurability Must Be Planned

Engineering managers can take one of three principal approaches to evaluating project status:

A. *Complete and Detailed Front-End Planning.* Establish realistic milestones through detailed front-end planning; use team-oriented management controls to move towards preestablished objectives.
B. *Milestone Planning.* Establish agreed-upon principal milestones prior to project start-up, but detailed measurements are to be defined *while* work is in process.
C. *Best Effort.* Manage the project as an ongoing effort toward principal milestones. Define progress in absolute terms of current accomplishments, without assessing the percentage of completion.

The budget realities and time constraints existing in today's organizations demand, however, that management be able to report progress and control the project status relative to specific preestablished overall project objectives. The consequences of not being able to measure project status are complex. Typically they affect business performance, customer relations, and, most crucially, they often affect the manager's ability to move the multidisciplinary effort toward a given end objective. The most serious but quite common result is project cancellation.

For those engineering managers who have great concern about the need for measurability, the choice of project performance measurement and control narrows down to either type A or type B. In fact, over 90% of all project managers interviewed in a recent field study rejected project management methods that did not rely on predefined, measurable milestones.[5] It is further interesting to note that managers of highly complex programs, such as research and development, advanced developments, and conceptual studies, emphasize the need for predefined measurable milestones most strongly. They further see themselves as being more successful in effectively establishing measurable parameters than their colleagues in more conventional and mechanistic situations.

The Quandaries of Measurability

Controlling engineering programs according to established plans is crucial to most organizations. Control depends critically on the manager's ability to measure the

[5]Over 250 engineering program managers were interviewed about (1) their methods and techniques in establishing measurable milestones, (2) conducting project performance measures, (3) the problems they experienced in the process, and (4) the solutions to these problems they had attempted. For details see H. J. Thamhain and D. L. Wilemon, "Project Performance Measurement: The Keystone to Engineering Project Control," *Project Management Quarterly,* January 1982; and "Effective Technical Project Reviews," *Convention Record, IEEE Annual Engineering Management Conference,* Atlanta, October 1987.

project status and performance with reasonable accuracy. Yet, at the time that the project plan is being developed, enough details often do not exist to enable definition of measurable milestones. In addition, pressures from upper management, the customer community, and the performing organization are usually strong "to get started" rather than spending more time up front with generating "paperwork'.

Typical for such a situation is the comment of an engineering manager who laments, "I know there will be problems with measurability later on, but I am under great pressure to open up some cost accounts so people can start working on the project and we can start billing the customer." This comment is echoed by many managers within profit and loss (P & L) responsibility. It also illustrates another reason for hastily started projects. The definition of measurable milestones requires the involvement and participation of personnel from all parts of the performing organization, an effort that takes time and funding. Unless this activity is planned and agreed upon with the sponsor organization, it is often difficult to get such a program definition phase funded as an afterthought to an already established contract.

Escaping the Quandaries

Most managers who find themselves in quandaries regarding operational efficiency do not ask for more funding for project definition; they resort to incremental planning while work is in progress. Project planning proliferates in a variety of organizational roles—coordinators, expeditors, staff engineers, committees, and task forces, for example. Many executives and analysts see these remedies as a form of project-focused centralization, which restricts resource managers in their freedom and power.

How Does Measurement Work in Practice?

Every engineering manager takes a different approach to establishing measurable milestones and has a reason for following that particular approach. However, the common theme among those who manage engineering programs successfully is the ability to measure *technical status* at preestablished review points.

At the time of the review, these milestones must be clearly defined, down to specific measurable parameters; however, it is often not necessary, and often is impractical, to establish these details at the outset of the program. What is required is a process that provides the discipline of predefining the major milestones at the beginning of the program *and* allows for incremental detailed development of the measurable parameters as the program progresses.

6.5 ESTABLISHING MEASURABLE MILESTONES

The process presented here has two key features: (1) it provides the discipline for establishing detailed measurable milestones, and (2) it creates involvement at all organizational levels, pervades project planning, and leads to improved communications and commitment.

The five-step procedure summarized in Table 6.3 relies on conventional project planning and control documents which constitute the project or program management plan. (See below for further details.) Project performance data usually evolve from the work breakdown structure and the statement of work into budgets, schedules, and specifications, which form the basis for describing the deliverable items at various points in the project life cycle.

One of the key criteria for establishing measurable milestones is the ability to define deliverable items, not only for the termination point of the program, but for all critical milestones through its life cycle, such as for reviews and tests. Figure 6.2 attempts to illustrate the multidimensional nature of measurability. Both the work breakdown structure and the statement of work are usually subdivided into task elements, each associated with specific (1) budgets, (2) schedules, and (3) specifications. If done properly, it is usually possible to establish one-to-one relationships among the three measures for each task module. Therefore with budgets, schedules, and specifications defined for each task element, down to the level of actual project control, the basic criteria for measurability are established. The deliverables can now be defined for groups of task elements or work packages.

Alternatively, deliverables can be defined first for various project phases and then coupled with those task elements necessary to complete them. In either case, the ability to measure the status of the deliverable depends on the ability to determine the

TABLE 6.3 Steps in Establishing Measurable Milestones

Planning Activity and Responsible Organization or Individual	Results
1. Start with customer and/or sponsor requirement and develop program management plan. (Customer and program office responsible)	Project/system specifications, statement of work, work breakdown structure, project management subplans, budget, schedule, project team roster
2. Define key milestones throughout the life cycle of the project. (Program office or engineering manager responsible)	Milestone schedule, dates for design review, tests, prototypes, installations, documentation, training
3. Define deliverable items for each milestone. (Engineering manager and task leader responsible)	List of deliverables (for example, milestone for design review: diagrams, tradeoff analysis, make–buy decisions, system specifications, bill of material, safety plan, test plans)
4. Define specific parameters for each deliverable item. (Task leader responsible)	Statement of work, task authorization, specifications, vendor test, sign-off, report, method
5. Establish modular cost budgets for each key milestone; try to establish cost accounts for each deliverable item. (Engineering manager and task leader responsible)	Budgets, elements of cost task budgets such as: Tasks A–D: deliverable 1, $12,000. Tasks E–K: deliverable 2, $50,000. Tasks L–P: deliverable 3, $8,000.

underlying task elements and their performance as defined by the corresponding project specifications, schedules, and budgets.

For most project situations, a combination of methods is used to establish measurable milestones. But in general, these milestones are defined by measurable task elements and by measurable deliverables.

The specific five-step procedure for establishing measurable milestones in detail, summarized in Table 6.3, is described below.

1. Develop the Program Management Plan. Starting with the customer requirements, a detailed program management plan is to be worked out in cooperation with the customer and the performing support organizations. Typically this plan contains the system specifications, statement of work, work breakdown structure, project management subplans, budgets, schedules, and the project team roster. The program management plan forms the basis for all subplanning, including establishing measurable milestones and subsequent engineering program direction and control.

2. Define Key Milestones. Prior to start-up, key milestones should be defined for all major phases of the program. The *type and number of milestones* must be carefully determined to assure meaningful tracking of the program. If the milestones are spread too far apart, continuity problems in program tracking and control can arise. On the other hand, too many milestones can result in unnecessary busywork, confusion, inappropriate controls, and increased overhead cost. As a guideline for multi-year programs, four key milestones per year seem to provide sufficient inputs for detailed program tracking without overburdening the system.

Selecting the *right type of milestone* is equally critical. Every key milestone should represent a checkpoint for a large number of activities at the completion of a major design phase. Design reviews, tests, prototype completions, and field installations are examples of milestones that encompass significant program segments with well-defined boundaries. Candidates for key milestones are usually easy to spot on a

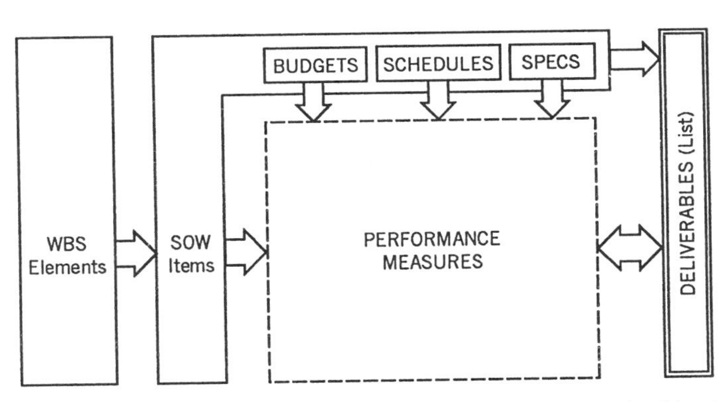

FIGURE 6.2 Budget, schedule, and technical performance measures should exist for each major work breakdown structure (WBS) and statement of work (SOW) item and for all deliverables.

program evaluation and review technique (PERT) network. They represent event nodes with a large number of dependencies.

The program office or engineering manager has the responsibility for defining key milestones in close cooperation with the customer and the supporting organization.

3. Define Deliverable Items. For each key milestone, a list of deliverable items should be defined. These deliverables represent tangible, measurable accomplishments describing the fulfillment of a milestone. For example, some of the deliverable items required to define a design review milestone are: (1) system diagram, (2) circuit diagram, (3) tradeoff analysis, (4) system simulation, (5) system specifications, (6) make–buy decisions, (7) bill of materials, and (8) test plans.

The responsibility for defining deliverable items for each key milestone rests jointly with the engineering manager and the task leader. Close cooperation with the supporting organizations is necessary to ensure meaningful items with measurable parameters.

4. Define Specific Parameters. Specific measurable parameters should be defined for each deliverable item. For example, some of the specific parameters that define the deliverable time "test plan" include: (1) the type of equipment to be tested, (2) specific test to be performed, (3) methods, (4) test apparatus, (5) parameters to be tested, and (6) documentation and reporting. Often these parameters refer to already-established components of the program plan, such as the specifications, statement of work, or other test plans.

The purpose of defining specific parameters is to provide the performing team and organization with the elements, measures, and criteria that describe each deliverable item. These parameters, together with the overlying specifications, provide the basis for the assessment of whether a deliverable item has actually been completed.

The responsibility for defining specific parameters rests with the task leader.

5. Establish Modular Cost Budget. Specific cost accounts should be established for each deliverable item. These accounts represent the cost modules for the program milestones and are necessary to relate the progress of a particular milestone or deliverable item to the established budget.

It is not usually necessary to go through another cost-estimating process. With some effort and intelligence, deliverable items can be related to the established task elements of the program plan—that is, the project tasks, derived from the work breakdown structure and the statement of work, should be partitioned to approximately fit these deliverables. This process connects the deliverables with the established task elements of the program for which budget, schedule, and performance measures are already defined. It enables a percent-complete assessment for each deliverable, and therefore for each key milestone, at any time by comparing the dollar value of completed task elements to the value of the corresponding deliverable item. The responsibility for establishing these cost accounts and budgets rests jointly with the engineering manager and the task leader. Budgets represent a commitment on the part of the company to provide resources. They also represent a commitment on the part of the responsible individual or manager to deliver the agreed-on results. The strength of these commitments depends on the uncertainties and contingencies associated with the original estimate. The underlying factors must be mutually understood on both sides to avoid conflict, mistrust, and game-playing. As an example, the operating budget for a training activity is likely to be developed with the mutual expectation that it will be met, while a budget projection for an advanced technology development or new product design might be less solid, with an implied understanding that additional resources might have to be negotiated at a later point in time.

The various factors that characterize a sound budget, one that people have bought into and feel comfortable with as a control too, are summarized in Table 6.4. Only through mutual trust and a clear understanding of the underlying assumptions can these budgets become an integrated part of the management control system for which they were designed. Once established, budgets are part of an operating plan and should be monitored by both the responsible activity leader and upper management. While an upward adjustment of a budget may be unavoidable or even acceptable in a specific situation, it is normally inexcusable to deplete the resource entirely and then try to negotiate an increase. Managers should keep a close eye on the expenditures in comparison to operational performance. If it becomes apparent that additional

TABLE 6.4 Characteristics of Effective Management Cost Control Systems

BUDGETS

- Budgets are broken into cost elements, such as activities broken into time phases, showing expenditure profiles
- They are estimated by responsible individuals
- Budgets are associated with known risk factors and uncertainties
- Budgets are agreed on between a responsible manager and upper management
- Budgets are made in constant dollars, hence providing for adjustment for inflation or overhead changes.

ACTIVITIES

- Activities are part of a clear and systematic cost model (e.g., the WBS)
- They are clearly defined in terms of the work to be performed, results, timing, and individual responsibilities
- Activities are agreed on by the individual responsible regarding the work, timing and budget
- There are measurable milestones and deliverables
- Activities are associated with a singular controlling authority, responsible for results
- Activities are visible throughout the project and the organization, and there is senior management involvement
- Activities are reflective of overall project objectives
- Activities are regularly reviewed by management
- Activities are monitored to detect early problems regarding task accomplishment and integration.

resources are needed beyond the agreed-on budget, the reason for such a budget variance should be assessed, management controls should be initiated to rectify the cause, and as a last resource, additional funds may be requested. It is hoped that all of these actions take place a long time before the budget runs out of money.

The Timing of Planning Activities

The process for establishing measurable program milestones is relatively simple. It depends on management's logic and the availability of relevant, detailed program data at the time of planning. The realities are, however, that these details are seldom available during the program definition phase. Therefore program managers should be willing to establish the framework for measurability early during program formation, but allow for the details to be worked out later, when they are actually available.

What we suggest is an incremental planning process, which follows the broad timing shown in Figure 6.3. The key milestones, such as design, reviews, tests, and deliverables, should be defined during the program definition phase. A mere statement of these milestones is sufficient at this stage. It does not require a detailed knowledge of the equipment or system to be built, but rather a commitment to establish logical checkpoints that are appropriately timed for effective management of the program.

The next step, defining deliverable items for each milestone, may require more detail than is available prior to program execution. However, many deliverables are of a generic nature (for example, diagrams, breadboards, tests, and management decisions) and should be defined during the program definition phase. Other items are not as obvious and need more program detail before they can be defined appropriately. In such a case, delineating the details of a particular milestone with its deliv-

		TIME →			
STEP	PLANNING ACTIVITY	PROGRAM DEFINITION PHASE	PROGRAM START-UP PHASE	PROGRAM EXECUTION PHASE	PROGRAM DELIVERY PHASE
1	OBTAIN CUSTOMER REQUIREMENTS, DEVELOP PROGRAM MANAGEMENT PLAN				
2	DEFINE KEY MILESTONES AND TASK STATEMENTS THROUGHOUT PROGRAM LIFE				
3	DEFINE DELIVERABLE ITEMS FOR EACH MILESTONE				
4	DEFINE SPECIFIC PARAMETERS FOR DELIVERABLES				
5	ESTABLISH MODULAR COST BUDGETS				

FIGURE 6.3 Timing of planning activities for measurability. Gray bars indicate completed work. Open boxes, work to be done.

erables, specific parameters, and costs can be defined as a specific task during the program definition phase. It becomes a contractual time—for example, part of the statement of work—with defined requirements, schedules, and budgets.

Figure 6.3 indicates that defining deliverable items for each milestone can range in timing from the program definition phase all the way into the execution phase. The same applies for defining specific parameters and modular cost budgets, a process that may extend as far as the program delivery phase. The only critical condition is that the particular milestone be fully defined, including underlying deliverables and parameters, by the time the program enters the phase where performance measurement and control toward reading the milestone become essential. The following are characteristics of a useable milestone:

- It is a clearly defined entity
- There are verifiable parameters for each deliverable item
- It is clearly related to the program management plan
- The responsible individual and organization are defined.

The Modular Approach Helps Break Down Complexity

Despite the relatively straightforward discipline described, the task of establishing measurable program milestones can become complex, especially for large, multidisciplinary programs. A simple method for breaking down the complexity of a milestone system is to divide the key milestones into their natural task categories and then, at the next level, into deliverable items. The concept is identical to the work breakdown structure used to describe the subdivision of the project effort, as discussed in Chapter 5. The charting method is illustrated in Figure 6.4. Not only does it simplify the complexity of the milestone system, but it also illustrates clearly the hierarchy of measures, overlaps, unchecked areas, and timing requirements. Furthermore the chart is an excellent tool to communicate the requirements for measurability among project team members. It leads to new insights into the requirements for effective project control, and it pervades the project planning process at all levels.

The concept described here can be extended to include the specific parameters for each deliverable item. A numerical listing of all milestone elements, with progressive indentation for successively lower levels, can provide an effective "dictionary" of all milestone measures, similar to the work breakdown structure dictionary.

Getting Team Support for Measurement

Making the project performance measurement system work requires more than just a procedure. It requires the involvement of all key project team members during the project planning phase and their ultimate commitment to the project objectives. This involvement, together with proper management support and project visibility, enhances the team's willingness to establish measurability. Project personnel throughout the organization must be convinced that project performance measurements are

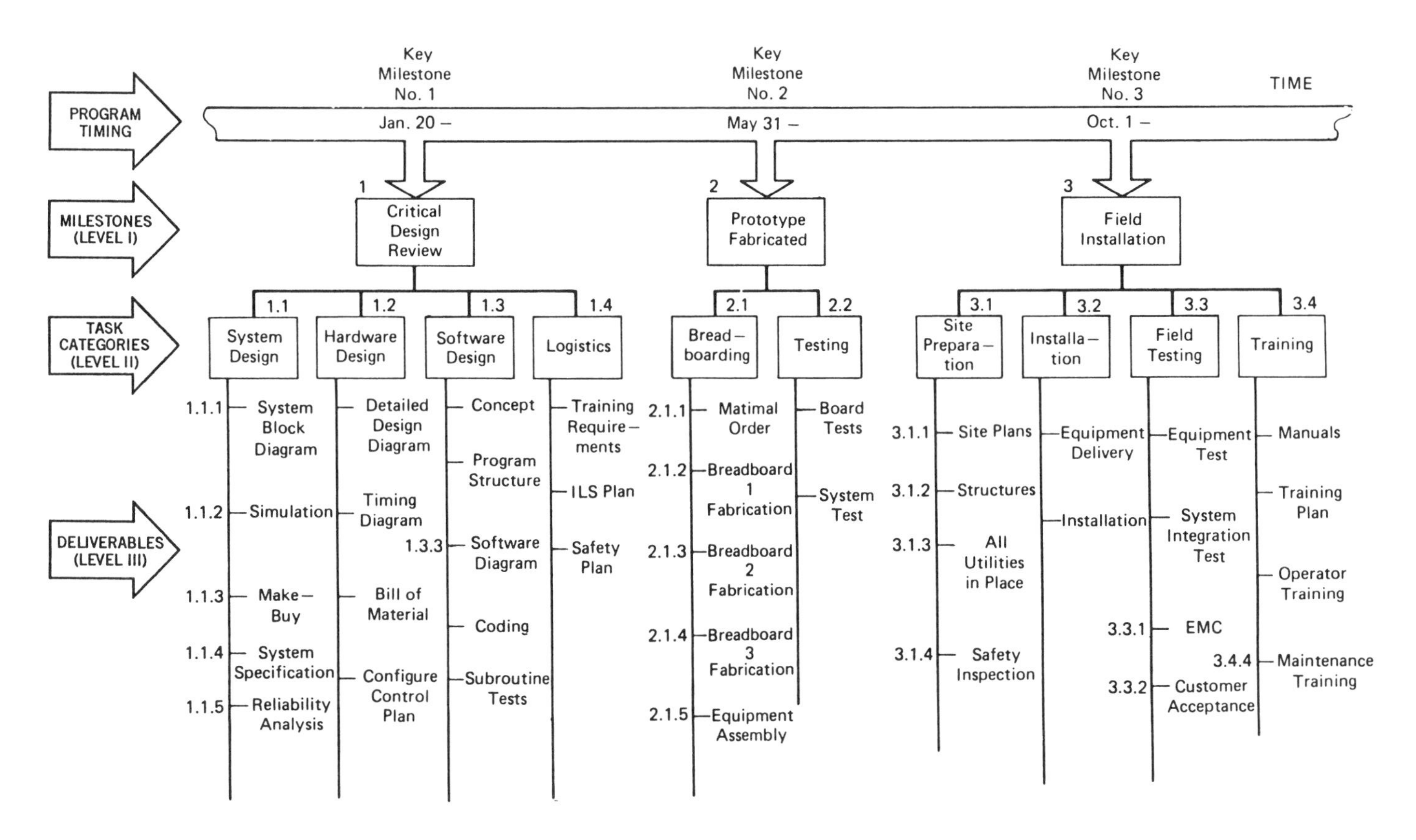

FIGURE 6.4 Breakdown of key milestones into task categories and deliverables.

helpful in their work in solving any potential problems. Moreover, team members must feel comfortable with the type of management control that they expect as a result of these measurements. If people feel threatened by the project performance measurement system, they are unlikely to participate in its implementation, therefore; management style, performance appraisal and reward system, and authority and reporting relations must be designed carefully to gain team support for the system. Management participation and a management style that fosters a professionally stimulating work environment, as well as the involvement of all project personnel, is also necessary.

6.6 CONTROLLING TECHNICAL PROGRAM PERFORMANCE

The Elements of Engineering Control

Engineering work is performed by people. It is the individual efforts of the personnel assigned to the project, not the project tracking system, that shape the results within the established parameters of technical performance, schedule, and budget. For the manager, controlling project performance means *helping the project personnel collectively to implement the agreed-upon plan.* It is a mistaken belief that engineering performance can be controlled simply by monitoring task schedules or resource expenditures against established spending profiles or budgets. Such monitoring is certainly an effective checking mechanism for major deviations. However, it indicates only that certain problems were encountered in the past, it does not predict them or detect small problems early on and rectify them.

Technically speaking, measuring budget performance, no matter how sophisticated, monitors only the derivative of an actual problem. This ripple effect, shown in Figure 6.5, can be explained as follows. First a technical problem surfaces, such as a difficulty in implementing a design requirement, a test failure, a parts shortage, or a personality conflict. The impact of such a technical problem often is that established technical performance parameters need to be downgraded in spite of increasing work effort. Tension, conflict, and fading interest can be undesired by-products of potential or actual work problems. Second, as a result of the technical problem, a schedule slip occurs. Often the reality of such a slippage becomes clear only after a certain time period past the problem. Furthermore, the schedule revision may be done in increments, because the full schedule impact unfolds with time. Third, as a result of the

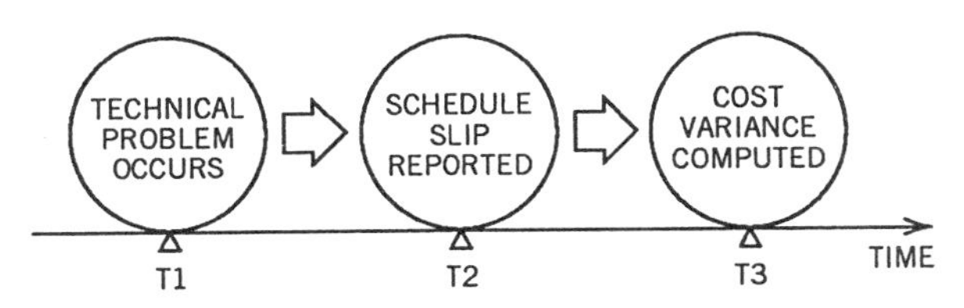

FIGURE 6.5 Ripple effect of technical problems on schedule and budget performance.

schedule variance, additional funding may be needed to reach a particular milestone, therefore increasing the projected cost at completion. The cost variance is first noticeable several time units after the problem has occurred; also, an additional time lag is caused by the time used to assess budget impact and to process and report the information.

Proper management leadership calls for staying on top of the technical work and identifying problems while they are small, or better yet, *before* they occur. This ability to predict problems in their early stages and to find multidisciplinary solutions is one of the leadership characteristics that seem to be a common ingredient of successful project management.[6]

Taken together, engineering management control rests on six support systems, which operates in concert with each other:

- Personal commitment
- Integrated program tracking
- Review meetings
- Project reports
- Management support
- Program leadership.

The prerequisites for any workable control system are (1) a functionally sound project team, (2) an agreed-upon project plan, and (3) the availability of the basic resources. Specifically, the following elements are critical for establishing and maintaining the integrity of a project plan and for providing the basis for any subsequent control:

1. Appropriate personnel, properly organized into a project team according to the tasks to be performed
2. Basic expertise and skills needed to perform the engineering tasks
3. Project and task leaders, clearly defined
4. Overall project requirements, results, and deliverables defined
5. Progress and performance measures
6. A project plan, agreed upon by all key personnel, defining the principal inputs and outputs for all major project subsystems
7. Communication channels linking all project activities, such as review meetings, status reports, and action requests
8. A special communications link to the customer or project sponsor, especially to communicate deviations from plan
9. Basic resources needed to perform the project; this includes the personnel, facilities, and materials for the life cycle of the project

[6]The gist of this philosophy was expressed by Macchiavelli in 1514: "Small problems are difficult to see but easy to fix. However, when you let these problems develop, they are easy to see but very difficult to fix." (Niccolo Macchiavelli, *Il Principle* [The Prince]).

10. Management's commitment to fully support the engineering project through its completion; this includes active participation in the management of the project, searching for solutions, helping to rectify problems, and continuous commitment to excellence and quality work, as well as a commitment to maintain a professionally stimulating work environment
11. A subcontractor and vendor control system that is clearly defined.

The appropriate configuration of the control system, its complexity and sophistication, depend on many factors, including the size and complexity of the engineering program. The 10% rule provides some guidelines as to how much overhead is reasonable for project administration and control: as a rule of thumb, the total administrative effort, including project tracking, direction, and control, should not exceed 10% of the total project budget. Indeed, we may find that smaller engineering efforts or subsets of a program need less formality in tracking and communicating.

Personal Commitment

More than any other variable, the commitment made by project team members to specific objectives provides a strong driving force toward established engineering performance. Commitment is an intrinsic drive which leads to self-actuating behavior and self-correction, as profoundly expressed by Sayles and Chandler[7]:

> Instead of relying soley on intense surveillance and monitoring by members of his staff, the manager tries to create a system that makes the organizations involved and the people in them want to do their utmost to cooperate and to achieve the sponsor's objective.

The personal commitment of engineering personnel to various project assignments grows with the personal involvement and the interest in the work and its applications.

Several specific suggestions can be put forth that will help enhance personal commitment and induce self-actuating, self-correcting project control.

1. Involve engineering personnel early, at the project formation. Active participation during project definition and requirement analysis leads to personal involvement and interest in the work itself and helps to build the team spirit. This involvement leads to an understanding of the requirements and more realistic assessment of the resources needed to perform. All of these ingredients are necessary prerequisites for personal commitment and self-guided control.

2. Create visibility and management interest for the project throughout the project life cycle. Management should explain the scope, objectives, applications, and benefits of the overall engineering effort. Frequent communiques such as memos, meetings, news articles, and exhibits issued by management and the project sponsor

[7]Leonard R. Sayles and Margaret K. Chandler, *Managing Large Systems: Organizations for the Future*, (New York: Harper & Row, 1971), based on studies of U.S. space programs.

assure the team of the importance of their efforts, help unify the team, and help blend individual interests with the overall project objectives.

3. Minimize personal insecurities and build confidence. Insecurity and anxiety often are major reasons for lack of commitment. Try to determine why these insecurities exist. They may come from several sources, such as insufficient understanding of the requirements and skills needed to perform, uncertain functional support, insufficient related experience, job insecurity, excessive competition among team members, distrust, and conflict. The personal interview prior to task assignment can be an important tool to identify the proper match of a potential team member's capabilities, desires, and needs with the job requirements and the overall project objectives. It is the responsibility of the project leaders to deal with these anxieties and build confidence on an ongoing basis throughout the life of the project.

4. Define individual responsibility clearly. Clearly defined roles, tasks, and responsibilities reinforce the individual commitments made and minimize role conflict and confusion over accountability. The principal management tools for defining these responsibilities are: (1) work packages, including task descriptions, statements of work, and lists of deliverables; (2) a task matrix; (3) charters; (4) organizational input/output statements; and (5) project personnel rosters. Furthermore the initial interview and regularly scheduled meetings during project start-up and later on for project review can help identify, negotiate, communicate, and clarify individual project responsibilities and adjust them to the dynamic needs during the project life cycle.

5. Define personal rewards and incentives. Personal rewards have many facets. They range from salary to bonuses on the financial side, but they also include professional rewards such as recognition, visibility, ability to learn, freedom to act, opportunity to travel, and possibility of advancement. Management should discuss the needs and desires of their engineering personnel and then delineate the basic rewards and incentives that will be considered. Care must be taken by management not to create an environment that is so ultra-competitive that it is threatening to people. Financial incentives to personal performance may stifle innovation and creativity, as people will likely focus on preestablished results—that is, the status quo—rather than adapting to changing requirements. On the positive side, many executives find a properly designed reward system, which carefully balances financial and professional awards, to be a very powerful tool, which is necessary for fueling personal motivation, drive, and commitment to engineering project success and overall productivity.

Taken together, it is the personal commitment to the successful completion of agreed-upon tasks that fosters a pervasive search for excellence and a unifying force among team members. It is this commitment that provides the basis for individual accountability and self-actuating engineering activity control.

Integrated Program Tracking

Program tracking and reporting is another important pillar in the engineering performance control system. The supporting elements such as computer-aided project management systems, meetings, and status reports provide a basis for measurability and eventually for integrated decision-making. They help identify problems and corrective actions. To be useful for management, it is crucial that the system have the ability to:

- Communicate true program status
- Integrate work status, schedule, and resource data
- Communicate program status to all involved parties, such as the task/or project team, functional support personnel, upper management, subcontractors, and the sponsor community
- Present the data in a clear format for easy interpretation, cross-validation, and management actions.

Every engineering organization that manages programs has its own tracking and reporting system, which is tuned to the specific needs of the activities at hand. But most of the successful managers, particularly for the more complex engineering efforts, rely on three specific tools for tracking their multidisciplinary activities: (1) cost tracking and analysis; (2) time and effort tracking and analysis; and (3) work status tracking and analysis. Often all three types of reporting systems are integrated into one set of documents. The information in one report frequently overlaps with that of other reports. For reasons of clarity, we will discuss each report separately in this chapter.

Cost Tracking and Analysis

The fundamental tool for tracking resource expenditures is the cost report, which provides a dynamic listing of expenditures for the various elements of costs and for the total program. For many programs, controlling costs is the bottom line. The cost report provides the starting point for analyzing, comparing, and controling program costs.

Format, Timing, and Distribution. Cost reports usually trail actual expenditures by eight to twelve days. Labor expenditures are commonly captured on the weekly time cards from the previous week. An additional two to five days are needed for data processing. While cost report data can certainly be hand-calculated and manually prepared, the number of calculations and the repetitive nature of analysis from one reporting period to the next make electronic data processing most cost effective, especially for large programs. Depending on the program size and complexity and the sophistication of the cost reporting system, the following reports are provided:

1. Activity Cost Report. A listing of weekly, monthly, and cumulative (total) expenditures for each program activity, as shown in Figure 6.6. The listing can also include worker-hour and percentage figures. Depending on the report's distribution,

ACTIVITY COST REPORT

WEEK ENDING: 23 JULY 1983

PROGRAM NAME: NEW MINICOMPUTER DEVELOPMENT

PROGRAM MGR : J.J.LEADER

			LABOR					MATERIAL				
					SPENDING PROFILE					SPENDING PROFILE		
COST ACCOUNT NO.	WBS ELEMENT NO.	ACTIVITY NAME	WEEKLY BUDGET $	WEEKLY ACTUAL $	TO DATE BUDGET $	TO DATE ACTUAL $	TOTAL BUDGET $	WEEKLY BUDGET $	WEEKLY ACTUAL $	TO DATE BUDGET $	TO DATE ACTUAL $	TOTAL BUDGET $
0	0	PROGRAM SUMMARY	117,500	103,200	506,700	483,450	1,000,000	80,000	56,776	145,500	134,987	255,000
1.	1.	EQUIPMENT DESIGN	20,000	18,300	100,100	83,000	305,000	22,000	22,500	85,000	97,300	125.000
1.1	1.1	MAINFRAME	5,000	5,800	18,500	22,000	70,000					
1.2	1.2	PRINTER	3,200	4,700	15,100	14,300	30,000					
1.3	1.3	TAPE UNIT	4,700	2,100	20,300	18,100	40,000					
1.4	1.4	GRAPHIC DISPLAY	2,800	2,800	13.700	12,000	80,000					
1.5	1.5	OTHER	4,300	2,900	32,500	16,600	85,000					
2.	2.	PROTOTYPE FAB	5,100	4,800	25,600	29,000	80,000	8,300	9.600	25,000	27,330	60,000
2.1	2.1	FABRICATION	2,200	2,900	12,300	14,600	45,000					
2.2	2.2	TESTING	2,300	1,100	10,200	10,400	25,000					
2.3	2.3	QUALITY ASSUR	600	800	3,100	4,000	10,000					

FIGURE 6.6 Sample activity cost report for minicomputer development example shown in work breakdown structure of Figure 5.1.

the data should be summarized for specific work packages or for levels of activities. The principal output data include:

- Activity name and cost account
- Budgeted versus actual cost per week
- Spending profile, planned versus actual costs to date
- Total budget per cost element
- Percent expenditure per cost element.

2. *Cash-Flow Report.* A weekly and cumulative total listing revenues and committed and actual expenditures. This report is especially helpful in managing the cash flow of externally funded programs with progress payment provisions. The principal output data of the cash-flow report include:

- Activity name and cost account
- Planned versus actual costs to date
- Percentage of work completed (input data/estimate)
- Estimated cost to complete
- Total cost at completion
- Earned value of work in progress
- Variance (estimated cost overrun)
- Cost-performance index.[8]

3. *Variance Report.* Provided that a percent-complete figure can be obtained for each activity listed in the report, a variance analysis can be performed, as discussed in the previous chapter, and its results can be summarized in the variance report. The principal output data include:

- Activity name and cost account
- Planned versus actual cost to date
- Percentage of work completed (input data/estimate)
- Estimated cost to complete
- Total cost at completion
- Earned value of work in progress
- Variance (estimated cost overrun)
- Cost performance index.

Cost reports can become very voluminous and cluttered with information that is irrelevant to project control if attempts are made to service too many people with the same report. For example, it is clear that the program manager needs a different report

[8]The Cost Performance Index (CPI) is the ratio of earned value (EV) divided by resources expended (SPENT), as defined in Chapter 5.

format with less detail and more perspective than the task leader responsible for the implementation of a particular program work package. If computers are being used for electronic data processing of the cost data, it is usually simple enough to request formats that summarize the cost report at the specific level of activity, such as for a particular work package or program level. The distribution of cost reports should be in accordance with the details needed at the respective user levels.

Where Do Cost Report Inputs Come From? Inputs for the cost analysis and report come primarily from the program plan for establishing the data base file, and the employees' time cards for updating the data file, as follows:

FOR ESTABLISHING THE COST DATA FILE

Item	Source
Activity description	WBS
Cost element number	WBS
Planned/budgeted costs	Budget
Payment schedules	Contract
Responsible individuals	Task matrix

FOR UPDATING THE COST DATA FILE

Item	Source
Tasks and hours worked/week	Time card
Employee name and department	Time card
Revised budgets	Management
Progress billings and sales	Contracts
Materials and subcontracts	Procurement
Percentage of work completed	Task leader

The employees' time cards are very convenient and effective instruments to collect program labor cost data. The only information to be provided by the employees is the task and hours worked on each task. The time card system will automatically reference the following data: department name, labor classification, hourly compensation and overtime pay, accrued costs, such as vacation, department overhead, other burden costs, and possible escalators.

Time and Effort Tracking and Analysis

Program schedule and work effort are highly interrelated. The ability to monitor and analyze these complex relationships provides another facet of engineering management control. Most schedule-related reports are formulated to present program evaluation and review technique/critical path method (PERT/CPM) type information. Many of these reports also include cost data. In general, electronic data processing (EDP) is a necessity for preparing these reports cost- and time-effectively. However, manual preparation of a summary report with few activities is workable and desirable in the absence of EDP facilities. The timing and distribution of the time and effort tracking report is similar to those of cost reports. The following types of time-related reports can be generated.

Time–Activity Reports. This is a listing of activities referenced to the work breakdown structure and associated with planned and projected schedule data. Depending on the sophistication of the reporting system, the following outputs can be generated for each activity, either in tabular or graphic format:

- Scheduled start and finish dates
- Work-force requirement
- Actual work-force loading
- Percentage of work completed
- Projected finish dates
- Projected schedule variances
- Project calendar, lists upcoming key activities
- Milestone analysis
- Work-force loading for each department
- Comparison of estimated and actual dates
- Impact analysis of slippages.

In essence, the time–activity report is based on an electronic bookkeeping operation, which sorts and groups activity and time data for effective display. The inputs for the time–activity report come from the project schedule and the work-force plan, and from a weekly update provided by the task leaders.

Time Analysis Reports. These reports have a PERT/CPM-type format, similar to the time–activity report, but with more analytical capability. Typically the following outputs are generated for each activity in tabular or graphic format:

- Activity and work breakdown structure reference
- Estimated time for completion
- Earliest and latest start and finish times
- Total slack and float
- Critical activities
- Ranking of most critical activities
- Trend analysis for selected variables
- Impact analysis of contingencies.

Work Status Tracking and Analysis

This is a listing of activities and their completion status. Typically, a work status report is generated on a weekly basis by each task leader and submitted to the next-highest project integrator level, such as the project manager. At that level, all task leader inputs are integrated into a single report, which is submitted to the next level, if such a level exists, or to senior management and the program sponsor. The integrated summary report also is distributed to all task leaders reporting into that program level. To limit the amount of time- and status-report writing and to keep the reports brief and informative, a specific format and page limitation should be specified. As a guideline, we suggest that the weekly work status report be limited to one page, addressing the following topics:

- Progress made during reporting week
- Changes introduced
- Problems and their impact
- Open items
- Important dates (milestones, meetings, etc.).

Using the Tracking System Effectively

The various reports that are generated by and for the project leaders are the working tools of engineering management. They provide the basis for measurability, indicate deviations from the plan, and report potential problems. These reports, if compiled properly, also provide a multidimensional overview and insight into the interrelationships among the many variables that come into play in the execution of a modern engineering program.

The tracking and reporting system can be used for management control for the following:

1. *Variance Analysis.* Weekly cost reports compare budgets with actual expenditures. Variances are analyzed to determine the cause. Problems are assessed and solutions are worked out with the performing personnel.

2. *Activity and Interface Problems.* Spending profiles of planned versus actual expenditures can provide clues to activity level and interface problems. For example, a new product development is overspending in the engineering department but underspending in manufacturing. This indicates potential interface problems between engineering and manufacturing. It could be that engineering is working hard toward moving the design into production, but problems have occurred and the schedule is slipping. As a result, manufacturing cannot get started. An underexpenditure against plan could be the result of superior performance or an easier-than-anticipated task, but most likely it indicates a slipping schedule because of "pipeline delays" or a low level of effort. The actual situation must be analyzed and problems must be dealt with on a one-to-one basis.

3. Additional Funds Needed Because of Scope Changes. A projected cost-at-completion increase should trigger an analysis of its cause, such as potential problems, customer-initiated changes, increasing project scope, or the reality of a clearer picture regarding the actual projected effort necessary. Problems must be dealt with, and scope changes should be negotiated with the project sponsor for additional funding.

4. Individual Department Performance. Work status reports can be used in conjunction with cost reports and the work breakdown structure to analyze the project performance of each functional support department and each subcontractor. Problems and variances can be analyzed on a department-by-department level. A search for solutions and schedule recovery should be initiated.

5. Schedule Performance. Project progress should be reviewed against established milestone schedules. PERT/CPM-type tracking systems help to determine project performance compared with established schedules. Causes of schedule slippage should be determined and schedule recovery plans should be worked out with the appropriate project personnel. Work schedules must be revised to reflect actual status and revised time estimates.

6. Insight, Commitment, and Involvement. The personnel involvement generated at all levels through the compilation and analysis of the reports provides additional insight, multidisciplinary understanding, and a reconfirmed commitment to the project objectives.

7. Specific Management Actions. An integrated program-tracking and reporting system helps management control the activities toward established parameters of technical performance, schedule, and cost. Specifically, it helps to:

- Isolate performance problems regarding their organizational origin and impact on technical, schedule, and cost performance.
- Detect technical problems early.
- Identify the cross-functional impact of deviations from the plan and the protective actions and contingency measures needed in affected project organizations.
- Identify unplanned shifts in work-force allocation and priorities.
- Identify needed adjustments to work-force schedules.
- Identify potential areas for relaxing technical specifications.
- Negotiate additional funding and time for changing scope and efforts.
- Update the project plan regarding schedule, budget, and performance specifications.
- Decide when to use a contingency plan.

Review Meetings

The program or project review meeting is considered one of the most important tools for controlling multidisciplinary activities. Its advantage over any other project control tool is the involvement of all key personnel at the time of the meeting. This participation and interaction of the team members leads to a comprehensive understanding of the project situation and its potential problems. It further cross-validates current data and helps to develop a true picture of the project status, including progress against plan, and potential problems and their impact on the various project operations. In this interactive mode, many of the potentially major problems can be identified and dealt with very early in their development. Moreover, project meetings are action-oriented: when potential problems surface, corrective actions can often be defined and assigned right during the meeting, since all the key decision-makers from both the project organization and the functional support departments should be present. Some other more subtle advantages relate to the synergism, risk-sharing, and unifying forces, which are part of the group process that takes place during the meeting. A summary of the features of project review meetings is provided in Table 6.5. Because of their importance to engineering managers for controlling technical activities, review meetings are discussed as a separate topic later on in this chapter.

Project Reports

Project reports are working tools of program management. Aside from the fact that specific reports are often requested by the customer as part of the contractual agreement, project reports serve many purposes that are vital to the effective overall management of a program or project.

1. *Recordkeeping.* Project reports document the status, results, changes, milestone dates, and other factors for immediate communication, future reference, and in support of other management actions such as review meetings.

TABLE 6.5 Characteristics of Effective Project Review Meetings

- Dynamic exchange of project status information
- Candor in identifying actual or potential problems
- Some structure, discipline, and agenda
- Limited group size
- Goal- and result-orientation
- Action orientation
- Senior management presence to assure decisions and closure
- People come prepared
- Continuity from previous review
- Low interpersonal conflict and power plays
- High visibility to total project team and support organizations
- Good team spirit, mutual trust

2. *Communication.* Reports are communication devices. If prepared properly, they integrate and unify the information from the various activities and make them available to the project team, the customer community, and top management.
3. *Control.* Project reports often become the basis for management actions. Presenting a particular situation, problem, or action request in writing often provides a more comprehensive picture to decision-makers and a clear justification for management action. In addition, the visibility generated by the report may foster a cooperative and supportive work environment.
4. *Contractual Protection.* Project reports form a chronological record of the program's progression through its various phases. These are often helpful in tracing program developments as a basis for change orders and in clarification of contractual requirements.

Setting up a proper project report system is the responsibility of the engineering manager. It is important that the various documentation efforts be properly coordinated and operate from the same data base, which should be the updated master program plan. Setting up the report system includes defining the type, detail, and frequency of the various reports required. If done formally, the program management plan is a convenient document to define the reporting requirements. Table 6.6 summarizes the basic types of project reports and thier characteristics.

Reporting Requirements Must Be Made Judiciously. In spite of all these benefits, project reports may have some unintended consequences. Obviously there is the element of cost. But in a more subtle way, reports have a tendency to legitimize the status quo; that is, certain problems, slippages, or overruns often become a permanent reality of the program once they are documented. Reports furthermore have the tendency to shift the responsibility of managing and rectifying a problem from the report writer to the report receivers. Yet another facet is that reports tend to proliferate. Team members may perceive the content of a report as criticizing their own activities, or being confusing or threatening. As a result, these team members write their own reports. These reports are often of the j-i-c category (just in case). They serve no purpose other than protection from potential future embarrassments or image problems, and can be a very costly administrative undertaking.

Taken together, reports are valuable in documenting results and providing continuity in changing organizations. They also provide a contractual basis for future negotiations and billings and serve as a communications device to remote members of the program community at large. However, the extent of reporting required and report contents, format, and distribution should be carefully considered to minimize the negative aspects of written reports.

Management Support

Engineering managers and project leaders are surrounded by a myriad of organizations that either support them or control their activities. Within the company, the

TABLE 6.6 Four Major Project Reports and Their Typical Characteristics

Report Type and Characteristics
STATUS REPORT
A comprehensive summary of the program status regarding its technical progress against schedule and budget. An outline of the required report content should be provided to the writer. Typical report length is between 10 and 100 pages but can be considerably more extensive depending on the type and magnitude of the program. *Timing:* end of program phase. *Prime responsibility:* program manager. Quarterly or less frequently, annually, more typical.
WEEKLY/MONTHLY PROGRESS REPORT
A brief summary of (1) progress, (2) changes, (3) problems, (4) action items, and (5) important dates. Such a summary report is being prepared by each task leader for each work package. These reports are then integrated by the program manager into a progress report for the total program. Ideal report length is 1 page. *Timing:* as needed. *Prime responsibility:* task leader and program manager.
OPEN-ITEM REPORT
A brief description of a problem or action item that needs attention. A form sheet can help provide uniformity and effective structure such as (1) administrative data for tracking, (2) open item, (3) impact, (4) suggested solution, and (5) person responsible for resolution. Typical length of this report is ½ page in free format or 1 page for form letter. *Timing:* as needed. *Prime responsibility:* any team member.
MINUTES OF MEETING
Brief summary of program review or status, stating (1) meeting highlights, (2) changes, (3) problems, and (4) action items. Minutes should be kept brief, ideally ½ page. *Timing:* as needed. *Prime responsibility:* meeting chairperson.

engineering manager has to elicit support from functional areas, such as manufacturing, finance, marketing, and research and development, each of which by itself is a highly complex organizational system. Externally, the engineering organization has to interface with the customer community, subcontractors, regulatory agencies, and financial and labor groups. All have different interests, present different risk factors, and have different lines of control and communication. Understanding these interfaces is important to engineering and program managers, as it enhances their ability to build favorable relationships with their support functions and with senior management. Management support is often an absolute necessity for dealing effectively with interface groups; therefore, a major goal for project leaders is to maintain the continued interest and commitment of senior management in their projects. Senior management should become an integral part of project reviews. It is equally important for senior management to provide the proper environment for the project to function effectively. Here the project leader needs to tell management at the onset of the program what resources are needed. Project organizations are shared-power systems with a tendency toward imbalance. Only a strong leader, backed by senior management, can prevent the development of unfavorable biases.

Four key variables influence the project manager's ability to develop favorable relationships with senior management: (1) ongoing credibility as program manager, (2) program visibility, (3) program priority relative to other organizational undertakings, and (4) the project manager's accessibility. All variables are interrelated and can be developed by the individual manager. Senior management can aid such developments significantly.

Program Leadership

Overall engineering leadership is an absolutely essential prerequisite for directing engineering activities and programs toward established goals. Leadership involves dealing effectively with individual contributors, managers, and support personnel within a specific functional discipline as well as across organizational lines, where engineering and program managers often have little or no formal authority. It involves information-processing skills, effective communication, and decision-making in a relatively unstructured environment. It also involves the ability to integrate individual demands, requirements, and limitations into decisions that benefit overall project performance.[8] It further involves the manager's ability to resolve intergroup conflicts, an important factor to overall program perfomance. Yet another facet of program leadership and control is the ability to interact effectively with the organizational culture and its value systems. The engineering project manager must understand the dynamics of organizational and behavior variables in order to foster a climate conducive to effective teamwork and to establish the basis for proper direction, leadership, and control.

Perhaps more than any other organizational function, quality leadership depends heavily on the manager's personal experience and credibility within the organization. Effective engineering management leadership does the following:

- Gives clear project leadership and direction
- Gives assistance in problem-solving
- Integrates new team members
- Has the capacity to handle conflicts
- Aids group decision-making
- Gives direction for proper planning
- Keeps clear communication channels
- Maintains high visibility for program efforts
- Recognition of accomplishments

[9]An example is juggling internal program schedules in an effort to compensate for one department's problems while holding the overall program to its predetermined timetable and budget.

- Actions toward commitment
- Maintains a professionally stimulating work environment
- Gives team members clear role and responsibility definition
- Shields team against outside pressures
- Represents project team to higher management.

With a strong engineering and program team, managerial leadership can focus on the early identification of problems and contingencies and on the direction of activities toward their resolution, rather than on administrative busywork. Problems may originate in a technical, organizational, personal, or environmental context, or in any combination of these. The manager who has a well-developed, committed team, which includes senior management and the sponsor, probably also has the ability to handle contingencies, with the support of all personnel involved. It is this type of leadership that facilitates self-actuating, self-correcting control of technical activities, a system that can cope with almost any kind of adversity.

6.7 THE TECHNICAL REVIEW—A SPECIAL MANAGEMENT TOOL

In today's complex, technology-based environment, project reviews, play an important role in facilitating the on-time on-budget integration of multidisciplinary engineering projects. Properly organized and conducted, these reviews provide real interaction among team members with resulting candor, feedback, action items, and on-the-spot decisions. They offer possibilities for real-time project control that no other management tool can match. Yet few project managers go into a project review meeting lightly. Their concerns range from the collective amount of time required to prepare and attend the meeting, to the conflicts, power plays, and justifications of poor performance that may occur.

Because of its great potential for both managerial benefits and problems, project review meetings have been the subject of considerable executive interest. Various articles have been written on the process of review meetings and their effectiveness in a general organizational framework.[10] Few studies, however, have investigated project review meetings in an engineering environment or considered the variables from the point of view of the broader organizational system, which includes project planning, work environment, and leadership style, an area that has become increasingly important to technology-based operations. Responding to these interests and needs, a field study was conducted analyzing the project review practices of technology-based companies. It has become, in part, the basis for the conclusions of this selection.[11]

[10]For discussion and literature review of this subject see H. J. Thamhain, "Effective Technical Project Reviews," *Proceedings, IEEE Engineering Management Conference,* Atlanta, Georgia, October 1987.
[11]For detailed discussion, see H. J. Thamhain, "Validating Technical Project Plans," *Project Management Journal,* October 1989.

Different Reviews, Similar Objectives

For the purpose of characterizing the large spectrum of project reviews, three different types of review meetings are recognized in this chapter: (1) the technical project or task review, (2) the management review, and (3) the design review. Table 6.7 provides a brief description for each type of project review. A sample agenda of a technical project review meeting is shown in Table 6.8. The basic differences among these reviews are the level of detail and technical content. The technically most detailed reviewed is the Technical project review, which is conducted, usually on a weekly basis, at the task level, while the program and design review are each integrated reviews of the total system, either from a business or engineering perspective.

However, in spite of the fundamentally different purposes of each type, the objectives and values of each review are similar and focus on effective communications toward accurate status, and early problem detection and solving.

These objectives also provide focus regarding the organization and conduct of these meetings. Examining these objectives may further help in auditing the structure and effectiveness of any particular review. The objectives of a well-run project review or design review meeting should include:

1. *Status and Progress Review.* Updating the task team on the integrated project status and progress compared with established project plans and cross-validating the findings. This is a platform from which many of the other meeting objectives, such as problem identification, status documentation, and recognition of accomplishments, evolve. An integrated status assessment can only be made collectively, with all task team members participating and cross-validating the findings.
2. *Communication to Others.* The results of the project or design review should be documented in a *brief report*, useful for communication with other task teams, project management, resource managers, and project sponsors. These reports also become part of the permanent record, for future reference.
3. *Change Management.* Review meetings are the vehicle for introducing necessary changes or potential changes to the project team for an integrated impact assessment and smooth implementation.
4. *Problem Identification.* Potential or actual problems often surface during status reviews discussing an apparently isolated problem. Well-run meetings should have the dynamics and introspection for defining and assessing interdisciplinary problems.
5. *Decision-making.* Review meetings can facilitate group decision-making, which is often necessary for resolving complex, multifunctional issues.
6. *Action Items.* Meetings are conducive to defining specific action times and choosing responsible individuals who are committed to their resolution.
7. *Project Integration.* The group meeting facilitates the integration of the project via interdisciplinary team involvement during the design, prototyping,

TABLE 6.7 Definition: Project Reviews, Three Principal Types

Definition: Technical Review

Technical reviews are regularly conducted at the task level, involving the people who actually perform the engineering work at the individual level. These teams are organized along WBS lines to implement a particular work package. Review meetings have the following characteristics:

FREQUENCY:

1 review per week

DURATION:

20–30 minutes

OBJECTIVES:

- Assess task status against plan. Communicate among team members
- Communicate plan changes and potential risks and contingencies
- Identify, evaluate, and deal with technical (or other) problems
- Establish basis for progress report
- Define action items
- Recognize accomplishments.

CHAIR:

Task leader

ATTENDEES:

All task team members

Definition: Management Review

Program or project reviews are regularly conducted at the overall program level, involving all senior project personnel from both the technical and administrative side, plus the functional managers who support the project and the senior personnel from those functions that will interface with the project activities later on, such as manufacturing, marketing, and, if applicable, the customer. Program reviews of the overall program activities have the following characteristics:

FREQUENCY:

1 review per month

DURATION:

30 minutes

OBJECTIVES:

- Integrated review of total program status against plan. Verify progress
- Deal with problems in an integrated functional and program-oriented way
- Assist and direct proper integration of multidisciplinary activities
- Direct/redirect resources
- Unify management team
- Establish data base for customer/sponsor liaison.

CHAIR:

Project manager

ATTENDANCE:

All senior design personnel, responsible functional department heads, lead personnel from interfacing departments such as manufacturing, product assurance, and marketing

Definition: Design Review

Design reviews are conducted at various subsystem levels as well as at the total system level. While technical reviews assess progress against preestablished objectives and try to minimize problems that stand in the way of achieving these objectives, design reviews assess (1) the validity and reality of these objectives regarding overall program success, (2) the integrity of the design approach; and (3) integrated progress as compared with the program plan. Design reviews are crucial milestones in an engineering program plan. They are tools for assessing integrated overall program performance and for directing and controlling the program toward the desired overall results. Design reviews have the following characteristics:

FREQUENCY:

Depending on design cycle, typical engineering programs may have three design reviews for each subsystem: preliminary design review (PDR), critical design review (CDR), and final design review (FDR).

DURATION:

4–8 hours

OBJECTIVES:

- Assess technical program status against plan
- Review/redirect technical objectives and requirements in relation to the evolving total system, its performance, and its integrity
- Assure doability
- Identify, evaluate, and deal with technical problems
- Assure cross-functional work integration
- Assure technical system integration
- Establish basis for renegotiating customer/sponsor requirements.

CHAIR:

Lead engineer

ATTENDANCE:

All senior design personnel, responsible functional department heads, lead personnel from interfacing departments such as manufacturing, product assurance, and marketing

---

testing, and production phases. It is therefore crucial to have the right mix of people from interfacing task groups participating in the meeting.

8. *Visibility and Team Spirit.* These can be developed by relating the accomplishments to the broader project objectives and recording the specific milestone results with reference to the individual team's efforts and their support groups. This visibility is further enhanced by upper management's presence

TABLE 6.8 Sample Agenda: Project Review

SAMPLE AGENDA: TECHNICAL PROJECT (OR TASK) REVIEW MEETING (WEEKLY)

No.	Agenda Item	Presenter	Time
1	Progress review	Project leader	5 minutes
2	Changes and contingencies	Project leader	5 minutes
3	Previous problems and resolutions	Project leader	5 minutes
4	New problems and discussion	Team	10 minutes
5	Action items	Team	5 minutes
		Total	30 minutes

in some of the meetings, as well as by notes published in the company newspaper.

9. *Recognition and Commitment.* Yet another objective of these reviews is the public recognition of accomplishments of the project reams and their management. Such recognition reinforces the individual commitment to the established project plan, as well as the resource commitment made by functional management. It further cross-validates the true work challenge associated with the project, one of the strongest professional drivers toward innovative, result-oriented behavior.
10. *Planning.* No project can be planned in detail for its total life cycle. Especially for longer projects, the meeting provides a forum of incrementally and dynamically updating the long-range plan.
11. *Clearinghouse.* Finally, the review meeting can be used as a clearinghouse for related problems and action items that originated outside the task group but impact on their work or that must be resolved by the task team.

Measures of Review Effectiveness

There are obviously differences among engineering managers regarding the characteristics of an effective project review meeting. These differences exist even among managers of the same company. They result from the existing differences in project type, size, and nature, as well as differences in organizational culture and managerial operating philosophy. However, there seems to be general agreement on a broad category of factors that are perceived to characterize effective project reviews, as shown in Table 6.5. In addition, some quantitative evidence exists from the calculation of Kendall tau correlations between each factor associated with effective reviews and some generally accepted project success factors: (1) perceived successful technical completion, (2) on-time project completion, (3) on-budget project completion, (4) innovative implementation, and (5) change orientation.[12] The statistics yield a positive correlation at a confidence level of at least 95% for all characteristics and success factors. This cross-validates the perception of project leaders that Table 6.5 characterizes not only effective project review meetings, but also seems to contain the ingredients important to project success.

Drivers and Barriers to Effective Project Review

We can probably learn from the collective experiences of engineering and project managers who identified several drivers and barriers that they perceived at being significant in managing project reviews effectively. Like the factors of Table 6.5, the drivers and barriers shown in Table 6.9 were statistically evaluated by calculating their rank-order correlation to the five project success factors of: (1) technical success, (2) being on time, (3) being on budget, (4) innovation, and (5) change orientation. All drivers had a statistically significant association at the 95% confidence level or better; all barriers were negatively associated with the project success factors. The factors in Table 6.9 have been rank-ordered by the strength of their association with project performance, with the most important drives and barriers listed first.

It is interesting to note the strong intercorrelation among drivers and intercorrelations among barriers. This supports the claim of many engineering managers that good review meetings have all factors going favorably in support of the review, while ineffective, unproductive meetings have a large number of barriers present.

Measurable Milestones and Checkpoints

Another important criterion for effective project review meetings, one pointed out consistently by engineering managers, is measurability. Progress can be reviewed only if the status of the project is measurable. Key milestones should be defined for all major phases of a project prior to its start-up. Each milestone should be defined by specific deliverables and results. During a project review, these deliverables become focal points of the discussion on progress, problems, and changes. The deliverables also help in defining what part of a milestone has been completed and where potential or actual problems exist. The details underlying a milestone, such as the specific tasks and deliverables, do not have to be defined at the beginning of the project, but can be worked out as the project is under way. In fact, the task details associated with a milestone, such as specifications and subplans, are often part of a formal work package that is executed as part of the overall project.

[12]The association was measured by utilizing Kendall's tau rank-order correlation. The Kendall tau coefficient is a measure of rank-order correlation, calculated with nonparametric statistical methods. For details see S. Siegel (1956). First, project review meetings were ranked by each of the characteristics shown in Table 6.4. Project team members were asked how strong each of the characteristic factors present in each meeting was. Then the projects were rank-ordered by each of the five performance factors, as perceived by senior management. Finally, tau coefficients and their confidence factors were calculated for each association.

TABLE 6.9 Drivers and Barriers To Effective Project Reviews

Drivers To Effective Project Reviews

- Agenda known in advance
- Detailed project plan
- Measurable activities and results
- Mandatory attendance of key project personnel
- Cross-functional representation
- Regularly scheduled meetings
- Short meetings
- Start on time, finish on time
- Adherence to agenda
- Project leader chairs meeting
- Focus on early problem identification
- Focus on problem-solving, not on analysis
- High-performance image
- High project visibility
- Upper management support and commitment
- Proper meeting preparation
- Suitable meeting room facilities
- Some guidelines, policies, and procedures

Barriers To Effective Project Reviews

- Mistrust, threats, fear
- Conflict among team members
- Indifference and disinterest toward project
- No clear project plan
- Unclear project objectives
- Milestones difficult to measure
- No upper management involvement
- Poor communication among team members during project work
- Role conflict
- Low team spirit
- Polarization of team, Groupits
- Compartmentalization
- Frequent personnel changes
- Rigid protocol
- No meeting leadership

Selecting the right type of milestone is critical. Engineering managers suggest that for a milestone to be effective for project reviews, the milestone should integrate a number of interrelated activities. Concept reviews, design kickoffs, design reviews, prototype tests, and software certifications are examples of milestones that integrate significant project content with well-defined boundaries. As discussed in Chapter 5, candidates for key milestones are usually easy to spot on a PERT network. They represent event nodes with a large number of dependencies. Table 6.10 lists a sample

TABLE 6.10 Examples of Project Milestones, Deliverables, and Results Needed to Assure Measurability During Reviews

Milestones	Deliverables and Results
Concept Review	Concept paper Preliminary budget and schedule Design objectives Product specifications Budget approval
Design Kickoff	Functional specifications Statement of work Schedules Budgets Task–Personnel roster Make–Buy analysis Work breakdown structure Sign-offs
Design Review	Functional specifications Product specifications Vendor purchasing specifications Quality plan Concept analysis Simulation Sign-offs
Prototype Tested	Test plan Test fixtures Test set-up and dry run Hardware operational Software operational
Software Certification	Software specifications User manual Flow diagrams Coding Test reports Software engineering maintainability evaluation Field engineering training plan Sign-offs

of five milestones, each with a short set of deliveries, just to illustrate the kind of detail necessary to define a project milestone in a meaningful way.

Yet another set of measurement points suggested by engineering managers consists of *checkpoints*. These are milestones with clear and simple status definition. Examples of such checkpoints are product announcements, sign-offs, and deliveries. Either they are complete or not. There is no complex evaluation necessary. They cannot easily be tampered with. Other good checkpoints include expenditures and staffing levels. Checkpoints help to cross-validate other more complex milestones and project status measures. They further provide a watchdog function against a serious project neglect or management oversight.

Progress and Design Validation

A special category of the project review is the project audit. It comes in various forms, such as design and progress validations. While the audits rely on the same management techniques as discussed so far and are subject to the same drivers and barriers toward effectiveness, the substance of the review audits is different. Engineering managers stress the importance of these validations as part of an integrated project review system. These reviews usually consider the total technological subsystem as an integrated part of its business environment. The reviews spend only part of their efforts assessing whether projects progress according to plan, but try to determine "Are we doing the right thing?" Does the plan still hang together and make sense in the current technological, market, and business environment? Because of their introspective nature and concern for objectivity, these audit-type reviews are usually organized and conducted by an interfunctional team that is impartial to regarding the project.

A selection of the most frequently mentioned audit validation methods is listed below:

1. Concept Analysis and Selection. A formal meeting reviewing the conceptual alternatives to the technical, managerial, and business issues involved. The outcome of these reviews is usually a management decision to select, rework, or cancel a particular engineering effort. These reviews are also used to freeze a design concept or to make a bid decision.

2. System or Project Audit. A formal process of analyzing the project and its future plan, with the intention of finding errors in the technical system, its operating logic and interfaces, and its business aspects.

3. Inspection. Often performed by a peer group, the process can be formal or informal. The objective is to identify hidden technical problems or oversights. The process relies on expert on-site observations and evaluation of documents and often relies on the cooperation of the project team also. The debriefing of the project team by the inspection team can be an additional source of information regarding hidden problems, innovative solutions and alternatives, and cross-validation of project status.

4. *Walkthroughs.* The term "walkthrough is often used interchangeably with "inspection." For those managers who make a distinction, the walkthrough refers to a highly systematic review of the product, design, etc. specification against the actual (customer) requirements. Usually both the designer and the independent reviewer "walk through" the data together, without any managers present. Debriefings can be held with the total project team in attendance or just as a one-on-one meeting.

5. *Verification and Validation.* Verification and validation are similar in process. Both reviews try to determine the integrity and correctness of the design. The *verification* checks the design against the specifications, including its interfaces (e.g., Are we building the system right?). The *validation* assures that each deliverable meets the precut and customer requirements (Are we building the right system?). Both verifications and validations can be conducted at the system or subsystem level. Top-down reviews conducted early in the design cycle can provide early feedback useful for redirecting the design, fine-tuning the product requirements, and improving the overall product and project performance.

6. *Sign-off.* This is a special method of validating progress or assuring correctness. If done properly, the sign-off process can be very effective in validating the correctness of the final deliverables. Interfacing work group leaders, such as those of engineering design and manufacturing, must sign off on the completion of an interfacing task (e.g., design). Since both leaders are aware of the sign-off requirement from the start of the project, there is pressure on both leaders to communicate and involve the other on work items that might affect the work interface or a subsequent work phase. Typically, engineering wants to involve manufacturing in their plans and reviews because they are the customer. Also manufacturing is interested in getting involved to assure that the design is producible and testable. Under the sign-off system, manufacturing has less of an excuse to say they did not know what engineering was doing. Sign-off fosters cross-functional involvement and cooperation.

The above discussion should help to break down the enormous complexities and intricacies involved in the management review of technical projects. Identifying the various components of these reviews, their drivers and barriers to effectiveness, and their managerial process can help the practitioner and scholar of engineering management to gain a better understanding of the dynamics involved. This may assist engineering managers to fine-tune their project review systems, their styles, techniques and mechanisms for effective decision-making and integrated control in their highly dynamic, multidisciplinary environment. This section concludes with a number of specific recommendations.

Recommendations to Increase Project-Review Effectiveness

A number of recommendations are stated to help increase the effectiveness of project review meetings and to aid engineering managers in controlling multidisciplinary tasks according to plan. The flow of recommendations follows the project planning and start-up sequence wherever possible.

The Project Plan Is The Foundation

1. *Plan ahead.* Plan your project in sufficient detail, including critical milestones and reviews. Project reviews must be made against an established plan.
2. *Assure Measurability.* Define measurable milestones on a monthly basis or in 10% project increments, whichever comes first. Project reviews can be useful tracking information and identify potential problems only if deviations from established plans can be measurably defined.
3. *Involve Project Personnel During the Planning* to build the basis for an agreement and commitment to the project objectives, and to develop a sense of ownership toward the project.
4. *Define Specific Task Responsibility* and designate responsible individuals for all work packages.
5. *Obtain Commitment* from project personnel to milestones and key results. This will foster some sense of ownership and a positive approach to problem identification, and encourage seeking out help and problem correction rather than covering up mistakes.

Make Meetings Work For You

6. *Recognize Barriers.* Project managers must understand the various drivers and barriers to effective project reviews. This is necessary to foster an environment and leadership style conducive to effective project meetings.
7. *Hold Regular Project-Review Meetings.* Depending on the type of review (see Table 6.7), these project reviews should be scheduled clearly and consistently, (e.g., *Tuesday mornings, 9:00–9:30* a.m.) This makes them easy to remember and to plan for, especially in a multiproject environment where many other things compete for attention.
8. *Assure Proper Representation* from functional support departments. This facilitates the participation of resource managers in problem identification and resolution. It also provides a link between the project team and upper management, which is necessary for proper management of the resource personnel, priority-setting, and resolving interfunctional problems.
9. *Keep Meetings Interactive* and on a mutual trust basis. Good review meetings are often noisy, with plenty of candor and broad involvement. It is this dynamic that helps to discover small problems at an early stage.
10. *Detect Problems Early and Resolve Them.* This is facilitated by the dynamics of meeting, the mutual trust relationships, the leadership style, and the action mentality of the project management team.
11. *Evaluate Interdisciplinary Effects* of any problem and try to minimize its impact. Try to isolate and confine the problem to a small subsystem if possible.

12. *Facilitate Interdisciplinary Integration.* Meetings can be effective vehicles for facilitating the coordination and integration of multidisciplinary tasks, by encouraging communications among various tasks groups and by providing visibility for the overall requirements and objectives.
13. *Make Accomplishments Visible.* This refuels the commitment to the established project plan and creates a pervasive result-oriented mentality, which helps the project team to concentrate on desired results.
14. *Don't Threaten.* Foster a fail safe environment. People who feel threatened, intimidated, or even just uncomfortable in a project review meeting will not be candid in identifying potential problems and concerns. The review meeting is not the place to reprimand.
15. *Get Involved.* Project leaders should be involved in the day-to-day operation of the project and should be intimately familiar with its status, challenges, and potential problems. This is a prerequisite for conducting an effective project-review meeting.
16. *Lead.* Project managers can influence the climate and logistics of the meeting by their own actions and leadership. Concern for people and the project, the ability to help and support the activities, and the ability to create enthusiasm for the project, the ability to help and support the activities, and the ability to create enthusiasm for the project can foster a climate of high involvement, motivation, and open, effective communication.
17. *Be Prepared.* Effective review meetings require homework and proper preparation by the leader and by the participants. A well-organized status summary, change announcement, or problem statement often requires detailed background work and considerable effort in organizing the presentation or discussion.

The Meeting Mechanics

18. *Send Agenda* with meeting invitation. Follow agenda during the meeting.
19. *Start On Time. Finish On Time.* Time overruns can be costly and disruptive to others. They also set bad examples for on-time performance.
20. *Send Reading Material and Homework Ahead of Meeting.* Don't take precious meeting time for this purpose.
21. *Don't Try to Solve a 30-Hour Problem in a 20-Minute Meeting.* Not only is this ludicrous, but it will also destroy your meeting.
22. *Assign Action Items* and follow up.
23. *Publish Major Milestones and Project Objectives.* This provides cross-functional visibility for other overall project roadmaps; it helps to unify the project team in striving for critical results and end dates.
24. *Keep Meetings Small.* Invite only those people who report directly to the task or project under review. Each invitee is free to bring any number of people

along, but only if they are needed to support the review. (The functional resource manager should always be encouraged to attend.)

25. *Project Leader Should Give Overview.* The project leader should give present project status, changes, and problems rather than each team member giving a progress review. A well-prepared leader can integrate among the various disciplines and state the project status more concisely than the team members. It also helps to cross-validate the status as seen by the leader and by the doers. Subsequent discussions on the "actual" project status often brings out potential and hidden problems, which would not have surfaced otherwise.
26. *Prepare Support Systems.* Depending on the size and nature of the project review, visual aids such as view graphs, flip charts, and wall-size master schedules should be available in the meeting room. The arrangement of the room must be conducive to the type of meeting intended.

A Final Note. In Summary, effective management of technical project reviews can be a critical determinant of engineering project success. These meetings can be organized and managed in a variety of formats and structures, depending on their primary purpose. They provide the mechanism for identifying problems and taking corrective actions. However, to be effective, the engineering manager must provide an atmosphere conducive to candor, interaction, and result orientation. Measurability of the project status and of all technical performance criteria must be the basis for any review. Together with the drive, direction, and leadership of the engineering manager and project leader, these are the key ingredients for effective project reviews and the basis for integrated decision-making and control in the multidisciplinary work environment of engineering management.

6.8 HOW TO MAKE IT WORK

Managing engineering organizations to fulfill established objectives and programs within given schedule and cost parameters requires more than just another plan. It requires the total commitment of all supporting organizations, plus the involvement and help of upper management and the customer or sponsor community. Successful engineering and program managers stress the importance of carefully designing the project planning and control system and the structural and authority relationships. All are critical to the implementation of an effective engineering project control system. Other organizational issues, such as management style, personnel appraisals and compensation, and communication, must be carefully considered in order to make the system self-actuating. Engineering personnel throughout the organization must feel that participation in a particular program is desirable for the fulfillment of their professional needs and wants. Engineering personnel must be convinced that management involvement in their work is helpful. Personnel must be convinced that identifying the true project status and communicating potential problems early will provide them with more assistance to problem-solving, more cross-functional support, and in the end will lead to project success and the desired recognition for their accomplishments.

In summary, effective control of engineering programs or projects involves the ability to:

- Work out a detailed project plan
- Involve the team during the program definition and feasibility phase.
- Define task responsibilities.
- Reach agreement on the plan with the project team, management, and the sponsor
- Obtain commitment from the project team members.
- Obtain commitment from upper management.
- Define measurable milestones and deliverables
- Attract, select, and sign on quality people
- Establish a controlling authority for each work package.
- Hold regular project and task reviews.
- Communicate true project status within team and to management and customer.
- Use integrated performance measurement system.
- Detect problems early and resolve them.
- Define, measure, and control specific checkpoints.
- Cope with changing customer requirements and environmental conditions.
- Make accomplishments visible.
- Manage and lead.

The implementation of engineering program controls starts with a detailed, doable, and agreed-upon program plan, which was developed through the participation of all program personnel and their managements. Implementing these controls is a complex managerial task, which encompasses all functions of management in all participating organizations. Because of its complexity, no one simple statement or procedural guideline can be sufficient to explain it. However, to provide focus on the key issues and responsibilities, Table 6.11 summarizes the major tasks and managerial responsibilities involved in the design and implementation of an integrated engineering management control system. Workable control further involves the continued adjustment of the plan according to the dynamics of the changing environment, both internally and externally to the company.

The five primary pillars that support a workable self-actuating control system for engineering management are:

1. Detailed program planning
2. Measurability of program status

TABLE 6.11 Preparation for Implementing Managerial Controls in Engineering

1. Program manager defines the global program objectives and requirements in terms of specifications, schedules, and resources.
2. Program manager and key personnel define the major program phases and key task categories. Use WBS.
3. Program manager and senior management determine who is responsible for each major task category; reach agreement.
4. Senior management and program manager determine the principal organizational structure and team configuration in which the program will be executed.
5. Program manager develops the detailed program plan, involving all key personnel, defining (1) specifications, (2) timing, (3) resources, and (4) responsibilities.
6. Program manager, task leaders, and sponsor or customer define deliverable items (results) for each program phase and subsystem.
7. Program manager and task leaders define measurable milestones and checkpoints throughout the program.
8. Engineering manager and program manager obtain commitments from all key personnel regarding the program plan, its measures, and results.
9. Project leader and engineering manager assure effective communications among team members, management, and sponsor. Regular status meetings, reviews, reports.
10. Project manager defines major work interfaces in advance. Require sign-offs from both sides.
11. Engineering manager, program manager, and task leaders determine needs and desires of program personnel and foster work environment conducive to fulfilling these personal objectives.
12. Senior management and engineering manager design personnel appraisal and reward system consistent with the employees' responsibilities.
13. Program manager and engineering manager determine interrelationships among people, organizations, and activities; determine appropriate communications and control channels.
14. Program manager and engineering manager define and install proper program tracking system, which includes the capture, processing, categorizing, and formatting of program performance data.
15. Program manager, task leaders, and engineering managers assure accurate measurement of program performance data, especially technical progress, compared with schedule and budget.
16. Engineering manager, senior management, and program manager develop program leadership to assure proper task coordination, problem identification, and corrective actions.
17. Engineering managers, program managers, and task leaders assure that there are involved and motivated personnel who find the project work professionally challenging and desirable.
18. Senior management assures their endorsement and support for the various program control measures.
19. Project leader and engineering manager develop a "can-do" atmosphere. Minimize threats, personal conflict, and power struggles.
20. Project leader and engineering manager lead project team toward final objectives. Focus on problem avoidance. Identify problems early, when they are small. Foster action-orientation.

3. Personnel involvement and commitment
4. Management support and leadership
5. Program visibility.

All five pillars are interrelated. If done properly, the process of establishing measurable milestones must involve both the performing and the customer organizations. This involvement creates new insight into the intricacies of the engineering activities or programs. It also leads to visibility of the activities at various organizational levels, and to management involvement and support. It is the involvement at all organizational levels that stimulates interest in the multidisciplinary activities and mission objectives' it also creates the desire for success. Taken together, this fosters a pervasive reach for excellence that unifies the engineering team and leads to commitment toward reaching each critical milestone. These are the characteristics of a self-actuating control system.

EXERCISE 6.1: PROJECT CONTROL (HIGH-TECH)

Situation[13]

You are a R&D manager at Hewlett-Packard. The date is April 15. You just have been given full responsibility for the Alpha software project, which consists of developing a new PC operating system.

Your predecessor quit his job this morning to take on a new assignment elsewhere. She was kind enough, however, to brief you on the project and furnish you with a complete set of project status data (see below). The former project manager told you that all was going fine so far. All activities are on schedule and the total cost is running below budget. To ensure proper communications, the former project manager discussed the assignment with each of the four task leaders individually prior to project kickoff, on January 3. The overall project plan had been worked out in detail by the corporate marketing group, which is also funding the new venture.

The next day you have lunch with two senior managers of your company who have been supporting your project via resource personnel since January. These managers voice some strong concerns over the handling of the project so far. They are greatly worried about activities, such as defining systems architecture, software standards, and hardware interfacing. There seems to be a general confusion over specific task responsibilities and deliverables and how the new project fits into the company's business plan. Back in your office, you begin to get nervous about your new responsibilities. This afternoon your boss stops in to inform you that a decision was made to formally review the project in three days. Senior managers from various divisions of your company are expected to attend the 1-hour review session.

[13]The lead-in situation is similar to Exercise 5.1. However the background data, the assignment, and learning objectives are different.

The Project Plan

The complete project plan for developing the new software system was worked out by the corporate marketing group and documented as follows: Originally the project core team consisted of four key individuals who were assigned various tasks as follows:

1. Sue was the project leader. She was also responsible for developing the project plan, including task definition, scheduling, cost modeling and estimating, budget planning, and management approvals.
2. Art was assigned to and is in charge of the software architecture. This includes a requirements analysis, software specs, system diagrams, protocols, and timing and controls.
3. Bob is responsible for all development activities for the new software up to a prototype demonstration. It includes design, coding, testing and debugging of 6 major software subsystems, rapid prototyping of all software, including integration and testing, and documentation.
4. Carl is in charge of providing liaison to the product development and marketing groups.

Some of the agreed-on milestone dates are: Project kickoff (1/3), budget proposal completed (2/10), system architecture defined (5/1), software documentation completed (11/15), project completed (12/30). Budgets are allocated to each task/project leader as follows: Art, $621 million; Bob, $65 million; Carl, $6.1 million; and Sue, $6.9 million. Some of the formal documentation is shown in Figure 6.7.

Your Assignment

As the new project manager how would you prepare for the upcoming review and its presentation to upper management? Specifically:

1. What challenges and problems do you anticipate in "controlling" the project according to plan?
2. What provisions have you made during the project definition phase to ensure adequate project control?
3. How would you measure "project status"? (By the end of May you will be asked to give a reasonably accurate "percent-complete" figure.)
4. How would you *manage* the project so that it will be completed according to plan?
5. Assume that it is now May 30. The project went well so far. However, in this morning's project review, a nasty design/development problem has surfaced which might have a major impact on your schedule and budget. Worse yet, it raises serious questions about the technical feasibility of the whole project.

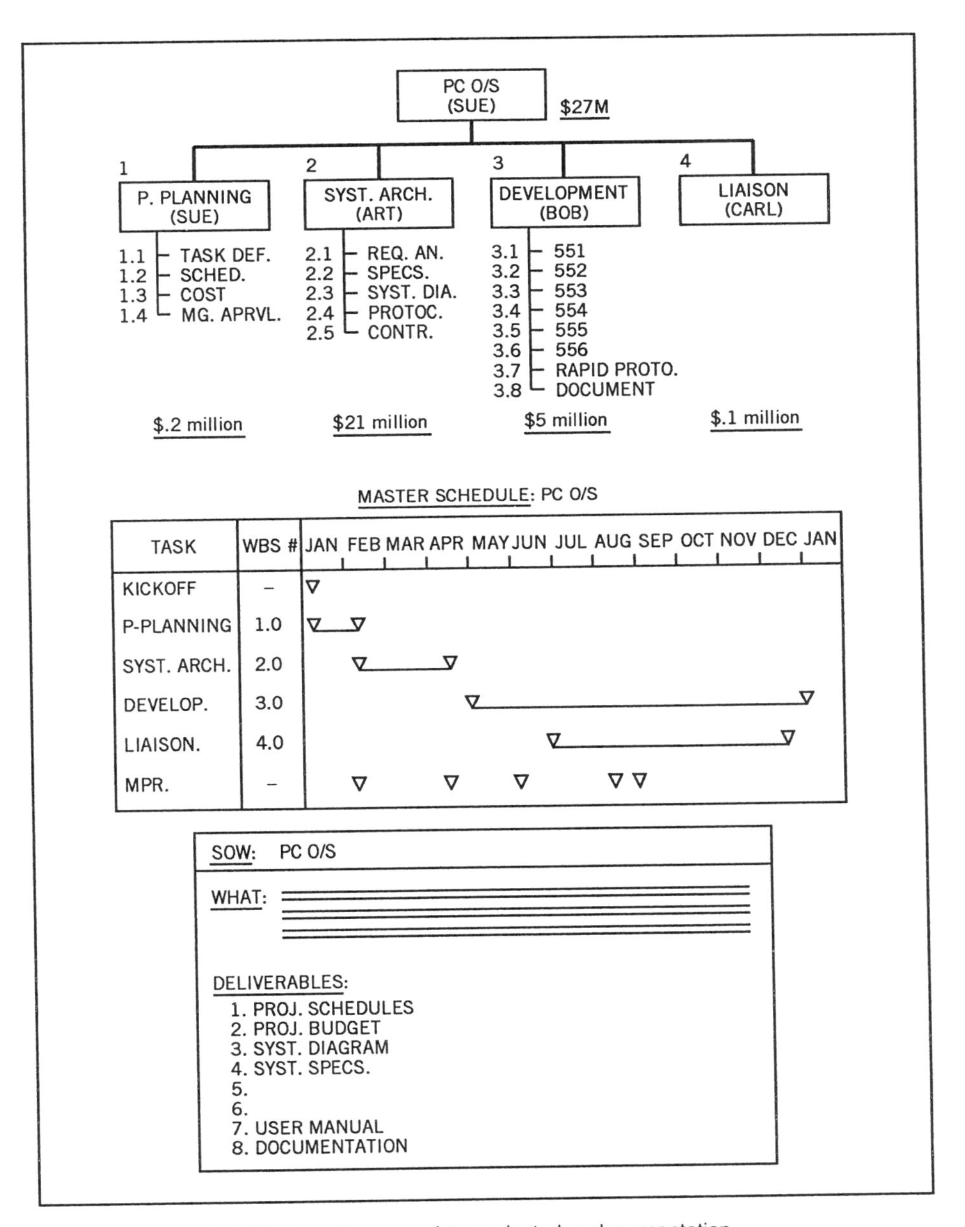

FIGURE 6.7 The complete project plan documentation.

How do you handle the situation? (Upper management has called you to a hastily organized meeting).

Delivery

The exercise can be completed in several modes: (1) individually as an exercise, to test the understanding of project control, (2) individually or in groups, as a homework assignment, take-home exam, or group exercise in a professional development seminar, and (3) in a multi-group workshop mode with specific roles assigned to various discussion groups, such as project leader, functional manager, sponsor/customer, and upper management.

APPENDIX: TERMS RELATING TO FINANCIAL CONTROLS

Activity accounting. Recording data by a specific organizational segment.

Activity base. A measure of an operating activity within a department, division, plant, or company; used to allocate indirect costs.

Actual costs. Costs measured by their cash equivalent value on an after-the-fact, arm's-length transaction basis. Also called *incurred costs and historical costs.*

Budget. An integrated plan of action expressed in financial terms. *Financial budgets* measure either (1) the *cash-flow* of incoming and outgoing resources or (2) the *capital expenditures* of major assets, such as plant, equipment, and property, or (3) the effect of a resource change on the *balance sheet. Operating budgets* express projected revenues and/or operating expenses.

Budget variance. The difference between actual costs incurred and the projected cost according to actual activity level. Also called the *cost variance.*

Capital budgeting. (A) The process of long-range planning involved with adding or reducing the productive facilities of a firm. (B) The process of long-range planning for specific projects.

Constraint. A restriction on the production process, limiting the amount of resources that can be committed.

Control. The process of measuring and correcting actual performance to ensure that a firm's objectives and plans are accomplished.

Controllable cost. A cost that can be regulated by a given manager in either the short run or the long run. A cost that is the responsibility of a specific manager.

Cost. (A) The cash or the cash-equivalent value of the resource obtained or the resource committed, whichever can be measured most objectively. (B) Value foregone to achieve an economic benefit.

Cost-benefit analysis. The systematic process of comparing costs and benefits between possible alternatives.

Cost centers. Organizational units where costs naturally come together. A natural clustering of costs by functional areas.

Cost effectiveness analysis. The measure of the relationship between the incurrence of cost and nonfinancial criterion.

Cost flow. The way accountants classify and process accounting data to show product and period costs.

Cost objective. Any focal point for which the measurement of costs is desired. Typically cost objectives include time periods, products, departments, and responsibility centers.

Differential cost. The difference in total costs between any two acceptable alternatives. Also called *incremental cost.*

Effectiveness. The accomplishment of a desired objective, goal, or action.

Efficiency. The accomplishment of a desired objective, goal, or action using the minimum resources necessary.

Engineering cost estimates. A method of separating costs into their fixed and variable components by direct estimates based on technical expertise.

Expense. (A) A cost that has been consumed in the production of revenue. (B) An expired cost.

Favorable variance (A) A variance where actual costs are less than budgeted or standard costs. (B) A variance where actual revenue exceeds budgeted revenue.

Labor rate (price) variance.. The variance that measures the ability to control wage rates and labor mix; the difference between actual wage rate and standard wage rate multiplied by the actual hours worked.

Lead itme. The interval between the time a purchase order is placed and the time materials are received and available for use.

Learning curve. A mathematical expression of the fact that labor time will decrease at a constant percentage over doubled output quantities.

Line-time budget. A budget in which accountability for expenditures of money is identified with specific expenditure lines in the budget.

Objectives. Specific quantitative and time–performance targets to achieve a firm's goals.

Overhead. All costs of operating the factory except those designated as direct labor and direct material costs. Also called *factory burden, manufacturing expenses, indirect factory costs, manufacturing overhead, indirect expenses, indirect manufacturing costs,* and *factory overhead.*

Overhead rate. A method of allocating the indirect factory costs to the products, creating an average overhead cost per unit of production activity.

Period costs. Costs that are not inventoried and are treated as an expense in the period in which they are incurred.

Performance budget. An adjusted budget prepared *after* operations to compare actual results with revenues and costs that *should* have been incurred at the actual level attained.

Present value. The concept that a sum invested today will earn interest and be worth more at a later date: a dollar in the hand today is worth more today than a dollar to be received (or spent) in the future.

Procedures. Detailed instructions specifying how certain activities are to be accomplished.

Profit centers. Organizational units where both revenue and costs naturally come together and income or contribution margin are used as control measures.

Risk. An exposure to loss because of inability to control conditions on which the firm is dependent.

Unfavorable variance. (A) a variance in which the actual costs are greater than the budgeted or standard costs. (B) A variance in which actual revenue is less than planned or budgeted revenue.

Variance. The difference between actual results and planned results.

Variance analysis (investigation). The investigation of the causes of *variances* in a *standard costing system.* (This term has a different meaning in statistics.)

BIBLIOGRAPHY

Allen, William F., Jr., "Project Control Engineering: Setting the Pace in the Nineties," *Cost Engineering,* Volume 32, Issue 10 (Oct 1990), pp. 9–13.

Anderson, Cindy and Mary M. K. Fleming, "Management Control in an Engineering Matrix Organization: A Project Engineer's Perspective," *Industrial Management,* Volume 32, Number 2 (Mar/Apr 1990), pp. 8–13.

Archibald, Russel D., *Managing High-Technology Programs and Projects,* New York: Wiley, 1976.

Babcock, Daniel L., *Managing Engineering and Technology,* Englewood Cliffs, NJ: Prentice-Hall, 1991.

Beidleman, Carl R., Donna Fletcher, and David Veshosky, "Using Project Finance to Help Manage Project Risk," *Project Management Journal,* Volume 22, Number 5 (June 1991), pp. 33–38.

Bergen, S. A., *R&D Management: Managing Projects & New Products,* Cambridge, MA: Basil Blackwell, 1990.

Beyer, Mark W., "A Project Control to Meet Project Change," *AACE Transactions* (1988), pp. P.5.1–P.5.7.

Blanchard, Frederick L., *Engineering Project Management,* New York: Marcel Dekker, 1990.

Blok, Frank G. and Johan A. Schuil, "Getting 'Integrated Project Control' to Work," *AACE Transactions* (1988), pp. D.6.1–D.6.8.

Bush, Dennis H., *New Critical Path Method,* Probus Publishing Company, 1991.

Christian, Peter H., "Capital Tracking and Project Control," *Computers & Industrial Engineering,* Volume 20, Issue 1 (1991), pp. 71–75.

Cleland, David I. and Karen M. Bursic, *Strategic Technology Management,* New York: AMACOM, 1992.

Cleland, David I., *Matrix Systems Management Handbook,* New York: Van Nostrand Reinhold, 1983.

Cleland, David I., *Project Management: Strategic Design and Implementation,* New York: McGraw-Hill, 1990.

Cleland, David I. and William R. King (eds.), *Project Management Handbook,* New York: Van Nostrand Reinhold, 1988.

Coulter, Carleton, III, "Multiproject Management and Control," *Cost Engineering,* Volume 32, Issue 10 (Oct 1990), pp. 19–24.

Drucker, Peter F., *Innovation and Entrepreneurship,* New York: Harper & Row, 1985.

Enrico, Jack F., "Project Control—Data, Information, and Computers," *AACE Transactions* (1991), pp. SK3(1)–SK3(7).

Henderson, Thomas R., "Comparative Cost Estimates for Project Control," *AACE Transactions* (1989), pp. C.4.1–C.4.7.

Kennedy, J. Alison, "Post Auditing and Project Control: A Question of Semantics?," *Management Accounting* (UK), Volume 66, Issue 10 (Nov 1988), pp. 53–54.

Kerzner, Harold, and Hans J. Thamhain, *Project Management for Small and Medium Size Businesses,* New York: Van Nostrand-Reinhold, 1984.

Kerzner, Harold, and Hans J. Thamhain, *Project Management Operating Guidelines,* New York: Van Nostrand Reinhold, 1986.

Kerzner, Harold, *Project Management for Executives,* New York: Van Nostrand Reinhold, 1982.

Kezsbom, Deborah S., Donald Schilling, and Katherine A. Edward, *Dynamic Project Management,* New York: Wiley, 1989.

Kunz, Gerald R., "Project Controls: Management's Decision-Making Tool," *Cost Engineering,* Volume 30, Issue 1 (Jan 1988), pp. 16–22.

Lennark, Raymond, "Grass Roots Project Control," *AACE Transactions* (1990), pp. P.7.1–P.7.6.

Love, Sydney R., *Achieving Problem Free Project Management,* New York: Wiley, 1989.

Lowery, Gwen, *Managing Projects with Microsoft for Windows,* New York: Van Nostrand Reinhold, 1992.

Mahler, Ed., "Project Administration Methodology: Achieving Schedule Control on a Large Project," *PM NETwork,* Volume 5, Number 6 (July 1991), pp. 9–33.

McKim, Robert A., "Project Control–Back to Basics," *Cost Engineering,* Volume 32, Issue 12 (Dec 1990), pp. 7–11.

McMullan, Leslie E., "Cost Control—The Tricks and Traps," *AACE Transactions* (1991), pp. 05(1)–05(6).

Middleditch, Mike, "Software Engineering: Flexibility Breeds Success," *Systems International* (UK), Volume 17, Issue 5 (May 1989), pp. 41, 44.

Moore, John M., "Effective Use of Management Control Systems," *AACE Transactions* (1990), pp. P.5.1–P.5.5.

Nicholas, John M., *Managing Business & Engineering Projects,* Englewood Cliffs, NJ: Prentice-Hall, 1990.

Obradovitch, Michael M. and S. E. Stephanou, *Project Management: Risks and Productivity,* Bend, OR: Daniel Spencer, 1990.

Postula, Frank D., "WBS Criteria for Effective Project Control," *AACE Transactions* (1991), pp. I6(1)–I6(7).

Pryor, Stephen, "Project Control—2: Measuring, Analysing and Reporting," *Management Accounting* (UK), Volume 66, Issue 6 (June 1988), pp. 18–19.

Pryor, Stephen, "Project Control—1: Planning and Budgeting," *Management Accounting* (UK), Volume 66, Issue 5 (May 1988), pp. 16–17.

Randolph, W. Alan and Barry Z. Posner, *Effective Project Planning and Management,* Englewood Cliffs, NJ: Prentice-Hall, 1988.

Rigg, Michael, "Breakthrough Thinking—Improving Project Effectiveness," *Industrial Engineering,* Volume 23, Issue 6 (June 1991), pp. 19–22.

Riggs, Henry E., *Managing High-Technology Companies,* Belmont, CA: Lifetime Learning Publication (Wadsworth), 1983.

Roussel, Philip A., Kamal N. Saad, and Tamara J. Erickson, *Third Generation R&D,* Boston: Harvard Business School Press, 1991.

Shina, Sammy G., *Concurrent Engineering and Design for Manufacture of Electronics Products,* New York: Van Nostrand Reinhold, 1991.

Siegel, Sidney, *Non-Parametric Statistics,* New York: McGraw-Hill, 1956.

Silverman, Melvin, *The Art of Managing Technical Projects,* Englewood Cliffs, NJ: Prentice-Hall, 1987.

Skimin, William E., Darrel E. J. O. Smith, John C. Krolicki, and Kenneth D. Zenner, "Scope Management on an Automotive Tool Project," *Project Management Journal,* Volume 22, Number 3 (Sept 1991), pp. 22–26.

Speed, William S., "Cost Scheduling Control of Paper Machine Rebuilds," *AACE Transactions* (1990), pp. I.1.1–I.1.3.

Tabucanon, M. T. and N. Dahanayaka, "Project Planning and Controlling Maintenance Overhaul of Power-Generating Units," *International Journal of Physical Distribution & Materials Management* (UK), Volume 19, Issue 10 (1989), pp. 14–20.

Thamhain, Hans J., *Engineering Program Management,* New York: Wiley, 1984.

Thamhain, Hans J., "Effective Technical Project Review," *Convention Record, IEEE Annual Engineering Management Conference,* Atlanta, (October 1987).

Thamhain, Hans J., "Phased Approaches to Engineering Program Management," Chapter in *Handbook of Technology Management,* D. Kocaoglu, (ed.), New York: Wiley, 1992.

Thamhain, Hans J., "Project Management in the Factory," Chapter 5 in *The Automated Factory Handbook,* New York: McGraw-Hill, 1990.

Thamhain, Hans J., "Validating Technical Project Plans," *Project Management Journal,* Volume 20, Number 4 (Oct 1989), pp. 43–50.

Thamhain, Hans J., "The New Project Management Software and its Impact on Management Style," *Project Management Journal,* (Aug 1987).

Thamhain, Hans J. and Harold Kernzer, "Commercially Available Project Tracking Software," *Project Management Journal,* (Sept 1986).

Thamhain, Hans J. and David L. Wilemon, "Criteria for Controlling Projects According to Plan," *Project Management Journal,* (June 1986), pp. 75–81.

Thamhain, Hans J. and David L. Wilemon, "Managing Projects According to Plan," *Project Management Journal,* (Sept 1986).

Thamhain, Hans J. and David L. Wilemon, "Project Performance Measurement," The Keystone to Engineering Project Control, *Project management Quarterly,* (January 1982).

Thamhain, Hans J. and David L. Wilemon, "Skill Requirements of Project Managers," *Convention Record, IEEE Joint Engineering Management Conference,* Denver, Colorado, (October 1978).

Trufant, Thomas M. and Robert H. Murphy "Comtemporary Planning in the '90s," *AACE Transactions* (1990), pp. H.3.1–H.3.5.

Urbaniak, Douglas F., "Integrated Program Management System is Key to Automotive Future," *Project Management Journal,* Volume 22, Number 3 (Sept 1991), pp. 17–26.

Wilkens, Tammo T., "Earned Value: Sound Basis for Revenue Recognition," *PM NETwork,* Volume 5, Number 8 (Nov 1991), pp. 28–32.

7
MANAGING INFORMATION

7.1 INFORMATION NEEDS FOR TECHNICAL MANAGEMENT

Over the last decade, U.S. organizations spent an estimated $85 billion annually for the acquisition, operation, and maintenance of data-processing systems. Engineering companies are among the top users of these systems. The significant growth in the number of computer installations, upgrades, and personal computer systems has been driven by four key factors: (1) the need for more accurate and timely information, (2) a more complex operating environment, which has become increasingly more difficult to manage without computer-supported methods, (3) the availability of increasingly more powerful computer systems and sophisticated application software, and (4) changes in managerial leadership style and decision-making.

If engineering managers are to carry out their responsibilities in the areas of research, design, product development, manufacturing, and engineering support services, they need ready access to the right information. The purpose of a computer-based information system for engineers is to integrate the collection, processing, and transmission of information so that engineering professionals can gain more systematic insight into the operations and functions they are managing. This will minimize guesswork and isolated problem-solving in favor of systematic, integrated problem-solving. It also significantly reduces the labor cost for generating and manipulating the data for necessary documentation, reporting, ordering or just record-keeping.

The type of information system required by an organization depends largely on its business and its people. It is this organizational environment that determines what information must be provided, how it is to be organized, formatted, and accessed to be useful for aiding decision-making and for the general management process. In today's Information Age, information is readily available to all players. Timely access to the right data, properly processed for decision-making, can provide a competitive edge. Secondly, information is a significant cost item for any business. The gathering, processing, transmission, and storing of information required to aid the administrative process and to fulfill the customer and legal requirements is very elaborate. The efficiency and cost effectiveness of all this information handling can often benefit from computerized processing. This is especially so in our technology-oriented businesses, where complex engineering activities require the coordination and integration of many subsystems, such as design, material handling, assembly, purchasing, and quality control, just to name a few. The tracking, controlling, and integration of these activities is facilitated via the various information systems installed throughout the organization.

In spite of all the great advances in computer and telecommunications technology, information processing and retrieval systems are not for free. In fact, information processing is associated with significant cost, depending on the accuracy, timeliness, and sophistication of the data systems requirements. Determining the optimum system for a specific organization is a difficult undertaking, therefore, it is not surprising that with its potential for great cost savings and increased business effectiveness on one side, and substantial capital investment and operating cost on the other side, systematic information management has become an increasingly important part of overall organizational control.

Focus on Engineering

Information has always been an important ingredient of engineering management, which relies on relevant data for effective management in all of its operations, from R&D to manufacturing and field services. What is new, however, is the increased pressure on managers to process information more effectively and to integrate their technical data with schedule, finance, market, contractor, and operational logistics data. Shorter product life cycles, better resource utilization, concurrent engineering, statistical process control, increased technical complexities and multidisciplinary programs, more demanding customers, and strong competition in all markets are only some of the drivers toward effective utilization of information.

Many books have been written on the subject of management of information. Any attempt to abstract the literature here would be improper and of little value. However, we hope to add some value to the book by bringing the subject into perspective regarding the vast number of options of data processing and application areas that exist today, and further to examine the impact of today's explosion of data processing on management style and engineering effectiveness.

What Is Available Today

Technological advances, especially in the areas of microelectronics and telecommunications, have created an avalanche of new systems and methods for processing information, which truly have reshaped organizational cultures and managerial processes. The evolution of end-user computing with its powerful personal computers and sophisticated software, in particular, has made data processing directly accessible to managers and professional specialists. With only a very basic knowledge of the computer operating system, the end user can access menu-driven software and process data ranging from electronic mail to manufacturing resource planning and computer-aided circuit layouts. Table 7.1 and Table 7.2 summarize the vast spectrum of computer applications and their capabilities available to managers and engineering professionals today. In many cases, the application software is available for either personal computers or larger, mainframe machines. Depending on the type of application, there are differences, pros and cons, between (1) dedicated software residing on a personal computer and (2) software that is accessed on a larger host computer, which is possibly shared among many users. Usually, the need for more customized application software increases with the complexity of user requirements such as accuracy, security, access time, special reporting features, and special applications. Customized software also frequently requires a more sophisticated management information system (MIS), with higher levels of integration, which translates into a more costly MIS. However, as shown in Figure 7.1, the level of customization often decreases for very complex, specialized situations, which may be handled best on a "local" system, such as a PC, to eliminate the additional complications that would come with attempting a higher level of integration. Once, the software is installed, however, the operational process is often identical on either system. Therefore, once a person is familiar with a particular application software, he or she can most likely work with that program in almost any setup or system.

Engineering Reporting and Documentation

Engineering is not exempt from the requirement of collecting, organizing, processing, and transmitting information for the sole purpose of "reporting" or "documentation." This is necessary to keep the customer, contractors, and top management informed, to fulfill legal requirements, to assist support functions in their planning and decision-making and to document engineering designs, just to name a few needs. Although engineering managers see the need for these reports, they are often very time-consuming and, in the mind of many engineering professionals, very boring and undesirable chores. Often, the computer can help in generating these reports more effectively. Software is available to assist in the formatting, condensing, and semi-automatic generation of reports. Software also helps in standardizing formats and assists in the distribution of the information into specific databases for additional processing. Examples of such chain processing are parts lists generated from design documents and then used for material ordering, prototyping, and production planning. Many other examples come from project-oriented activities, which require periodic reporting of schedule and resource information against technical progress and deliverables. Typical examples of such reports are shown in Figure 7.2.

Integrating Text, Graphics, and Analysis

Engineering reporting most commonly combines several types of information: narrative reports, tables, graphs, and pictures. To be effective, these reports must be for-

TABLE 7.1 Areas of Application for Computer-Aided Engineering Activities

Project Management

A large variety of software products assist project professionals, managers, and customers in planning, tracking and controlling interdisciplinary activities. *Features* include assistance to scheduling, budgeting, resource allocation and leveling, project calendar, performance analysis, reporting, and computer-aided decision-making.

HOST SYSTEMS

Personal computers, minicomputers, mainframe computers

INTEROPERABILITY

Good for mini- and mainframe-based host systems, very limited for PCs.

Word Processing

A large variety of software products assists in the efficient generation of quality documents such as letters, reports, memos, and manuscripts. *Software features* include text, input/output, storage and retrieval; editing, formatting, thesaurus and spell-check, moving, cutting and pasting of text, list processing and mail merging, data/text transfer, different font styles, proportional spacing, search-and-replace functions, and data validation.

HOST SYSTEM

Mostly personal computers.

INTEROPERABILITY

Some, via local area networks (LANs) interconnections with mini- or mainframe computers, which function as file servers or interconnect with other application packages, such as spreadsheet or database programs.

Graphics

Specialty software designed to assist in the evaluation and graphical presentation of information. *Software features* include statistical functions that convert numbers into pictures; graphical data display options (line, bar, overlapped bar, pie, exploded pie, x/y scatter, trend, area, curve, 3-D); graphical and text annotation; and overhead and 35-mm slide generation.

HOST SYSTEMS:

Primarily personal computer.

INTEROPERABILITY:

Many programs can read directly from text processing, spreadsheet, and other data bases. Limited LAN ability with some products.

Spreadsheet

Programs assist in the statistical evaluation and manipulation of data, usually structured in an x/y matrix format. *Features* include flexible matrix structure expandable to several hundred columns and rows, extensive math functions for calculating new data from selected column and row data, move and copy cell data or ranges, limited text processing, graphic display and data base features often integrated with analytical spreadsheet, user-defined macros.

(continued)

TABLE 7.1 Areas of Application for Computer-Aided Engineering Activities *(continued)*

HOST SYSTEMS

Personal computers.

INTEROPERABILITY

Data exchange with other spreadsheet and graphics programs.

Data Base

Data Base Management Systems (DBMS) assist users in the creation and management of recordkeeping systems. It tracks the information and its interrelationships. *Features* include: create and manage electronic data files; fast retrieval and analysis; query/search; data validation; sorting; formatting; math functions; mailing lists/labels; some text processing; and spreadsheet capability.

HOST SYSTEMS

Personal computers (special installation on minis and mainframes is available for most software products).

INTEROPERABILITY

Data exchange with other data bases, graphics, text processing, and spreadsheet programs.

Communication

Programs allow retrieving and transmitting of electronic data between terminals, central computers, and systems, outside the primary database. Electronic mail, local area networks (LAN), and distributed data processing (DDP) are examples of communication-oriented software applications.

HOST SYSTEMS

Personal computers, minicomputers, mainframe computers; often software is integrated among all systems.

INTEROPERABILITY

As designed for.

CAD/CAM

Computer-aided design (CAD) and computer-aided manufacturing (CAM) software assists engineers in the design and production of new products. Instead of traditional paper-and-pencil designs, engineers create new designs electronically. Included are computer-assisted analysis, validation, documentation, transfer data generation, and change management.

HOST SYSTEMS

Personal computers, minicomputers.

INTEROPERABILITY

Currently, very limited. In future, trends for strong LAN and DDP applications.

MRP

Materials requisition planning (MRP) is a computer software system that integrates several production-related information systems to assist production scheduling and material management, semi-automatically: production planning, master production schedules, bill of materials, material purchasing, inventory control, operations control, resource planning, and capacity planning, cost accounting, and cash-flow planning.

HOST SYSTEMS

Mainframe computers, minicomputers.

INTEROPERABILITY

As designed for within MRP system.

CAM

Computer-integrated manufacturing (CIM), one of the most exciting new developments in operations management, links CAD and CAM systems to assist in the automated planning and control of all manufacturing processes. Ultimately integrated with the company's information and control systems, manufacturing will be linked to the operations of the entire organization.

HOST SYSTEMS

Mainframe computers.

INTEROPERABILITY

As designed for.

Expert System

A special class of software products that can be user-programmed to simulate expert decision-making, based on given operational criteria and rules. Expert systems can be designed and programmed to determine the best move in a chess game or, in an engineering environment, to determine optimal designs, investments, bid decisions, or development schedules. Further, the system can help to predict the effects of management decisions and engineering decisions on complex real-world situations. Mostly designed for interactive use on high-speed microprocessors.

HOST SYSTEMS

Most personal computers.

INTEROPERABILITY

Very limited.

TABLE 7.2 Capability of Various Application Software Packages

Type of Software Package	Text Processing	Graphics	Spreadsheet	Database	Communi-cations	Project Managemt	Engineerg. Design	Mfg. Eng.	Production Control
Project management	2	3	2	2	1	4	—	2	2
Word processing	4	2	2	2	1	1	—	—	—
Graphics	2	4	1	—	1	2	2	1	1
Spreadsheet	2	3	4	2	1	2	2	2	2
Database management systems	2	2	1	4	1	2	2	2	2
Communications	2	1	—	4	4	1	—	2	2
Computer-aided design (CAD)	2	2	1	—	1	2	4	4	2
Computer-aided mfg. (CAM)	2	2	1	2	1	2	2	1	1
Matl. requirements planning (MRP)	2	2	2	1	2	1	1	4	4
Computer-integrated mfg. (CIM)	2	2	2	2	2	2	2	2	4
Expert systems (ES)	2	1	—	1	1	2	2	2	2

Key: —, No capability. 1, Some capability. 2, Limited capability for selected applications. 3, Strong capability. 4, Primary capability (Principal Application).

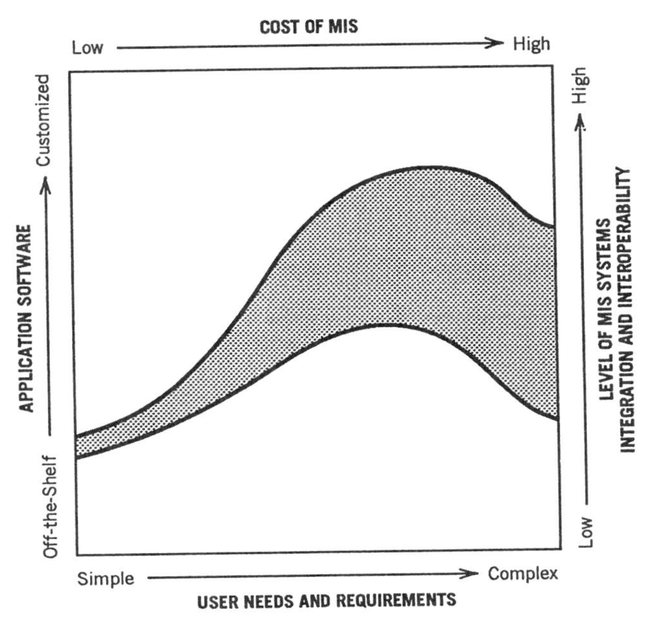

FIGURE 7.1 Relationship between MIS user needs and applications software customization.

matted for effective communication. That is, they must highlight the important information, be structured for easy reading and logical data flow, and formatted for decision-making. In addition, the operating environment is usually in transition—the data to be analyzed and reported this week evolved from last week's operation and report.

Advances in high-speed computers and space-age telecommunications have fostered a large variety of software products to assist engineering professionals in generating the necessary reports effectively and efficiently. One of the largest variety of end-user software available today is aimed at text processing, graphics, and spreadsheet applications (see Table 7.1). When the text processing, graphics and spreadsheet applications are combined with the specialty reporting available via project-management software, they give engineering managers a very powerful set of tools to assist in their communications and decision-making.

Word-Processing Software. Word-processing software, combined with dedicated fast end-user computing, gives managers ready access to information and makes them less dependent on clerical or technical staff. Managers with personal computers can use text-processing programs for the preparation or updating of correspondence, can schedule meetings with the help of electronic calendars, and can access specific databases prior to decision-making. Another major advantage is the ease of updating, revision, and text transfer, which is a great benefit for the generation of major reports, proposals, or manuscripts. Revisions can be made quickly and neatly with a minimum of additional retyping. Yet another feature is the portability

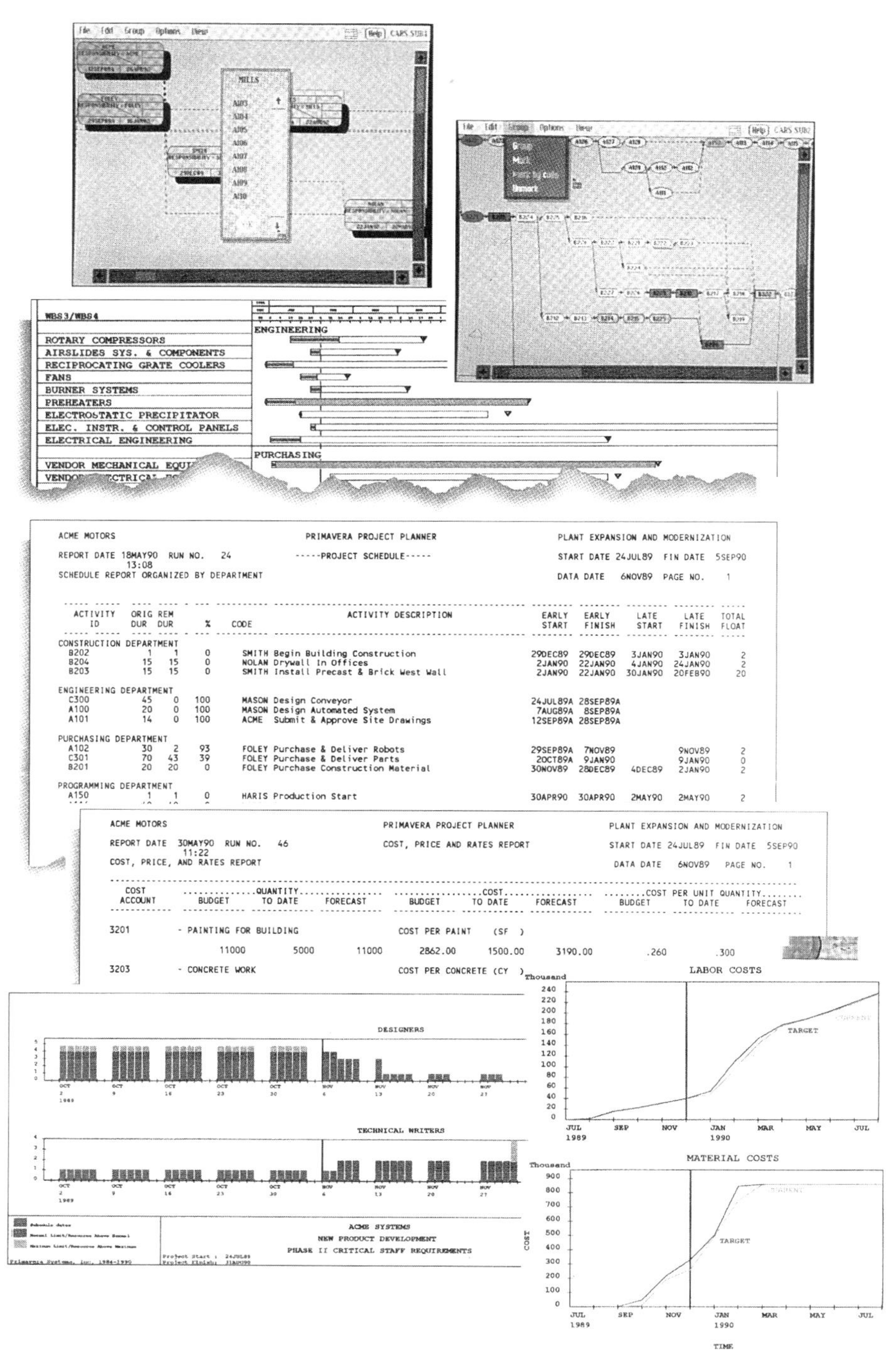

ACME MOTORS — PRIMAVERA PROJECT PLANNER — PLANT EXPANSION AND MODERNIZATION

REPORT DATE 18MAY90 RUN NO. 24 13:08 — -----PROJECT SCHEDULE----- — START DATE 24JUL89 FIN DATE 5SEP90

SCHEDULE REPORT ORGANIZED BY DEPARTMENT — DATA DATE 6NOV89 PAGE NO. 1

ACTIVITY ID	ORIG DUR	REM DUR	%	CODE	ACTIVITY DESCRIPTION	EARLY START	EARLY FINISH	LATE START	LATE FINISH	TOTAL FLOAT
CONSTRUCTION DEPARTMENT										
B202	1	1	0		SMITH Begin Building Construction	29DEC89	29DEC89	3JAN90	3JAN90	2
B204	15	15	0		NOLAN Drywall In Offices	2JAN90	22JAN90	4JAN90	24JAN90	2
B203	15	15	0		SMITH Install Precast & Brick West Wall	2JAN90	22JAN90	30JAN90	20FEB90	20
ENGINEERING DEPARTMENT										
C300	45	0	100		MASON Design Conveyor	24JUL89A	28SEP89A			
A100	20	0	100		MASON Design Automated System	7AUG89A	8SEP89A			
A101	14	0	100		ACME Submit & Approve Site Drawings	12SEP89A	28SEP89A			
PURCHASING DEPARTMENT										
A102	30	2	93		FOLEY Purchase & Deliver Robots	29SEP89A	7NOV89		9NOV89	2
C301	70	43	39		FOLEY Purchase & Deliver Parts	2OCT89A	9JAN90		9JAN90	0
B201	20	20	0		FOLEY Purchase Construction Material	30NOV89	28DEC89	4DEC89	2JAN90	2
PROGRAMMING DEPARTMENT										
A150	1	1	0		HARIS Production Start	30APR90	30APR90	2MAY90	2MAY90	2

ACME MOTORS — PRIMAVERA PROJECT PLANNER — PLANT EXPANSION AND MODERNIZATION

REPORT DATE 30MAY90 RUN NO. 46 11:22 — COST, PRICE AND RATES REPORT — START DATE 24JUL89 FIN DATE 5SEP90

COST, PRICE, AND RATES REPORT — DATA DATE 6NOV89 PAGE NO. 1

COST ACCOUNT	QUANTITY BUDGET	QUANTITY TO DATE	QUANTITY FORECAST	COST BUDGET	COST TO DATE	COST FORECAST	COST PER UNIT QUANTITY BUDGET	COST PER UNIT QUANTITY TO DATE	COST PER UNIT QUANTITY FORECAST
3201 - PAINTING FOR BUILDING				COST PER PAINT (SF)					
	11000	5000	11000	2862.00	1500.00	3190.00	.260	.300	
3203 - CONCRETE WORK				COST PER CONCRETE (CY)					

FIGURE 7.2 Examples of computer-generated engineering project reports. (Courtesy of Primavera Systems, Project Management Software.)

of the text file. Rather than sending a final paper draft for integration into a report or proposal, one can send the data on disk for merging with the other text files.

Spreadsheet Programs. Spreadsheet programs assist in the analysis and preparation of quantitative information, such as project resources and timing, manufacturing, planning, and engineering manpower allocation. Once the data have been input or transferred from another source in a prescribed format, they can easily be manipulated by row and column, statistically summarized with extensive mathematical functions and, with most spreadsheet programs, graphically displayed. A bonus benefit occurs in case of a repeat analysis of data, such as updated production schedules or what-if analysis. Here a comprehensive analysis or report can be generated with a minimum of new data input. Many of the spreadsheet programs also include some graphics and database features.

Graphics Capability. Perhaps one of the strongest growth areas in computer software, business graphics capabilities are provided as an integral part of many reporting packages. Graphic reporting is an effective way of communicating information in a clearer and more concise form than text or tables alone. With the right software, the effort of obtaining the graphic picture is minimal, and in case of a repeat graph with the same format but different data, the effort is virtually zero. Sophisticated peripherals can generate the graph in many colors on paper or directly on overhead transparencies or 35-mm slides; however, a regular dot-matrix printer can produce very nice artwork that looks good in a report or proposal.

7.2 MANAGEMENT INFORMATION SYSTEMS IN ENGINEERING

Modern engineering organizations carry out a wide variety of business transactions which are interrelated in many of their internal functions as well as within their external interfaces. Information systems can help in an integrated cross-functional processing of these data. Two general categories of data can be noted:

1. *Business Data Processing.* These are by and large semiautomated transactions, such as payroll accounts payable, invoicing, and recordkeeping. The information system processes specific data from predefined sources into required records for further transactions, as needed. Business data processing requires minimal management interaction decision inputs or control. It relies on a preprogrammed process executed by the computerized system.
2. *Management Information System (MIS).* Defined as an interactive system of people, machines, and procedures, MIS is designed to generate relevant information from the organization's internal and external environment for management decision-making.

The above classification provides a simple partitioning between the two information systems, which can be helpful for breaking down conceptual complexities, defining systems requirements, and developing an information system. For most operational situations, however, both categories overlap a great deal. An example might be the type of information needed to optimize the design of a new communication system design regarding total life cycle cost. Managers must integrate information from various functional areas such as design, manufacturing, field operations, and engineering services, which might involve thousands of details about design concepts, materials, processes, product interfaces, customer operating costs, and reliability data, just to name a few.

Types of Management Information

Yet another useful classification of management information is according to its level of managerial use within an organization:

1. *Strategic Business Information* is used by senior managers to aid in strategic planning and decision-making that affects long-range business aspects. It may involve the acquisition, allocation, and disposition of resources, personnel policies, and product and market decisions. Often the strategic information required is so unique that the existing MIS can only provide part of it. Many companies rely extensively on outside consulting firms, specialized in this area, to narrow that information gap.
2. *Management Control Information* is used in optimizing organizational performance compared to established objectives. The needed information comes mostly from multifunctional sources and is aimed at objectives such as quality, cost control, safety, and training. Typical subsystems of this category are project management systems, production management systems, marketing information systems, and financial information systems.
3. *Operational Control Information* is used in effectively managing specific operations, such as production scheduling, engineering design to specific requirements, payroll operation, and inventory control.

Levels of MIS Capability

Sophisticated decision support requires sophisticated data processing capabilities. On the other hand, many important clerical functions can be adequately supported with very basic data-processing systems. A five-point classification of MIS capability depending on the level of sophistication has been suggested by Robert Kreitner[1]:

Level 1: Storage of Raw Data Only. Manual recordkeeping is replaced by computer processing. Does not provide any data manipulation. Examples: Appointment calendar, employee absenteeism record, bill of material.

Level 2: Selective Data Retrieval. Data are stored, organized, and selectively displayed, in a way that is useful for decision-making. The system can respond

[1]See Robert Kreitner, *Management,* Boston, MA: Houghton Mifflin, 1986, pp. 577–579.

to selective requests such as providing list of suppliers qualified for a particular contract or listing of components rank-ordered by lead time.

Level 3: Elementary Computation. Data can be arithmetically and logically manipulated, summarized, and compared as is done in a simple spreadsheet. Examples: Calculation of all critical paths on a development project and their rank ordering, or a cash-flow analysis.

Level 4: Advanced Computation. More advanced mathematical calculations and logic functions can be performed on the data in storage. Examples include statistical process control information, computer-aided design, and computer graphics.

Level 5: Mathematical Modeling: The computer system simulates specific management situations, generating diagnostics and predictions for given scenarios, including what-if analysis. Mathematical modeling is extensively used in programs such as computer-aided design (CAD), computer-aided manufacturing (CAM), computer-aided engineering (CAE), computer-integrated manufacturing (CIM), or any system that provides artificial intelligence (AI).

Depending on the data-processing capabilities, the sophistication of the management information system varies. As shown in Figure 7.3, the MIS functions range from aiding clerical work, to providing information, decision support, and programmed decision-making.

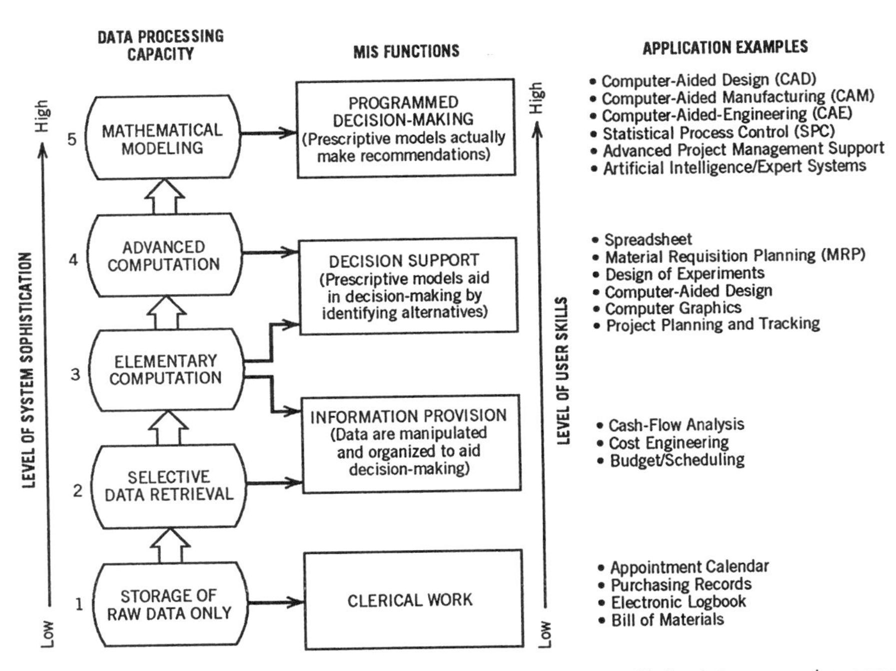

FIGURE 7.3 The functional sophistication of an MIS increases with its data-processing capabilities and user skill level.

Central Data Processing and the Personal Computer

Managers who wish to automate data processing and use computerized information systems have basically three options, centralized data processing, end-user computing, and distributed data processing.

1. Centralized Data Processing. Centralized data processing is the traditional approach to data processing. The computer hardware and its support staff are located in one central location and perform most or all of the data processing for the company's various systems. The central facilities can perform batch or real-time processing via remote online terminals. Reports and data summaries are sent back and distributed as requested. *The advantages* of centralized processing include the following: (1) availability of powerful, sophisticated mainframe systems to all users as needed, (2) cost sharing of one facility among all users, (3) highly specialized and skilled support staff, (4) assistance in data management and problem-solving, (5) enhancement, upgrade, and maintenance of system and end-user training are ongoing central functions, and (6) interoperability with other systems and subsystem integrations can be centrally managed. *Disadvantages* of centralized processing include: (1) potential time delays in gaining access to the central system and slow turnaround, (2) limited interaction, such as for "what-if analysis," (3) access limited to specific locations, (4) it is slow in application software upgrades, (5) there is poor interoperability with "local" systems such as personal computers, and (6) it has limited ability to accommodate special user needs.

2. End-User Computing. In 1990, U.S. businesses used an estimated 13 million personal computers (PCs). With the increased availability of user-oriented hardware and software, many managers have their personal data-processing facility right in their offices. Managers do not have to rely on other functions and personnel located in a central facility, but can perform a considerable amount of data processing that is under their immediate control on their own laptops, desktops, or personal computers. End-user applications range from computer-aided design to project management information systems and statistical analysis. In addition, these engineering managers often have secretaries and administrative staff who are highly experienced in the use of personal computers and able to assist the manager in working with the local computer information system. It is estimated that today 95% of the nation's engineering managers have access to personal computers. *The advantages* of end-user computing include: (1) having a dedicated system for fast, inexpensive processing, (2) quick, flexible software and hardware upgrading, (3) portable application software and data for other departments or customer sites, (4) having a user-interactive system, (5) tailoring of information and formats to specific needs, and (6) having high confidence in information output as a result of the hands-on data interface. *Disadvantages* of end-user computing include: (1) limited data-processing and storage capability by comparison to central mainframe computers, (2) lack of integration with other users outside the local area, (3) managers' temptation to spend considerable time as computer operators, (4) proliferation of computer hardware and software throughout the organization with no coordination and limited compatibility,

(5) potential information-security problems, and (6) system maintenance and repair problems.

3. Distributed Data Processing. As indicated by the large number of personal computers in use, engineering managers did not wait until the ultimate centralized management information system arrived, but bought their own local system and used it. End-user computing is here to stay. Yet centralized data processing offers great additional features and advantages and should be integrated with the local systems. *Distributed data processing* (DDP) is the technology link that connects both central and end-user computing. Shown schematically in Figure 7.4 *(bottom)* DDP is defined as an interactive system of computers that are geographically dispersed and connected by telecommunications. Each computer is able to process data independently and to transfer data to other systems. Personal computers that are interconnected with a DDP system can use the centralized data-processing system to access files, transfer application programs, perform large computational tasks, use special application programs, and tie into the central management, information system. At the same time, the user can perform any data-processing locally that his or her machine is capable of. In this mode, the user has a fully dedicated, interactive computer available, with the central processing facility as a time-shared extension of the personal system. Such decentralized operation is especially advantageous in organizations that are managerially decentralized, requiring relatively few interactions among operational units and headquarters. DDP can keep communications costs low and ease the demand on the central system. *The advantages* of DDP include: (1) increased utility of all organizational computer equipment and lower systems costs, (2) reduced complexity of the central facility's systems, its operations, and data management, (3) adaptability of DDP to the structure and needs of an organization, (4) increased manageability of overall information systems, and (5) combined central processing and end-user processing. *Disadvantages* include (1) difficulties associated with the complexities of computer networking, (2) high cost of developing and installing a distributed data-processing system, (3) high communications costs (for most DDP systems, communications accounts for more than 50% of its operating costs) and (5) the organization may lose overall control over the information system.

In looking at the three options, it should be clear that they do not represent an either–or choice. Most organizations are sufficiently complex to encourage a combination of one, two, or all three systems to coexist, depending on the specific application under examination. With the enormous increase in personal computer users, it is not surprising that the trend towards decentralized systems is currently continuing. Local area networks (LANs), which will be discussed next are yet another derivative of decentralized computing. Each approach has its pros and cons.

Local Area Networks (LANs)

With the growing trend toward decentralized computing, local area networks (LANs) have become increasingly important vehicles for data-system integration. A local area

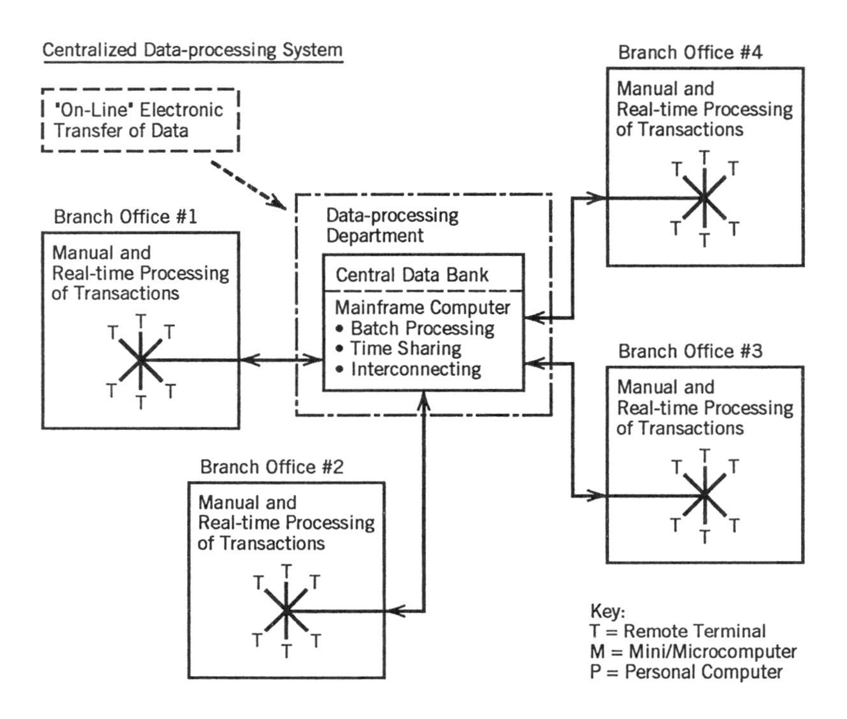

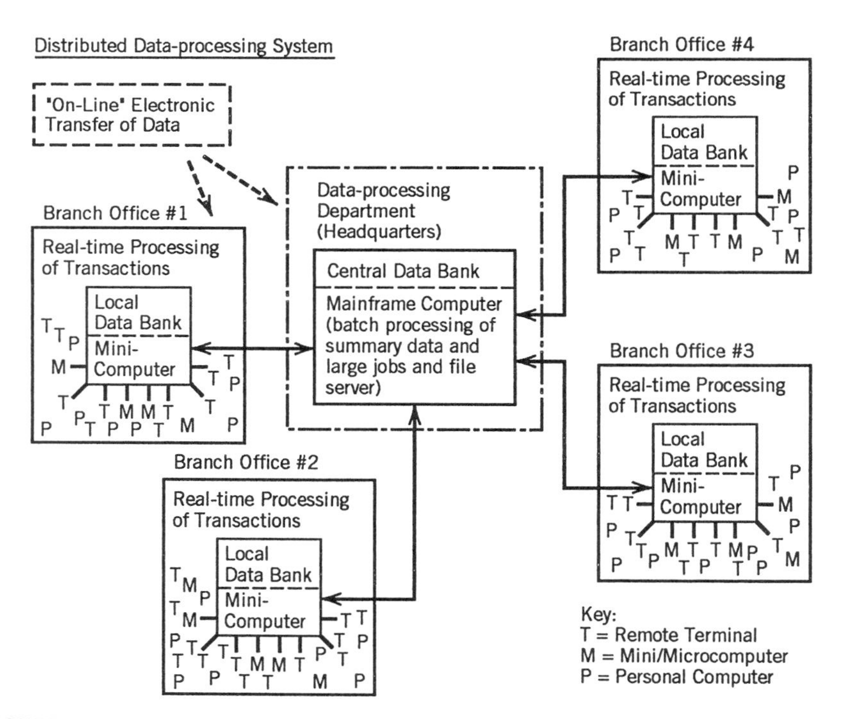

FIGURE 7.4 Schematic diagrams of centralized and distributed data-processing systems.

network (LAN) is a special version of a distributed data processing (DDP) system. It provides for the integration of a cluster of workstations with a host computer, featuring limited interoperability and communications among the end users connected with the LAN system. More specifically, a local area network links a variety of data-processing equipment such as personal computers, minicomputers, and mainframe computers. These links may also connect to graphics equipment, facsimile (fax) machines and CAD/CAM systems.

Noting the advances in local area network technology and their availability throughout many companies, engineering managers have begun to take advantage of these new systems for communicating information. In the United States, an estimated 20 million LAN users existed in 1990.

The earlier LAN systems were hard-wired, commonly using telephone wires within a limited "local area," such as one building. However, recent software development, high-speed packet switching, and the establishment of some formal standards and protocols such as Ethernet, today enable large numbers of personal computers and other data-processing equipment to be networked. In addition, special hardware/software known as bridges and routers are now available for interconnecting LANs, providing the infrastructure for true global data networks.

The importance of LAN technology to the engineering community is reflected by the development of several specialized products, such as Auspex Systems Network Server, NS–5000, and the strong participation of engineers in LAN user groups, such as Caucus and the Advanced Networking Group (ANG) formed in 1989.

Providing More System Interoperability

Information system integration involves more than the type of intercommunication provided by local area networks. It involves free access to data banks and the application subsystems of an organization. Total data-system integration is a very complex, costly, and risky undertaking. However, with the current technological advances and the managerial need for effective data processing and communications, we see a steadily increasing upgrading of information systems toward total interoperability and integration.

7.3 CHARACTERISTICS OF MANAGEMENT INFORMATION SYSTEMS TODAY

With the powerful software and hardware systems readily available to engineering managers, combined with the need for quick information and reaction time, most engineering managers do not wait until a nicely integrated management information system becomes available in their companies. These managers typically introduce new applications software to their engineering organizations if they feel that it is operationally advantageous. This often leads to piecemeal, spotty information-processing and isolated problem-solving. It also deprives other managers outside the immediate application area from access to this information. In addition, it most likely requires completely new data-entry and data-system setup when a work package or information travels across organizational lines. Examples are product developments that require data transfers from engineering design to manufacturing or a contract project that is tracked via different systems in engineering, cost accounting, and contract administrations.

Therefore, the global objective of a management information system is to link the various data sources and information-processing subsystems into a comprehensive information system. It is a computer-based network that integrates the collection, processing, and transmission of information, conducive to the needs of the user. Some of the desirable characteristics of a MIS are:

1. Easy Access to Appropriate Information. An effective MIS responds to the manager's needs for information by indicating if and where the requested data are available. In the event that the requested information is insufficiently available from the system's database, the MIS becomes a data-gathering system, which can tap auxiliary data sources, including the broad spectrum of online databases commercially available today.

2. Easy Information-Processing for Management Actions. The information retrievable from the database is seldom in a form directly useful for managerial decision-making or reporting. It must be processed. The MIS should provide such information-processing with a minimum of user interface. In many cases, special applications software, integrated into the company's MIS, provides effective processing of the information for managerial decision-making, control, and reporting. Examples of such application subsystems are project management, production planning, financial planning, cash-flow, change management, and engineering design and analysis programs.

3. Integration of Various Subsystems. Although the MIS subsystems and their data banks are usually maintained by their user departments, an effective management information system is centrally coordinated. It ensures that the various data banks and information-processing systems, including special applications software, can be operated from various points of the system and that data can be transferred back and forth among the subsystems as needed. Such integration reduces intermediate data-processing and the need for repetitive data entry and processing by different departments. In addition, such systems integration enables senior management to obtain an integrated cross-functional overview of the business.

4. Cost Effectiveness. The cost of acquisition and maintenance of the MIS and its subsystems must be affordable and cost-effective. Determining the value of information is difficult. However, four primary factors contribute to it: accuracy, access time, reporting formats, and special processing for decision-making. Another cost dimension is customizing software to specific organizational needs. All variables are dynamically interrelated. No simple formula guides a cost-effective MIS design. However, as shown in Figure 7.1, with increasing user requirements, the need for more customized application software and the level of system integration increases up to a point. Beyond a certain level of user requirement, particularly for specialized applications, the trend often reverses. Many of the sophisticated applications software are difficult to customize and to integrate fully with the MIS.

5. *Expandability.* The MIS should be designed for future growth. The system should permit the integration of new and existing subsystems, to grow with the information needs of the business and to take advantage of new software developments and their enhancements.

6. *Enhancing Communications.* An effectively designed MIS facilitates communications across functional lines as well as up and down the organization. The key to such communication is the system's ability to transfer information back and forth among subsystems and among different users. The other necessary prerequisites are the skill development and attitudes of the system users. User training at all levels of personnel is often an absolute necessity in order for such communications to flourish.

7.4 TOWARD TOTAL INTEGRATION

If engineering managers are to effectively perform their roles as coordinators and integrators of multidisciplinary activities, they need access to multifunctional information. They also must communicate data to others, who are often not part of the same "local" information system. Managers are often frustrated to find that the needed information already exists in the same data bank and could be used for immediate decision support or further analysis, except that a proper data link does not exist between the two systems. The manual transfer and reformatting of the information is costly and time-consuming; it also leads to errors, misinterpretation, and security problems.

Here is a typical example. An engineering manager wants to retrieve statistical quality-control data stored at the product-assurance MIS. The data would be further processed and used for decision support of electronic component selections in a new product design. Later on, when the manufacturing department starts detailed produceability planning, manufacturing engineering personnel want direct access to the design, parts, and test data kept in the CAD database.

What we need is the ability to interact among individual management information systems. The key to such interaction is free access to the various data banks, including applications programs and controls. Ultimately this would result in total interoperability of all information subsystems, providing format flexibility, reporting, and interfaces to other computer systems. These are the characteristics of a totally integrated management information system, which is diagrammed in Figure 7.5.

Database Management

The foundation of any MIS is its data bank. It contains the raw data that will be accessed by the MIS user, either directly or indirectly via analytical method or models. The data bank must be maintained for accuracy, timeliness, proper format, and security. An added challenge for the designers of an MIS is to provide for data

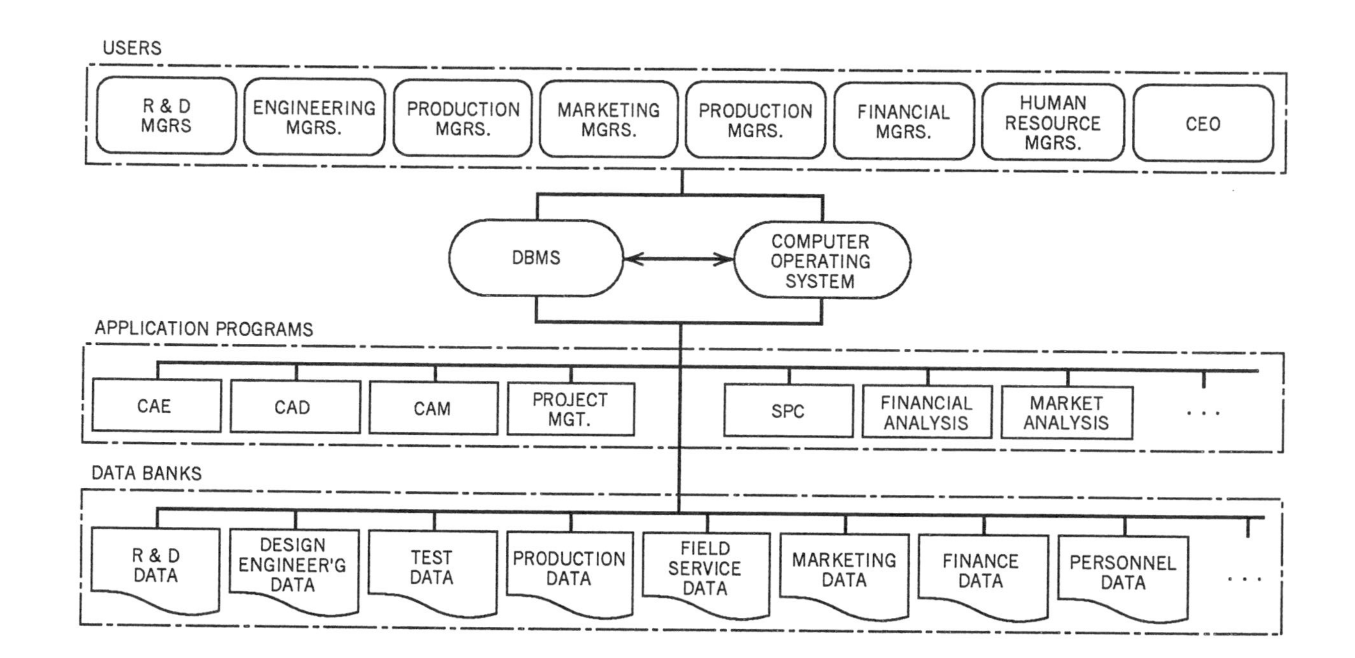

FIGURE 7.5 Conceptual view of integrated MIS using database management system (DBMS).

sharing among different area users. An engineering manager might access the data bank, or database, of the product assurance MIS to retrieve statistical quality-control data for deciding on a vendor component needed for her new product design. Later on when the manufacturing department starts detailed produceability planning, manufacturing engineering personnel want direct access to the design, parts, and test data kept in the CAD database.

What is needed is a Database management system (DBMS) conceptually shown in Figure 7.5, which permits simultaneous access to the many different data banks that exist throughout the company. Ideally, such an integrated MIS would greatly enhance the power and sophistication of information and decision support available to managers. At the current state of systems technology, totally integrated MIS do not exist. However, advances in the area of database management systems provide opportunities toward database sharing, especially if DBMS concepts are considered at the time of the design or reconfiguration of an MIS.

The importance of systems integration and data-sharing to managers is underlined by the popularity of commercially available software such as dBase III and Paradox, which sold over 2,000,000 copies by 1990. Some advanced software, such as dBase III Plus, is designed for PC users to access and share information from different databases within an organization. These software developments are important milestones toward total MIS integration, enabling individuals to network their PCs across functional lines and different organizational levels. The ultimate objective is to provide free access to data banks of all information subsystems from all end-user points. This would eliminate the problems of sharing common information resources and central data-processing facilities among all users.

Where Are We Heading?

With the technology toward miniaturization expected to continue, we will also see faster, more powerful system components such as memory, processor, and display units. The development of upgraded, standardized software for operating systems and their interfaces, together with well-defined standards such as the IEEE's "Futurebus" Specification 896.1 and protocols for open system interconnection (OSI), will lead to a more effective interface of personal computers with mainframe-based information systems.[2] As a result, we will see a growth in local and wide-area networks. The infrastructure for global data networking is beginning to emerge. High-speed packet switching has pushed data rates already beyond 100 mega-bids per second, while optical fiber networks hold the potential for even the most bandwidth-hungry application, such as mixing high-speed data transfers among supercomputers with high-definition television (HDTV). At the heart of the engineering office of the future will be the managerial workstation, which is expected to be integrated into the daily engineering activities. The engineering system of the future will have the following characteristics:

[2]Open system interconnection (OSI) refers to a computer interface environment which can support application software and communications based on a variety of operating systems. The computer interface becomes just another communications node in the user network.

1. Changes in Future Workstations. Future workstations will be even smaller, lighter, more compact and mobile than today's. Increasingly, the bulky cathode tube display will give way to flat panels with increased resolution and possibilities for 3-dimensional perspectives. Increased computer power, expressed in speed, memory, and operating system sophistication, will lead to more programs and information storage for user-specific work situations at the workstation level.

2. Changed Office Environment and Tasks. The computer workstation will replace much of the need for traditional paper-and-pencil design, analysis, recordkeeping, and information-sharing. Engineering professionals will spend a great deal of their working time in front of the computer. Tomorrow's office furniture and layout will reflect the increased need for physical comfort and privacy. More people will work at home and in satellite offices, without pressures to share workspace, and will be able to utilize resources over a wide range of work hours in an extension of the 9:00 to 5:00 workday.

3. Changes in Engineering Labs. Engineering experimentation, prototyping, testing, and laboratory studies will be more directly linked with the data-processing systems via instruments and sensors. The laptop computer has already replaced the traditional lab notebook for recordkeeping in many situations. More rapid analysis, access to cross-functional information from other team members, and sophisticated analytical applications software will enable individual engineers and technicians to perform more complex, more integrated, higher-level work, which traditionally had to be done by teams of people. In addition, information networking will provide rapid access to research and design data, with the possibility for their integration. This will help in the integrated management of engineering research and development, including design validations, project status assessments, and integrated systems level decision-making.

4. Changes in Systems Interface and Human Interactions. In response to the desire to increase engineering productivity, the physical layout of the personal computer system, its work environment, and interfaces will be designed with continued emphasis on human factors and ergonomics. The keyboard will remain the primary interface mechanism between the user and the machine: however, we expect to see more applications that feature touch panel displays, pointers, mice, and voice recognition. Interconnection with other systems within the organization and with outside services will be simplified to a level that eliminates the need for special training and protocol memorization to use them. The computer workstation will become an instrument for effectively interconnecting to all communication channels and systems that were traditionally separate entities, such as the telephone, calendar, electronic mail, stock market report, time cards, vendor lists, catalogs, trade journals, inventories, and work status and site security reports. The quest for increased interoperability among these information systems and total integration will continue, with increased pressures from the engineering community.

5. Changes in Training. At the present time most people entering the engineering profession have working knowledge of and skills in end-user computing. This is in sharp contrast to previous generations of professionals, who could function technically effectively with only limited exposure to computers. Changes in job content, new technologies, changing market and vendor situations, and outright changes in a person's job or career are just a few of the factors that drive the need for providing basic computer user training as a continued professional development in the present and future. The continued growth of special applications software and open system interconnection models requires considerable specialized end-user training and upgrading. It also requires a great deal of "people skills" and management to direct and unify data-processing toward having an integrated information system. Extensive training in MIS-oriented team-building, leadership, communication, decision-making, and conflict management will be necessary to accomplish the necessary goals.

7.5 THE HUMAN SIDE OF MIS

In spite of the cutting-edge technology available and in use for information systems, hardware, and software, MIS success is not guaranteed by technology alone. People are the most important ingredient in the formula for successful MIS implementation and use. This contention is supported by many studies that argue that user attitude and organizational ambiance are the most crucial factors in the successful design and implementation of any management information system.[3]

Whether we look at the development of a new MIS, its site implementation, or its actual operation, it involves people interacting with people. Technical tools and the intrinsic features and capabilities of a system are important; however, information specialists generally agree that success or failure of a MIS has little to do with the technical attributes of the system.[4] As nicely stated by Kreitner[5]: "Even the best-designed MIS can be brought to its knees by overt and covert user resistance. Enthusiastic support, on the other hand, can greatly enhance a MIS's development and use." When the system fails to produce the intended results, the failure is usually due to people resisting the new system. The reasons could range from fear to incompatibility of the MIS with operational processes to "not-invented-here" syndromes. Managers may also see the MIS as competing with their traditional roles of integrating and communicating information. For example, the MIS may provide for collecting, integrating, summarizing, analyzing and distributing cost schedules and performance data of engineering projects on an ongoing basis. This eliminates the need of

[3]Michael J. Cerullo, in a study of 122 of the 1,000 largest U.S. firms, found that user attitude is the most important factor that determines success of a new MIS implementation. See "Information Systems Success Factors," *Journal of Systems Management*, December 1980. See also F. Warren McFarlan, "Portfolio Approach to Information Systems," *Harvard Business Review*, September–October, 1981, pp. 142–150, and G.W. Dickson and John K. Simmons, "The Behavioral Side of MIS," *Business Horizons*, Vol. 13, August 1970.

[4]See Kate Kouiser and Ananth Srinivasan, "User–Analyst Differences: An Empirical Investigation of Attitudes," *Academy of Management Journal*, September 1982, pp. 630–646. Also see H. C. Lucas, Jr., *Why Information Systems Fail*, New York: Columbia University Press, 1975, and Daniel Robey, "User Attitudes and MIS Use," *Academy of Management Journal*, September 1979, p. 527.

[5]Robert Kreitner, *Management*, Chapter 16, Boston, MA: Houghton Mifflin, 1986, p. 593.

the manager to perform these data-processing functions and to present the data at staff meetings. Although the system very effectively assists the manager, he or she may resist the MIS because of perception of lower status, power, and managerial influence. The type and level of resistance varies, and it often comes in subtle forms. On the soft side of resistance, people may just ignore the system, make excuses for their inability to work with the MIS, or may criticize various aspects. More outraged forms of resistance include blaming all sorts of problems on the system, building cases in incompetence, and engaging in outright sabotage. Therefore, it is important for managers responsible for the design, implementation, and effective use of information systems to understand the complex interaction of organizational and behavioral variables. Most importantly, these managers must understand the reasons why people resist the MIS and must build a work environment conducive to their professional and personal needs. This involves technical MIS expertise, information and training sessions, and, equally important, diplomacy, persuasiveness, tact, and sensitivity to people's needs and concerns.

Barriers to MIS Use

Problems most commonly encountered in situations of resistance to use the information system include the following:

- Poor understanding of management objectives for MIS
- Insufficient user skills and training
- Fear of job loss or status, and power reduction
- Information overload
- Role conflict and power struggle
- Lack of clear management direction, policies, and procedures
- Weak top-management support
- Users not involved in the requirements analysis, system planning, and development
- Lack of commitment from interfacing user groups
- Natural resistance to change.

Recent field studies[6] have found some quantitative evidence of the unfavorable effects of these factors on MIS performance. Using rank-order correlation techniques, all of the above problems were found to be negatively associated with effective MIS use at a 97% confidence level or higher. The stronger and more frequently these

[6]In a recent field study, Thamhain and Wilemon found quantitative evidence that many of the barriers and resistances toward MIS use are generally present during the execution of technical projects. Thus the lessons learned from project management regarding effective planning, active involvement, management support, team building, communications, and leadership, apply also to the implementation of MIS. For detailed discussion see "Building High-Performing Technical Project Teams," *IEEE Transactions on Engineering Management*, Vol. 34, No. 3, (August 1987), pp. 130–137.

problems occurred, as perceived by a work team, the lower was their performance rating as effective MIS users by upper management. Determining the statistical association of barriers to MIS performance has two uses. First, it offers some clues as to what an effective MIS work environment looks like. This can stimulate thoughts of how to foster a work environment responsive to the needs of people and conducive to user acceptance. Second, the findings may help in designing specific on-site investigations of user environments, with the objective of diagnosing and correcting problems related to both the system and the people.

Neutralizing User Resistance to MIS

One of the most interesting findings of management research is that 85% of all user-perceived MIS problems can be linked to people's attitudes rather than to technical attributes of the system. These perceived problems ultimately lead to resistance to using the MIS. Middle- and lower-level managers have been found to offer the strongest resistance to using established MISs. Kreitner found their resistance was related to the influx of personal computers, which were favored, together with self-generated databases, in lieu of the organization's central MIS.[5] Regardless of the specific user group and the type of information system, a number of recommendations can be advanced to potentially increase user effectiveness in using the MIS.

1. Establish Clear Objectives. Management should establish and communicate clear objectives for any new information system under consideration or for any extension of existing MIS operations. If the users understand the advantages, objectives, and potential benefits of the new system, they may be more likely to get interested and participate with a positive attitude and desire for success. In addition to the short-term objectives, the importance of the system in meeting the organization's needs should be clear to all potential users. Senior management can help develop a priority image by communicating the basic objectives and providing management guidelines and active involvement.

2. Plan Ahead. Effective planning early in the life cycle of an MIS project should involve all user organizations and systems specialists, as well as top management. The information system should be developed within the context of long-range business planning, considering the broad needs of the entire organization in meeting its strategic goals. Part of this planning effort is a detailed project definition, which defines the objectives, requirements, and scope of the proposed system and is responsive to the needs and concerns of the future users. Proper interactive planning can be very pervasive in stimulating interest and support for the new system throughout the organization and in unifying the many local interests toward the new integrated information system.

3. Involve All Stakeholders. One of the side benefits of proper system planning is the involvement of people at all organizational levels and functions. This will lead to a better understanding of the overall objectives and benefits and a realistic assessment of the system requirements and support and training needs. Most importantly, it will stimulate interest and desire for the new system, which not only lowers the initial anxieties and minimizes resistance to change, but leads to cooperation among the various user groups and commitment toward successful implementation. A systems steering committee or user advisory committee can be an effective vehicle for facilitating active involvement and establishing linkages among user groups, systems specialists, and management.

4. Build a Favorable Image. Management, through their actions, involvement, and communications, should build a favorable image for the MIS project in terms of priority, importance to the organization, visibility, and potential benefits to all participants. Such a favorable image leads to intrinsic motivation of the people in the organization to work toward the established MIS goals. It is also a pervasive process, which fosters a climate of active participation at all levels, helps to unify the organization behind the MIS, and minimizes dysfunctional conflict.

5. Have Regular Reviews and Briefings. Management should hold regular status-review meetings for all people who are involved in the system's implementation as well as for the future users. This is a good tool for communicating objectives and progress, and for reassuring management commitment. These meetings prepare the future user for the new system and its potential benefits, help to alleviate many concerns, stimulate active involvement and a sense of ownership, and help to unify the MIS community behind the new system.

6. Minimize Threats. Management must understand the potential barriers to MIS implementation and build a work environment that shows sensitivity and concern for the personal and professional needs of the people. Proper communications and active involvement of the MIS community will minimize the natural anxieties and enhance the desire to participate in the new system's development and in its applications.

7. Humanize the Information System. Assure the users that the system is here to support their professional activities and to help them in planning, tracking, and controlling their activities more effectively. The objective for the new system *is not* to control the users or to restrict their freedom of decision-making, but to help professionals to control more effectively and innovatively.

8. Fit the System to the User. The best information system is bound to be rejected if it does not meet the needs of its users. Proper matchmaking is to some degree technical. However, the human side of *perceived* relevancy and fit of the system to the user's needs is also crucial to successful implementation. The active involvement of the users during the conceptual phases of the project and during informational and training sessions can help in selecting, designing, and fine-tuning the technical aspects of the system, as well as assuring the users that the system is built around their needs, thus fostering a sense of ownership, commitment and desire to make it work.

9. Provide User Training. Extensive education and training are often required with a new system. The benefits are both technical and psychological. Of course, it is important that the users be familiar with the basic operational details of the system so that they can utilize its features. In addition, user training has major psychological benefits. Familiarity with the system's features and benefits, along with the user's new operational skills, all help in building a "can-do" image in the mind of the user. This reduces potential anxiety and increases the desire for further involvement, skill development, and application. Training sessions also help to identify potential problems regarding personal interfaces, attitudes, and organizational conflict, which often can be diagnosed and resolved *before* the system goes into operation. Finally, these training sessions are excellent organization development tools for team-building, thus helping to unify the MIS community behind the new or upgraded information system.

A Final Note

Preparing the organization and its people for change is one of the primary responsibilities of management. The implementation of a new or enhanced information system involves new ways of conducting business in many forms, affecting design, communications, decision making, recordkeeping and performance evaluations. It involves the whole spectrum of management skills to identify and integrate the organizational needs toward meeting the functional system's requirements. The more actively management can involve all future user groups in the conceptual phases, keep them informed, and provide user training, the greater the trust and interest in the new system and the desire to use it. Such active involvement also helps to build and unify the user team behind the overall objectives. It results in a higher quality of information exchanges, including candor, sharing of ideas, and innovation. It is this professionally stimulating environment which also has a pervasive effect on the organization's ability to cope with change, conflict, shared power and decision making. Facilitating the acceptance of a new information system is the shared responsibility of the functional managers in cooperation with the company's senior management. In order to identify the critical drivers to success, the engineering manager must be a social architect who understands the organization, its people, and their needs. This involves effective leadership, including people skills. Finally, engineering managers can influence the climate toward MIS effectiveness by their own actions. Their concern for their people, sensitivity to their specific needs, commitment to the new system, providing of resources for training and development, and ability to stimulate personal enthusiasm for the new MIS resource can foster a climate of high motivation, work involvement, open communication, and ultimately successful transition to the new information system.

BIBLIOGRAPHY

Adler, Paul S., M. Donald, D. William, and Fred MacDonald, "Strategic Management of Technical Functions," *Sloan Management Review*, Volume 33, Issue 2 (Winter 1992), pp. 19–37.

Benjamin, Robert I., "Information Technology in the 1990's: A Long Range Planning Scenario," *MIS Quarterly*, 6:2 (June 1982), pp. 11–32.

Blackwell, Gerry, "MIS Acceptance at Standstill as Apple Strives to Win Fortune 500 Accounts," *Computing Canada*, Supplement (Sept 1991).

Bucken, Mike, "Open Systems—Positioning Architectures to MIS," *Software Magazine*, Volume 11, Issue 15 (Dec 1991), pp. 23–24.

Cerullo, Michael J., "Information Systems Success Factors," *Journal of Systems Management*, (December 1980).

Cusack, Sally, "Keeping Centralization on Track," *Computerworld*, Volume 24, Issue 26 (June 25, 1990), pp. 67, 70.

Davis, Gordon B. and Margrethe H. Olson, *Management Information Systems*, New York: McGraw-Hill, 1985.

Davis, G. B. "Strategies for Information Requirements' Determination," *IBM Systems Journal*, Volume 21, Number 1 (1982).

Deitz, Daniel, "Project Management: Pulling the Data Together," *Mechanical Engineering*, Volume 112, Issue 2 (Feb 1990), pp. 56–57.

Dickenson, G. W. and John K. Simmons, "The Behavioral Side of MIS" *Business Horizons*, Volume 13, (August 1970).

Donation, Scott, "Technology Marketing: Magazines Defend Their Niches," *Advertising Age*, Volume 62, Issue 48 (Nov 11, 1991), pp. S17–S18.

Drigani, Fulvio, *Computerized Project Management*, New York: Marcel Dekker, 1989.

Dunn, Richard L. and Dick Johnson, "Getting Started in Computerized Maintenance Management," *Plant Engineering*, Volume 45, Issue 7 (Apr 4, 1991), pp. 55–58.

Eldred, William A., "A Proposed Approach to Computer-Supported TQM in Maintenance Work Induction and Accomplishment," *Computers & Industrial Engineering*, Volume 21, Issue 1–4 (1991), pp. 123–127.

Ezerski, Michael, "Simple Doesn't Mean Limited in SNMP Nets," *Computer Technology Review*, Volume 11, Issue 15 (Dec 1991), pp. 10–11.

Fairweather, Virginia, "Managing a Megaproject," *Civil Engineering*, Volume 59, Issue 6 (June 1989), pp. 44–47.

Farid, Foad, and Roosbeh Kangari, "Knowledge-Based System for Selecting Project Management Microsoftware Packages," *Project Management Journal*, Volume 22, Number 3 (Sept 1991), pp. 55–61.

Feigenbaum, E., and P. McCorduck, *The Fifth Generation*, Reading, MA: Addison-Wesley, 1983.

Franklin, M. Steven, "Combining PC Flexibility and Mainframe Muscle," *AACE Transactions* (1990), pp. N.3.1–N.3.5.

Friedlander, Philip, and Sandra Lehde, "The CASE Generation Gap," *Software Magazine*, Volume 12, Issue 2 (Feb 1992), pp. 106–105.

Galletta, Dennis F. and Ellen M. Hufnagel, "A Model of End-User Computing Policy: Context, Process, Content and Compliance," *Information & Management*, Volume 22, Issue 1 (Jan 1992), pp. 1–18.

Guimaraes, Tor, and Jayant V. Saraph, "The Role of Prototyping in Executive Decision Systems," *Information & Management*, Volume 21, Issue 5 (Dec 1991), pp. 257–267.

Hazzah, Ali, "IBM's Information Model: Rosetta Stone for Developers?" *Software Magazine*, Volume 10, Issue 9 (July 1990), pp. 87–96.

Hubbard, Craig, "Maintenance Now Seen as an Engineering Process," *Computing Canada* (Canada), Volume 16, Issue 22 (Oct 24, 1990), pp. 36, 38.

Kirby and East, *A Guide to Computerized Scheduling,* New York: Van Nostrand Reinhold, 1990.

Kouiser, Kate and Ananth Srinivasan, "User-Analyst Differences: An Empirical Investigation of Attitudes," *Academy of Management Journal* (September 1982), pp. 630–646.

Krietner, Robert, *Management,* Chapter 16, Boston, MA: Houghton Mifflin, 1986.

Levine, Harvey A., "How Do Graphical User Interfaces Affect the Usability and Power of PM Software?" *PM Network,* Volume 5, Number 2 (Apr 1991), pp. 19–22.

Levine, Harvey A., "Latest PM Software Releases," Volume 5, Number 3 (Oct 1991), pp. 19–22.

Livingston, Dennis, "Engineering Giant Upgrades to 3-D CAD," *Systems Integration,* Volume 23, Issue 10 (Oct 1990), pp. 64–70.

Lorenzen, A. M., F. Giambelluca, and M. J. Stagliano., "Factors Leading to Success in Introducing A New Generation of Project Management Information Systems," *1985 Project Management Institute Proceedings,* Volume 2.

Lowery, Gwen, *Managing Projects with Microsoft for Windows,* New York: Van Nostrand Reinhold, 1992.

Lucas, H. C., Jr., *Why Information Systems Fail,* New York: Columbia University Press, 1975.

Mahoney, John and David P. Kenney, "End-User Computing for Project Controls," *AACE Transactions* (1990), pp. M.3.1–M.3.6.

McDonald, Donald F., Jr., "Getting Personal: Mainframe-Mini-Micro Links Versus a Common User Database," *Cost Engineering,* Volume 33, Issue 7 (July 1991), pp. 30–31.

McFaarlan, F. Warren, "Portfolio Approach to Information Systems," *Harvard Business Review,* (Sept/Oct 1981), pp. 142–150.

McNair, C. J. and Ronald Teichman, "Fast Forward: The Challenges Facing Controllers in a Global Market," *Corporate Controller,* Volume 4, Issue 1 (Sept/Oct 1991), pp. 34–38.

Owings, Clinton, Charles Sirro, and John A. Smith, "Integrated and Corporate-Wide Project Management Information Systems," *1982 PMI Proceedings,* Toronto, Canada.

Paz, Noemi, William Leigh, Jim Pullin, and Jim Ragusa, "Information Systems Configuration by Expert System: The Case of Maintenance Management," *Computers & Industrial Engineering,* Volume 19, Issue 1–4 (1990), pp. 553–556.

Puckett, James C., Ronald L. Strauss, and Cynthia Wilson, "Cradle-to-Grave Material Management," *AACE Transactions* (1991), pp. 06(1)–06(9).

Radding, Alan, "Management Software: Project Management," *Bank Management,* Volume 66, Issue 5 (May 1990), pp. 62–67.

Raymond, Louis, "Information Systems Design for Project Management: A Data Modeling Approach," *Project Management Journal,* Volume XVIII, Number 4 (Sept 1987).

Robey, Daniel "User Attitudes and MIS Use," *Academy of Management Journal* (Sept 1979), p. 527.

Robins, Gary, "Information Engineering: Flexibility Is Key," *Stores* (Section I), Volume 73, Issue 2 (Feb 1991), pp. 50–52,

Rowen, Robert B., "Software Project Management Under Incomplete and Ambiguous Specifications," *IEEE Transactions on Engineering Management,* Volume 37, Issue 1 (Feb 1990), pp. 10–21.

Ruhl, Janet, "Teach Yourself to Be a Good Manager," *Computerworld,* Volume 24, Issue 47 (Nov 19, 1990), pp. 123.

Savage, Charles M., *Manufacturing Engineering,* Volume 102, Issue 1 (Jan 1989), pp. 59–63.

Scott, George M., *Principles of Management Information Systems,* New York: McGraw-Hill, 1986.

Snell, Ned, "Three Ways to Plan Network Upgrades," *Systems Integration,* Volume 24, Issue 12 (Dec 1991), pp. 63–67.

Smith, Todd, "Users Compete in QIC Data Management," *Computer Technology Review,* Volume 11, Issue 15 (Dec 1991), pp. 33, 35.

Spirer, Herbert F. and Jack A. Opiola, "An Approach to Selecting A Computer-Based Project Management-Institute Seminar/Symposium 1983 (Oct 1983).

Stevenson, James J., Jr., "Project Management Information System," *AACE Transaction,* (1990) pp. D.2.1–D.2.5.

Taylor, Allen G., Robb Ware, and Duncan William, "Project Management Software," *Computerworld,* Volume 25, Issue 10 (Mar 11, 1991), pp. 55–64.

Teresko, John, "What MIS Should be Telling You About CASE," *Industry Week,* Volume 239, Issue 7 (Apr 2, 1990), pp. 82–85.

Thamhain, Hans J., "The New Project Management Software and its Impact on Management Leadership Style," *Project Management Journal,* Volume 18, Number 3 (Aug 1987).

Thamhain, Hans J. and David L. Wilemon, "Building High-Performing Technical Project Teams," *IEEE Transactions on Engineering Management,* Volume 34, Number 3 (Aug 1987), pp. 130–137.

The, Lee, "MIS Buys into Mail Order," *Datamation,* Volume 37, Issue 25 (Dec 15, 1991), pp. 27–29.

Walwyn, Mel, "Mac Makes Headway but Still Needs Push into LAN Markert," *Computing Canada,* Supplement (Sept 1991), pp. 10, 14.

Winston, Alan, "Full-Function Tools Trade Off Flexibility," *Software Magazine,* Volume 11, Issue 1 (Jan 1991), pp. 59–65.

Woodward, Charles P. and Linda M. Ramme, "Micro to Mainframe Multiproject Scheduling," *AACE Transactions* (1988), pp. G.10.1–G.10.6.

Yau, Chuk and Jimmy Liu, "A Generalized Software System Shell for Decision Support System Development," *Computers & Industrial Engineering,* Volume 16, Issue 2 (1989), pp. 313–319.

8

ENGINEERING'S JOINT RESPONSIBILITIES AND INTERFACES WITH MARKETING

8.1 THE NEED FOR INTEGRATED MANAGEMENT

A company or institution does not exist for the purpose of providing work for its engineering department. Sometimes engineers like to think that the company is in business so that they can develop new products and technological capabilities. Obviously, it is the other way around; engineering supports the overall mission of the company. This requires a great deal of market orientation and integrated, organized team efforts among all the technical, business, and marketing functions of a company.

There is often a great deal of tension and conflict at the boundaries of marketing and the various engineering functions. This tension is quite normal in traditionally structured companies, which have separate engineering and marketing departments, each headed by a more or less autonomous director, with specific and independent charters. But this tension can also be productive, as it provides some checks and balances and reminds all individuals of organizational realities.

Among the many overlapping areas, six specific activities are discussed in this chapter that require special cooperation and responsibility-sharing between the engineering and marketing organizations:

1. Forecasting
2. Technology development
3. New product and service development
4. Bid proposal efforts
5. Technical product selling
6. Field engineering and service.

Almost every business decision in a technology-oriented company needs critical inputs from both engineering and marketing. Disasters often happen when either engineering or marketing suffers from myopia and fails to give adequate attention to both the market realities and engineering capabilities. Examples of such mismatches include the Supersonic Transport (SST) and Hovercrafts, which were engineering marvels but commercial disasters. Other examples are instant movie film systems, which lacked customer acceptance, and the cars produced by the marketing myopia of the American automobile industry in the 1970s, which did not anticipate the demand for small cars although the technical ability to make small cars was clearly available to them.

The challenge faced by management today is to closely integrate engineering and marketing scenarios into planning and strategic decisions. Corporate goals and subsequent plans and decisions must be consistent with the capabilities of the organization and the realities of the market.

8.2 FORECASTING

The ability of an organization to conjointly forecast technological trends and market needs that are consistent with their own organizational strength and desires is crucial to the development of effective business strategy. Most engineering managers find this process difficult, as technologies and markets develop interactively. (Typical examples are personal computers and home video equipment.) But, because of the interactive nature between market and technology, it is important that a company's technological direction and leadership be based on a shared vision between the technologists and the markets. This will assure that technological developments are aimed at the right market needs and time frame. It will further assume that the long-range objectives and plans that evolve from these forecasts are consistent with organizational goals, charters, and global directions. There seems to be a continuous, often overlapping process of global goals, long-range forecasts, and operational planning. It is important that an organization not only generate quality forecasts, but also forecast the right type of data, consistent with and integrated with the organization's market, financial, and operational strength.

A number of tools and techniques are available to assist us in forecasting the various parameters needed in our plans. A selected number of forecasting techniques is summarized in Table 8.1. For most practical situations, a combination of several techniques is used to forecast. For example, the forecasting of a certain technological capability can be aided by a combination of qualitative and quantitative techniques such as S-curves, experience curves, scanning, and Delphi techniques. Then all the results are integrated in a scenario analysis. The fourteen techniques are briefly discussed below.

Time Series Analysis. This quantitative method tries to approximate a historic data series by a mathematical formula, such as a regression line, sine wave, or any other form that may be described as a mathematical function. Then the formula is

TABLE 8.1 Summary of Selected Forecasting Methods

Method Name and Description	Areas of Application
Time series: Analysis of a variable over time	• Short-range forecasting
Regression analysis: Analysis of two interdependent variables	• Forecasting • Trend analysis
Mathematical modeling: Describes dependencies among various variables mathematically	• Forecasting • Feasibility • What-if analysis
Simulation: Simulation of actual situations via computer, mechanical model, or group	• Forecasting • Feasibility • What-if analysis
Experience curve: Cost/price trends of a selected product or service over time in constant dollars across industry lines	• Long-range cost forecasting • Cost/product comparisons
Learning curve: Production cost of a specific product compared with cumulative production quantity for one organization	• Production cost forecasts • Production strategy • Market strategy
S curve: Graphic display of technological development (one variable) over time	• Technology forecasting • Comparison of technological advances for different approaches • Prediction of new technological advances
Forcefield analysis: Monitoring of related variables to determine trends of broad sets of variables such as needs, technology, social factors	• Forecasting • Driver (barrier assessment)
Scenario analysis: A situational analysis with specific scenarios. Integrates various techniques such as brainstorming and analytical methods	• What-if analysis • Strategy formulation • Strategy assessment • Situational assessment • Concept summaries
Delphi Method: Survey of expert opinions	• Forecasting • Feasibility assessment
Content analysis: Scanning and monitoring of relevant factors through media	• Trend analysis • Social or political forecasting • Technology prediction
Scanning techniques: Broad assessment of observations in the environment relevant to a situation under analysis	• Trend analysis • Broad predictions of changes and breakthroughs
Focus groups: Expert monitoring of a user group discussing applications and needs for a particular product or service	• Need assessment • Demand forecasting • Feasibility
Brainstorming: Situational analysis by experts in a group meeting, based on current perceptions	• Situational assessment • Strategy formulation

used to calculate future data points. A typical application for time series analysis is for short-range forecasting based on historic sales figures. Time series methods include: moving averages, exponential smoothing, nonlinear modeling, and autocorrelation methods such as Box–Jenkins. This technique is usually better for making short-range rather than long-range forecasts. Its accuracy depends largely on whether the factors that influenced the variable, such as sales, in the past, will also be present in the same magnitude and direction in the future.

Regression Analysis. Regression analysis tries to determine the mathematical relationship between the primary variable, such as sales, and one or several dependent variables, such as interest rate, relative price level, and product life stage. Once such a relationship has been established, the value of the primary variable can be calculated for the forecasting period if the dependent variables can be estimated for this period.

Mathematical Modeling. Mathematical modeling attempts to describe functional relations among a set of environmental variables. Formulating these input–output relations for any real-world situation in engineering or marketing usually is highly complex and its outcome depends on many assumptions and is associated with many uncertainties.

Simulation. Simulation refers to modeling a real-world situation with the aid of computers, mechanical models, or actual user groups.[1] Hence, simulation can range from mathematical models to focus group approaches. It is used for analyzing complex scenarios that cannot be described by simple mathematical relations. Special categories of simulation are artificial intelligence (AI) and expert systems (ES), which simulate the logic and decision-making process of expert personnel in particular situations.[2]

The Experience Curve. The experience curve graphically shows the price development of a given product or service over time. The price, given in constant dollars, usually decreases for almost every standard product and service as a result of technological advances and experiences across companies and industry lines worldwide. The experience curve is similar to a learning curve, but on a more global scale. As shown in Figure 8.1 for semiconductor memory, the price curve, plotted on a log–log scale, declines with time and cumulative production (experience). The trend for a technological product, such as shown in Figure 8.1, is a direct result of the technological advances that may be recorded with the S curve (see below); it does not depend on the experiences of one particular company, but on the industry as a whole.

The Learning Curve. The learning curve is similar in concept to the experience curve, except it measures the production cost of *one* product produced by *one* particular organization. The cost is expected to decrease with the cumulative produc-

[1]For some detailed application of computer simulation to business, including specific models, see Einsheff.
[2]A small sample of references are included in this book.

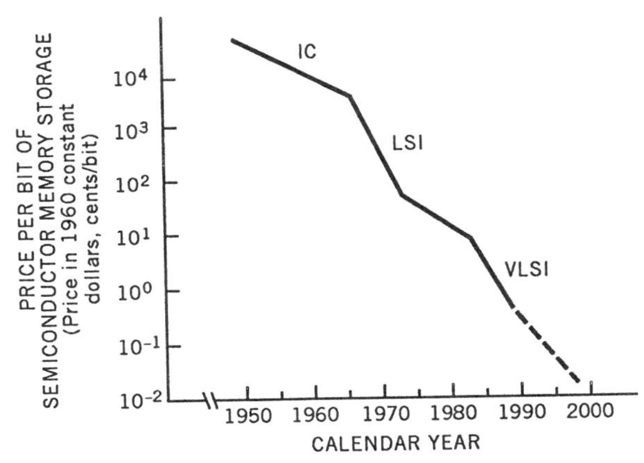

FIGURE 8.1 Experience curve showing on log–log scale price trends of semiconductor memory over the past 40 years: IC, integrated circuits; LSI, large-scale integrated circuits; VLSI, very large scale integrated circuits.

tion volume. The learning curve is a good tool for forecasting unit production cost once a cost basis has been established.

The S Curve. The S curve is a technique originally developed by McKinsey[3] to record technological progress over a period of time. It is designed to predict future trends. To be useful, the scales must be carefully chosen; various application-oriented technologies are then combined in one graph, as illustrated in Figure 8.2. The curve in Figure 8.2 depicts the development of the silicon-integrated circuit (IC) from the invention of the germanium transition. After a slow start, the technology went through major advances in the 1960s, resulting in a sharp increase in chip density (transistors/chip). Then the exponential growth leveled off until the new metal oxide semiconductor (MOS) technology (Curve B) enabled the fabrication of even denser chips. By 1980, integrated circuit technology went through two additional advances involving several technologies, which resulted in large-scale integrated circuit (LSI) and very large-scale integrated circuit (VLSI) capabilities (Curve C). With today's capabilities of packing over one million transistors on one chip, one might wonder where the capabilities are heading and what the technological limitations are. Not to worry. The S curve will take care of it, at least for a while. What is happening is that new technologies with new capabilities will be "pulled" by applications. The reason that LSI exploded over the last 20 years by an exponential factor of 10^6 is not because we went through a particular calendar period, but because we had continued needs for higher levels of integration within our electronics industry and ultimately within our societies. This confirmed need drove the past developments and will continue to drive the future advances. Thus, in 1970 one could have predicted the LSI density for

[3]The S curve concept, developed by the McKinsey Consulting Company, quantifies technology-based progress by showing product performance over time.

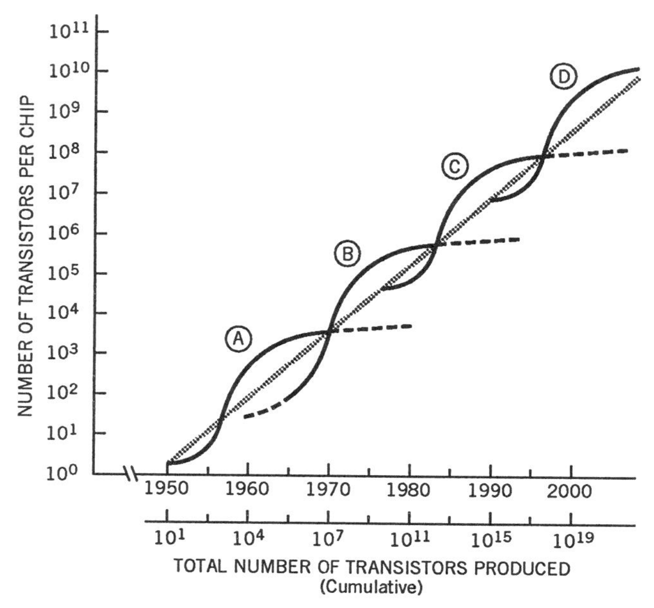

FIGURE 8.2 S curve of solid state/LSI technology, 1950–2000.

1980 or 1990, using the global trend of the S curves, regardless of the specific emerging technologies that made it actually possible, if one made the assumption that the demand for LSI was going to continue exponentially. Therefore, the same predictions are possible for future years, given the same assumptions.

Forcefield Analysis. "Forcefield Analysis" refers to a process of examining countervailing forces to predict primary trends in technology, market needs, or competitive behavior.[4] As an example, the forcefield process can be used to determine cost–performance trends of microprocessors by identifying and analyzing the major drivers and barriers to more advanced processor technology and production methods. Factors may include future trends in VLSI and factory automation. This very common concept of forcefield analysis is similar to the Japanese Ishikawa Diagram Technique.

Scenario Analysis. Scenario analysis, one of the most common and effective methods of forecasting markets and technologies, involves a broad analysis of the current situation and its future in the light of specific scenarios. These scenarios could focus on demographics, technology, or cost trends. Or it could include a "what-if" analysis assessing the impact of anticipated happenings on given strategies or established operations. One example of a scenario analysis is shown below in Table 8.2,

[4]The formal study of these forces, known as forcefield analysis is based on the pioneering research by Kurt Lewin.

TABLE 8.2 Scenario Analysis Example for Assessing Market for a New Low-Cost Personal Computer (PC)

Features Needed and Wanted	Discriminating Success Factors	Ghost Factors
	Scenario: Future Market Needs (+5 years.)	
Lightweight, true portable, under 6 pounds, low cost	Performance comparable to HP 7000 and IBM 6000, weight under 5 pounds	Unreliable operation and service
Application compatible with popular laptops	8 hours battery operation, rechargeable	Etc.
Features and performance factors . . .	Price under $800	Etc.
Etc.	Etc.	Etc.
	Scenario: Competitor's Response (+5 yrs.)	
COMPETITOR A		
Lightweight, true portable	Performance compatible to xyz at under 10 pounds	No standard o/s
COMPETITOR B		
Lightweight, true portable	Networking capability	Disk drive
COMPETITOR C		
Etc.	Etc.	Limited display capability
Etc.	Etc.	Etc.
	Scenario: Our Strategy (+5 yrs.)	
Lightweight, true portable, 10-hour battery, rechargeable	Under 5 pounds, open architecture	No brand recognition
40 MIPS, 4 MB memory, Graphical user interface	Under $800.00, color monitor, LAN capability	No service network
Fully xyz compatible	Etc.	Etc.
Etc.	Etc.	Etc.

which has the objective of forecasting demand volume for a new low-cost portable personal computer in an established market.

The results of the scenario analysis are often derived by a variety of methods, which can include systematic analytical approaches as well as more open, intuitive methods such as brainstorming. The crucial point is the delineation of the specific scenarios, which then become focal points of the analysis.

Delphi Method. The Delphi Method is a survey of expert opinions. It is similar to brainstorming, but the expert opinions are collected independently without any discussion or feedback from within the Delphi group. The process is particularly effective if a simple measure of a complex scenario is to be obtained; for example, the answer to "What random-access memory capacity will be attainable on a single chip within the next five years?" The sample data are then statistically evaluated. In their simplest form, the statistical averages provide measures of the forecast and the variances give measures of the sample's quality or confidence level.

Content Analysis. By monitoring media content, such as trade journals, we can try to predict trends in technology, markets, and social factors. The content analysis concept, which was made popular by John Naisbitt,[5] relies on the fact that more popular, trendy, or state-of-the-art news will fill more media space. Therefore, a collective analysis of media sources may permit the forecasting of certain long- and or short-range developments in a manner similar to the Delphi approach.

Scanning. Scanning refers to an open-minded observation of developments in the current environment. Through deductive reasoning, many broad-based trends can be forecasted.

Focus Groups. A focus group is a group of field experts who typically are users of a particular product or service, discussing the pros, cons, future applications, and potential improvements of that product or service.[6] The focus group is monitored by specialists who try to determine from the group discussion and the actual use of a particular product or service what improvements are needed, and what future applications, limitations, and markets are. The focus group concept is used most extensively for predicting markets, applications, and improvement needs; however, it can also be used for technological forecasting, design reviews, feasibility assessments, and value analysis.

Brainstorming. Brainstorming relies on collective thinking, discussion, and reasoning. the quality of its output depends strongly on the expertise of the participants on the subject area, as well as the group's interpersonal compatibility, which is conducive to such a process. While the process usually lacks analytical detail and

[5] *Megatrends* by John Naisbitt, New York: Warner Books, 1982.

[6] The basic concepts, analytical models, and processes of focus group discussions are described by B. J. Calder, "Focus Groups and the Nature of Quantitative Market Research," *Journal of Marketing Research*, August 1977, pp. 353–364.

quantitative precision, it can integrate a wide spectrum of variables and inputs and analyze decisions that are too complex for other forms of analysis and forecasts.

The Importance of Joint Reviews

Forecasts are organizational team efforts. Because of the enormous complexities and intricacies of the real-world system we are dealing with, forecasts should not be undertaken by a single person or single organizational unit, such as marketing or engineering. No matter how competent a particular functional group is and how appropriate the forecasting technique, the generation of the models, inputs, and procedures should be conducted jointly, involving all parties that can contribute expert knowledge to the scenario to be analyzed.

Regardless of the type of model or method used, systematic reviews of the forecasts and their underlying assumptions are crucial to a quality forecast and subsequent business strategy. Senior personnel from all major functions, such as engineering, manufacturing, marketing, field services, product assurance, finance, and the legal department should participate and contribute to the formulation of forecasts and their subsequent reviews.

Enhancing Multifunctional Cooperation and Performance in Forecasting

Forecasting in engineering requires the integrated assessment of many variables from the various environmental subsystems such as technology, markets, and the economy. It also requires a cross-functional integration of these variables in order to assemble a meaningful picture that is relevant to a particular company. A number of suggestions, derived from empirical studies of engineering organizations, can potentially increase multifunctional cooperation during forecasting efforts and ultimately increase engineering management performance. They are briefly discussed below.

Use a Systematic Process. Forecasting in the business environment requires a disciplined process regarding the type of variables, global areas scanned, and data collection, evaluation, and integration. Some procedural and technical formality is necessary to assure meaningful and useful results.

Broad-based Inputs and Flexibility. Within a disciplined framework of forecasting, managers must assure broad-spectrum monitoring of the environment from various viewpoints, such as technology user, competitor, looking at internal strengths, demographics, etc. Open and unconventional thinking should be encouraged and multifunctional involvement should be assured. People from many organizational areas contribute and look at different sources for data and scan the environment with different perspectives. This requires active involvement of key personnel from all areas of the organization, a process that must be facilitated and orchestrated by top management.

Use Many Data Sources. Many data channels and sources exist for accessing the outside world. A significant amount of data can be obtained by expert observations, which can be integrated via Delphi and brainstorming approaches. However, other sources of data are often needed to make proper forecasts, such as market studies, technical conferences, and advice of consultants requiring costly and time-consuming efforts. Management must make long-term decisions and resource commitments to establish an appropriate capability for environmental monitoring and for forecasting on an ongoing basis.

Link Forecasts. The predictions made for the various business components, the economy, market, and technology, must be linked together and presented as an integrated forecast.

Assess Measures and Trends. Environmental trends can often be predicted more accurately than absolute measures. Further, trends are a more integrated measure, which normalize many peripheral effects such as inflation and cycles. Trend analysis can also help in identifying the onset of radical changes in the market, technology, or other domains.

Use a Combination of Techniques. No single technique will produce meaningful forecasts. Methods should be used in combination. When several techniques produce data that point in the same direction, the confidence level in the production increases. The techniques used should include a balance of quantitative and qualitative approaches.

Process Qualitative Inputs. Many of the most valuable, highly integrated variables come through qualitative channels, for example, expert opinions. Using a systematic approach, such as Delphi, brainstorming, or scenario analysis for processing this information can facilitate the integration of the data across various channels and greatly enhance the quality and confidence level of the forecast.

Have Systematic Reviews. Management must assume that forecasts are not developed in an isolated environment. Forecasting should be an ongoing process. Systematic, regular reviews, requiring the active participation of all relevant functions, help to integrate the multidisciplinary inputs and assure the forecast's relevancy to the specific company situation. These reviews also help to build coalitions among the many organizations that are eventually required to work together in order to produce results consistent with the forecast.

8.3 TECHNOLOGICAL DEVELOPMENT

Technological development is the lifeblood of many companies. However, activity itself does not guarantee business success. Anticipating technological needs consistent with future market needs and internal business capabilities is truly a challenge, because technologies evolve within a system of mutually supporting developments and evolve interactively with their market needs. As an example, microprocessor technology, which led to personal computers, automatic cruise control, and hundreds of daily utensils, would not have been possible without the enormous advances in semiconductor/LSI technology and its supporting systems in the areas of chemistry, mechanics, optics, and materiel. Thus technological developments require intense interaction among all major functions of an organization, especially between engineering, R&D, and marketing. This interaction is required for defining corporate technology strategies and plans, as well as for developing new technologies and transferring them into production and the market. This shared vision assures that market sensitivity and production capability are properly considered by research and development, and that engineering personnel are most creative about the right opportunities.

Strategy Formulation

Technological strategy cannot be developed in isolation. We must consider the three components: the organizational desires, capabilities, and market needs. Technological strategy formulation must consider all phases of the "value chain" of an organization, a concept made popular by Michael Porter in 1985 and summarized in Table 8.3.[7] The reason is that change in technology affects the value added in every

TABLE 8.3 A General Approach to Formulating Technology Strategy as Suggested by Porter*

Formulating technology strategy must involve a broad assessment of the company's needs and capabilities relevant to global technology trends. The Strategy formulation requires an integrated set of management activities, as shown below:

1. Identify all technology categories in the value chain of your company.
2. Identify relevant technologies and their trends worldwide.
3. Determine which technologies and their potential changes are most significant to your business.
4. Assess your company's capabilities in the technology areas relevant to your business.
5. Assess the resource requirements, costs, risks, and alternatives for making improvements in these relevant technology areas.
6. Develop an integrated technology strategy that reinforces the company's overall competitive position.
7. Apply this technology strategy consistently throughout the company.

*Summarized from Michael E. Porter, *Competitive Advantage,* New York: Free Press, 1985.

[7]Michael Porter, *Competitive Advantage,* New York: Free Press, 1985.

operation of an organization. Technological strategy formulation should therefore be an integrated part of overall strategic business planning activities. Frederick Betz poses three fundamental questions that should become the focus of any technology plan and should link business and technology strategy:[8]

1. In the future, what business should the firm engage in?
2. How should the firm be positioned in these businesses?
3. What research, production and marketing will be necessary to attain this position?

The scenario analysis technique can provide a simple and effective format for analyzing the impact of technological change on a company's business, including threats to the business and opportunities for it. Among the various researchers in this area, Donald Pyke proposed a mapping technique, in which technological change can be systematically analyzed in three hierarchically organized categories: Level 1, environmental scanning and analysis; Level 2, technological capability analysis; and level 3, product line analysis.[9]

The birth of the personal computer and microprocessor industry in the 1970s is a typical example of organizations predicting environmental and technological changes, such as the need for more effective service, automation, and household conveniences; sensing business opportunities; and incorporating them into their business strategies and operating plans.

To accomplish an effective technology plan, companies must take a systematic approach, as well as an approach that involves all organizational components of the value-added chain. At the minimum, the approach must involve key personnel from R&D, engineering, production, field services, and marketing. Often, special multi-functional task groups, such as suggested by Bitondo and Frohman, can be useful in brainstorming, and formulating or reviewing technological strategy.[10] It is unfortunate that even today many companies use the budgeting process as the primary vehicle for technology planning. Logic, desire, commitment on the part of the individual manager and the company as a whole to certain goals, and an understanding of capabilities and resource requirements must come first, before the budgeting process can be conducted in a meaningful way.

While strategy formulation is a long-range process, it is also continuous. The established long-range strategy should be updated periodically and adjusted to changing conditions and assumptions. Key personnel from the contributing functions should review the established technology plan and make the necessary adjustments. This strategy review can often be combined with a tactical review of schedules, results, resource acquisitions, and budgets of the ongoing technology program.

[8]Frederick Betz, *Managing Technology,* New Jersey: Prentice-Hall, 1987.

[9]Donald Pyke, "Mapping—A System Concept for Displaying Alternatives," in J. Bright and M. Schoeman, *A Guide to Practical Technological Forecasting,* New Jersey: Prentice-Hall, 1973.

[10]Dominic Bitondo and Alan Frohman, "Linking Technological and Business Planning," *Research Management,* Vol. 14, No. 6, 1981, p. 19.

Enhancing Multifunctional Cooperation and Performance in Technological Development

The implementation of technology strategy as an integrated part of the company's business strategy involves techniques of effectively leading and integrating the multidisciplinary activities. The process of creating desired technological systems usually begins with the innovation of a new device or system, or even an invention, which marks the start of a new S curve. The progress and ultimate success of a new technological system depends on many factors, mostly related to the supporting infrastructure, standards, applications, user psychology, and ultimate market demands. A long time may elapse between the early success of a new technology development and its commercialization. A number of suggestions, derived from empirical studies of engineering organizations, can potentially increase multifunctional cooperation during the technological development phase and ultimately increase engineering management performance. They are listed below.

Plan Ahead. An agreed-on plan for technological development establishes the roadmap for this complex, multidisciplinary effort. Clear goals, objectives, and endorsement by management help to unify the multifunctional team and to enhance collaboration.

Have Regular Reviews. In a changing environment, regular reviews, involving personnel from all contributing areas, are necessary to fine-tune the efforts and to refuel the commitment to collaborate and to assist in the ultimate transfer of the new technology.

Use a Phased Approach. Break the technology development program into its natural phases. Make people from all phases part of the team. Joint reviews, recognition of mutual accomplishments, overall program visibility, and active management involvement will help to unify the team and gain cross-functional cooperation.

Have Management Involvement and Visibility. Active involvement of senior management in the planning and subsequent execution of technological development is one of the best catalysts for organizational collaboration and downstream participation. It assures personnel that the effort is important to the company, that resources are allocated, and that success will be rewarded.

Designate Transfer Agents. The success of a technology strategy depends on the company's ability to transfer and integrate the technology into the market. Organizational interfaces and key individuals responsible for successful technology transfer should be identified early in the development cycle and committed to it.

Use No Threats. Management must foster an atmosphere of mutual trust and a fail-safe environment, so that collaboration may occur. Unrealistic demands, job-security problems, risks of failure, and power struggles drive people to parochial and protective efforts, which are a major barrier toward cooperation.

8.4 NEW PRODUCT AND SERVICE DEVELOPMENT

The development of a new product or service usually requires very intense involvement of four functions: engineering; production or operations; marketing; and field services. While conventional wisdom tells us that each function must optimize the value of the new product or service based on its specialty, with minimal interference from others, modern management practices suggest integrated cooperation among all departments throughout the project development life cycle.

For example, let us look at a new-product feasibility analysis early in the product development cycle. We need market definition, demand forecasts, functional specifications (specs), feature catalogs, production cost estimates, etc. Who should do what first? This is not a simple question, since all components of the feasibility analysis are interrelated. What is needed is an interfunctional team effort, where each organization has its prime responsibility (such as marketing has for the market analysis), but cross-functional team efforts are used for brainstorming, conceptualizing, and reviewing.

Similar cross-functional team efforts are necessary to design products that can be economically produced, can be effectively serviced in the field, and meet quality and reliability standards.[11] As shown in Figure 8.3, these efforts increase with decreasing unit production cost. Low-cost products typically required for competitive high-volume production demand large and costly up-front design and manufacturing engineering efforts and capital equipment.

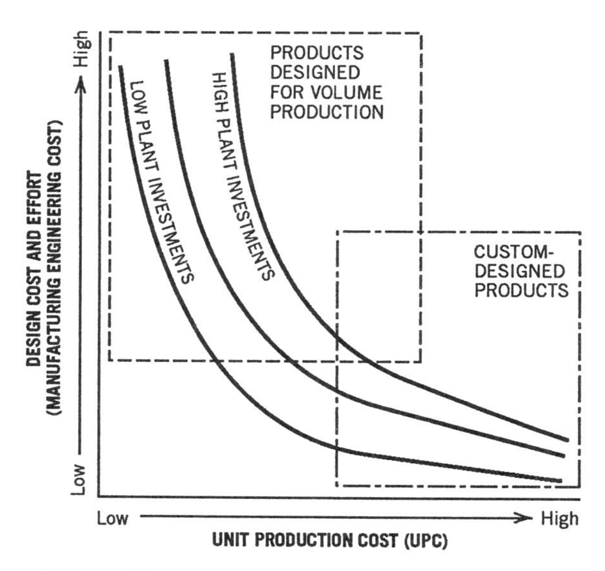

FIGURE 8.3 The high investment needed for low-cost production.

[11]Design to unit production cost (DTUPC) procedures are formal procedures that involve cross-functional team efforts, especially between design engineering and production engineering.

The phased approach to product development, as discussed in Chapters 4 and 5, is a simple and effective method for identifying critical interfaces and facilitating cross-functional cooperation. The procedure requires that the product development activities are broken down into discrete phases, such as concept development, engineering design, prototyping, testing, manufacturing, etc. For each phase, the key deliverable items including documentation are defined, together with the key individuals responsible for moving the project from one phase into the next. Often these individuals are requested to sign off on the completion and acceptability of the work. The awareness of mutual dependence leads to early and continuous cross-functional involvement and cooperation among the people who are ultimately responsible for making the project successful. Table 8.4 summarizes the key task responsibilities that come into play during a typical product design life cycle and lists the management tools that help to facilitate interdisciplinary task interfacing in the area of new product development. Some additional discussion regarding field engineering participation in new product developments is presented later in this chapter under the "Field Engineering" section.

Enhancing Multifunctional Cooperation and Performance in New Product Development

A number of suggestions, derived from empirical studies of engineering organizations, can potentially increase multifunctional cooperation during new product development efforts and ultimately increase management performance. The same suggestions made for enhancing technological development also apply here: *Plan ahead, have regular reviews, use a phased approach, have management involvement, designate a transfer agent,* and *use no threats.*

In addition, eight guidelines were provided by William E. Souder,[12] summarized here, to help alleviate interface problems and to enhance multifunctional cooperation; they are summarized below:

1. *Break Large Projects into Smaller Ones.* Souder found that task teams in excess of eight people experience interface problems. Therefore it is suggested to break new product development projects into small task groups to assure increased face-to-face contact and personal involvement with support departments, which leads to better cooperation and multifunctional commitment.
2. *Take a Proactive Stance Toward Interface Problems.* Team leaders should identify potential interface problems and openly discuss them and seek resolutions.
3. *Eliminate Small Problems Early.* Project leaders must stay close enough to the work and the people to identify problems during their early stages. Conflict, distrust, anxieties, and priority problems all start small and are more manageable than their fully developed versions.

[12]William E. Souder, "Managing Relations Between R&D and Marketing in New Product Development Projects," *Journal of Product Innovation Management,* Vol. 5, No. 1, March 1988, pp. 6–19.

TABLE 8.4 Key Responsibilities Throughout the Product Design Life Cycle

Task Item or Phase*	R&D	Engineering	Manufacturing	Marketing
Feasibility analysis	Technology impact, technology forecast, technology plan, concept paper	Concept, features, system specs, risk analysis, system justification	Produceability study	Market analysis (demand, competition, price-performance), product plan
Product or project planning	R&D product plan	System definition, functional specs, Make-buy plan, Documentation Plan, Reviews and approvals	Produceability plan	Product business plan
Design and development	R&D technology review, special design	System design, detail design and development, prototype, test and integrate, diagnostics plan, alpha site test, value engineering	Produceability plan (detail), manufacturing plan (quality plan, material plan, test plan, assembly plan), design assurance engineering	Updated business plan, release schedule, promotional plan
Manufacturing	Advanced manufacturing techniques	Technology transfer liaison, engineering changes	Manufacturability reviews, tooling, material requirements planning (MRP), vendor selection, manufacturing process design, pilot built, volume production	Product strategy, updated market forecast and plan, cost reviews
Marketing	Liaison	Product support plan, reliability and quality review, performance verification	Manufacturing cost update, quality improvement program	Distribution channels, sales literature, sales training, promotion and market development, selling
Field service	Liaison	Beta site test reports, product support plan, product enhancement	Review quality assurance audits, fine-tune production process, product enhancements	Liaison, bid proposals

*Phases and their corresponding responsibilities can overlap during the product life cycle.

Field Services	Product Assurance	Finance	Project Management	Primary Interfaces
Operations and maintenance concept, serviceability analysis	Preliminary quality plan	Preliminary cost estimates and schedules, target budgets, pro forma cash flow analysis, ROI analysis	Product/project objectives, feasibility coordination, preliminary project plan	R&D engineering, manufacturing, marketing
Field service plan	Quality plan, product safety plan, qualified vendor list	Risk analysis, capital investment plan, financial reviews	Project plan, project organization, management approvals	Engineering, manufacturing, marketing, field services, product assurance finance
Warranty plan, installation plan, serviceability reviews	Design assurance, engineering plan, design reviews	Budget reviews, cost tracking	Project coordination, interface to top management, customer, sponsor, and vendors	Engineering, manufacturing, marketing, product assurance
Product support plan, warranty and installation cost, field support documentation	Pilot run testing, statistical process control (SPC), vendor sourcing approval	Product cost update, capital acquisition request reviews, financial analysis of product	Product transfer coordination	Engineering, manufacturing, product assurance
Reliability and repairability statistics, field service readiness training	Design assurance engineering certification, product ship authorization	Final product cost analysis, pricing review	Final documentation (coordination), final report, transfer of project organization	Engineering, marketing, field services, product assurance
Field performance report	Quality performance monitoring and reporting	Liaison	Liaison	Engineering, marketing, field services, product assurance

4. *Involve All Parties Early in the Project Life Cycle.* R&D, engineering, manufacturing, marketing, and field services in particular should be joint participants in planning and decision-making of the new product development early in its life cycle.
5. *Promote and Maintain Dyadic Relationships.* Dyads are fostered any time people with complementary skills are assigned to work together and given significant autonomy.

   These powerful interpersonal alliances can become the kernel of larger circles of responsibility such as between R&D and engineering or marketing. For example, a successful dyad formed by one R&D and one marketing person will draw other R&D and marketing people into it and will grow into an R&D–marketing network.
6. *Assure Open Communications.* Open communications should be the explicit responsibility of every team member. It also must be encouraged and facilitated by management through their examples, policies, and actions.
7. *Use Interlocking Task Forces.* Create a multifunctional, top-level steering committee consisting of key personnel from all major functions that will be involved during the new product development life cycle, such as R&D, engineering, production, marketing, finance, field services, product assurance, and legal departments.
8. *Clarify the Decision-Making Authorities.* Each department participating in the new product development should have a clearly defined charter, which clarifies its specific responsibilities, decision-making powers, controls, and interfacing roles. These charters provide the foundation for managing a multifunctional product development project with a minimum of dysfunctional conflict and optimum interdepartmental cooperation.

8.5 BID PROPOSAL EFFORTS

Bid proposal efforts require one of the highest degrees of cooperation among the technical, marketing, and business functions of an organization. Starting with the identification and evaluation of a new business opportunity for the customer, proposal development and pricing must involve people from those functions that collectively represent the company in the most competent and credible way possible. This requires more than a random collection of talents from the appropriate functions; what is needed is a carefully orchestrated team of professionals who can present an integrated solution consistent with the company's business and marketing objectives, and who are responsive to the specific customer needs.

An extensive treatment of the bid proposal process is provided in the next chapter. Section 8.5 discusses the important aspects related to joint engineering responsibilities with other organizations. Table 8.5 summarizes some of the management tools available to facilitate cooperation and task integration during a bid proposal effort. The bid proposal process will be briefly discussed below in chronological order.

TABLE 8.5 Management Tools to Facilitate Task Interfacing During Bid-Proposal Activities

Co-location: Establishing the proposal team in one dedicated office area, to enhance communications and work integration

Customer Presentation: Presentation of concepts, ideas, and potential solutions to the customer in support of bid proposal. Often used during preproposal phase to clarify or define requirements

White Paper: A well-written report explaining a potential solution or capability to the customer. Can be submitted during preproposal phase or as a no-bid response, or as an alternative to a formal bid response

Incentive System: Rewards given to employees for identifying new business leads; recognition, visibility, performance reviews, bonus

Cross-Functional Committee: Ad-hoc committees of experts from selected functional areas, who evaluate or audit new bid opportunities, strategies, or proposals

Acquisition Plan: Formal document summarizing the bid opportunity and win strategy, including resource requirements, schedules, and competition

Management Review: Formal review of a bid opportunity as a basis for a preliminary or final bid decision, submission, or pricing

Red Team: Impartial expert team, who evaluate a plan or proposal and suggest improvements

Proposal Support Group: Permanent functional resource group specializing in bid proposal developments

Proposal Specialist: Person specialized in proposal developments; consults to proposal team or leads the proposal effort

Storyboarding: Technique to facilitate group writing via incremental text development and reviews, using office wall for display of text and illustrations in development

Incremental Text Development: Text and illustrations are developed in steps: outline, synopsis, storyboard, first draft, second draft, final proposal

Joint Presentations: Joint presentations to user community by various organizations such as marketing, engineering, manufacturing, and often the customer

Opportunity Identification and Evaluation

While the initial lead to a new business opportunity can come from any source, it takes multifunctional resources to analyze and evaluate the opportunity regarding its realistic business potential. Simple forms and collection methods, such as via superiors, together with some *incentive system,* can enhance the frequency and quality of new business leads. It also may enhance the variety of bid sources and customer contacts. Screening and evaluating of these leads is the first truly multidisciplinary effort. It should be conducted by a cross-functional committee or task force on a regular basis. This committee should have senior representatives from all major functions who, collectively, can assess the value of a new opportunity quickly and economically. Typically such a committee consists of representatives from engineering, manufacturing, field services, and of course, marketing. Members from other functions such as legal, finance, and quality may have to be consulted as needed.

Bid Decision

The decision to pursue a specific business opportunity actively is a resource commitment, which requires careful analysis in order to answer the following questions: (1) Do we have the capability and resources to perform according to customer requirements? (2) Can we develop a winning proposal in time? (3) Can we win? (4) Is it desirable from an overall business and financial perspective? Especially for the larger bid decisions, an acquisition plan should be carefully prepared by the organization ultimately responsible for the contract performance, but with inputs from all functions needed to support the contract eventually. This plan becomes the basis for the final management reviews and bid decision. Ideally, the members of the cross-functional committee, who help develop the acquisition plan, have also been selected and are committed to key responsibilities on the proposal development and contract project, if the opportunity matures to these phases. While some managers are concerned about losing objectivity by having evaluators no longer impartial to the bid decision, we find the benefits derived from the continuity, desire to succeed, commitment to developing a winning proposal, and the ownership of the contract to be far greater than the potential problems of bias toward bidding. In any case, the final judgment must be made by the responsible business manager, who must weigh the costs and risks of the undertaking, together with its impact on other operations, against the potential benefits and business plans. Sometimes an independent, impartial cross-functional team, called a *red team,* is formed to review the acquisition plan and to state an unbiased opinion.

Proposal Developments

Proposal developments of any size are highly multifunctional efforts that need the participation and cooperation of many organizational departments, such as R&D, engineering, manufacturing, product assurance, cost accounting, technical publication, field services, and marketing. Ideally, the proposal development is a continuation of the bid evaluation and acquisition planning effort. In addition, a special proposal support department can be a valuable asset in organizing, coordinating, and integrating the effort and in leading the proposal through the dense logistical jungle of producing a winning document from the customer request for proposals (RFP). Group writing techniques, such as storyboarding facilitate multidisciplinary task integration. The technique, described in Chapter 10, relies on incremental text development. Each author starts with an outline, then writes a synopsis, then a summary, and eventually the final text. Each module evolves incrementally from the previous text. For each phase of such a proposal development, all authors display their modules (storyboards) sequentially on a wall. In essence, each display represents a complete draft proposal. All members of the proposal team can review the entire proposal as it evolves, which is helpful in fine-tuning the text and illustrations with regard to content and interfaces. Yet another technique for facilitating integration employs cross-functional review teams, who carefully read the storyboards at each phase and give written and oral feedback to the writers on how they can strengthen the proposal or implement the win strategy more effectively. Similar to review teams previously discussed in Chapters 4 and 5, the members of cross-functional review teams are senior representatives of the various operations affected by the proposal. These review teams are often called by various colors, such as the "red team."

The integration of a bid proposal development is usually helped considerably by collocating the team in one location during the time of the proposal effort. It further helps if the senior management of the company gets involved and actively participates in the strategy development, storyboard reviews, and management discussions.

Customer Presentations and Negotiations

Many proposal efforts involve technical presentations to the prospective customer and most proposals involve negotiations before a contract is awarded. Both activities are multidisciplinary and require careful preparation to be effective. The same techniques used earlier in the proposal cycle to enhance integration, cooperation and use checks and balances can also be applied here with benefit.

Enhancing Multifunctional Cooperation and Performance in Bid Proposal Efforts

Taken together, an effective bid proposal is a well-integrated multidisciplinary document that is responsive to a specific customer need. The quality of the proposal seems to be directly correlated to the involvement of top management, the priority image and visibility of the acquisition, and the enthusiasm, drive, and commitment of the people responsible for the proposal development.

Some specific suggestions are made below for increasing multifunctional cooperation during bid proposal efforts.

Start Early Involvement of All Functions. Preproposal efforts of requirements analysis, customer meetings, and internal capability assessment should involve personnel from those functions that will be responsible later on for the proposal development. This creates understanding and interest in the work itself; it develops confidence, and establishes communication channels across functional lines.

Have Multifunctional Bid Decision Participation. Senior managers from all functions that eventually contribute to the proposal development should actively participate in the bid decision process. This helps to break down functional boundaries and to unify the management team behind the multifunctional effort.

Use Effective Planning. Bid proposal efforts involve many organizational disciplines. Proper planning early in the proposal or preproposal stage should involve the key contributors from each supporting function, including subcontractors. These planning activities help to unify the proposal team by solidifying common objectives, establishing communication channels, and by fostering a work environment of common interests and commitment.

Collocate Proposal Team. The most effective way of creating multifunctional cooperation at the team level is by collocating the entire proposal team in one area. Although this approach often meets resistance from team members and requires a higher resource commitment, it is often the best way to unify the team for a highly concentrated project such as a bid proposal effort.

Get Top Management Support. Active involvement and support of senior management is crucial to fostering interfunctional cooperation. It sends a clear message to all parties that the bid effort is important and desirable for the company, holding future rewards to those who participate and cooperate.

8.6 SELLING TECHNOLOGY-DRIVEN PRODUCTS

Differences Between Selling Technology-Driven Products and Others

"We sell solutions, not products," a statement coined by Buck Rogers of IBM, describes clearly the challenges of selling technology-based products. Winning a sales contract requires more than reading a spec sheet to a customer. It requires significant homework in analyzing the customer requirements and defining the system that is best suited for solving the specific customer problem at hand. In contrast to more basic products, selling of high-technology products and services is more challenging because their proper application and performance depend largely on the detailed technical understanding of the product, available alternatives, and knowing how to use it skillfully. Whether we look at word processors, medical electronics, or laser tools, the customer often does not understand the intricacies of the product or its workings; nor does he or she want to learn them. All the customer is usually interested in is solving a specific operational problem.

Bridging the gap between a technology-driven product and a problem/solution-driven customer community requires a technically highly skilled and applications-oriented team, who can suggest simple and elegant solutions to complex operating problems. Moreover this need for multifunctional subject experts increases with the complexity of the system to be purchased and its organizational impact, as shown in Figure 8.4. This requires often strong engineering–marketing interaction, involving joint responsibilities in five primary areas: requirements analysis, building credibility to perform, proposal development, pricing, and contract negotiations.

Traditional approaches to product marketing and selling assume that we can identify the customer group (market segment) well in advance of going to market. They also assume that the product or service can be clearly defined in the promotional literature, so that the customer can easily evaluate the suitability of the product against his or her needs. The traditional sales function focuses primarily on the

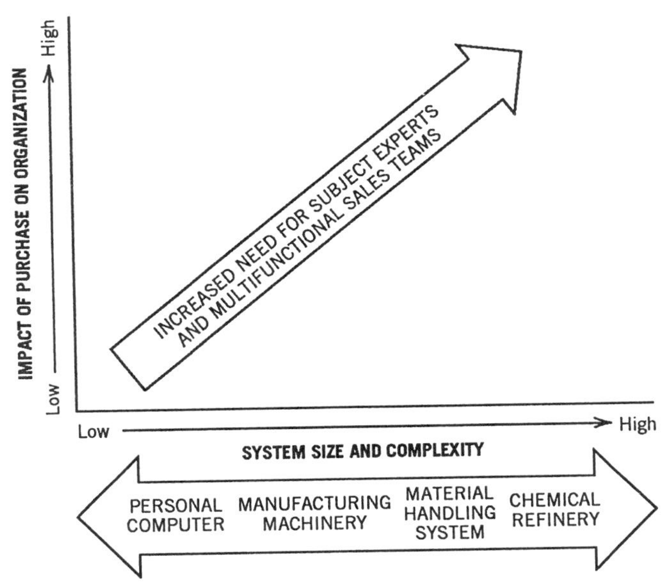

FIGURE 8.4 The need for subject experts increases with system size and organizational impact of the system to be purchased.

interpretation of promotional and contractual literature and the facilitation of order processing and delivery. Usually a minimal amount of technical understanding of the product or customer application is needed by the sales force in the traditional approach. This is very different, however, for technology-driven product selling, which often requires the sales person to work directly with the customer in identifying solutions to the problem. The requirements analysis and specific proposal development often is too intricate and multidisciplinary for a single seller–buyer format, but requires a team approach of highly specialized professionals from both the selling organization and the customer community. A typical technology-driven sales situation is characterized as follows:

- Focus on customer problem, impact, and constraints;
- Presence of subject experts on both the customer and seller side;
- Solutions based on existing time and budget constraints;
- The seller has credibility with the customer and there is mutual trust;
- A search for alternative solutions is necessary;
- There is a focus on product features relevant to application;
- Buying motives are largely rational and purchasing decisions are strongly influenced by technical needs, specifications, costs, and schedules;
- There is concern for broad product liabilities;
- There is concern for after-sales support;

- Most products are perceived as commodities;
- There is long-term commitment to the product and customer.

Specific Joint Responsibilities

Meeting the above challenges requires skills and support from many organizations outside the sales/marketing group, which are summarized in Table 8.6. Cross-functional collaboration ranges from brief counseling to in-depth, funded participation such as a bid proposal development or negotiation. Engineering participation is necessary for effective role performance of the technical sales function and should be planned, encouraged and rewarded by senior management. In fact, it is suggested that this joint responsibility be formally acknowledged as part of the charter or mission statement of the various operating functions, be supported via the budgeting process. Table 8.7 lists several forms of cross-functional technology-driven sales support, which are briefly discussed below.

Needs Assessment. Especially for larger systems or multiple components, the sales professional has to work together with the customer in assessing the specific customer needs and requirements. This often requires the support of technical spe-

TABLE 8.6 Multifunctional Knowledge and Skills Needed for Driven Technology-Sales Support

Effective role performance of the technical sales function requires:

UNDERSTANDING OF:

- Product and its application
- Technology involved
- Customer's environment
- Customer's technical and business needs
- Customer's constraints and risk factors
- Product alternatives and competition

SKILL CATEGORIES NEEDED

- Analytical skills
- Technical skills
- Financial skills
- Communication skills
- Sales and negotiation skills
- Planning and organizing skills
- Multifunctional integration skills
- Conflict-resolution skills

TABLE 8.7 Technology-Driven Sales Support, by Category

Support Category and Its Mission	Primary Knowledge and Skills Necessary
Needs Assessment: Determine specific customer needs and requirements	Product, technology, customer application, market and competition
Technical Liaison: Maintain and build customer relations, assist customer, position for future sales	Customer environment, product application, technology/product trends
White Paper: Get potential customer interested in product or service	Customer environment, customer needs, product application
Journal Article: Advertise company/product capabilities, build credibility	Product application
Promotional Literature: Advertise specific product or service capability	Product application, market knowledge
Trade Shows: Advertise product or service capability, contact potential customers, learn customer needs	Product knowledge, communication skills
Customer Meetings and Entertainment: Learn customer needs, show potential solutions, build "can-do" image	Product knowledge, communication skills
Bid Proposals: Communicate specific solution to a customer's requirements	Customer requirements, product knowledge, customer and market, communication skills
Negotiations: Contractual agreement	Negotiation skills, communication skills, product knowledge

cialists from the R&D and engineering functions. To be effective, these technical support people must be well briefed on the basic customer problems and constraints. They also must be knowledgeable about the product, its functions, applications, and competing alternatives, and be able to communicate well with a nontechnical audience.

Technical Liaison. This is a common form of maintaining customer relations used by the sales department. It involves regular visits to existing customers, providing information on current product upgrades and developments as well as customer assistance on how to use products more effectively or replace them with new ones. Sales personnel often must team up with technical experts to consult effectively on a customer problem.

White Paper. In many situations, the customer is not accessible for informal discussion, counseling, or missionary selling, but operates behind closed doors, carefully selecting the contacts and channels for solicitation. A well-written report or "white paper" may get the customer interested in a particular product or service and open the door for further sales dialogue. A white paper usually states and diagnoses the customer problem, as perceived, and proposes an attractive potential solution. Such a paper is usually a joint effort between the marketing and technical people and presents a crafty blend of technical, economical, and business considerations.

Journal Article. Similar to a white paper, a journal article can help in communicating product capabilities and possible applications to prospective customers. If written for promotional purposes, the article should be jointly developed by technical, marketing, and sales people, hence being a true multifunctional joint effort.

Promotional Literature. Promotional literature, such as product bulletins, application notes, and space advertisements, requires joint functional efforts to be effective. Once developed, these promotional pieces can become modules in a variety of marketing efforts.

Trade Shows. Trade shows are yet another area that requires joint efforts between marketing and the technical operations. Although trade shows are usually organized by the marketing or sales department, technical personnel also should be present to interact with the customer community. Proper preparation for participation in trade shows requires substantial homework and multifunctional involvement.

Customer Meetings and Entertainment. These can be very effective promotional vehicles. With the right mix of people present in the right setting, many complex, multifunctional issues can be dealt with in an integrated, interactive manner, hence conducting a miniature needs-assessment, requirements analysis, and feasibility study in the very short timeframe of a customer meeting.

Bid Proposal Efforts. For selling larger systems or a large number of components, the bid proposal is an established vehicle for formal contract bidding. Bid proposals are projects all by themselves, requiring an intensive joint effort among many company functions and outside organizations, as discussed in detail in Section 8.5 and in the next chapter.

Negotiations. Negotiations of a technology-based contract often become very technical, requiring the joint participation of people from R&D, engineering, manufacturing, product assurance, field services, and other functions that eventually will contribute to the successful execution of the contract and its after-sale support.

8.7 FIELD ENGINEERING AND SERVICES

Cooperation among the various engineering and field operations is critical to the effective functioning of the overall company. Among the many interfaces, four specific ones and their activities are selected for discussion.

New Product Development

Total product life cycle considerations often require intense integration of field service and quality requirements with design and development efforts. Repairability, reliability, interoperability, and cost of ownership must be designed into the product. Specific participation of senior field personnel early in the design cycle can inject reality into the requirements analysis and conceptual design phases. Participation can range from committee and task force activities to detailed product-user needs assessments, which involve market studies, focus groups, and test marketing. In addition to the highly focused joint efforts at the beginning of the product design cycle, a continuous involvement of selected field engineering and service personnel can be beneficial. Examples of such involvement include: participation in design reviews; soliciting feedback on product concepts, specifications and models; assigning joint responsibility for serviceability; independent design verification and validation; and advanced test marketing.

Customer Problem-Solving

Many field engineering activities involve solving customers' problems. Aside from routine maintenance and repair, solutions to customer problems require detailed technical analysis and system engineering work, which must be supported by engineering specialists. This cooperation does not just happen; it must be planned for and endorsed by upper management, and the organization must be designed for power- and resource-sharing to facilitate such multifunctional cooperation.

Field Installation

The installation of a new system at the customer site, including its start-up, operator training, and initial maintenance procedures, is usually the responsibility of the field engineering organization. However, depending on the special nature of the new system, its engineering complexities and customization, engineering specialists from the original design team may need to get involved to assure smooth and proper operation of the new system at the customer site. Such a team effort must be carefully planned as part of the overall project to assure timely resource availability as well as proper integration and coordination of the many activities that come into play in a technology-based system installation.

Field Education and Training

Today's technology-based products and customized systems have created a strong need for specialized customer education and training. They range from specialized onsite operator training to more general skill development, such as the need of bank tellers for computerized information processing, and to broad-based public seminars for gaining knowledge and skills in a particular product category, such as minicomputers or software packages. The responsibility for coordinating and delivering such training programs rests with the field engineering department; however, the highly specialized technical nature of the training and specialized customer environment often require hands-on participation of engineering and marketing specialists. This joint effort often includes the needs and requirements analysis, course design, material development, and presentation. Field education and training should be a critical component in any product and project plan. It is often a deliverable end item that significantly determines the overall project success. In addition, training can be an important source of revenue as well as a promotional platform for new business.

Enhancing Multifunctional Support for Field Engineering

Like other operations of a company, field engineering is an autonomous department. It is often chartered as a profit center. Interdepartmental cooperation does not happen easily with such an organizational structure, but needs to be fostered by senior management through their interest, actions, and policies. It further requires an understanding by the management of all interfacing groups that such interfunctional cooperation is necessary and mutually beneficial. The incentives for cooperation must be supported by the budgeting process. The ability to allocate or withhold funds for support, and the personal rewards of the people for their performance, are powerful drivers toward cooperation, which also legitimize other nonfinancial incentives, such as recognition, visibility, priority, and professional interest. These multifunctional efforts must be jointly planned and agreed to, as well as jointly reviewed and managed, like any other project. In addition, fostering a cooperative organizational culture involves more than just structural issues, but encompasses some of the most complex organizational and leadership components, as is discussed in the next section.

8.8 FOSTERING A COOPERATIVE ORGANIZATIONAL CULTURE

The effective role performance of an engineering manager often requires efforts that reach far beyond idea generation, concept development, and prototyping. It usually requires linking these efforts to the market via the manufacturing department and other transfer organizations. To be successful, the engineering organization must lower the strong impedance across organizational boundaries. This impedance is very natural and predictable because of the different organization objectives, cultures and values; it is also very disruptive to transferring technical ideas into the marketplace and the user environment. One of the true challenges faced by engineering managers is to foster a work environment conducive to interorganizational communications.

The most innovative, creative people are not natural teamplayers.[13] They enjoy conceptual work, have ideas, and are great individual contributors. Most of these

[13]This is one of the observations made by John Jewkes in his well-known studies on innovation. See J. Jewkes, D. Sawyers and R. Stillerman, *The Source of Innovation*, London: Macmillan, 1985.

people are not very good, however, in integrating the concepts and ideas of other individuals. They are often useless when their efforts involve joint activities that require strong communication links to other departments or organizations outside the company.

Because of the great need to communicate and cooperate on one side, and natural barriers that exist on the other, engineering managers must design their organizations so that the barriers to interorganizational teamwork are minimized. Specifically, an organizational environment conducive to interdisciplinary cooperation has the following:

A Unified Engineering Team

Interdisciplinary cooperation is enhanced if the people from various organizations see themselves as working on the same project with common overall objectives. Other drivers toward cooperation come from the pride and desire to belong to a particular project team and from the interest in the work itself. Management can help in unifying the engineering team by communicating the significance of the project to the company and by clearly stating global objectives and project goals. Organization interdependencies should be stressed and downstream–upstream project involvement should be encouraged early in the project life cycle.

Gatekeepers

These are professionals who understand the technology, functional disciplines, culture and values of other interfacing organizations. The more radical the technology, the newer and the more threatening the techniques that are employed, and the more diverse the functional cultures of the interfacing departments, the more critical are effective gatekeepers in fostering a cooperative interdisciplinary environment. Some people are natural gatekeepers, with interests and empathy for other organizations or activity areas. Provided they also have the specific job skills, these people can become valuable candidates for lead engineer or task manager in an activity area that requires integration and cooperation with other areas. Gatekeepers can also be developed. Job rotation, training programs, and project assignments can help to build cross-functional skills and understanding. In fact, one excellent way to develop gatekeepers is to let a person migrate with a project through its various life cycle phases. Such a directed job rotation not only develops an understanding of the functioning, values, and organizational requirements of other departments, but it also cultivates cross-functional contact personnel, credibility, and trust.

Power-sharing

Interdisciplinary cooperation can only exist if managerial powers are equitable, at least to some degree. There must be incentive for cooperation in the form of cross-functional budgeting, performance evaluations, and organizational charters. Management must design organizations to assure that power is balanced among the various departments and that there is enough flexibility to cope with the dynamics of the various projects and organizational requirements. The matrix design is usually a good model, which can be fine-tuned for various organizational requirements. (See Chapters 1 and 2 for discussion of matrix environments.)

Resource Availability

Resources, in the form of personnel, facilities and funding, must be available according to established plans, if cross-functional cooperation is to be assured. In fact, cooperation can be channeled as well as interrupted via resources allocation. Proper project planning, careful resource allocation, and continuous management commitment are necessary to assure that the resources are available when needed and that cooperation is enhanced rather than disrupted.

Proper Planning

Good planning practices can foster interdisciplinary cooperation in several ways. First, the participation of the various support groups in the program plan development creates early involvement and interest, which helps to unify the team behind the plan and its global objectives. Second, the program plan becomes the vehicle for linking the various organizational support groups during the program's execution, defining the interfaces and integration points as well as the timing and resource requirements. Third, the plan makes the engineering activities visible to other resource managers, enhancing top-down cooperation and commitment.

Global Goal

Cross-functional cooperation is enhanced by a common goal that is recognized by all parties involved in the project. This helps to lower organizational barriers and interests, while promoting unified efforts toward the overall project objectives. The project manager and upper management can help in identifying and promoting such a common goal by relating the project and its significance to broader business and market objectives, as well as assuring the importance of the project to the company. The company's newspaper, progress reports, and staff meetings are effective vehicles for promoting such a common goal value system.

Upper Management Support

The visible interest and active involvement of upper management is a strong catalyst for cross-functional cooperation. A clear message from the top regarding the significance of a project helps to unify the multidisciplinary team effort behind the overall goals. It also puts pressure on the team members through the formal chain of command and reward system to perform well, which lowers particular barriers and enhances cooperation toward the overall project objectives. Taken together, active upper management support is pervasive in developing a cooperative organizational environment. Engineering managers and project managers can enhance upper management interest by keeping the project and its progress visible and by involving management in the reviews, decisions, and milestone demonstrations.

BIBLIOGRAPHY

Akiyama, Kaneo, *Function Analysis,* Cambridge, MA: Productivity Press, 1991.

Beckert, Beverly, A., "Concurrent Engineering: Changing the Culture," *CAE,* Volume 10, Number 10 (Oct 1991), pp. 51–56.

Betz, Frederick, *Managing Technology,* Englewood Cliffs, NJ: Prentice-Hall, 1987.

Bitondo, Dominic and Alan Frohman, "Linking Technological and Business Planning," *Research Management,* Volume 14, Number 6 (1981), p. 19.

Brown, D. H., "Product Information Management: The Great Orchestrator," *CAE,* Volume 10, Number 5 (May 1991), pp. 60–66.

Calder, B. J., "Focus Groups and the Nature of Quantitative Market Research," *Journal of Marketing Research,* (Aug 1977), pp. 353–364.

Coombs, Gary, Jr. and Luis R. Gomez-Megia, "Cross-Functional Pay Strategies in High-Technology Firms," *Compensation & Benefits Review,* Volume 23, Number 5 (Sept/Oct 1991), pp. 40–48.

Dean, James W. Jr. and Gerald I. Sussman, "Organizing for Manufacturable Design," *Harvard Business Review,* Volume 67, Number 1 (Jan/Feb 1989), pp. 28–36.

Editorial, "Concurrent Engineering, Global Competitiveness, and Staying Alive: An Industrial Management Roundtable (Part 1)," *Industrial Management,* Volume 32, Number 4 (July/Aug 1990) pp. 6–10.

Editorial, "CAD/CAM Planning: Building an Engineering Team," *CAE,* Volume 10, Number 7 (July 1991), pp. CC4–CC6.

Emshoff, James R. and Roger L. Sisson, *Computer Simulation Models,* New York: Macmillan, 1970.

Forrest, Janet E. and Michael J. C. Martin, "Strategic Alliances: Lessons from the New Biotechnology Industry," *Engineering Management Journal,* Volume 2, Number 2 (Mar 1990), pp. 13–21.

Gupta, Ashok K. and Everett M. Rogers, "Internal Marketing: Integrating R&D and Marketing within the Organization," *The Journal of Services Marketing,* (Spring 1991), pp. 55–68.

Gupta, Ashok K. and David L. Wilemon, "Accelerating the Development of Technology-Based New Products," *California Management Review,* (Winter 1990), pp. 24–44.

Hartley, John R., *Concurrent Engineering,* Cambridge, MA: Productivity Press, 1991.

Hunsucker, John L., Japhet S. Law and Randal W. Sitton, "Transition Management—A Structural Perspective," *IEEE Transactions on Engineering Management,* Volume 35, Number 3 (Aug 1988), pp. 158–166.

Hunsucker, John L., Shaukat A. Brah, and Daryl L. Santos, "How NASA Moved from R&D to Operations," *Long-Range Planning,* Volume 2, Number 6 (Dec 1988), pp. 38–47.

Hyder, Debra A. and John Lew Cox, "Industrial Engineering Computer Tools—System User/System Implementer: A Joint Responsibility For Success," *Computers & Industrial Engineering,* Volume 15, Number 1–4 (1988), pp. 450–455.

Jewkes, John D. Wawers and R. Stillerman, *The Source of Innovation,* London: Macmillan, 1962.

Krouse, John, Robert Mills, Beverly Beckert, Laura Carrabine, and Lawrence Berardinis, "Building an Engineering Team," *Industry Week,* Volume 240, Number 14 (July 15, 1991), pp. CC4–CC6.

Kumar, Sanjoy and Yash P. Gupta, "Cross Functional Teams Improve Manufacturing at Motorola's Austin Plant," *Industrial Engineering,* Volume 23, Number 5 (May 1991), pp. 32–36.

Maccoby, Michael, "Move from Hierarchy to Heterarchy," *Research-Technology Management,* Volume 34, Number 5 (Sept/Oct 1991), pp. 46–47.

Massimilian, Richard D. and Laura Pedro, "Back from the Future: Gearing Up for the Productivity Challenge," Volume 79, Number 2 (Feb 1990), pp. 41–43.

McKeown, James J., "New Products from New Technologies," *Journal of Business & Industrial Marketing,* Volume 5, Number 1 (Winter/Spring 1990), pp. 67–72.

Millson, Murray R., S. P. Rag, and David Wilemon, "A Survey of Major Approaches for Accelerating New Product Developments," *Journal of Product Innovation Management,* Volume 9, Number 1 (Mar 1992), pp. 53–69.

Moad, Jeff, "Navigating Cross-Functional IS Water," *Datamation,* Volume 35, Number 5 (Mar 1, 1989), pp. 73–75.

Naisbitt, John and P. Aburdene, *Megatrends 2000,* New York: William Morrow, 1990.

Naisbitt, John, *Megatrends,* New York: Warner Books, 1982.

Nicholas, John M., *Managing Business & Engineering Projects,* Englewood Cliffs, NJ: Prentice-Hall, 1990.

Ohmae, K., *The Borderless World: Power and Strategy in the Interlinked Economy,* New York: Harper, 1990.

Orsini, J. and N. Karagozolu, "Marketing/Production Interfaces in Service Industries," *SAM Advanced Management Journal,* (Summer 1988), pp. 34–38.

Porter, Michael, *Competitive Advantage,* New York: Free Press, 1985.

Pyke, Donald, "Mapping-A System Concept for Displaying Alternatives,"in J. Bright and M. Schoeman (eds), *A Guide to Practical Technological Forecasting,* Englewood Cliffs, NJ: Prentice-Hall, 1973.

Roussel, Philip A., Kamal N. Saad, and Tamara J. Erickson, *Third Generation R&D,* Boston: Harvard Business School Press, 1991.

Scott-Morgan, Peter B., "Managing Interfaces: A Key to Rapid Product and Service Development," *Prism,* (Second Quarter 1991), pp. 55–69.

Shina, Sammy G., *Concurrent Engineering and Design for Manufacture of Electronics Products,* New York: Van Nostrand-Reinhold, 1991.

Siegel, Bryan, "Organizing for a Successful CE Process," *Industrial Engineering,* Volume 23, Number 12 (Dec 1991), pp. 15–19.

Smilor, Raymond W. and David V. Gibson, "Technology Transfer in Multi-Organizational Environments: the Case of R&D Consortia," *IEEE Transactions on Engineering Management,* Volume 38, Number 1 (Feb 1991), pp. 3–13.

Souder, William E., "Managing Relations Between R&D and Marketing in New Product Development Projects," *Journal of Product Innovation Management,* Volume 5, Number 1 (Mar 1988), pp. 6–19.

Souder, William E. and Suheil Nassar, "Choosing an R&D Consortium," *Research Technology Management* (Mar/Apr 1990), pp. 35–41.

Williams, Frederick and David V. Gibson, *Technology Transfer: A Communications Perspective,* Newbury Park, CA: Sage Publications, 1990.

9 BID PROPOSAL DEVELOPMENT

9.1 MARKETING TECHNICAL PROJECTS

New contracts are the lifeblood of many engineering organizations. The techniques for winning these contracts follow established bid proposal practices, which are highly specialized for each market segment. They often require intense and disciplined team efforts among all organizational functions, especially engineering and marketing, plus significant customer involvement.

Bid proposal efforts come in many different forms, sizes, and complexities. There are considerable differences among the various types of proposals, such as bidding in the private versus government sector, or bidding for project versus product business. However, the principal challenges, critical factors, and management practices and processes for developing winning proposals are similar and will be addressed in this chapter, with a focus on technical project proposals.

What Makes Project Marketing Different?

Marketing projects requires the ability to identify, pursue, and capture one-of-a-kind business opportunities, which entails unique managerial challenges, listed below:

1. *Systematic Effort.* A systematic effort is usually required to develop a new program lead into an actual contract. The program acquisition effort is often highly integrated with ongoing programs and involves key personnel from both the potential customer and the performing organization.

2. *Custom Design.* Traditional businesses provide standard products and services for a variety of applications and customers; projects are custom-designed to fit specific requirements of a single customer community.
3. *Project Life Cycle.* Project-oriented businesses have a beginning and an end and are not self-perpetuating. Business must be generated on a project-by-project basis, rather than by creating demand for a standard product or service.
4. *Marketing Phase.* Long lead times often exist between project definition, start-up, and completion.
5. *Risks.* Risks are present, especially in the research, design, and production of programs. The program manager has not only to integrate the multidisciplinary tasks and program elements within budget and schedule constraints, but also to manage inventions and technology.
6. *Technical Capability to Perform.* This capability is critical to the successful pursuit and acquisition of a new project or program.

Why Sell Projects?

One may wonder why companies pursue project business in spite of the risks and problems, and usually very modest contract profits. Indeed there are many reasons why projects and programs are good business, especially in engineering, as listed below:

1. Although immediate profits as a percentage of sales are usually small, the return on capital investment is often very attractive. Progress payment practices keep inventories and receivables to a minimum and enable companies to undertake programs many times larger in value than the assets of the total company.
2. Once a contract has been secured and is managed properly, the program is of relatively low financial risk to the company. The company has little additional selling expenditure and has a predictable market over the life cycle of the program.
3. Program business must be viewed from a broader perspective than motivation for immediate profits. It provides an opportunity to develop the company's technical capabilities and to build an experience base for future business growth.
4. Winning one program contract often provides attractive growth potential, such as (1) growth with the program via additions and changes, (2) follow-on work, (3) spare parts, maintenance, and training, and (4) being able to compete effectively in the next program phase, such as nurturing a study program into a development and finally a production contract.

In summary, new business is the lifeblood of an organization. It is especially crucial in project-oriented businesses, which lack ongoing, conventional markets.

New business developments are complex undertakings. The management of a new program acquisition is different from that for commercial marketing practices. it requires special skills, tools, techniques, and most importantly, competent and experienced people from all parts of the organization to carry the effort successfully through the various stages of acquisition planning, marketing, proposal development, and contract negotiation. This fundamental set of management techniques is discussed in this chapter. But first we will define the basic proposal types and markets of a project-oriented business.

9.2 PROPOSAL TYPES AND FORMATS

The majority of proposals that are prepared by companies are based on inquiries received from prospective clients. The inquiry stipulates the conditions under which the clients wish the work to be done. The responses we make to inquiries received from clients are termed "proposals," even though in many cases no commitment is proposed. Accordingly, proposals can be classified broadly in two major categories: qualification proposals and commercial bid proposals.

Qualification Proposal

The qualification proposal generally gives information about company organization, qualifications, working procedures, or information for a specific area of technology. Qualification proposals make no offer to perform services and make no commitments of a general or technical nature. These are also called *informational proposals* if the contents relate to company organization, general qualifications, and procedures. They are sometimes called *white papers, technical presentations,* or *technical volumes* if technical and economic data are provided for a specific area of technology. A special form of the qualification proposal is the *presentation.*

Commercial Bid Proposal

The commercial bid proposal offers a definite commitment by the company to perform specific work or services, or to provide equipment in accordance with explicit terms of compensation. A commercial bid proposal may also contain the type of information usually found in qualification proposals.

Both qualification and commercial proposals may be presented to the client in various forms under a wide variety of titles, depending on the situation, for example, the client's requirements and the firm's willingness to commit its resources under the specific circumstances involved. The most common forms are:

- Letter proposals
- Preliminary proposals
- Detailed proposals
- Presentations.

There are no sharp distinctions among these on the basis of content. Differentiation is mainly by the extent of the work required to prepare them. Included in the following list are definitions of the most common forms.

1. *Letter Proposals.* Letter proposals are either qualification or commercial proposals. They are brief enough to be issued in letter form rather than as bound volumes.
2. *Preliminary Proposals.* Preliminary proposals are either qualification or commercial proposals and are large enough to be issued as bound volumes. They may be paid technical and/or economic studies, bids to furnish services, or other offerings of this kind.
3. *Detailed Proposals.* Detailed proposals are most often commercial bid proposals, generally including the preparation of a detailed estimate. They are the most complex and inclusive proposals. Because of the high cost of preparation and the high stakes involved in the commitments offered, the organization and contents of the documents are defined and detailed to a much greater degree than in other kinds of proposals.
4. *Presentations.* Presentations are generally in the nature of oral qualification proposals. Selected personnel, specialized in various areas, describe their subjects verbally to client representatives in time periods varying from an hour to an entire day. To aid in the success of a presentation, audiovisual aids are encouraged. Many companies maintain a library of photographic slides developed just for this purpose.

9.3 DEFINING THE MARKET

Customers come in various forms and sizes. Particularly for large programs with multiple-user groups, customer communities can be very large, complex, and heterogeneous. Very large programs, such as military or aerospace undertakings, are often sponsored by thousands of key individuals representing the user community, procuring agencies, Congress, and interest groups. Selling to such a diversified, heterogeneous customer is a true marketing challenge, which requires a highly sophisticated and disciplined approach.

The first step in a new business development effort is to define the market to be pursued. The market segment for a new program opportunity is normally in an area of relevant past experience, technical capability, and customer involvement. Good marketers in the program business have to think as product line managers. They have to understand all dimensions of the business and be able to define and pursue market objectives consistent with the capabilities of their organizations.

Market Predictability. Program businesses operate in an opportunity-driven market. It is a mistaken belief, however, that these markets are unpredictable and unmanageable. Market planning and strategizing is important. New program opportunities develop over periods of time, sometimes years, for larger programs. These developments must be properly tracked and cultivated to form the basis for management actions such as bid decisions, resource commitment, technical readiness, and effective customer liaison.

The strategy of winning new business is supported by systematic, disciplined approaches, which are illustrated in Figure 9.1 and discussed below in five basic steps:

1. Identifying new business opportunities
2. Planning the business acquisition
3. Developing the new contract opportunity
4. Developing a winning proposal and pricing
5. Negotiating and closing the contract.

9.4 IDENTIFYING NEW BUSINESS OPPORTUNITIES

Identifying a new program opportunity is a marketing job. During the initial stages one does not evaluate or pursue the opportunity—that comes later. Furthermore, identifying new opportunities should be an ongoing activity. It involves the scanning of the relevant market sector for new business. This function should be performed by all members of the project team in addition to marketing support groups. There are many sources for identifying new business leads such as: customer meetings on ongoing program, professional meetings and conventions, trade shows, trade journals, customer service, advertising of capabilities, and personal contacts. All one can expect at this point is to learn of an established or potential customer requirement in one of the following categories:

- Follow-on to previous or current programs
- Next phase of program
- Additions or changes to ongoing programs
- New programs in your established market sector or area of technological strength
- New programs in related markets
- Related programs, such as training, maintenance, or spare parts.

For most businesses, ongoing program activities are the best source of new opportunity leads. Not only are the lines of customer communication better than in a new market but, more importantly, the image as an experienced, reliable contractor has been established, it is hoped, giving a clear competitive advantage in any further business pursuit.

The target result of this analysis is an acquisition plan and a bid decision. Analyzing the new opportunity and preparing the acquisition plan is an interactive effort.

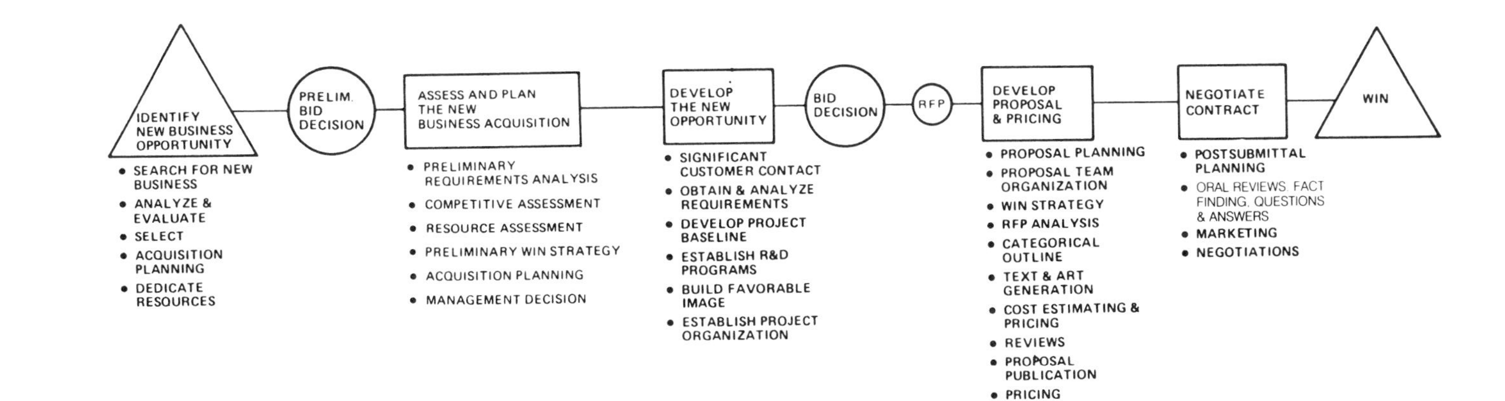

FIGURE 9.1 Major phases and milestones in a bid proposal life cycle.

Often many meetings are needed between the customer and the performing organization before a clear picture emerges of both customer requirements and contractor's capabilities. A major fringe benefit of proper customer contact is the potential for building confidence and credibility with the customer. It shows that your organization understands their requirements and has the capability to fulfill them. This is a necessary prerequisite for eventually negotiating the contract. The new business identification phase concludes with a formal analysis of the new project opportunity, which is summarized in the following section.

9.5 PLANNING THE BUSINESS ACQUISITION

The acquisition plan provides the basis for the formal bid decision and a detailed plan for the acquisition of the new business. The plan is an important management tool, which provides an assessment of the new program opportunity as a basis for appropriating resources for developing and bidding the new business. Typically, the new business acquisition plan should include the following elements:

1. *Brief Description of New Business Opportunity.* This states the requirements, specifications, scope, schedule, budget, customer organization, and key decision-makers.
2. *Why Should We Bid?* The answer gives a perspective with regard to establishing business plans and lists desirable results, such as profits, markets, growth, and technology development.
3. *Competitive Assessment.* This is a description of each competing firm with regard to their past activities in the subject area, including (a) related experiences, (b) current contracts, (c) customer interfaces, (d) specific strengths and weaknesses, and (e) potential baseline approach.
4. *List of Critical Win Factors.* List the specific factors important to winning the new program, including their rationales. (Example: low implementation risk and short schedule are important to the customer because of the need for equipment in two years.)
5. *Evaluation of Ability to Write a Winning Proposal.* This addresses the specifics needed to prepare a winning proposal, including (a) availability of the right proposal personnel; (b) understanding of customer problems; (c) company's unique competitive advantage; (d) expected bid cost to the under customer budget; (e) special arrangements, e.g., teaming, license model; (f) engineering's readiness to write proposal; and (g) the ability to price competitively.
6. *Win Strategy.* The win strategy is a chronological listing of critical milestones guiding the acquisition effort from its present position to winning the new program. It should show those activities critical for positioning your company uniquely in the competitive field. This includes noting the timing and responsible individuals for each milestone. For example, if low implementation risk

and short schedules are important to the customer, the summary of the win strategy may state:

(a) build credibility with the customer by introducing key personnel and discussing baseline prior to request for proposal;
(b) stress our related experience on ABC program;
(c) guarantee 100% dedicated personnel and list the program personnel by name;
(d) submit a detailed schedule with measurable milestones and specific reviews;
(e) submit XYZ module with proposal for evaluation.

7. *Capture Plan.* The capture plan is a detailed action plan in support of the win strategy and all business plans. It should integrate the critical win factors and specific action plan. All activities, such as timing, budgets, and responsible individuals should be identified, and it should have measurable milestones. The capture plan is a working document to map out and guide the overall acquisition effort. It is a living document, which should be revised and refined as the acquisition effort progresses.
8. *Ability to Perform Under Contract.* This is often a separate document, but a summary should be included in the acquisition plan stating (a) technical requirements, (b) work force, (c) facilities, (d) teaming and subcontracting, and (e) program schedules.
9. *Problems and Risks.* This is a list of problems critical to the implementation of the capture plan, such as (a) risks to techniques, staffing, facilities, schedules, or procurement; (b) customer-originated risks; (c) licenses, patents, and rights; and (d) the contingency plan.
10. *Resource Plan.* The resource plan summarizes the key personnel, support services, and other resources needed for capturing the new business. The bottom line of this plan is the total acquisition cost.

There are many ways to present the acquisition plan. However, an established format, which is accepted as a standard throughout the organization, has several advantages. It provides a unified standard format for quickly finding information during a review or analysis. Standard forms also serve as checklists. They force the planner not only to include information conveniently obtainable, but also to seek out the other data necessary for winning new business. Finally, a standard format provides a quick and easy assessment of the new opportunity for key decision-makers.

The Bid Board

Few decisions are more fundamental to new business than the bid decision. Resources for the pursuit of new business come from operating profits. These resources are scarce and should be carefully controlled. Bid boards serve as management gates for the release and control of these resources. The bid board is an expert panel, usually convened by the general manager, which analyzes the acquisition activities to determine their status and also to assess investment versus opportunity in acquiring new business. An acquisition plan provides the major framework for the meeting of such bid boards.

Major acquisitions require a series of bid board sessions, starting as early as 12 to 18 months prior to the request for proposal. Subsequent bid boards reaffirm the bid decision and update the acquisition plans. It is the responsibility of the proposal manager to gather and present pertinent information in a manner that provides the bid board with complete information for analysis and decision. This requires significant preparation and customer contact. A team presentation is effective, as all disciplines should be involved.

The Bid Decision

After the proposal inquiry is received and logged in, it should be screened as soon as possible to facilitate the bid/no-bid decision. Because most inquiries have a short response time, the sooner a decision can be made, the longer you have to prepare your proposal. As part of the screening process, it is important that the sales representative or proposal manager review the document thoroughly to determine its total value to the organization. It also allows the decision-makers to determine if they have the capabilities required to bid on the job.

The technical nature of the current marketplace is such that organizations must compete in specialized areas of an industry. In many cases, you may find anywhere from 10 to over 50 organizations bidding on a project. However, often only a handful are serious competitors capable of performing the work. The rest are just wasting time and money. Therefore, an organization must concentrate its proposal efforts in areas that further their success by satisfying the objectives of the organization. Careful screening of the inquiry may also reveal that more information is needed to prepare a responsive proposal properly. A bid/no-bid decision is usually made based upon a set of criteria judged to be important in selecting projects that contribute to an organization's continued growth and success. A typical set of questions to be answered in support of a bid decision is shown in Table 9.1. In any event, the decision to bid or not to bid is very important to resource allocation and should be made by senior management. In making the decision to bid, management must be convinced that the project is good for the company, that they have the necessary resources and capabilities, and that they have a good chance of winning. Management must be realistic; if they cannot be competitive or responsive in meeting the requirements of the project, then they should not waste their time preparing a proposal. In some instances, an organization may feel obligated to submit a proposal to the customer to show interest and to develop a rapport for future work. If a decision to bid is made, a commitment to provide full resources in preparation of the proposal must follow.

All proposal efforts must have the endorsement of the commercial or sales department and, where needed, of the legal and operations departments. It is the sales representative's or proposal manager's responsibility to obtain this clearance. For

TABLE 9.1 Checklist in Support of a Bid Decision

1. Does the company have capabilities and resources to perform the work?
2. Can the company phase in the work to meet client's schedule?
3. What is the company's technical position?
4. What is the company's approach to project execution?
5. Is the project of special importance to the client?
6. Would doing the project enhance our reputation?
7. What has been our past experience and contractual relationship with the client?
8. What is the company's commercial approach and price strategy?
9. What are the client's future capital expenditures?
10. Who is the competition and do they have any special advantages?
11. Does the client have preferred contractor and if so, why?
12. What is the probability of the project going ahead?
13. Does the project meet the immediate or long-range objectives of the company?
14. Are there other work prospects for the company in next six months? In one year?
15. Are there other special factors and considerations?

proposal efforts budgeted under a certain amount, say, less than $500 or $1000, no further management authorization is needed. The proposal manager must, however, have the agreement of either the general sales manager or the operation's general manager to proceed with each such proposal to assure compliance with the group's profit plan and staffing.

Proposal efforts identified as costing in excess of this break point must have the approval of an inquiry review committee (IRC). It is the sales representative's responsibility to complete a request for approval of bid preparation, as shown in Table 9.2. The proposal manager obtains inputs on costs and staffing from the selected group and provides this to the sales representative for inclusion in the request for approval of bid preparation.

The IRC usually meets once a week. All inputs must be in the marketing departments on the day prior to the meeting. The IRC is usually chaired by the senior vice-president of marketing. The members of the committee may include the vice-presidents of operations, engineering, research and development, legal and financial areas, and the proposal manager. The proposal manager discusses the workload of his department so members of the IRC can make the proper decision on resource allocation. In addition to providing proposal budget authorization, the agreement to spend money on preparing a proposal, the IRC essentially provides a review to:

- Provide management guidance and support for the sales efforts
- Ensure that day-to-day activities are consistent with long-term objectives
- Reduce the risks of making premature, conflicting, or nonproductive commitments.

TABLE 9.2 Request for Approval of Bid Preparation

Date: ____________
No: ____________

1. Client ____________
2. Project ____________
3. Units and capacities ____________
4. Location ____________
5. Scope of work ____________
6. Value of project ____________
7. Type of contract ____________
8. Type of proposal ____________
9. Proposal cost ____________
10. Due date of proposal ____________
11. % probability to go ahead ____________
12. % probability for award to company ____________
13. Other factors
    a. Time of award (if long range) ____________
    b. Financing (if financing involved) ____________
    c. Source of knowhow (if other than company) ____________
    d. Special contract conditions (if not company standard) ____________
    e. Manpower availability (if not readily available) ____________
    f. Secrecy conditions (if any) ____________
    g. Competition (if known) ____________
    h. Unusual factors (if any) ____________

Management Review Committee
Approved By: ____________ Date: ____________

Input from the sales group is required during the inquiry and bid-decision stage. The sales representative may have additional information, which should be included in the inquiry. In many cases, the inquiry is the end point of a long series of thinking and acquisition planning activities, which may or may not have involved outside sales help. The sales representatives, through their normal contact with the client, have a feel for the customer's thinking on such areas as: technical approach; management organization; operational systems; plans and schedules; related and future work; and dollars budgeted for the project. The sales representative may also find out such information as strengths and weaknesses of the competition and customer bias toward a particular competitor. In general, you want to know as much as possible about the customer's total requirements, so that you can present the best possible proposal to him. In some instances, a visit by the customer to your facilities and a discussion with the project team will become part of the proposal effort.

Once a decision is made not to bid on a project, the customer should be notified in writing. Some organizations respond with a form letter, but such a response could cause the customer to interpret this as lack of interest and could result in loss of future work. A specific letter for that inquiry should be prepared, explaining to the customer why your organization could not respond.

9.6 TYPICAL PHASES OF CONTRACT DEVELOPMENT

Selling a new program is often unique and different from selling in other markets. It requires selling an organization's capability for a custom development—something that has not been done before. This is different from selling an off-the-shelf product that can be examined prior to contract. It requires establishing your credibility and building confidence in the customer's mind so that your organization is selected as the best candidate for the new program. Building such a "can-do" image usually has six phases, which often overlap.

Significant Customer Contact

Early customer liaison is vital in learning about the customer's requirements and needs. It is necessary to define the project baseline, the potential problem areas, and the risks involved.

Establishing meaningful customer contact is no simple task. Today's structured customer organizations involve many key decision-making personnel, conflicting requirements and needs, and biases. There rarely is only one person responsible for signing off on a major procurement. Technical and marketing involvement at all levels is necessary to reach all decision-making parties in the customer community. Your new business acquisition plan will be the roadmap for your marketing efforts. The benefits of this customer contact are that you:

- Learn about the specific customer requirements
- Obtain information for refining the baseline prior to proposal
- Participate in customer problem-solving
- Build a favorable image as a competent, credible contractor
- Check out your baseline approach and its acceptability to the customer
- Develop rapport and a good working relationship with the customer.

Prior Relevant Experience

Nothing is more convincing to an engineering customer than demonstrated prior performance in the same or a related area of the new program. It shows the customer that you have produced on a similar task. This reduces the perceived technical risks and associated budget and schedule uncertainties. Therefore it is of vital importance to demonstrate to the customer that your organization understands their new requirements and has performed satisfactorily on similar programs. The image of an experienced contractor can be communicated in many ways:

- Field demonstration of working systems and equipment
- Listing of previous or current customers, their equipment, and applications
- Model demonstrations
- Technical status presentations
- Product promotional folders
- Technical papers and articles
- Trade show demonstrations and displays
- Slide or video presentation of equipment in operation
- Simulation of the system, equipment, or services
- Specifications, photos, input–output simulations of the proposed equipment
- Advertisements.

Demonstrating prior experience is integrated with and interactive with customer liaison activities. To be successful, particularly on larger programs, requires both leadership and discipline. Start with a well-defined customer contact plan as part of your overall acquisition plan. This requires well-planned involvement at all levels in order to make contacts with relevant personnel in the customer community. One major benefit received from the intensive marketing efforts is that you create an image with the customer of an experienced, sound contractor. Second, you are learning more about the new program, its specific requirements, the risks involved, and the concerns and biases of the customer, which will make it easier to respond effectively to a formal or informal request for proposal.

Establishing Readiness to Perform

Once the basic requirements and specifications of the new program are known, it is often necessary to mount a substantial technical preproposal effort to advance the baseline design to a point that permits a clear definition of the new program. These efforts may be funded by the customer or borne by the contractor. Typical efforts include (1) feasibility studies, (2) systems designs, (3) simulation, (4) design and testing of certain critical elements in the new equipment or the new process, (5) prototype models, or (6) any developments necessary to bid on the new job within the desired scope of technical and financial risks.

Development prior to contract is expensive, has no guarantee of return, and keeps the company from pursuing other activities. Then why do organizations spend their resources for such development? It is often an absolutely necessary cost for winning new business. These early developments reduce the implementation risks to an acceptable level for both the customer and the contractor. Furthermore, development might be necessary to catch up with a competitor or to convince the customer that certain alternative approaches are preferable.

As preproposal developments are costly, they should be thorough and well-detailed and should be approved as part of the overall acquisition plan. The plans and specific results should be accurately communicated to the customer. This will help to build a quality image for your firm while giving the potential contractor additional insight into the detailed program requirements. One should not overlook two sources of funding for development activities prior to contract:

1. Customer funding. Often the customer is willing to fund contract-definition activities because they may reduce the risks and the uncertainties of contractual performance.
2. Activities' inclusion with other ongoing developments. The program manager may find that a similar effort is underway in a corporate research department, or even within the customer's organization.

Another element of credibility is the contractor's organizational readiness to perform under contract. This includes facilities, key personnel, support groups, and management structure. Reliability in this area is particularly critical in winning a large program relative to your company size. Often a contractor goes out on a limb and establishes a new program organization to satisfy specific program or customer requirements. This may require major organizational changes.

Establishing the Organization

Few companies go into reorganization lightly, especially prior to contract. However, in most cases it is possible to establish all the elements of the new program organization without physically moving people or facilities and without erecting new buildings. What is needed is an organization plan exactly detailing the procedures to be followed as soon as the contract is awarded. The new program organization can be defined on paper, together with its proper charter and all structural and authority relationships. This should be sufficient for customer discussion and will give a head start once the contract is received. Usually it is not the moving of partitions, people, or facilities that takes time, but determining where to move them and how to establish the necessary working relationships.

As a checklist, the following organizational components are listed; they should be defined clearly and discussed with the customer prior to a major new contract:

- Organizational structure
- Charter
- Policy-management guidelines
- Job descriptions
- Authority and responsibility relationships
- Type and number of offices and laboratories
- Facilities listing
- Floor plans
- Staffing plan
- Milestone schedule and budget for establishing the organization.

A company seldom needs to reorganize completely to accommodate a new program. Reorganization uses resources and has risks for both the contractor and the customer. Most likely the customer and program requirements can be accommodated within the existing organization by redefining organizational relationships, authority, and responsibility structures without physically moving people and facilities. Matrix organizations in particular have the flexibility and capacity to handle large additional program business with only minor organizational changes.

The Kickoff Meeting

Soon after the decision to bid has been made, a preliminary-proposal kickoff meeting or proposal strategy planning meeting should be held. This meeting, chaired by the proposal manager, is attended by the heads of the various contributing departments, the sales representatives, and possibly senior management. Because there is a limited amount of time available for preparation of a proposal, it is mandatory that the proposal effort be planned in all aspects to make the most use of that time. A plan produces the best effort and response to a client's request that the time period allows. To be time effective, no work should commence without first having a proposal strategy planning meeting.

After the strategy planning meeting and development of a preliminary proposal plan, the proposal manager calls a kickoff meeting. Attendance is limited to those who actually contribute to the definition and initiation of the proposal effort, such as the following persons:

- Sales representatives responsible for the inquiry
- Heads of departments and groups responsible for any of the proposal work
- Key personnel from departments and groups responsible for any of the proposal work, if they have been selected
- The general manager of operations and a member of the legal department responsible for proposal review, who are notified of the meeting so they may attend if they wish.

The proposal manager sets the meeting time and place and writes a kickoff memo or notice to inform the participating departments what needs to be discussed at the meeting regarding the proposal effort. The kickoff memo and meeting are also referred to as the *proposal coordination memo and meeting.* The purpose of the kickoff meeting is to inform all participants about the proposal plan and its objectives. The kickoff meeting gives you an excellent opportunity to brief and obtain feedback from a large number of contributing people at one session. It is crucial to success that all personnel contribute to the proposal so that it meets the client's project needs. The proposal manager is responsible for preparing the minutes of the kickoff meeting, which are issued no later than the second working day after the meeting. The contents are divided into five main sections, in the same fashion as the kickoff meeting.

Topics that should be covered in every kickoff meeting or memo to ensure a good start on the work are listed below:

1. *Project Scope.* The type of plant or unit, the client, the project location, the order of magnitude of total installed cost (if known), the job or proposal number, and any other general identification, plus designation of proposal manager and other key personnel.
2. *Commercial Objectives.* The type of proposal required is discussed, as well as the management philosophy to be incorporated into the proposal. The sales representative provides background on the client's requirements, exceptions to be taken, and on the client's budget, if known.
3. *Proposal Staffing.* Three factors—the proposal schedule, the man-hour budget, and the type of proposal desired—determine which departments will participate and also set the proposal staffing requirements. Staffing should be settled soon after the kickoff meeting or after issuance of the kickoff memo.
4. *Assignments of Participants.* Work assignments with clear areas of responsibility will be made. Relationships between participating departments should be spelled out. Because the proposal manager may be on loan from another department, the role of coordinator of the work must be defined clearly.
5. *Proposal Dates, Schedules, and Budget.* The key dates for issuance of proposal documents and completion of important portions of the work should be discussed, and the participants should be advised of the available man-hour budget. Cautionary measures should be taken to be sure that the schedule is realistic. If not, the sales representative has to notify the client and seek an extension of the due date.
6. *Qualifications.* On occasion, the proposal may have to be tailored to sell the client on your qualifications and capabilities in some area covered by the proposed project. The desired slant to accomplish this should be considered and made known to all involved.
7. *Type of Estimate.* The type of estimate to be submitted with the proposal already may have been decided on the basis of the client's requirements. If not, the proposal manager clearly must define the requirements. The type of estimate determines the quality and quantity of estimating information that must be generated by the participants in the proposal effort. The proposal manager specifies what is required for proper support of the estimate.
8. *Final Proposal Contents.* The type of proposal determines to some extent what must be included in the final documents sent to the client. Specific requirements for the contents should be brought to everyone's attention. The proposal manager develops and includes a tentative table of contents for the proposal documents under discussion. Definite assignments for preparation of the written drafts for each section of the proposal are also made.
9. *Working Information.* Copies or a summary of the inquiry documents are distributed before or with the kickoff memo. Any other information from the client that is to be used in preparing the proposal is also distributed and discussed. If the information is inadequate, the proposal manager arranges to obtain additional information from the client. The technical basis for the design should be summarized, preferably in a standard project design questionnaire.

Useful information from other proposals or projects should be called to everyone's attention. Its validity for use in the proposal work should also be settled.

The final phase of the bidding effort is usually the development and submission of the formal proposal document, which will be discussed in the next section.

9.7 DEVELOPING A WINNING PROPOSAL

Bid proposals are payoff vehicles. They are one of the final products of your marketing effort. Whether you are bidding on a service contract or on an engineering development, a government contract or a commercial program, the process is the same and, in the end, you must submit a proposal.

Yet with all due respect to the importance of the bid proposal as a marketing tool, many senior managers point out that the proposal is only one part of the total marketing effort. The proposal is usually not the vehicle that sells your program—the proposal stage may be too late. The program concept and the soundness of its approach, the alternatives, your credibility, and so on, must be established during the face-to-face discussions with the customer. So why this fuss about writing a superior proposal? Because we still live in a competitive world. Your competition is working toward the same goal—winning this program. They, too, may have sold the customer on their approaches and capabilities. Hence among the top contenders, the field is probably very close. More importantly, beating most of the competition is not good enough. As in a poker game, there is no second place. Therefore while it is correct that the proposal is only part of the total marketing effort, it must be a superior proposal. Proposal development is a serious business in itself.

Most people hate to work on proposals. Proposal development requires hard work and long hours, often in a constantly changing work environment. Proposals are multidisciplinary efforts of a special kind, but, like any other multifunctional program, they require an orderly and disciplined effort that relies on many special tools to integrate the various activities of developing a high-scoring, quality proposal. This is particularly true for large program proposals, which require large capital commitments. Smaller proposals often can be managed with less formality. However, at a minimum, the proposal plan should include the following to ensure a quality bid proposal:

- Proposal team organization
- Proposal schedule
- Categorical outline with writing assignments and page allocation
- Tone and emphasis and win strategy
- Request for proposal (RFP) analysis
- Technical baseline review
- Proposal writing assignments

- Reviews
- Art and illustration development
- Cost estimating
- Proposal production
- Final management review

For each activity or milestone, the plan should define the responsible individual(s) and schedules.

Storyboarding Facilitates Group Writing

Most bid proposals are group writing efforts. Organizing, coordinating and integrating these team efforts can add significantly to the complexities and difficulties of managing proposal developments, especially for the larger engineering development bids. Storyboarding is a technique that facilitates the group writing process by breaking down its complexities and integrating the proposal work incrementally. It is based on the idea of splitting up proposal-writing among the various contributors and then developing the text incrementally via a series of writing, editing, and review phases. The number and type of phases indicated in Table 9.3 together with the timing, might be typical for a major bid proposal development with a 30-day response time.

Table 9.3 can also serve as a guide for larger or smaller proposals by scaling the effort up or down. For larger efforts, more iterations are suggested among phases 5 through 9, while smaller efforts can be scaled down to 7 phases, including only phases 4 through 10. The actual timing of proposal developments should be scheduled like any other project. A bar-graph schedule is sufficient and effective for most proposal efforts. Each phase of storyboarding is briefly described below.

TABLE 9.3 Typical Phases of Storyboarding for Bid Proposal

Phase	Day of Completion
1. Categorical outline	Day 1
2. Synopsis of approach	Day 3
3. Roundtable review	Day 4
4. Topical outline	Day 5
5. Storyboards	Day 10
6. Storyboard review	Day 11
7. Storyboard Expansion	Day 22
8. Staff review	Day 24
9. Editing	Day 28
10. Printing and delivery	Day 30

Phase One: Categorical Outline. The first step in the storyboarding process is the development of a categorical outline. This is a listing of the major topics or chapters to be covered in the proposal. For larger proposals, the categorical outline might form the first two levels of the table of contents, as is shown in Figure 9.2. The categorical outline should also show the responsible author, a page estimate, and references to related documents for each category. The categorical outline can often be developed *before* the receipt of the RFP, and should be finalized at the time of proposal kickoff.

Phase Two: Synopsis of Approach. A synopsis of approach is developed for each categorical topic by each responsible author. The synopsis is an outline of the approach, which addresses three questions related to the specific topic:

1. What does the customer require?
2. How are we planning to respond?
3. Why is the approach sound and good?

The format of a typical synopsis is shown in Figure 9.3. As an alternative, the proposal manager can prepare these forms and issue them as policy papers, instead of having each author develop them.

In preparation for the roundtable review, the completed synopsis forms and categorical outline can be posted on a wall, in sequential order. This method of display facilitates open group reviews and analysis in a very effective way.

CATEGORICAL OUTLINE

SHEET 1 OF 1
DATE 7/2/92

PROGRAM: RADAR SYSTEM

PROPOSAL ADDRESS	TITLE	AUTHOR	PAGE ESTIMATE TEXT	ART	SPEC REFERENCE
I	INTRODUCTION	Al	3	1	- - -
II	PROPOSED SYSTEM	Beth	42	25	xyz 799
III	RELIABILITY AND MAINTAINABILITY	Carlos	27	8	QRT 09944
IV	REQUIREMENTS	David	12	4	VX-99; MX-334
V	PROPOSED TESTING AND TEST FACILITIES	Ellen	15	11	- - -
VI	MATERIAL IDENTIFICATION	Frank	9	2	AVRR 33
VII	PROGRAM SCHEDULE	George	7	5	7000.2
VIII	PROCUREMENT, FABRICATION, AND ASSEMBLY	Harry	19	10	DFMA-123

FIGURE 9.2 Typical categorical outline form for storyboarding a major proposal.

SYNOPSIS OF APPROACH

PROPOSAL NAME:	VOL:	PROPOSAL ADDRESS	WORK PACK:
RADAR SYSTEM	II	II.A.1 Configuration	#4

RFP REFERENCE: 04.99.22

AUTHOR: Mark	EXT: 333
BOOK BOSS: Beth	EXT: 111

UNDERSTANDING OF REQUIREMENTS:

RFP OPERATION BY NON-COMMUNICATION MILITARY PERSONNEL. OPERATIONAL FEATURES INCLUDING GROWTH CAPABILITY, DEPLOYMENT, DEGRADED OPERATION, FLEXIBILITY. TIME TO DISPERSE AND DEPLOY.

APPROACH & COMPLIANCE:

POINTS TO BE MADE

- EASE OF OPERATION
- GROWTH CAPABILITY
- DEPLOYMENT
- DEGRADED OPERATION
- FLEXIBILITY
- SET UP & STAND DOWN TIMES
- COMMUNICATIONS DURING TRANSIT
- NET CONTROL
- DISPERSAL TIME

SOUNDNESS OF APPROACH:

OUR SUBSYSTEMS ACCOMMODATE ALL DESIRABLE FEATURES, IS EASY TO OPERATE & DEPLOYS RAPIDLY

RISKS:

FIGURE 9.3 Typical synopsis form, which is used to expand each categorical topic from Figure 9.2 on storyboarding.

Phase Three: Roundtable Review. During this phase, all synopsis of approach forms are analyzed, critiqued, augmented, and approved by the proposal team and its manager. It is the first time that the total proposal approach is continuously displayed in summary form. Besides the proposal team, technical resource managers, marketing managers, contract specialists, and upper management should participate in this review, which usually starts four days after the proposal kickoff.

Phase Four: Topical Outline. After the review and approval of the synopsis forms, the categorical outline is expanded into the specific topics to be addressed in the proposal. The topical outline thus created forms the table of contents for the bid proposal. As for the categorical outline, each topic has the responsible writer, page estimates, and document references designed. An example of a topical outline form is shown in Figure 9.4.

Phase Five: Storyboard Preparation. The storyboard preparation is straightforward. Typically, a one-page storyboard is prepared for each topic by the writer assigned. As shown in Figure 9.5, it represents a detailed outline of the author's approach to the writeup for that particular topic. Often the storyboard form is divided into three parts: (1) topic and theme section, (2) text outline on the left side of the form, and (3) summary of supporting art to be prepared on the right side of the form.

The storyboard represents a summary of the key messages which should be articulated under a particular proposal section. The topic heading, theme (if any), problem statement, and exit or conclusion must be written out in full, just the way they might appear in the final text. Expression of these key sentences is important. They must be relevant, responsive, and comprehensive.

The composition of text and art is arranged on the storyboard in this format for convenience only. As shown in Figure 9.6, the format chosen can be either modular or nonmodular. For the modular concept, the storyboard format is copied into the proposal layout, text left, illustrations right; in nonmodular form the art becomes an

TOPICAL OUTLINE

SHEET 1 OF 6
DATE 7/09/92

PROGRAM: RADAR SYSTEM

PROPOSAL ADDRESS	TITLE/TOPIC	AUTHOR	PAGE ESTIMATE TEXT	PAGE ESTIMATE ART	SPEC REFERENCE
I	INTRODUCTION	Al	3	1	---
II	PROPOSED SYSTEM	Beth	42	25	XYZ 799
A.	Mechanical Design	Mark	9	4	RRM-44
1.	Configuration	Mark	2	1	
2.	Construction	Nancy	1	2	
3.	Control and Output	Paul	5	–	CCC-88
4.	Thermal Considerations	Ralph	1	1	
B.	Antenna Design	Stanley	11	10	ANT-777
1.	Transmit Antenna	Stanley	5	4	
2.	Receive Antenna	Stanley	6	6	
C.	Filter Design	Mark	9	6	FFF-XQC-3
1.	Transmit Filter	Mark	2	–	
2.	Receive Filter	Willy	2	1	
3.	Receive Pad	Mark	5	5	
D.	Transmitter Design	Victor	11	6	XM-001
E.	Frequency Source, System	Walter	2	5	FS-007
1.	Frequency Source				
2.	System Clock and Timing . . .				
III	RELIABILITY AND MAINTAIN . . .	Carlos	27	8	QRT 09944
A.	Organization	Marie	4	–	
B.	Program Plan	Carlos	12	5	
1.	Related Activity Interface . . .	Carlos	4	2	
2.	Subcontractor and Vendor . .	Carlos	8	3	

FIGURE 9.4 Topical outline sample form, which expands the categorical outline of Figure 9.2.

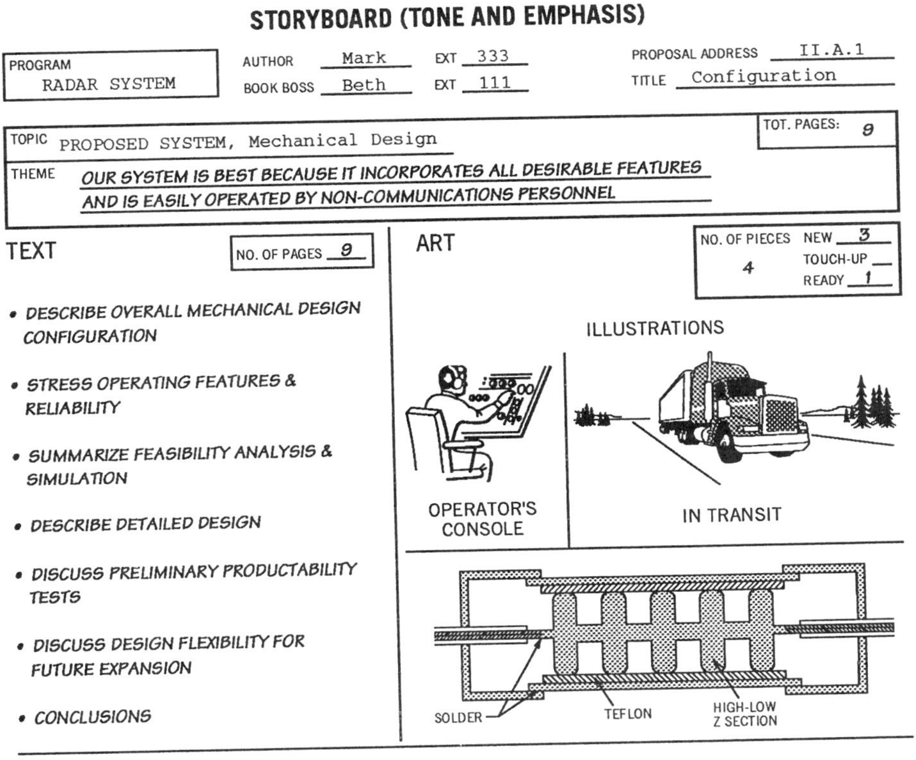

STORYBOARD (TONE AND EMPHASIS)

PROGRAM RADAR SYSTEM

AUTHOR Mark EXT 333

BOOK BOSS Beth EXT 111

PROPOSAL ADDRESS II.A.1

TITLE Configuration

TOPIC PROPOSED SYSTEM, Mechanical Design

TOT. PAGES: 9

THEME *OUR SYSTEM IS BEST BECAUSE IT INCORPORATES ALL DESIRABLE FEATURES AND IS EASILY OPERATED BY NON-COMMUNICATIONS PERSONNEL*

TEXT

NO. OF PAGES 9

- *DESCRIBE OVERALL MECHANICAL DESIGN CONFIGURATION*
- *STRESS OPERATING FEATURES & RELIABILITY*
- *SUMMARIZE FEASIBILITY ANALYSIS & SIMULATION*
- *DESCRIBE DETAILED DESIGN*
- *DISCUSS PRELIMINARY PRODUCTABILITY TESTS*
- *DISCUSS DESIGN FLEXIBILITY FOR FUTURE EXPANSION*
- *CONCLUSIONS*

ART

NO. OF PIECES 4 NEW 3 TOUCH-UP READY 1

ILLUSTRATIONS

OPERATOR'S CONSOLE

IN TRANSIT

SOLDER TEFLON HIGH-LOW Z SECTION

FIGURE 9.5 Storyboard format and example.

integral part of the text. The final layout should, however, be of no concern to the authors at this point in the proposal development cycle.

Phase Six: Storyboard Review. The completed storyboard forms are pasted on the walls of the review room in logical sequence, presenting the total story as we want to tell it to the customer.

Storyboard reviews should start within 5 working days after the synopsis reviews. The reviews are held in a special area and attended by the author, the proposal team, and key members of the functional organization. The storyboard review permits a dialogue to take place between the proposal team and its management and the author. All proposal writers are notified of the review times by such media as the daily bulletin and microschedule, so that they can participate in the reviews of their storyboards.

The review permits the proposal team to insert, modify or correct any approach taken by an author. An additional output of the storyboard review is a final proposal outline. This proposal outline includes categorical headings, topics, and art log numbers, provides the proposal management team with an activity overview, and

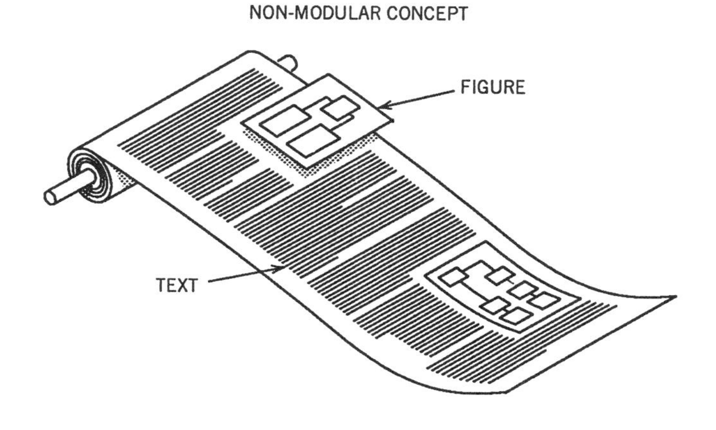

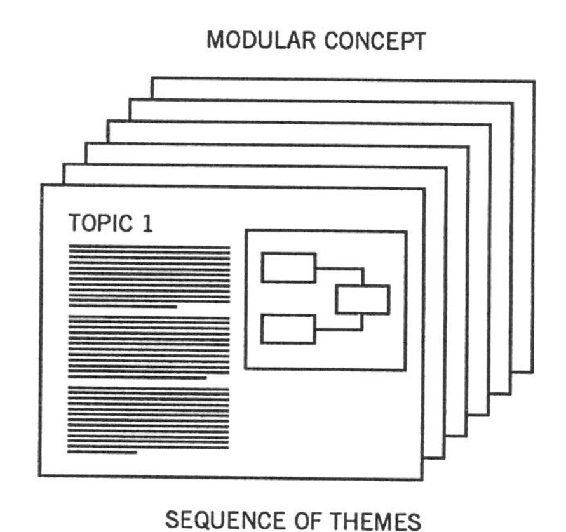

FIGURE 9.6 Nonmodular versus modular concept of text and art presentation.

serves as a control document. The storyboard review provides the team with the single most important opportunity to change direction or change approaches in the proposal preparation.

Like the synopsis review, storyboarding is an interactive process. During the reviews, a copy of the latest storyboard should always be on display in the control room.

Phase Seven: Final Text Generation by Expanding Storyboards. After storyboard approval, each author prepares a storyboard expansion. It is the expansion of information on each topic into about a 500-word narrative. As part of the storyboard expansion, all authors should finalize their artwork and give it to the publications specialist for processing. The completed storyboard expansions represent the first draft of the final proposal. The material is given to the technical editor, who will edit them for clarity. Each responsible author must review and approve the final draft, which might cycle through the editing process several times.

This final text generation is the major activity in the proposal development process. All other prior activities are preparatory to this final writing assignment. If the preparations up to the storyboard review are done properly, writing the final text should be a logical and straightforward task without the hassles of conceptual clarification and worries about integration with other authors.

As a guideline, 10 working days out of a 30-working-day request-for-proposal response period may be a reasonable time for the final text generation. Because of its long duration, it is particularly important to set up specific milestones for measuring intermediate progress. The process of final text generation should be carefully controlled. The proposal specialist, if available, will play a key role in the integration, coordination, and controlling of the final text generation and its output.

The final text should be submitted incrementally to the publication department for editing, word processing or typing, and media storage for future retrieval.

Phase Eight: Staff Review—A Final Check. The final proposal review is conducted by the proposal team and its management group plus selected functional managers, who will later provide contractual services. In addition, a specialty review committee may be organized to check the final draft for feasibility, rationale and responsiveness to the RFP. For a given chapter or work package, the staff review is accomplished in less than a day. The comments are reviewed by the original authors for incorporation.

Phase Nine: Final Edit. After the review comments are incorporated, the entire proposal is turned over to the publications department for final format editing, completion of art, word processing, and proofreading of the final, reproducible copy. The proposal is then returned to the publication specialist or proposal specialist, with all of the completed art, for a final check. The authors are then given a final opportunity to look at the completed proposal. Any major flaws or technical errors that may have crept into the copy are corrected at that time.

Phase Ten: Printing and Delivery. The proposal is now ready for pasteup, printing, and delivery to the customer.

9.8 PROPOSAL CONTENT AND ORGANIZATION

Although every individual proposal is different, the basic composition of each proposal has similar features. In general, the three principal elements of a proposal are the technical section, the management section, and the cost section. Each of these sections is equally important, and each must stand on its own because the client may evaluate each separately.

The technical component describes the approach that the company will take to meet the client's project requirements, including the techniques, system, and tools used to do the job as well as the schedule developed for the project.

The management component describes the organization, staffing, and special capabilities of the firm to perform the work satisfactorily. This section also describes the company's qualifications and background, its structure and overall capabilities and experience, including its policies, procedures, and financial status.

The cost component describes the details used in estimating the cost of doing the work and includes enough backup to give the client a high degree of confidence in its accuracy. Past cost performance can also be included, together with the terms and conditions under which the company will perform the work, including contract agreement, contract comments, secrecy agreements, proprietary data, and other matters.

Depending upon the type or size of the proposal required the proposal elements can be arranged in different ways. If the proposal is a small effort, the technical, management, and cost sections can be combined within one cover, but are separated physically by organization. If a large proposal effort is required, each individual module can be organized into separate volumes. The usual organization of proposals is two volumes, one technical and one commercial or cost and management.

To provide clarity and continuity of approach, the technical, management, and cost sections of the proposal should follow the same general format, and each should summarize the pertinent points of the overall proposal. Further, by standardizing the format, much general information, such as company background, qualifications, management approach, and plant layout, can be copied from previous proposals. This provides expediency and uniformity among bid proposals.

Proposals Come in Various Sizes

Depending on the project, type of solicitation and customer, bid proposals can range from a simple business letter to multivolume documents containing thousands of pages of text, illustrations, and other support material. However, regardless of its size, a bid proposal should contain the three important components: (1) technical, (2) managerial, and (3) cost. Table 9.4 presents an example of the type of specific information that would be included in each of the three components for a large project or complex industrial product. Sometimes the client requests a specific format to aid in his own evaluation of the proposal. This also aids your organization when putting the proposal together. Below are examples of outlines for two proposal types: (1) a single-volume proposal, and (2) a multi-volume proposal, discussing the technical, managerial and cost volume.

The Single-Volume Proposal

For both small and large projects, a single-volume proposal may suffice. In such a case, the *frontmatter* should contain:

- The transmittal letter
- Title page
- Disclosure protection statement, if any
- Table of contents

TABLE 9.4 Bid Proposal, Contents of Technical Proposal, Management, and Cost Proposal Volumes

1. TECHNICAL PROPOSAL	2. MANAGEMENT	3. COST PROPOSAL
Introduction and background of contractor's company	Process schedule	Price for services offered in proposal
Organization of contractor's company	Process description	Breakdown of price (materials, labor, etc.)
Schedule of professional personnel	Operating requirements	Escalations (lump-sum contract)
Resumé of key personnel or a resumé summary	Plot plans and elevations	Amount for subcontract work
Project management policy or philosophy	Process flow diagrams	Amount of off site facilities
Description of contractor's engineering department	Engineering flow diagrams	Taxes
Description of contractor's procurement department	Utilities flow diagrams	Royalty payment
Description of contractor's financial controls department	Heat and materials balance	Alternative systems
Experience list of similar plants built	Equipment list	Optional equipment
Experience list of large complexes built	Equipment data sheets	Prior adjustments (labor, efficiency, etc.)
Experience list of all plants built by contractor	General facilities, such as piping, instrumentation, electrical, civil, construction, etc.	Schedule of payments
Experience list of using a client's process	Contractor's or client's specifications or standards	
Photographs of plants built by contractor	Services provided by contractor	
Draft contract	Services provided by client	
	Model and/or rendering of proposed plant	

- List of illustrations and tables
- Summary, abstract, and/or introduction.

The *main body* of the proposal should contain:

- The opening statement
- Reference to inquiry document
- Description of the project
- Statement of work plan
- End result of proposed plan
- Participation in related activities
- Company objectives and qualifications.

If multivolume proposals are required, a standard 3-volume breakdown is shown below which suggests some typical content elements as a guideline for individual proposal layout and development.

The Technical Volume

The technical volume is usually the most difficult to prepare in both time needed and complexity. As a minimum, the technical volume should contain the following:

- *Statement of Problem.* This should be the company's interpretation of the statement of the problem, in order to show the customer that the company understands fully the work to be accomplished.
- *Technical Discussion.* This is a detailed discussion of the intended technical approach to achieve the objectives. This section can also include the work breakdown structure, a specification list, detail and summary schedules, hedge positions, potential risks, and methodology for tradeoffs.
- *Program Plan.* The program shows the logical steps that your company plans to follow to achieve the objective. For complex projects, you may wish to include a detailed program plan for such items as procurement, quality assurance, and so on. The program plan can also include resumes of key technical personnel, assignment of key technical personnel, linear responsibility charts, and manpower planning.
- *Description of Facilities.* This section describes the facilities that will be used for the project, unless it is already included in the management volume.
- *Exceptions, Deviations, and Assumptions.* This section delineates any exceptions, deviations, or assumptions that may differ from the customer's contract, statement of work, schedules, or specifications.
- *Background or Supplemental Information.* This section should contain background or supplemental technical information and should not be a duplication of the management volume. Related technical experience may be included here.

The Management Volume

The following sections appear frequently in management volumes; many subtopics are standard so-called *boiler plates.* However, all material must be properly integrated and responsive to the customer's formal request.

- *Administrative Management Capability.* This includes such items as the organizational structure of the company, the financial stability, the accounting system, employee compensation policies, Affirmative Action plans, employee safety and health programs, small business participation, quality assurance and control plan, and security. This section can also include a total staff profile, total company turnover data, total company manpower availability, and talent inventory.

- *Program Management.* This describes your company's exact method for managing the project, including the organizational charts for the project, a definition of the project manager's authority and responsibility, and interface mechanisms with the customer. Additional information includes resumés of key personnel and methodology for managing subcontracts and joint ventures.
- *Facilities.* This describes briefly the facilities that will be utilized on the project and may also include the utilization rate of the facilities for all ongoing activities.
- *Cost and Schedule Controls.* This is a management summary of the detailed sections of the cost proposal, assuming that the cost proposal contains an in-depth description. It includes the methods for authorizing costs, tracking costs, comparing actual to planned expenditures, doing variance analyses, and reporting.
- *History and Experience.* This describes briefly the history of your company and your related and nonrelated experiences (with emphasis, of course, upon the related activities).

The Cost Volume

The actual contents of the cost volume depends upon the type of contract and the nature of the product or services rendered. The following items may be included as part of the cost volume:

- *Basic Material.* This includes the abstract or title page, table of contents, introduction, and perhaps a summary.
- *Statement of Work.* This may be customer-furnished or developed in-house and may simply be rewritten in the contractor's own language to show that the contractor understands the work required fully.
- *Cost Summary.* This is an overall picture of the total cost for the project, perhaps broken down to one level of reporting.
- *Supporting Schedules.* These may very well be the same schedules that appear in the technical and management proposals.
- *Fee or Profit Statement.* This is the supporting data to justify either the fee or profit.
- *Government-or Customer-Furnished Equipment.* The proposal must state the equipment that you expect to be furnished to your company, usually free of charge (except for refurbishment).
- *Elements of Cost and Cost Breakdown.* This is the basis against which costs are measured, controlled, and accumulated.
- *Cost Format.* This is a further clarification of the cost breakdown, identifying the exact methods by which the costs are shown to the customer, either before or after accumulation.

- *Cost Estimating Techniques Used.* This describes the company's policy and procedures for estimating the work, whether customized or by estimating manuals.
- *Supporting Data.* This section contains the supporting data for labor, material, overhead estimates, if required by the customer.

9.9 NEGOTIATING AND CLOSING THE CONTRACT

Sending off the bid proposal signals the start of the postsubmission phase. Regardless of the type of customer or the formalities involved, even for an oral proposal, the procurement will go through the following principal steps:

1. Bid proposals received
2. Proposals evaluated
3. Proposal results compared
4. Alternatives assessed
5. Clarifications and new information from bidders
6. Negotiations
7. Award.

While bidders usually have no influence on proposal evaluation or the source selection process, they can certainly prepare properly for upcoming opportunities for customer negotiations. Depending on the procurement, these opportunities for improving the competitive position come in the following forms:

- Follow-up calls and visits
- Responding to customer requests for additional information
- Fact-finding requested by customer
- Oral presentations
- Invitations to field visits
- Sending samples or prototypes
- White paper
- Supportive advertising
- Contact via related contract work
- Plant or office visits
- Press releases
- Negotiations.

The bidder's objective on all postsubmission activities should be improving the competitive position. For starters, the bidder must assess the proposal relative to customer requirements and competing alternatives. In order to do this realistically, the bidder needs customer contact. Any opportunity for customer contact should be utilized. Follow-up calls and visits are effective in less formal procurements, while fact-finding and related contract work are often used by bidders in formal buying. These are just a few methods to open officially closed doors into the customer community. Only through active customer contact is it possible to assess realistically the competitive situation and organize for improvement and winning negotiations. Table 9.5 provides a listing of steps for which the bidder should organize. Customer contact and interaction are often difficult to arrange, especially in more formal procurements, which may require innovative marketing approaches.

While the proposal evaluation period is being used by the customer to determine the best proposal, the bidder has the opportunity to improve his position in three principal areas, (1) clarification of proposed program scope and content, (2) image building as a sound reliable contractor, and (3) counteracting advances made by competing bidders.

The proposal evaluation period is highly dynamic in terms of changing scores, particularly among the top contenders. The bidder who is well organized and prepared to interact with the customer community stands the best chance of being called first for negotiations, thus gaining a better basis for negotiating an equitable contract. Table 9.5 provides some guidelines for organizing and preparing for the postsubmission effort.

9.10 PROPOSAL BUDGET AND COST CONTROL

All new business development activities, especially proposals, should be accounted for by a cost budget. The sum of these individual budgets approximates the total budget allocated to the bidding effort by upper management. The proposal manager, assisted at times by a project manager, is responsible for determining the cost or budget for each proposal. Once that is established, the proposal manager is responsible for monitoring and controlling the expenditures logged against the budget.

The proposal manager should prepare the proposal budget and maintain cost control records in terms of man-hours rather than dollars. This method provides a quick means of evaluating the status of the effort without unnecessary dissemination of confidential or sensitive information.

As with any work effort, if you can control the end product, then you can control the work (including the cost) required to produce that product. Relating this to the proposal preparation effort, the first step toward producing the end product (the proposal) is to establish the format and content and then to develop a comprehensive, detailed plan. To implement the plan, it is essential that all members of the project team understand the proposal plan. A substantial effort should be made before starting on a proposal to define the work required from each participant. As discussed earlier, the major means of informing the entire team is the kickoff meeting. All team members must attend this meeting so that they can be briefed properly. This meeting ensures that each member of the team fully understands what is required. This results in the development of a better man-hour estimate and there-

TABLE 9.5 Organizing the Postsubmission Effort

1. *Reassess Your Proposal.* Study your proposal and reassess (1) strengths, (2) weaknesses, and (3) compliance with customer requirements.
2. *Plan Action.* Develop an action plan listing all weak points of your proposal, potential for improvement, and actions toward improvement.
3. *Open Communications.* Establish and maintain communications with the customer during the proposal evaluation period. Determine the various roles people play in the customer community during the evaluation.
4. *Find Your Score.* Try to determine how you scored with your proposal. Find out what the customer liked, disliked, and perceived as risks, credibility problems and standing against the competition. To determine your score realistically and objectively requires communications skills, sensitivity, and usually a great deal of prior customer contact. Determining your proposal score is an important prerequisite for being able to clarify specific items and to improve your competitive position.
5. *Seek Interaction.* Seek out opportunities for interacting with the customer as early as possible. Such opportunities may be had by presenting additional information for clarifying or enhancing specific proposed items. The meeting or presentation should be requested by the customer. It is a great opportunity for the bidder to "sell" a proposal further. This includes clarifications, modifications, options, and image-building.
6. *Prepare for Formal Meetings.* Be sure you are well-prepared for meetings, fact-finding sessions, or presentations requested by the customer. This is your opportunity not only to clarify but also to strengthen your proposal, show additional material, and introduce new personnel if needed.
7. *Reassess Cost and Price.* Cost and proposed effort are often fluid during the initial program phases. Many times the discussion of the proposal narrows down the real customer requirements. This provides an opportunity to reassess and adjust the bid price.
8. *Obtain Start-Work Order.* It is often possible to obtain a start-work order before the program is formally under contract. This provides a limited mutual commitment and saves time.
9. *Stay on Top Until Closure.* From the time of bid submission to obtaining the final contract, the bidder must keep abreast of all developments in the customer community that affect the proposal. Try to help the customer to justify the source selection and be responsive to customer requests for additional information and meetings. Frequent interaction with the customer helps to build trust, mutual respect, and credibility.
10. *Conduct Formal Negotiations.* Negotiation comes in many forms. Program contract negotiations mostly center on technical performance, schedules, and costs. They should be conducted among the technical and managerial personnel of the contracting parties. If, in addition to the technicalities, the contract covers extensive legal provisions, terms, and conditions, the bidder should seek the interpretation and advice of legal counsel.

fore a better and more realistic budget. The plan is published as the kickoff or coordination memo.

Immediately following the kickoff/coordination meeting, the proposal manager and/or the project manager assigned to the effort develops a budget for the bid proposal activity. A budget should be developed for each major step. For example, if you need to make a presentation to qualify for a project before submitting a proposal, you should have separate budgets prepared for the presentation and the proposal effort. The budget for the proposal should then include, as well as can be determined, the actual cost of the presentation. Typically, the proposal budget is prepared by the proposal manager. However, it usually requires some man-hour input from the participating departments.

Once a detailed task breakdown has been developed for each participating department, the proposal manager has each department provide a man-hour budget estimate for their portion of the work. If the proposal extends over several weeks, the department will also provide an estimate of personnel assignments, week by week. This information is used by the proposal manager in setting up a proposal progress chart. On large proposals, it is necessary to turn the departmental man-hours over to the estimating department, so that they can apply appropriate factors and other support costs to develop the final total proposal cost. Proposals of this size require upper-management approval of the final proposal budget.

The approach to controlling costs of proposal activities must allow for a wide range of both costs and time of performance that can be encountered on such activities. Effective cost control begins at the very start of a proposal effort and is continued throughout its life until the proposal is submitted to the client or until conclusion of all chargeable activities. To accomplish this, it is necessary to use a formal system or procedure. One method of effective control is to record the time charges applied to a proposal effort from start to finish. This is normally done by opening or assigning a charge number at the beginning of a proposal activity and closing it out at its completion. All persons working on the proposal can charge their time to the charge number; therefore, all time charged against the proposal is recorded. Charge numbers can be opened and closed as many times as necessary, but to keep down overhead costs, this should be done only when required for project continuity.

Control of labor costs requires real-time reporting of time charges, either on a daily report, giving man-hours charged by each person, or on a weekly report that provides total hours charged by each department or section. Other information that can be reported is weekly payroll cost and cumulative cost-to-date on a weekly or monthly basis. The proposal manager and the project manager (or a designer) should monitor both reports for all proposal activities. Another report is the monthly cost report, which covers payroll, travel, long-distance telephone calls, telex, and out-of-pocket expenses. Control of out-of-pocket, travel, and other miscellaneous charges is more difficult than controlling labor charges, because of the much longer lag time for their reporting.

Certain charges for other costs are subject to prior approval of the manager of proposals or operations manager, as follows:

- Travel by all personnel working on the activity (approval of expense accounts)
- Costs invoiced by other offices (approval of invoice)
- Costs invoiced by subcontractors or outside consultants (approval of invoice).

All questionable charges should be reviewed and appropriate action immediately taken on all mischarges. The principal tools for controlling proposal labor costs are:

- The man-hours budget for the effort
- The proposal schedule
- The weekly reports of man-hours charged against the proposal effort.

From these principal tools, the proposal manager can prepare a proposal man-hour report and/or progress chart to provide a unified record and control for the proposal effort. The proposal manager sets up the record and chart for the proposal as soon as the budget and schedule are established. The record and chart enable the proposal manager or others to make quick comparisons of progress, budgets, costs, and schedule, including dates for key proposal events.

Each week, the proposal manager can update the report and chart, using information from the accounting department as input. With the dates for key events as guideposts, progress, budget, costs, and schedule can be reviewed to determine whether or not the effort is proceeding proposal manager any actual or potential problems.

In reviewing the accounting department reports, the proposal manager makes sure that individuals who charge time to the proposal have actually been working on it. If mischarges are evident, the department concerned will transfer the charges to other accounts.

To provide for the practical control of costs over the wide range of expenditures that may be encountered, especially for large proposal activities, it may be worthwhile to set up separate budgets at the task level. The accounts are then negotiated with individual task leaders who are accountable for controlling their budgets according to the agreed on plan.

9.11 RECOMMENDATIONS TO MANAGEMENT

Winning a bid proposal depends on more than the pricing or right market position, or luck. This is true for any type of a contract, whether project or product, whether consumer, commercial, or military customer. Winning a piece of business depends on many factors, which must be carefully developed during the preproposal period, articulated in the bid proposal, and fine-tuned during contract negotiations. Success is the result of hard work, which starts with a proper assessment of the bid opportunity and a formal bid decision. To be sure, a low price bid or at least a price-competitive bid, can help in many situations. However, it may be interesting to note some research findings. A low price bid is advantageous toward winning only in contracts with low complexity, low technical risk, and high competition. In most other situations, price is a factor toward winning only in the context of all competitive components such as: (1) compliance with customer requirements, (2) best-fitting solution to customer problem, (3) real demand, (4) relevant past experience, (5) credibility, (6) long-range commitment to business segment, (7) past performance on similar programs, (8) soundness of approach (9) cost credibility, (10) competitive price, (11) delivery, (12) after-sale support, and (13) logistics. The better you understand the customer, the better you will be able to communicate the strength of your product relative to the customer requirements.

Some specific recommendations are made to summarize the discussion in this chapter and to help business managers responsible for winning new contracts and professionals who must support these bid proposal efforts to better understand the complex interrelationships among organizational, technical, marketing, and behavioral components, and to perform their difficult roles more effectively.

1. Plan Ahead. Develop a detailed business acquisition plan that includes a realistic assessment of the new opportunity and the specific milestones for getting through the various steps needed for bidding and negotiating the contract.

2. Involve the Right People, Early. In order to get a realistic assessment of the new opportunity compared with internal capabilities and external competition, form a committee of senior personnel early in the acquisition cycle. These people should represent the key functional areas of the company and be able to make a sound judgment on the readiness and chances of their company to compete for the new business effectively.

3. Closeness to the Customer. Especially for the larger contracts, it is important for the bidding firm to be closely involved with the customer prior to the RFP and, if possible, during the bid preparation. A company that has been closely involved with the customer in helping to define the requirements, in conducting feasibility studies with the customer, or in executing related contract work will not only understand the customer requirements better, but also will engender higher trust and credibility regarding the company's capacity to perform in comparison to a company that just submits a bid proposal.

4. Select Your Bid Opportunities Carefully. Bid opportunities are usually plentiful. Only qualified bidders who can submit a competitively attractive proposal have a chance of winning: submitting "more" proposals does nothing to improve your win ratio; it only drains your resources. Each bid opportunity should be carefully assessed to determine whether you really have the ingredients to win.

5. Make Bid Decisions Incrementally. A formal bid decision, especially for the larger proposals, requires considerable homework and resources. By making these decisions in several steps, such as initial, preliminary, and final, management can initially quickly screen a large number of opportunities and narrow them down to a shorter list without spending a lot of time and resources. Management can then concentrate the available resources on analyzing those opportunities that really seem to be most promising.

6. Be Sure You Have the Resources to Go the Full Distance. Many bid proposal activities require large amounts of resources and time. In addition, resources may be needed beyond the formal bidding activities for customer meetings, site visits, and negotiations. Further, the customer may extend the bid submission deadline, which will cost you more money as you continue to refine your proposal. Serious consideration should be given at the very beginning of any potential bid whether or not you truly have the resources and are willing to commit them. Develop a detailed cost estimate for the entire proposal effort.

7. Obtain Commitment from Senior Management to Make the Necessary Resources Available When Needed. This includes personnel and facilities.

8. Do Your Homework. Before any proposal writing starts, you should have a clear picture of the strengths, weaknesses, and limitations of (1) the competing firms and (2) your company. In addition, you need to fully understand the customer requirements and constraints, such as budgets, and the customer's biases. This requires intense customer contact and market research.

9. Organize the Proposal Effort Prior to the RFP. Run the proposal development like any other project. You need a well-defined step-by-step action plan, schedules, budgets, team organization, and facilities. You also have to prepare for support services, such as editing and printing.

10. "Grow" a Proposal Specialist. The efficiency and effectiveness of the proposal development can be greatly enhanced with a professional proposal specialist. This is an internal consultant who can lead the team through the proposal development process, including providing checks and balances via reviews and analysis.

11. Know Your Competition. Marketing intelligence comes in many forms. The marketer who is in touch with the market knows his competition. Information can be gathered at trade shows, bidder's briefings, customer meetings, professional conferences, from the literature, and via special market service firms.

12. Develop a Win Strategy. Define your niche or "unfair advantage" over your competition and build your win strategy around it. Only after intensive intelligence-gathering from the competition and the customer and careful analysis of these market data against your strengths and weaknesses can a meaningful win strategy be developed. Participation of key personnel from all functions of the organization is necessary to develop a meaningful and workable win strategy for your new contract acquisition.

13. Develop the Proposal Text Incrementally. Don't go from RFP to the first proposal draft in one step. *Use the storyboard process.*

14. Be Fully Compliant to the RFP. Don't take exceptions to the customer requirements unless absolutely unavoidable. A formal RFP analysis, listing the specific customer requirements, helps to avoid unintended oversights and also helps in organizing your proposal.

15. Demonstrate Understanding of Customer Requirements. A summary of the requirements and brief discussions helps to instill confidence in the customer's mind that you understand the specific needs.

16. Demonstrate Ability to Perform. Past related experiences will score the strongest points with your customer. But showing that your company performed well on similar programs, has experienced personnel, and has done analytical homework regarding the customer's requirements may rate very favorably with the customer too, especially when you have other advantages such as innovative solutions, favorable timing, or good pricing.

17. Progress Reviews. As part of the incremental proposal development, assure thorough reviews that check (1) compliance with customer requirements, (2) soundness of approach, (3) effective communication, and (4) proper integration of topics into one proposal.

18. Red-Team Reviews. For "must-win" proposals, it may be useful to set up a special review team that evaluates and scores the proposal, in a process similar to that used by the customer. Deficiencies that might otherwise remain hidden can often be identified and dealt with during proposal development. Such a red-team review can be conducted at various stages of the proposal development. It also is important to budget enough time for revising the proposal after red-team review.

19. Use Editorial Support. A competent editor should work side by side with the technical proposal writers. A good editor can take text at a rough-draft stage and "finalize" it regarding logic, style, and grammar. However, the proper content has to come from the technical author. Therefore, for the process to work, text often cycles between the author and editor several times until both agree on it. The professional editor not only frees the technical writer for the crucial innovative technical proposal development, but also provides clarity and consistency to the proposal. We further find that having professional editors increases the total writing efficiency by a factor of 2; that is, using professional editors reduces the total proposal-writing budget to one-half of what it would cost otherwise.

20. Price Competitively. Pricing is a complex issue. However, for most proposals, a competitively priced bid has the winning edge. Knowing the customer's budget and some of the cost factors of the competition can help in fine-tuning the bidding price. Further, cost-plus proposals must have cost credibility, which is built via a clearly articulated cost model and a description of its elements of cost. Pricing should start at the time of the bid decision.

21. Prepare for Negotiations. Immediately after proposal submission, work should start in preparation for customer inquiries and negotiations. Responses to customer inquiries regarding clarifications on the original bid can be used effectively to score additional technical points and build further credibility.

22. Conduct a Post Mortem. Regardless of the final outcome, a thorough review of the proposal effort should be held and the lessons learned should be documented for the benefit of future proposals.

A Final Note

Winning new contract business is a highly competitive and costly undertaking. To be successful, it requires special management skills, tools, and techniques, which range from identifying new bid opportunities to bid decisions and proposal developments.

Companies that win their share of new business usually have a well-disciplined process. They also have experienced personnel who can manage the intricate process and lead a multifunctional team toward writing a unified, winning proposal. Their management teams use good logic and judgment in deriving their bid decisions; they also make fewer fundamental mistakes during the acquisition process. These are the managers who position their companies uniquely in the competitive field by building a quality image with the customer and by submitting a responsive bid proposal that is competitively priced. These are the business managers who target specific opportunities and demonstrate a high win ratio.

9.12 CONTRACT INFORMATION SOURCES

Handbooks and Reference Material

A Guide for Private Industry, U.S. Department of Defense, U.S. Government Printing Office, Washington, DC 20402.

A Guide to Doing Business with the Department of State U.S. Department of State, U.S. Government Printing Office, Washington, DC, 1990.

Contract Planning and Organization, United Nations Publications, United Nations, LX2300, New York, NY 10017 (1974).

Directory of Federal & State Business Assistance: A Guide for New and Growing Companies, U.S. Department of Commerce, Washington, DC 20230.

Doing Business with the Federal Government, U.S. General Service Administration, U.S. Government Printing Office, Washington, DC, 1989.

DOD Supplement to Federal Acquisition Regulation, U.S. Government Printing Office, Washington, DC 20402.

DOD Small and Disadvantaged Business Utilization Specialists (SADBUS), List, U.S. Department of Defense, Washington, DC.

Federal Acquistion Regulation, U.S. Government Printing Office, Washington, DC 20402.

Government Contracts Guide, Commerce Clearing House, 4025 West Peterson Avenue, Chicago, IL 60646.

Government Contracts Directory, U.S. Government Data Publications, 422 Washington Building, Washington, DC 20005, (Annual).

Guide to Contracting Regulations, U.S. Government Printing Office, Washington, DC 20402.

Guide for Drawing up Contracts for Large Industrial Works, United Nations, LX2300, New York, NY 10017 (1972).

Handbook for Small Business Programs of the Federal Government, U.S. Senate Committee on Small Business, Washington, DC.

How to Seek and Win NASA Contracts, National Aeronautics and Space Administration, Washington, DC.

Selling to NASA, National Aeronautics and Space Administration, U.S. Government Printing Office, Washington, DC, 1990.

Selling to Navy Prime Contractors, U.S. Government Printing Office, Washington, DC 20402.

Selling to the Military, U.S. Department of Defense, U.S. Government Printing Office, Washington, DC, 1990.

Selling to United States Government, United States Small Business Administration, Washington, DC 20416 (1973).

Small Business Guide to Federal R&D Funding Opportunities, National Science Foundation, Washington, DC.

The National Directory of State Agencies, Cambridge Information Group, Cambridge, MA.

United States Government Purchasing and Sales Directory, U.S. Small Business Administration, U.S. Government Printing Office, Washington, DC 20402.

Women Business Owners: Selling to the Federal Government, U.S. Small Business Administration, U.S. Government Printing Office, Washington, DC, 1985.

Periodicals and Newspapers

Briefing Papers, Federal Publications, 1725 K Street NW, Washington, DC 20006. Bimonthly.

Commerce Business Daily, U.S. Department of Commerce, Office of Field Services, U.S. Government Printing Office, Washington, DC 20402.

Federal Register, U.S. Government Publication, daily, U.S. Government Printing Office, Washington, DC 20402.

Federal Register Index, U.S. Government Publication, monthly, U.S. Government Printing Office, Washington, DC 20402.

Forms of Business Agreement, Institute of Business Planning, IPB Plaza, Englewood Cliffs, NJ 07632. Monthly.

Government Contractor, Federal Publications, 1725 K Street NW, Washington, DC 20006. Biweekly.

Government Contracts Reports, Commerce Clearing House, 4025 West Peterson Avenue, Chicago, IL 60646. Weekly.

NCMA Newsletter, National Contract Management Association, 675 East Wardlow Road, Long Beach, CA 90807. Monthly.

On-Line Databases

Defense Market Measurement System, Frost and Sullivan, 109 Fulton Street, New York, NY 10038.

Federal Register, Capitol Services, 511 Second St. NE, Washington, DC 20002.

Associations and Societies

Electronic Industries Association (EIA), 2001 I Street NW, Washington, DC 20006.

National Contract Management Association, 2001 Jefferson Davis Highway, Arlington, VA 22202.

National Institute of Government Purchasing, 1001 Connecticut Avenue NW, Washington, DC 20036.

BIBLIOGRAPHY

Ammon-Wexler, J. and Catherine Carmel, *How to Create a Winning Proposal,* Santa Cruz, CA: Mercury Communications.

Asner, Michael, "Management: Winning Proposals," *Computing Canada* (Canada), Volume 15, Issue 14 (July 6, 1989), p. 18.

Asner, Michael, "Creating a Winning Proposal," *Business Quarterly,* Volume 55, Number 3 (Winter 1991), p. 36(4).

Barakat, Robert A., "Storyboarding Can Help Your Proposal," *IEEE Transactions on Professional Communication,* Volume 32, Issue 1 (Mar 1989), pp. 20–25.

Beveridge, Jim M. and E. J. Velton, *Creating Superior Proposals,* Talent, OR: J. M. Beveridge Associates.

Bond, David F., "USAF Systems Command Reforms Maze of RFP Procedures," *Aviation Week & Space Technology,* Volume 132, Issue 2 (Jan 8, 1990), pp. 24–25.

Boughton, Paul D., "The Competitive Bidding Process: Beyond Probability Models," *Industrial Marketing Management,* Volume 16, Issue 2 (May 1987), pp. 87–94.

Buchman, Matthew L., "The RFRP: Winning the Race of Technology," *Journal of Systems Management,* Volume 41, Issue 9 (Sep 1990), pp. 6–9.

Budish, Bernard Elliott, and Richard L. Sandhusen, "The Short Proposal: Versatile Tool for Communicating Corporate Culture in Competitive Climates," *IEEE Transactions on Professional Communication,* Volume 32, Issue 2 (Jun 1980), pp. 81–85.

Bush, Don, "Proposal Writing: A Profusion of Panaceas," *Technical Communication,* Volume 37, Number 3 (Aug 1990), p. 261(1).

Clauser, Jerome K., "PODs: Pump Primers for Proposal Writers," *Technical Communication,* Volume 36, Number 2 (Apr 1989), p. 114(5).

DeRose, Louis J., "Negotiations: Clouding the Ethics Issue with Pseudo-Ethics," *Purchasing World,* Volume 30, Issue 9 (Sept 1986), pp. 34–36.

Edelman, F., "The Art and Science of Competitive Bidding," *Harvard Business Review,* Volume 43 (July/Aug 1965).

Falleder, Arnold, "The Making of a Proposal Writer," *Nonprofit World,* Volume 7, Issue 4 (Jul/Aug 1989), pp. 26–27.

Fulton, William, "To Bid or Not to Bid," *Planning,* Volume 54, Issue 12 (Dec 1988), pp. 15–19.

Gould, Lee, "Ten Tips for a Good RFP," *Modern Materials Handling,* Volume 45, Issue 14 (Dec 1990), pp. 55.

Hamilton, Ogden J., "Pros and Pitfalls of Creative RFPs," *Travel Weekly,* Volume 49, Number 93 (Nov 19, 1990), p. 32(1).

Holtz, Herman, "The $650 Billion Market Opportunity," *Business Marketing,* Volume 71, Issue 10 (Oct 1986), pp. 88–96.

Knapp, Bonnie Ogram, "Writing a Proposal? No Problem," *Training,* Volume 25, Issue 3 (Mar 1988), pp. 55–58.

Kunz, Gerald R., "Project Controls: Management's Decision-Making Tool," *Cost Engineering,* Volume 30, Issue 1 (Jan 1988), pp. 15–22.

Lilley, Vic, "Choice Proposals," *Reactions* (UK), (Dec 1990), pp. 33–34.

Livingston, Dennis, "Federal Systems Integration—Part V: The Envelope, Please," *Systems Integration,* Volume 22, Issue 9 (Sept 1989), pp. 65–70.

Livingston, Dennis, "Federal Systems Integration—Part IV: Romancing the Feds," *Systems Integration,* Volume 22, Issue 9 (Sept 1989), pp. 61–63.

Loew, Gary Wayne, "Designing an Effective Request for Proposal," *Systems/3X World,* Volume 15, Issue 2 (Feb 1987), pp. 52–60.

McIntyre, Kathryn J., "Don't Pick Vendor Solely on Price: Kloman," *Business Insurance,* Volume 23, Issue 23 (Apr 24, 1989), pp. 20–21.

Nevers, David, "Negotiating Contracts: Coming to Terms with Your Vendor," *Business Communications Review,* Volume 17, Issue 6 (Nov/Dec 1987), pp. 10–11.

Shaw, Abigail, "Putting Technology in the Background," *Inform,* Volume 4, Issue 6 (Jun 1990), pp. 34–40.

Skimin, William E. Smith, Darrel E. J. O. Smith, John C. Krolicki, and Kenneth D. Zenner, "Scope Management on an Automotive Tooling Project," *Project Management Journal,* Volume 22, Number 3 (Sept 1991), pp. 22–26.

Thamhain, Hans J., "Marketing for Project-Oriented Businesses," *Project Management Quarterly* (Dec 1982), pp. 29–39.

Ware, Robb, "A View of Alternatives," *Journal of Systems Management,* Volume 41, Number 2 (Feb 1990).

White, Leslie, "Negotiating with Suppliers," *Computing Canada* (Canada), Volume 16, Issue 13 (June 21, 1990), pp. 18, 47.

10
MOTIVATION AND LEADERSHIP

10.1 NEW REALITIES IN ENGINEERING MANAGEMENT

An understanding of motivational forces and leadership skills are essential for effective management of multidisciplinary engineering activities. The ability to build project teams, motivate people, and create organizational structures conducive to innovative and effective work requires sophisticated interpersonal and organizational skills. Specifically, today's engineering management environment is characterized by:

- Complex multidisciplinary tasks
- Strong competition for scarce resources
- Authority that is largely perceived by the members of the organization to be based on earned credibility, expertise, and perceived priorities.
- Dual accountability of most personnel, especially in project-oriented environments
- Shared power between resource managers and project on task managers
- Individual autonomy and participation greater than in traditional organizations
- Weak superior–subordinate relationships in favor of stronger peer relationships
- Subtle shifts of personnel loyalties from functional to project lines
- Engineering performance that depends on teamwork
- Group decision-making that tends to favor the strongest organizations
- Reward and punishment power travel via both vertical and horizontal lines in a highly dynamic pattern

- Influences of reward and punishment from many organizations and individuals
- Individual accountability and self-actuating behavior
- Changing technology, uncertain operating environment
- Dynamic organizational relations
- Need for innovation and creativity
- Multiproject involvement of support personnel and sharing of resources among many engineering activities.

There is no single magic formula for successful engineering management. However, most senior managers agree that effective engineering managers need to understand the interaction of organizational and behavioral elements in order to build an environment conducive to their teams' motivational needs, a prerequisite for effective leadership of complex multidisciplinary engineering undertakings.

What engineering executives find is that the management of technical complexities in an often unorthodox organizational environment with ambiguous authority and responsibility relations requires a more sophisticated leadership style than that needed for traditional management situations. In fact, engineering performance is based on individual accountability, commitment, and self-actuating behavior. It is also based on the unification of various individual efforts into an integrated team effort, consistent with the overall goals of the department. All these factors point toward a more open and adaptable management style, which is often described as a "team-centered" style. Such a management style is based on the thorough understanding of the motivational forces of people and an understanding of their interaction with the organizational environment. The basis for such a team-centered leadership style is the manager's ability to foster a work environment that is professionally stimulating and interesting. People like what they are doing and unify behind the task objectives, becoming committed to making them happen. Of course, effective engineering leadership requires more than just an understanding of human behavior. However, motivated people are self-directing; and like riding horses, they are easier to lead in the direction they are already going!

10.2 MOTIVATIONAL FORCES IN ENGINEERING

Understanding people is important for the effective management of today's technical programs and engineering activities. The breed of manager that succeeds within these often unstructured work environments faces many challenges. Within the company they must be able to deal effectively with a variety of interfaces and support personnel over whom they often have little or no control. Outside the company, managers have to cope with constant and rapid changes in technology, markets, regulations, and socioeconomic factors. Moreover, traditional methods of authority-based direction, performance measures, and control are impractical in such contemporary environments.

Finding ways of gaining the desired level of support is a true challenge for managers and project leaders in engineering organizations. Typically these organizations have unconventional relationships between superiors and subordinates, which often violate the unity-of-command principle, but accept managerial power-sharing and dual accountability. To be effective, managers in these situations require a large degree of interpersonal and organizational skills, plus an understanding of the motivational bases and needs of project personnel. The ability to recognize these needs and to build a work environment conducive to their fulfillment is crucial for effective teamwork and managerial performance.

Sixteen Specific Professional Needs

In field research on engineering personnel, 16 specific professional needs were identified.[1] These studies showed that the fulfillment of professional needs can drive engineering personnel to higher performance; conversely, the inability to fulfill these needs may become a barrier to teamwork, personal drive, commitment, and creativity, and ultimately may negatively affect overall engineering performance. The rationale for this important correlation is found in the complex interaction of organizational and behavioral elements.

Effective engineering management performance involves three primary issues: (1) people skills, (2) organizational structure, and (3) management style. All three issues are influenced by the specific task to be performed and the surrounding environment. The same three issues surface again in satisfying professional needs of engineering professionals. The degree of satisfaction of any of the needs is a function of:

- The right mix of people with appropriate skills and traits
- The right organization and resources, according to the tasks to be performed
- The right leadership style.

The sixteen specific professional needs that have been identified for engineering personnel are listed below:

1. Interesting and Challenging Work. Interesting and challenging work satisfies various professional esteem needs. It is oriented toward the intrinsic motivation of the individual and helps to integrate personal goals with the objectives of the organization.

2. Professionally Stimulating Work Environment. This leads to professional involvement, creativity, and interdisciplinary support; it also fosters team-building and is conducive to effective communication, conflict-resolution and commitment to organizational goals. The quality of the work environment is defined through its

[1]See H. J. Thamhain, "Managing Engineers Effectively," *IEEE Transactions on Engineering Management,* Vol. 30, No. 4, November 1983, pp. 231–237.

organizational structure, facilities, and management style. It includes a very complex set of variables that perhaps are most closely resembled by Ouchi's Theory Z Organization.[2]

3. Professional Growth. Professional growth is measured by promotional opportunities, salary advances, learning of new skills and techniques, and professional recognition. A particular challenge exists for management in a limited-growth or zero-growth businesses—to compensate for the lack of promotional opportunities by offering more intrinsic professional growth in terms of job satisfaction.

4. Overall Leadership. This involves dealing effectively with individual contributors, managers, and support personnel within a specific functional discipline as well as across organizational lines. It involves technical expertise, information-processing skills, effective communication, and decision-making skills. Taken together, leadership means satisfying the need for clear direction and unified guidance in reaching established objectives.

5. Tangible Rewards. These include salary increases, bonuses, and incentives, as well as promotions, recognition, better offices, and educational opportunities. Although extrinsic, financial rewards are necessary to sustain strong long-term efforts and motivation. Furthermore, they validate more intrinsic rewards, such as recognition and praise, and reassure people that higher goals are attainable.

6. Technical Expertise. People need to have all necessary interdisciplinary skills and expertise available within the engineering team in order to perform the required tasks. Technical expertise includes an understanding of the technicalities of the work, the technology and underlying concepts, theories and principles, design methods and techniques, and functioning and interrelationships of the various components that make up the total system.

7. Assistance in Problem-Solving. Assistance in problem-solving, such as facilitating solutions to technical, administrative, and personal problems, is a very important need. If it is not satisfied, it often leads to frustration, conflict, and poor quality engineering work.

8. Clearly Defined Objectives. Goals, objectives, and outcomes of an engineering effort must be clearly communicated to all affected personnel. Conflict can develop over ambiguities or missing information.

9. Management Control. Management control is important to engineering professionals for effective performance. The managers must understand the interaction of organizational and behavioral variables in order to exert the direction, leadership, and control required to steer the engineering effort toward established organizational goals without stifling innovation and creativity.

10. Job Security. This is one of the very fundamental needs that must be satisfied before people consider higher-order growth needs.

[2]In his book, *Theory Z,* (Reading, MA: Addison-Wesley, 1983) William Ouchi describes a management style that is a hybrid of American and Japanese styles. In organizations using Theory Z, managers rely largely on networks for command, control, and communications, even though traditional hierarchical structures exist and form the basic organizational framework.

11. Senior Management Support. Senior management support should be provided in four major areas: (1) financial resources, (2) an effective operating charter, (3) cooperation from support departments, and (4) provision of necessary facilities and equipment. It is particularly crucial to larger, more complex undertakings.

12. Good Interpersonal Relations. These are required for effective engineering teamwork, since they foster a stimulating work environment with involved, motivated personnel, low conflict, and high productivity.

13. Proper Planning. Proper planning is absolutely essential for successful management of complex engineering work. It requires communication and information-processing skills to define the actual resource requirements and administrative support necessary. It also requires the ability to negotiate resources and commitment from key personnel in various support groups across organizational lines.

14. Clear Role Definitions. This helps to minimize role conflict and power struggles among team members and/or supporting organizations. Clear charters, plans, and good management direction are some of the powerful tools to facilitate clear role definitions.

15. Open Communications. This satisfies the need for a free flow of information, both horizontally and vertically. It keeps personnel informed and functions as a pervasive integrator of the overall project effort.

16. Minimizing Changes. Although the engineering manager has to live with constant change, project personnel often see change as an unnecessary condition that impedes their creativity and productivity. Advanced planning and proper communications can help to minimize changes and lessen their negative impact.

Implications for Organizational Performance

The significance of assessing these motivational forces lies in several areas. First, the above listing provides insight into the global needs that engineering professionals seem to have. These needs must apparently be satisfied *continuously* before engineering personnel can reach high levels of performance. The emphasis is on *continuously* satisfying these needs, or at least providing a strong potential for such satisfaction. This is consistent with findings from other studies, which show that in technical environments a significant correlation exists between professional satisfaction and organizational performance.[3] From the above listing we now know more specifically on what areas we should focus our attention. In fact the above listing provides a model for managers to monitor, define, and assess the needs of their people in more specific

[3]See B. N. Baker, D. C. Murphy, and D. Fisher, "Factors Affecting Project Success," in D. I. Cleland and W. R. King, eds., *Project Management Handbook,* New York: Van Nostrand Reinhold, 1983, pp. 670–671, for a discussion of determinants of project success as perceived by customers, management, and project personnel. Also, H. Thamhain found a strong correlation between professional satisfaction of team personnel and project performance; see "Managing Technologically Innovative Project Teams Toward New Product Success," *Journal of New Product Innovation Management,* Vol. 7, No. 1, March 1990.

ways. With their awareness of professional needs, managers can direct their personnel and build a work environment that is responsive to these needs. As an example, from top-down the total work structure and organizational goals might be fixed and not negotiable; however engineering managers have a great deal of control over the way the work is distributed and assigned. The same degree of operational and environmental control exists in most of the other identified need areas.

Second, the above listing of needs provides a topology for actually measuring organizational effectiveness as a function of these needs and the degree at which they seem to be satisfied. In fact, I recently completed a field study of 300 engineering professionals and their management, which provides some interesting insight into the effects of professional need satisfaction on engineering performance.[4] We measured the degree of satisfaction for each of the 16 needs, as perceived by individual contributors. Then we obtained measures of engineering group performance from the immediate supervisor. These measures included the group's degree of:

- Innovation and creativity
- Communication effectiveness
- Commitment by individuals to objectives
- Work involvement and energy
- Interfacing with other organizations
- Willingness to change
- Capacity to resolve conflict
- Overall organizational effectiveness.

The degree of association between professional need satisfaction and organizational performance is shown in Table 10.1. Specifically, the table indicates the strength of association between each need and performance factor, using Kendall's tau rank-ordered correlation measures. Although all need satisfactions are positively correlated with group performance, certain factors have a particularly strong association. A summary of conclusions from this analysis, and their management implications, is given below.

1. Professional Needs Perception. The sixteen needs listed in Table 10.1 were perceived by engineering personnel as particularly important to their ability of working professionally, efficiently, and in a personally satisfied manner.

2. Performance Correlates. All sixteen needs correlated positively to overall engineering performance and other indirect measures of organizational effectiveness, such as innovation, communications, commitment, personal involvement and energy,

[4]For a detailed discussion of the field research study, see H. J. Thamhain, "Managing Engineers Effectively," *IEEE Transactions on Engineering Management*, Vol. 30, No. 4 (May 1983), pp. 231–237, with follow-up studies published in "Managing Technology: The People Factor," *Technical & Skill Training Journal*, Vol. 1, No. 2, August/September 1990.

TABLE 10.1 The Effect of Professional Need Satisfaction on Engineering/Technology Group Performance

	Engineering Group Characteristics and Performance							
Team Member Needs	Innovation and Creativity	Communications	Commitment	Involvement and Energy	Organizational Interface	Willingness to Change	Conflict Handling	Overall Effectiveness
Work Challenge	3	3	3	3	2	2	2	3
Stimulating Environment	2	3	2	3	2	1	2	2
Growth	1	2	2	2	1	2	1	2
Leadership	3	2	2	2	2	2	3	3
Rewards	1	1	2	2	1	2	1	2
Expertise	2	2	2	2	1	1	1	2
Assistance and Help	1	1	2	1	2	2	2	2
Clear Objectives	3	1	2	2	2	1	1	2
Control	2	1	1	2	2	2	2	2
Job Security	2	1	1	1	1	1	1	2
Management Support	2	2	3	2	2	2	2	2
Good Personal Relations	1	2	1	2	2	2	3	3
Good Planning	2	2	2	3	2	1	1	2
Clear Role Definition	1	1	2	1	2	2	2	2
Open Communications	2	3	2	2	2	3	3	2
Minimum Changes	1	1	2	2	1	2	1	2

1 = $+\tau$, Positive Kendall's tau Correlation at significance level of $\propto = 85\%$.
2 = $+\tau$, Positive Kendall's tau correlation at significance level of $\propto = 95\%$.
3 = $+\tau$, Positive Kendall's tau correlation at significance level of $\propto = 99\%$.

organizational interface, willingness to change, and capacity to resolve conflict. The stronger the need satisfaction of individuals in an engineering group, the higher was their group's performance.

3. Most Significant Factors. As indicated by their strong positive association, all needs seem to be important to group performance. However, several factors appear to be very important. For instance, having professionally interesting and challenging work and having overall leadership seemed to influence group performance more strongly than any other factors. Having a professionally stimulating work environment and open communications also can be included with the set of factors that are statistically most strongly associated with group performance measures.

4. Intercorrelations. As one might expect, all need factors cross-correlate to some degree. That is, people who find their work professionally challenging also find the organizational environment stimulating, perceive good potential for professional growth, find their superiors having sufficient leadership qualities, etc. In fact, this cross-correlation has been tested via a Kruskal-Wallis analysis of variance by ranks at a confidence level of $p = 96\%$, which indicates a very strong cross-correlation.[5]

It is clear, however, that there are costs associated with satisfying these needs and managing necessary changes. What we are lacking is a good method for comparing the relative costs of an organizational change with the gain in engineering productivity; this is clearly an area for future research. The challenge is for managers to develop enough understanding of the dynamics of their organizations that they are able to diagnose potential problems and the need for change. Only by understanding those variables that influence engineering performance can one fine-tune the organization toward maximum long-range engineering productivity and overall cost-effectiveness. In an engineering environment, one of these variables is the degree of need-satisfaction perceived by professional personnel.

Motivation as a Function of Need Satisfaction

So far we have identified a number of needs that appear important to engineering professionals. Their fulfillment leads to satisfaction and can be considered a reward. The motivational process can be explained with a simple model, as shown in Figure 10.1. Let's take the example of an engineer named Carol Edwards, who has as a goal advancing in her career. This would satisfy her needs for more money and prestige, interesting work, job security, etc. These satisfiers are by and large just perceptions on part of the individual of what the outcome of reaching the goal would give. Such perceptions may, for instance, include more money, which may not be guaranteed. They may include even erroneous ideas, such as getting a reserved parking spot or

[5]Kruskal and Wallis developed a nonparametric test in 1952 based on ranks, as an extension of the more commonly known Mann-Whitney U-test. It allows to test differences and similarities of samples and thus allows one to measure the degree of cross-correlation among these samples. For specific methods and formulas, see Roger E. Kirk, *Experimental Design & Procedures for the Behavioral Sciences*, Belmont, CA: Wadsworth Publishing Company, 1968.

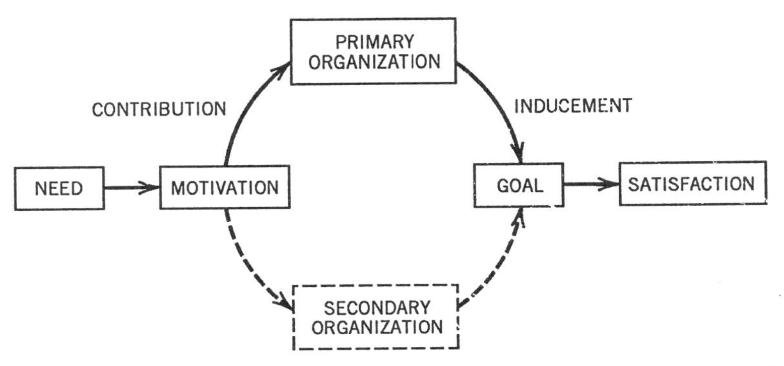

FIGURE 10.1 The inducement–contribution model of motivation. Explaining the role of needs and probability of success as a function of motivation and contribution to the organization.

the key to the executive washroom (which do not exist in Carol's company). However, regardless of their reality, it is the *perceived* outcome that motivates the engineer to work toward this goal of professional growth. Furthermore, since the employer's organization provides the opportunity for the goal, the engineer is induced to contribute to the organization. Therefore the model is called the inducement–contribution model.

Management's Reaction to Employee's Needs. Why should management pay attention to an employee's goals? We found from the needs analysis discussed earlier in this chapter that professional people perform at higher levels of efficiency if they can satisfy their needs within the organizational environment and in their work itself. If the probability of reaching a given goal becomes very low, the person's inducement to contribute to the organization by working toward the goal is low. In such a situation, the person in our example, Carol, is likely to exercise one of several behavioral options or a combination of them. *Option One:* Carol may give up on her original goal and manifest either the "sour grapes" syndrome or rationalizes that her goal is unattainable. In either case, the inducement toward organizational contribution disappears. *Option Two:* Carol terminates her activities with her current organization and seeks out employment with another organization, where the opportunity for reaching the desired goal appears better. *Option Three:* Carol may stay with the current organization, but may try to satisfy her needs by working toward her goals through an "outside" organization. She may involve herself in volunteer work or recreational activities. While this may help to satisfy the employee's need for interesting and stimulating activities, her efforts and energy are directed within a secondary organization, which provides inducement and receives her contributions, which is being "leaked away" from her employer.

From the above scenario, we can see that the employee's need satisfaction is in the long term as important to the functioning of the organization as it is to the employee. A good manager should be closely involved with the staff and their work. The manager should identify any unrealistic goals and correct them. The manager should work to change situations that impede the attainment of realistic goals, so that employees' energies aren't channeled elsewhere. The tools that help the manager to facilitate professional satisfaction are work sign-on or delegation, career counseling and development, job training and skill development, and effective managerial direction and leadership.

Motivation as a Function of Risks and Challenges

Additional insight into motivational drive and its dynamics can be gained by considering motivational strength as a function of the probability and desire to achieve the goal. We can push others or ourselves toward success or failure because of this mental precondition. Our motivational drive and personal efforts increase or decrease relative to how likely the expected outcome is to happen. Personal motivation toward reaching a goal changes with the perception of doability and challenge. Figure 10.2 expresses the relationship graphically. A person's motivation is very low if the probability of achieving the goal is very low or zero. As the probability of reaching the goal increases, so does the motivational strength. However, this increase continues only up to a certain level; then motivation decreases as the probability of success approaches 100%. This is possibly an area in which the work is perceived as routine, uninteresting, and holding little potential for professional growth.

Success as a Function of Motivation. The saying, "The harder you work, the luckier you get" expresses the effect of motivation in pragmatic terms. People who have a "can-do" attitude, who are confident and motivated, are more likely to succeed in their mission. Winning is in the attitude. This is the essence of the self-fulfillment prophecy. In fact, high-performing engineering teams often have a very positive image of their capabilities. As a result, they are more determined to produce the desired results, often against high risks and odds. This has been demonstrated via many studies, such as the classic study at UCLA in the early 1960s. A class of high

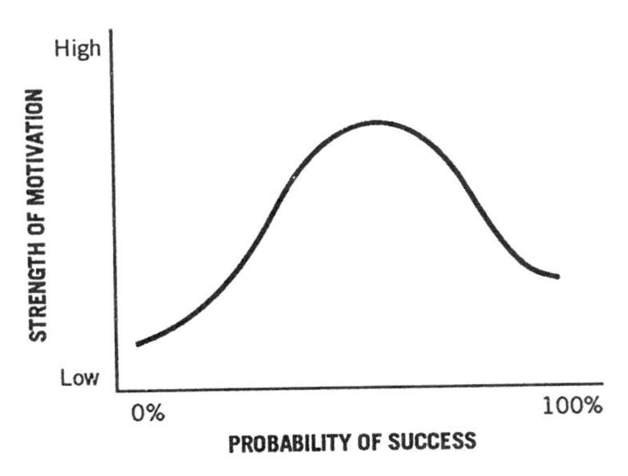

FIGURE 10.2 The Pygmalion Effect, relating strength of motivation to the probability of success.

school students was randomly divided into two groups. Group A was as bright and academically prepared as Group B. Two teachers were given the assignment to teach a given subject as effectively as possible. Teacher A was told that Group A is an extremely bright group of students, who were selected specifically for this experiment. Teacher B was given to Group B and told that this group was a particularly slow group, but that the directors expected that the teacher would do the best she could do with this slow group. At the end of a given time period, both groups were tested on their knowledge of the subject. Maybe it does not surprise the reader that Group A tested considerably higher in the knowledge gained in the subject. The experiment was repeated thousands of times with different groups, different teachers, different subjects, different schools. The outcome remained the same Group A was ahead of B. *What you expect from others is more likely to happen.* Also, what you expect from yourself is more likely to happen. This is a self-fulfilling prophecy.[6]

In another recent study, Dennis Waitley investigated the 1984 Olympic athletes. He found that most of the athletes who broke a record knew ahead of time that that was the particular event where they could do it. This is an example of the Pygmalion effect. The name goes back in history to the saga of King Pygmalios of Cyprus, who sculptured a beautiful woman out of marble. He then fell in love with his own creation and hoped that the statue could be brought to life. He prayed intensely to Aphrodite. And, indeed the statue came to life as Galatea, who then married Pygmalios. Hence if you believe in your ability you can even bring a stone to life, at least according to the saga.

Manage in the Range of High Motivation

This simple Pygmalion-effect model can provide considerable insight into the dynamics of motivation and can help in guiding the manager's efforts toward building a productive work force. Long before managers used the model, many school systems paid attention to motivation, when they tried to group students of each grade into classes of various levels of aptitude homogeneous grouping. The intent was to place the pupil in a class that matched his or her academic ability and background, because if work was too challenging, it became threatening and mentally overwhelming, resulting in the student having a low motivational level. Similarly, if the work was too easy, it appeared boring and dull, also resulting in low motivation to study.

Applying this scenario to the engineering workplace suggests that managers should ensure that their people are properly matched to their jobs. Further, managers must foster a work environment and direct their personnel in a manner that promotes "can-do" attitudes and facilitates continuous assistance and guidance toward successful task completion. The process involves four primary issues, as discussed below.

[6]The impact of superiors' expectations of subordinates was studied by S. Livingston. See "Pygmalion in Management," *Harvard Business Review*, July–August 1969, pp. 81–89. The study concluded that poorer subordinates' performance is associated with superiors having lower personal standards for themselves and lower expectations of their subordinates.

1. Work Assignments

- Explain the assignment, its importance to the company, and the type of contributions expected.
- Understand the employee's professional interests, desires, and ambitions and try to accommodate to them.
- Understand the employee's limitations, anxieties, and fears. Often these factors are unjustly perceived and can be removed in a face-to-face discussion.
- Develop the employee's interest in an assignment by pointing out its importance to the company and possible benefits to the employee.
- Assure assistance where needed and share risks.
- Show how to be successful. Develop a "can-do" attitude.
- If possible, involve the employee in the definition phase of the work assignment, for instance via up-front planning, a feasibility analysis, needs assessment, or a bid proposal.

2. Team Organization

- Select the team members for each task or project carefully, assuring the necessary support skills and interpersonal compatibility.
- Plan each engineering project properly to assure clear directions, objectives, and task charters.
- Assure leadership within each task group.
- Sign-on key personnel on a one-on-one basis according to the guidelines discussed in item number 1.

3. Skill Development

- Plan the capabilities needed in your engineering department for the long-range. Direct your staffing and development activities accordingly.
- Encourage people to keep abreast in their professional field.
- Provide for on-the-job experiential training via selected work assignments and managerial guidance.
- Provide the opportunity for some formal training via seminars, courses, conferences, and professional society activities.
- Use career counseling sessions and performance reviews to help in guiding skill development and matching them with personal and organizational objectives.

4. Management

- Develop interest in the work itself by showing its importance to the company and the potential for professional rewards and growth.

- Promote project visibility, team spirit, and upper-management involvement.
- Assign technically and managerially competent task leaders for each team, and provide top-down leadership for each project and for the engineering function as a whole.
- Manage the quality of the work via regular task reviews and by staying involved with the project teams, without infringing on their autonomy and accountability.
- Plan your projects up front. Conduct a feasibility study and requirements analysis first. Assure the involvement of the key players during these early phases.
- Break activities or projects into phases and define measurable milestones with specific results. Involve personnel in the definition phase. Obtain their commitment.
- Try to detect and correct technical problems early in their development.
- Foster a professionally stimulating work environment.
- Unify the task team behind the overall objectives. Stimulate the sense of belonging and mutual interdependence.
- Refuel the commitment and interest in the work by recognizing accomplishments frequently.
- Assist in problem-solving and group decision-making.
- Provide the proper resources.
- Keep the visibility and priority for the project high. No interruptions.
- Avoid threats. Deal with fear, anxieties, mistrust, and conflicts.
- Facilitate skill development and technical competency.
- Manage and lead.

10.3 FORMAL MODELS OF MOTIVATION

Employee motivation has been a major concern of managers throughout history. However, it was only with the beginning of the industrial era at the turn of the century that researchers and practitioners began to formalize concepts and theories of human behavior. These theories can be used to model motivation and leadership behavior and therefore help to understand human behavior.

No one theoretical model, no matter how complex, can describe human behavior accurately for any situation, and certainly not for all situations. However, each model allows us to look at human behavior in a unique way, exploring different facets of the intricate and complex human behavioral system. It is especially the simple model that provides powerful insight and helps us, in concert with other models, to understand people and manage them effectively. This section discusses a selected number of concepts that should provide an integrated perspective on motivation and leadership. They also provide the basis for applying the specific management techniques that are discussed toward the end of this chapter.

The Nature of Motivation

Motivation is an inner drive that transforms activities into desired results. More specifically, people have needs and choose goals that they believe will satisfy these needs. In the effort to reach the goal, the individual is motivated (inducement) to engage in certain activities, which in turn will make contributions to the host organization. After attaining part or all of the goal, the individual consciously or unconsciously judges whether the effort has been worthwhile. Depending on the conclusion reached, the individual will continue, repeat, or stop the particular behavior. As discussed earlier, Figure 10.1 illustrates this basic process graphically. Although somewhat simplistic, the model illustrates the basic concept of motivation and its dynamics. Over the years, more elaborate and detailed models have been developed to reflect the intricacies of human motivation more realistically.

Early Theories of Motivation

Early motivational theories, such as those developed by Frederick Taylor at the beginning of this century, assumed that work is inherently unpleasant and that incentives such as money must be provided.[7] In these early theories money often is seen as the primary goal in the motivational process. These early theories further assumed that managers knew much more about the job than the workers, who needed to be told exactly what to do and how to do it.

The Hawthorne Studies, conducted by Elton Mayo at Western Electric between 1927 and 1932, provided additional insight to motivation.[8] The studies gave birth to the *human relations approach,* complementing the earlier scientific theories. The human relations approach assumes that people want to participate in management decisions and get involved with the organizational process. They enjoy belonging to their organization and want to feel important and useful. These assumptions led to the recommendations that managers should keep their people informed and involved, allow them self-direction and control, and utilize the informal social process for motivating employees toward organizational goals.

More recently, the so-called human resource approach to motivation evolved. It focuses intensely on human needs. The approach is based on the belief that people not only enjoy the feeling of being part of the organization and its decision-making process, but that specific contributions and accomplishments are highly desired and valuable to the worker. This work itself can be a principal motivator and inducement to higher productivity.

Each theory makes many assumptions, which underlie its validity. Modern concepts of motivation and leadership recognize the situational nature and the dynamics of each case. Many of the specific concepts deal only with a particular aspect of the overall model, such as the content, process, or reinforcement of motivation. Let us explore some of the most popular concepts.

[7]See Frederick Taylor, *Principles of Scientific Management,* New York: Harper and Brothers, 1911.

[8]See Elton Mayo, *The Human Problems of an Industrial Civilization,* New York: Macmillan, 1933.

Maslow's Hierarchy of Needs

Abraham Maslow is known as the father of the need hierarchy[9], a theory that focuses on content. As shown in Figure 10.3, there are five hierarchical levels of human needs.

1. Physiological needs are most important. They include the basic human needs for survival and biological functioning, such as food, air, and sex.

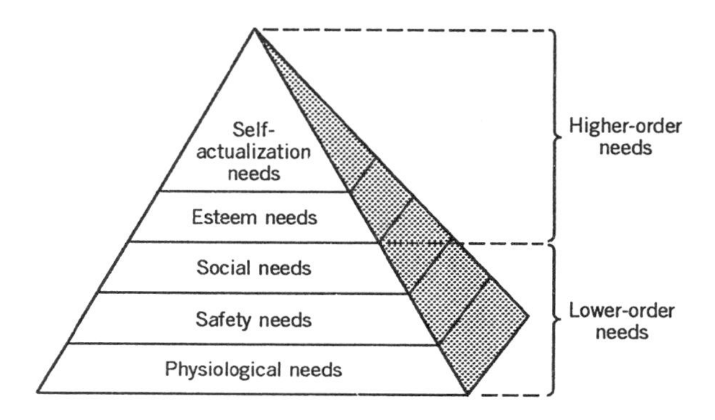

Need Category	Management Influence Areas
Self-actualization	Challenges in job. Creative work. Provide advancement opportunities. Encourage participation in decisionmaking. Provide flexibility, autonomy. Encourage high achievement.
Esteem	Recognize and publicize good performance. Significant job activities. Respectful job title. Merit-pay increase. Responsibility.
Social	Permit social interaction. Sponsor activities. Keep groups stable. Encourage cooperation.
Safety	Safe working conditions. Job security. Fringe benefits. Good base salary/wages.
Physiological	Fair/minimum salary or wages. Comfortable working conditions. Heat, lighting, space, air conditioning.

FIGURE 10.3 Maslow's hierarchy of needs.

[9]The concept of the needs hierarchy was first advanced by W. C. Langer, but Abraham Maslow is best known for refinement of the theory. For details see W. C. Langer, *Psychology of Human Living*, New York: Appleton, Century, Crofts, 1937, and A. H. Maslow, "A Theory of Human Motivation," *Psychology Review*, Vol. 50, 1943, pp. 370–396.

2. Safety needs come next. They include needs for clothing, shelter, and a secure physical and emotional environment. They also include job security and money.
3. Social needs are the needs of belonging, including love, affection, peer acceptance, information-sharing, discussion, participation, and general peer-group support.
4. Esteem needs represent ego needs, which include recognition, respect from others, self-image, and self-respect.
5. Self-actualization needs represent the highest order needs. They include high-level achievements.

The lower-order needs are often referred to as "hygiene needs," while the higher-order needs are called "growth needs." Maslow's model suggests that lower-order needs must be fulfilled first before the next higher-order need can occur. Furthermore, people are motivated only if that need is not already fulfilled; only deficit needs are motivators. For example, people must have their physiological needs fulfilled first before they have safety needs. A person who makes so little money that he and his family are starving is not motivated by the opportunity to negotiate a no-layoff agreement. Because of its intuitive logic, Maslow's need hierarchy enjoys a certain popularity and acceptance among managers. However, researchers have found many flaws in the theory. The five levels may not always be present and may occur in different order, including the recurrence of lower-order needs after higher needs were filled.[10] Yet, because of its simplicity, the model provides a very important framework for guiding the management practitioner and for more detailed analysis and research regarding human behavior.

Herzberg's Two-Factor Theory

Another popular content-oriented theory of motivation was developed by Frederick Herzberg in the 1950s. Based on his interviews with accountants and engineers, Herzberg concluded that people can be motivated by two sets of factors[11]:

1. *Satisfiers,* which motivate intrinsically. Their presence increases satisfaction
2. *Hygiene factors,* which motivate extrinsically, or superficially; their presence decreases dissatisfaction.

[10]One of the best known modifications of Maslow's theory was developed by Clayton Alderfer at Yale University. Called the ERG theory, the model recognizes three levels of needs: (E) existence, (R) relatedness, and (G) growth. The theory suggests that several levels of needs can cause motivation at the same time. It further suggests that frustration at one level may cause regression to a lower level, which becomes a renewed motivator, while the higher level need disappears, at least for some time, before resurfacing again.

[11]For a good summary of Herzberg's two-factor theory and its managerial applications, see Frederick Herzberg, "One More Time, How Do You Motivate Workers," *Harvard Business Review,* January–February 1968, pp. 53–62.

Table 10.2 shows a sample of factors for each of Herzberg's two categories. Notice that the true motivators or satisfiers are derived from the work content, while the hygiene factors are related to the work environment.

The management implications of Herzberg's theory are in several areas. One, employees are motivated in two stages. Managers must provide hygiene factors at an appropriate level. Working conditions must be satisfactory and safe; proper rules, guidelines, and procedures must exist to conduct business in an orderly way; the manager must provide sufficient technical supervision; and the pay and job security must be fair and acceptable. However, none of these hygiene factors really excites and stimulates the employee; they merely assure that his or her dissatisfaction is being kept to a minimum. After the hygiene factors have been provided for, the manager should concentrate on stage two, fostering a work environment rich in intrinsic motivators, such as professionally interesting and stimulating work, helping people to achieve desired results, recognize their accomplishments, and facilitating personal and professional growth. Managers also should try job enrichment to provide higher levels of motivation.

Implication number two is that intrinsically motivated people are less concerned about hygiene factors; that is, people who have work challenges, achievements and recognition and enjoy their work may need less supervision, procedures, and even pay and security to minimize their level of dissatisfaction.

A third implication is that motivational factors can be enhanced or even created by developing a sense of ownership in the organization and pride and value for the work itself. Japanese-style management is a typical example of fostering such an environment. Many examples in our own country, such as the turnaround in attitudes and values in the automotive industry, show an application of Herzberg's theory. Another facet of this enhancement concept is that motivation can be stimulated at all levels of the work force, not only at the professional or highly skilled level. However, this requires some magic and creative thinking on the part of management.

Yet, for all its conceptual insight and managerial applications, Herzberg's theory is under considerable fire. His critics argue that the theory works only with people who have at least the potential for intrinsic motivation, an assumption that is probably correct for engineering professionals, but doesn't always hold in general. Further,

TABLE 10.2 Herzberg's Motivational Factors

MOTIVATION FACTORS	HYGIENE FACTORS
Achievement	Technical supervision
Recognition, praise	Working conditions
The work itself	Interpersonal relationships
Job enrichment	Personal life
Responsibility	Pay and security
Advancement	Company policy and administration
Personal growth	Status

some people argue that the emphasis is too much on how people feel and it assumes that productivity will follow automatically when they feel motivated, which may not be true. However, Herzberg's theory contributed considerably to managerial understanding of motivation and had a major impact on modern management practices and research. Especially in the management of engineering personnel, Herzberg's concepts are very stimulating and provide some guidance for effective management practices.

Vroom's Expectancy Theory

Victor Vroom developed a framework for evaluating a person's level of motivation.[12] As illustrated in Figure 10.4, the model focuses on how people choose their goals and what factors influence their motivational levels. The Vroom model predicts the level of motivation, M, based on three factors: (1) the probability of reaching a goal, called *expectancy*, E, (2) the probability that the goal will actually produce the desired results, called *instrumentality*, I, and (3) the probability that the results are indeed desirable and satisfying, called *valence*, V. Mathematically this relationship is expressed as: $M = E \times I \times V$. An example may help to clarify. Our engineer Carol Edwards reads an advertisement that your company is looking for a new vice-president of engineering. Carol knows that certain results would be associated with the position, such as more money, status, etc. She also values these results highly. Therefore, both instrumentality and valence are high. However, her expectancy of getting the job, that is, of reaching the goal, is very low. As a result, her motivation is very low. She does not apply for the job.

A different situation exists when Carol's boss asks her to rerun a lab test because the original test results got lost. Carol knows that she can repeat the test and recreate the data. Both her expectancy and instrumentality are high. However, the outcome does not create any satisfaction. On the contrary, the work interferes with her desired activities. Valence is low and Carol's motivation to do the assignment is low.

Yet another situation exists when Carol discusses a new, challenging engineering assignment with her boss. The boss assures Carol that the project is very important

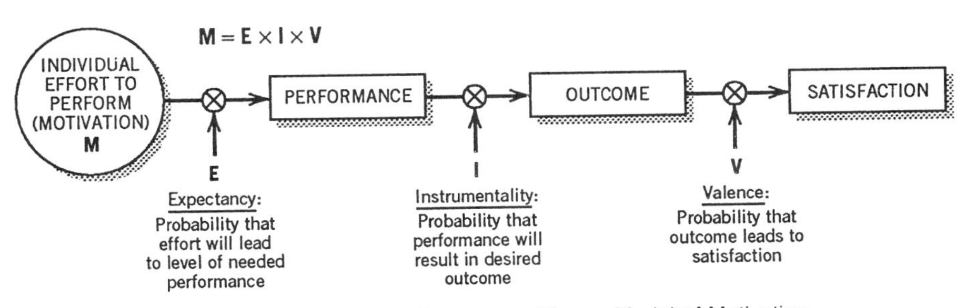

FIGURE 10.4 Vroom's Expectancy Theory Model of Motivation.

[12]See Victor Vroom, *Work and Motivation*, New York. Wiley, 1964.

to the company. Its successful completion will be highly visible and lead to company-wide recognition and further growth potential. Carol finds the potential outcome very desirable, has reasonable assurances that the outcome will actually occur, and feels confident that she can handle the job. All three probabilities are high. Carol is highly motivated.

In summary, Vroom's model suggests that for motivation to occur, three conditions must be met. The individual must expect that his or her effort will actually lead to the goal, that the goal will actually produce results, and that these results are desirable.

The assumptions underlying the Vroom's model are: (1) people have a choice among alternative plans and activities, and they exercise that choice, (2) people consciously or unconsciously evaluate the outcome of their future efforts and adjust the level of effort accordingly, (3) only three factors are the primary determinants of motivation, and (4) people can actually separate various goals and outcomes in real-world situations and analyze them one at a time, rather than orchestrating all of their goals and needs as an integrated system.

Vroom's theory provides a useful insight into the dynamics of human motivation. The following suggestions for engineering managers are derived from the broader context of Vroom's theory. Engineering managers should:

1. Be aware of their employees' desires and ambitions
2. Assign tasks in a way that emphasizes the correlation between organizational goals and personal goals
3. Make sure that subordinates have the ability and organizational support to succeed on the job. Develop a "can-do" mentality
4. Provide a positive picture regarding the possibility of successfully achieving goals and list the outcomes that will follow
5. Catch technical problems and personal limitations early in their development and rectify them
6. Recognize accomplishments on an ongoing basis; provide visibility
7. Provide tangible rewards to validate the perception of success.

Equity Theory

Simply put, equity theory states that people will be motivated toward certain efforts only if they perceive the outcome to be fair relative to others. Equity theory is applied by Shakespeare who says through one of his characters, "He is well paid that is well satisfied." Obviously, the outcome from a job is more than just pay. It includes recognition, skill development, promotional potential, social relations, freedom of action, and the esteem that follows achievement. Equity theory assumes that people evaluate their job performance and rewards by comparison to others. Accordingly, a person is motivated in proportion to the perceived fairness of the rewards received.[13]

[13]For further discussion see R. C. Huseman, J. D. Hatfield, and E. W. Miles, "A New Perspective on Equity Theory," *Academy of Management Review*, November 12, 1987, pp. 232–234.

The challenge to managers is to administer rewards fairly and equitably. Since most engineering work is based on group efforts, people continuously compare their efforts and rewards to those of others. They have their own biases regarding fairness and may value rewards differently than the boss thinks. Managers must be close to their people to know what makes them tick and address and administer awards accordingly. Managers must also understand the dynamics of the team and know how the individual compares himself or herself to others. When it comes to salary adjustments, managers should realize that their people are likely to be sufficiently biased, hence to perceive inequities even though the manager does not. This is one of the reasons that management usually tries to keep salary actions confidential, although it doesn't always succeed. Distributing salary adjustments over different times for different people makes a direct comparison more difficult. A one-person salary adjustment is a less newsworthy event than a unified salary action for the whole team.

Equity theory provides another framework for analyzing human behavior and formulating management actions. It provides an especially valuable insight for managing team efforts, which are so important in engineering. Every model and theory has its limitations. Seasoned managers acknowledge their shortcomings but use those parts that are relevant to their situations.

Reinforcement Theory

Reinforcement theory is based on the assumption that behavior that results in a desired outcome (reward) is most likely to be repeated, but that a behavior that is punished will be discontinued. This theory was originally tested on animals, but researchers such as B.F. Skinner have shown that it also applies to humans.[14] Although many critics argue the relevancy of animal experiments to humans, the inherent logic of the theory and its simplicity led to a wide acceptance of the theory by managers in building some of their actions. According to reinforcement theory, desired managerial results can be obtained by using one or several of four specific methods described in Table 10.3: positive reinforcement, avoidance, negative reinforcement (or) punishment, or extinction.

Operant Conditioning

In addition to providing a stimulus to people after they have engaged in a desired or undesired behavior, the manager can *condition* his or her people by telling them ahead of time what reward or punishment they can expect. For example, the boss needs the results from a field test urgently by Wednesday at noon. The boss tells the engineer of the urgent need. The boss provides incentives for the on-time delivery by offering the afternoon off from work if the employee delivers on time. In addition the boss points out, however, that the boss would expect the engineer to work overtime without pay if the engineer runs into difficulties. Hence we condition the employee with a combination of stimuli—one positive and one negative reinforcement.

[14]For a comprehensive treatment of this subject, see B. F. Skinner, *Science and Human Behavior*, New York: Macmillan, 1953.

TABLE 10.3 Methods of Reinforcement

Method and Description
POSITIVE REINFORCEMENT (REWARD)
Provide a reward when desired results have been achieved. Rewards can include money, promotion, praise, or the work itself.
NEGATIVE REINFORCEMENT (PUNISHMENT)
Managers may choose to reprimand, hold pay, etc. to punish an employee who has engaged in an undesirable behavior such as tardiness. The employee might modify his or her behavior to avoid future punishment.
AVOIDANCE
People may work toward a goal to avoid punishment. For example, an employee may be motivated to come to work on time to avoid a reprimand.
EXTINCTION
The manager can take actions to eliminate an undesired behavior or the cause of it. Examples may be the firing of an employee whose tardiness is a negative influence on others, or closing the cafeteria at certain hours to eliminate undesired coffee breaks.

How to Manage

Motivation is of great concern to managers as it directly effects the level of performance, including innovative, creative behavior, risk-taking, personal drive and effort, and the willingness to collaborate with others. Behavioral models provide some insight to the manager as to why people behave in a particular way, what energizes them, and how their behavior can be channeled to produce desired results. Yet, none of the models or theories really tells task leaders and engineering managers what to do in specific situations. There are, however, certain methods that seem to produce more favorable overall behavior in an engineering environment where we cannot always define specific objectives, don't have all the answers, must deal with frequent changes, and want to encourage risk-taking to achieve innovative results. What seems to work best in such an environment are approaches that fulfill the esteem needs of the professional person, leading to encouragement, pride, work challenge, recognition, and accomplishments, while minimizing personal risks, threats, fear, and conflict. Specific recommendations will be made at the end of this chapter. (See Section 10.6)

10.4 LEADERSHIP IN ENGINEERING

Leadership is the most talked about, written about, and researched topic in the field of management. Yet, in spite of the great deal of research, management theorists cannot come up with a perfect model that explains the situational nature of leadership and helps to guide managers toward effective leadership behavior.

So why should managers pay attention to leadership theories? Although they may not be useful in defining a cookbook approach to effective management in a given situation, leadership theories may help in analyzing why, in a given situation, a particular management behavior produces certain results, so we can learn from our past experiences and the experiences of others, and hence fine-tune our style.

Early leadership concepts focused on particular traits such as intelligence, vocabulary, confidence, attractiveness, height, and gender. Literally hundreds of studies were conducted during the first two decades of this century. For the most part, the results were disappointing. These studies could not define a common set of traits for successful leaders and effective management. Neither could these studies explain the enormous trait and style differences that exist among successful leaders. The traits, styles, values, and contributions to society differed considerably from each other.

Based on the lack of success in identifying personal traits of effective leaders, research began to focus on other variables, especially the behavior and action of leaders.

Rensis Likert and the Michigan Studies

In the late 1940s, studies at the University of Michigan under Rensis Likert identified two basic styles of leader behavior[15]:

1. *Job-Centered Leader Behavior.* The leader relies primarily on authority, reward, and punishment power to closely supervise and control work and subordinates. This style follows the characteristics of McGregor's Theory X behavior.[16]
2. *Employee-Centered Leader Behavior.* The leader tries to achieve high performance by building an effective work group. The leader pays attention to the worker's needs and helps to achieve them. Focus is on team development, participation, individual accountability, and self-direction and control. McGregor's Theory Y would best characterize this behavior.

In assessing the relevance of these two concepts for engineering management situations, it might be tempting to discard the job-centered style as inappropriate because it appears too simplistic. In addition, an authoritarian style seems to be inappropriate to engineering and technology management. However, a closer look

[15]Originally Likert's studies identified four systems of management: (1) autocratic; (2) benevolent autocratic, both relying on job-centered styles of management; (3) consultive and (4) participative group styles of management, which are employee-centered. For details see Rewsis Likert, *New Patterns of Management,* New York: McGraw-Hill, 1967.

[16]Douglas McGregor classified people for motivational purposes into two categories: (1) those who follow "theory X" dislike work, are lazy, must be controlled, and prefer clear direction and established norms. (2) those who follow "theory Y" enjoy their work, need professional challenges, freedom, and growth potential; however they are also willing to take responsibility and exercise leadership. For details see Douglas McGregor, *The Human Side of the Enterprise,* New York: McGraw-Hill, 1960.

shows that many effective engineering managers actually pay considerable attention to the planning, tracking, and controlling of the work, in addition to staying close to their people, building a coherent task team, and fostering a work environment that emphasizes individual responsibility, recognition, and accomplishments. Hence, the manager's style can combine both dimensions, as concluded by the well-known Ohio State University research on leadership behavior[17] and later on presented in the form of the managerial grid.

The Managerial Grid®

Developed by Robert Blake and Jane Mouton, the Managerial Grid has become a widely recognized framework for examining leadership styles.[18] As illustrated in Figure 10.5, the Grid displays leadership qualities in a two-dimensional matrix. Its horizontal axis represents concern for production, namely:

- Desire for achieving better results
- Cost effectiveness, resource utilization
- On-time performance
- Satisfaction of organizational objectives
- Getting the job done.

The second dimension, shown on the vertical axis, represents concern for people, which is defined as:

- Promoting cooperation and friendship
- Helping people to achieve their personal goals
- Building trust, friendship, and respect
- Minimizing conflict
- Facilitating to achieve results.

By scaling each axis from 1 to 9, Blake and Mouton created a framework for measuring and identifying leadership behavior, which they classified into five major styles:

1,1 Style: Impoverished Management. Minimal concern for either productivity or people.

[17]The Ohio State University studies on leadership behavior go back to 1940 and have many published research papers. For an informative summary of these studies, see Edwin A. Fleishman, "Twenty Years of Consideration and Structure" in E. A. Fleishman and J. G. Hunt, *Current Developments in the Study of Leadership*, Carbondale, Illinois: Southern Illinois University Press, 1973.

[18]See Robert R. Blake and Jane S. Mouton, *The Managerial Grid*, Houston, Gulf Publishing, 1964, and Robert R. Blake and Anne Adams McCanse, *Leadeship Dilemmas—Grid Solutions*, Houson: Gulf Publishing, 1991 which refers to the Managerial Grid as the "Leadership Grid."

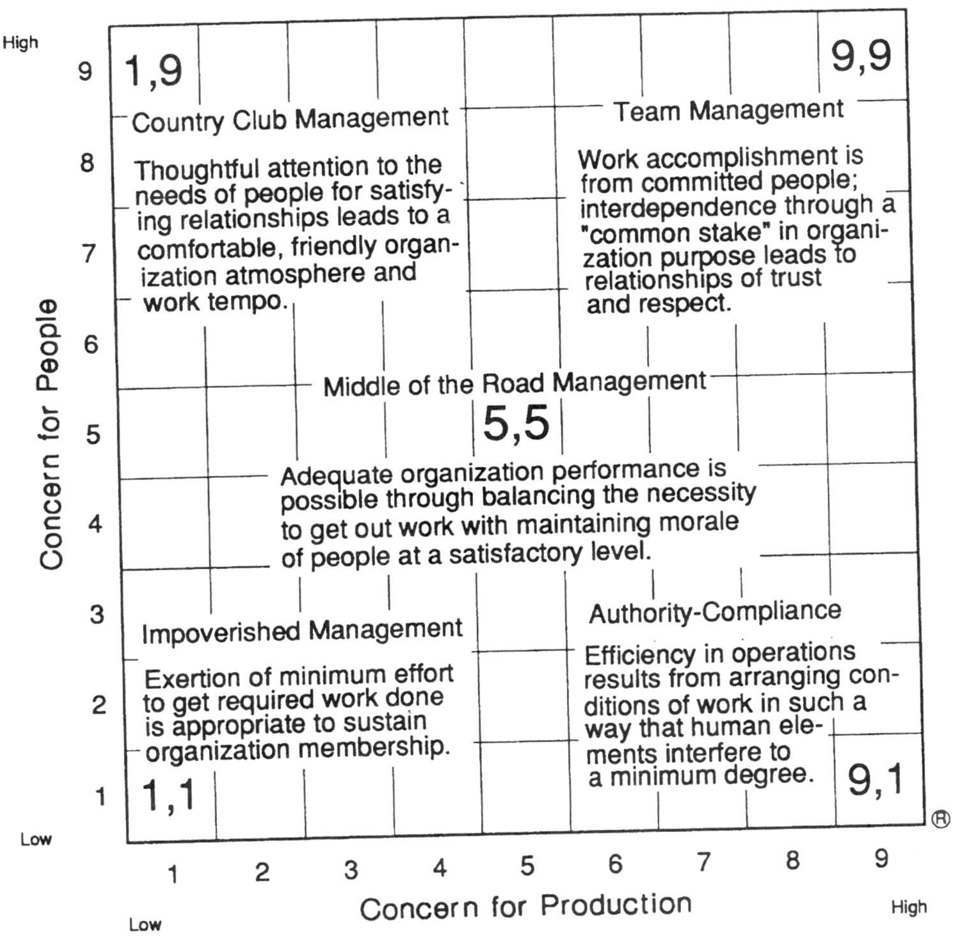

FIGURE 10.5 The Leadership Grid® Figure from *Leadership Dilemmas—Grid Solutions,* by Robert R. Blake and Anne Adams McCanse. Houston: Gulf Publishing Company, p. 29. Copyright © 1991, by Scientific Methods, Inc. Reproduced by permission of the owners.

5,5 Style: Middle of the Road Management. Moderate concern for both productivity and people, to maintain status quo.

9,1 Style: Authority-Compliance Management. Primary concern for productivity; people are secondary.

1,9 Style: Country Club Management. Primary concern for people, productivity is secondary.

9–9 Style: Team Management. High concern for both productivity and people.

According to Blake and Mouton, 9,9 is the ideal style of management for most work environments. This finding seems to be especially relevant in an engineering environment where managers have to lead people towards innovative results in an uncertain environment. In addition the engineering manager often has to step across functional lines and build a work team consisting of people over whom the manager has limited authority and control. To get results, the manager must build trust and respect and direct the activities via team commitment and self-control. Many training programs have been developed that focus on the Leadership Grid with the objective of assisting managers in achieving a leadership style that is most effective for their specific work environment.

Situational Leadership Theories

Fascinated by the results of effective leadership, researchers have attempted to identify the traits and ingredients of successful leadership behavior for a long time. Yet after decades of formal studies, both management scholars and practitioners are convinced that there is no one best behavior profile or style that fits all situations. Leadership effectiveness seems to be a very intricate function of many variables. At the minimum it depends on the complex integration of variables from the leader, the followers, and the work situation. *The contingency theory of leadership,* first described by Fred Fiedler,[19] recognizes these complexities and provides a fresh insight to leadership effectiveness. Some of other popular situational leadership theories include path goal leadership theory; Vroom–Yetton model; the attributional model; and the operant conditioning approach.

Fiedler's Contingency Theory of Leadership. Many situational leadership theories have been proposed to date. However, Fiedler's theory is the one that has been most thoroughly tested so far. It is the result of thirty years of research by Fred E. Fiedler and his associates into the situational effectiveness of task-oriented versus relationship-oriented leadership. Fiedler characterizes the *work situation* using three variables with the following ranges:

1. *Leader–Member Relations.* Range: Good to poor
2. *Task Structure.* Range: Structured to unstructured
3. *Position Power of the Manager.* Range: High to low

Fiedler concluded that a task-oriented leadership style is most effective when all three situational variables are favorable that is, leader–member relations are good; the task is structured; and the manager's position power is high. Fiedler also found that a task-oriented leadership style was the most effective one when leader–member relations were poor; the task was unstructured; and the manager's power was low (i.e., all three variables were unfavorable). He further found that a relationship-oriented leadership style was most effective when the manager faced a mixed situation, in which some variables were favorable but others were unfavorable.

[19]See Fred Fiedler, *A Theory of Leadership Effectiveness,* New York: McGraw-Hill, 1967.

While critics point to the small number of variables and their descriptors in Fiedler's theory, as well as the difficulty of measuring situational components, as severe limitations of the theory, the theory does provide important lessons for the engineering manager, as noted below:

- Engineering managers typically work in situations with moderate task structure in which they have low position power but good relationships with their team members. Fiedler's model would suggest a relationship-oriented style of leadership as optimal for such situations. This is consistent with many other research studies of technical and innovation management.
- The start-up phase of a new technical task is often either all favorable or unfavorable regarding member relations, task structure, and power. In that case, a directive, more task-oriented style would be most effective, a contention that is supported independently by other research.
- A contingency, change, or problem that suddenly surfaces is likely to weaken the manager's position power and strain member relations. If this contingency, change, or problem is combined with an unstructured task, the manager might have to resort to a more task-oriented or directive style to maintain effectiveness. If prior to the contingency all the situational variables were highly favorable (and therefore the manager did not need much people-orientation) more attention must be given to the staff as a result of the contingency, changes, or problem, and the leader must give more attention to facilitating conflict-resolution, rebuilding trust, motivating people, and refueling their task-commitment.
- The model may also be useful for a situational "what-if analysis", in which it may be used for making managers aware of potential situational changes and assessing whether the particular situation calls for more relationship or direction.

While scholars may argue that Fiedler's theory assumes leadership behavior is a personality trait that cannot be changed to fit a particular situation, the model provides an important framework for assessing leadership effectiveness in situations involving many components. Such a topology has been used in various field studies for determining drivers and barriers to engineering management performance, as well as for determining specific professional training and development needs.[20]

The Life Cycle Model. Another popular situational leadership theory was developed by Paul Hersey and Kenneth Blanchard.[21] As an extension of the managerial

[20]For specific methods and results, see H. Thamhain and D. Wilemon, "Leadership, Conflict, and Project Management Effectiveness," *Sloan Management Review,* Vol. 19, No. 1, Fall 1977, and "Managing Technologically Innovative Team Efforts Towards New Product Success," *Journal of Product Innovation Management,* Vol. 7, No. 1, March 1990.

[21]See P. Hersey and K. Blanchard, *Management of Organization Behavior and Utilizing Human Resources,* 4th ed., Englewood Cliffs, NJ: Prentice-Hall, 1982.

grid, the model shown in Figure 10.6 provides insight into leadership effectiveness as a function of work group maturity. *Maturity* is defined as the team members' ability and willingness to assume responsibility for their group assignment. This includes many variables, including skill level, experience, knowledge, confidence, motivation, mutual trust and respect, and commitment to agreed-on results.

Leadership style is defined in terms of managerial grid dimensions: X for task-oriented behavior and Y for relationship-oriented behavior. However, unlike the two-dimensional approach suggested by Blake and Mouton, the life cycle situational

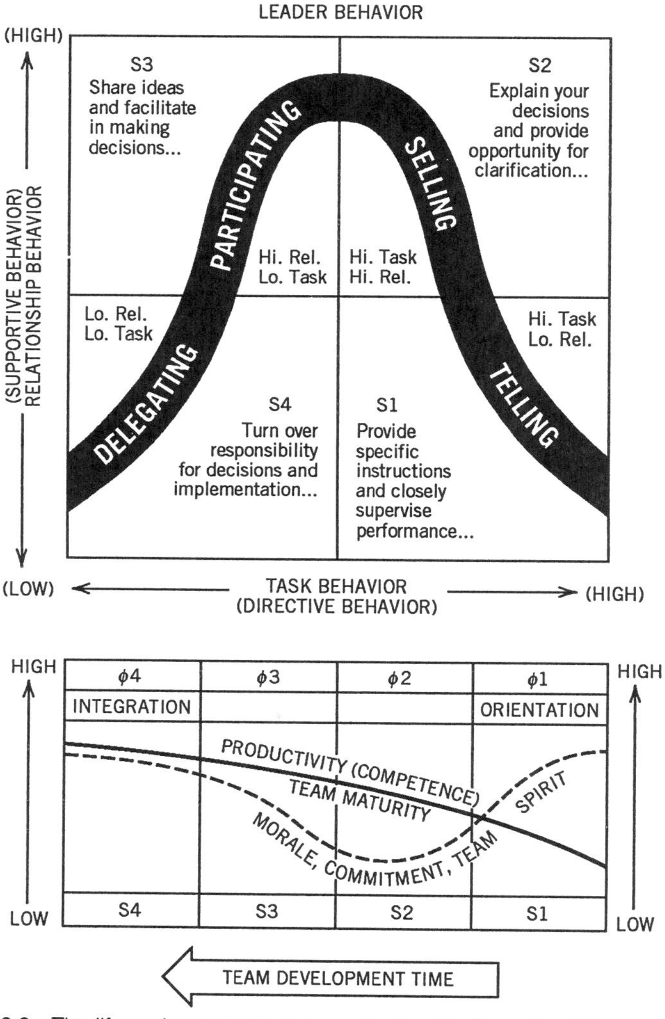

FIGURE 10.6 The life cycle approach to team leadership. From Paul Hersey and Ken Blanchard, *Management of Organizational Behavior:* Utilizing Human Resources, 5th ed., Englewood Cliffs, NJ: Prentice-Hall, 1988, p. 287. Reprinted by permission.

model provides a more flexible interpretation of the grid components, resulting in four styles of management, which are related in their effectiveness to group maturity:

1. Telling Style. This high task-oriented, low relationship-oriented style focuses on goal-setting, planning, organizing, guiding, directing, and controlling. It is suggested as an effective style with groups of low maturity, such as a newly formed project team, or during the initial phases or project definition. The manager provides specific instructions and close supervision.

2. Selling Style. This high task-oriented, high relationship-oriented style (equivalent to a 9–9 style in grid jargon) focuses on gaining team acceptance and buy-in to the assignment. It further facilitates interaction among team members, communication with other departments, and feedback. This type of leadership is suggested for work groups that are well into their formation stage, are beginning to behave as a team, but still have a lot of concerns about their specific roles and missions and are still unsure of their ability to perform the given task as a team.

3. Participating Style. This low task-oriented, high relationship-oriented style focuses on team-building by providing encouragement, support, visibility, and recognition to team members. The managerial objective is to strengthen team integration and self-actuation. The leader is a facilitator who provides guidance with a minimum of direct supervision and control. The participative leadership style is recommended for moderately to highly mature teams. Most project teams and engineering task groups are expected to be in this maturity state during the majority of the activity phases.

4. Delegating Style. This low task-oriented, low relationship-oriented style relies on a fully developed and integrated team for self-management and self-actuation of their activities, according to agreed-on plans. The manager's role is reduced to delegation, observing, and monitoring, while the team is fully committed to and capable of producing the desired results and achieving self-development.

The life cycle model of leadership is particularly interesting and useful in project-oriented situations where managers have to deal with the formation and development of multidisciplinary teams throughout the project life cycle. Also, the nature of engineering work does not always guarantee a linear progression of team maturity toward the "high" end. Work interruptions, interfunctional transitions, problems, and changing requirements can make it necessary to alter the team's composition, which affects the team's characteristics regarding maturity, readiness, and ability. Situational leadership is a reality in engineering management. The effective engineering manager is a social architect who understands the dynamics of his or her organization, the work, and the people. These managers are able to diagnose potential problems and the need for change.

Perspectives on Engineering Management

Engineering management is risky business in more than one sense. Activities are highly complex, and their precise outcome often is unpredictable. Organizational structures do not always follow classical concepts of clear lines of authority, communications and command, the type used in traditional theories of motivation and leadership. Technical people are a special breed of characters who do not always follow the conventional motivation and leadership models. In addition, managers of technology-based organizations must deal with social, ethical, and environmental challenges.

Thus, managers of research, development, and engineering organizations must maintain and integrate two cultures, the corporate–organizational–business culture and the scientific–technical culture. For effective role performance, engineering managers must be skilled to deal with complex administrative tasks, people, and technical challenges. We are entering a far more dynamic era requiring greater professionalism in research, design, and engineering management. Tomorrow's manager, in addition to being technically qualified, must be respected, skilled in dealing with people, and comfortable with the latest business practices and administrative techniques. Hence, the role of the engineering manager is to:

1. Translate long-range strategic business objectives into operational engineering objectives and program areas for research, development, and engineering.
2. Find, hire, and encourage creative people to pursue new technical knowledge and to make technical inventions
3. Ensure a free and creative atmosphere for research focused on economic benefits
4. Implement research and development resulting in product, process, and service creation and improvement, consistent with the business of the corporation.

To perform this role effectively, engineering managers must exhibit a style of leadership that balances managerial and technical values. Specific aspects of effective technical leadership are listed below from a major field study:[22]

- Clarity of management direction
- Understanding of the technology leaders are working with
- Understanding of the organization and its interfaces
- Understanding of personnel's professional needs
- Ability to satisfy professional needs and motivate
- Good planning and administrative skills
- Credibility within the organization and its support groups
- Clear written and oral communication channels
- Delineation of goals and objectives
- Providing methods to aid group decision-making

[22]Hans J. Thamhain, "Managing Engineers Effectively," *IEEE Transactions On Engineering Management*, Vol. 30, No. 4, November 1983.

- Providing assistance in problem-solving
- Helping to resolve conflict.

Many of the above leadership components become *substitutes* for traditional leadership influences, such as authority and position power. These leadership substitutes were formally studied by Kerr and Jermier,[23] who characterized them in seven areas: (1) team member ability, (2) experience, (3) training, (4) knowledge, (5) need for independence, (6) professional orientation, and (7) indifference to organizational rewards. For example, an engineer who has skills and abilities to perform her job and also has a high need of independence and professional satisfaction may not need or in fact may resent a leader who provides direction and structure. By a similar argument, leadership substitutes come into effect for an engineer who performs some routine, highly structured tasks, but finds the process and the results professionally stimulating and challenging. Here the task itself provides the subordinate with an adequate level of intrinsic motivation, so that close supervision is unnecessary and may even be counterproductive.

In summary, the engineering manager must span the boundaries between the technical and the business worlds. The engineering functions in a modern corporation cross boundaries both within and outside the organization. Because their managers are trying to maintain and integrate two cultures, the corporate economic culture and the scientific–technical culture, it is important that the engineering manager exhibit a style of leadership that balances managerial and technical values. Engineering personnel do value technical work and expertise and expect certain things from management—such as providing leadership and support for technical creativity. The potential gains from increased engineering productivity are great. Management research has provided some insight into the dynamics of leadership in high-technology situations. However, no management model is perfect, and probably none ever will be. We must work with the tools and knowledge that exists today, seeking better ways to manage the resources available to us. Improving engineering productivity requires awareness, skills, commitment, ingenuity, action, and perseverence. Maybe these are the ingredients of the highly praised transformational leadership style.[24]

10.5 THE POWER SPECTRUM IN ENGINEERING MANAGEMENT

Motivation and Managerial Power

Why do people comply with the requests or demands of others, for example, their superiors? One reason is that they see the other people as being able to facilitate the

[23]For additional discussions that also include a study on managerial style effectiveness with engineers and technicians, see S. Kerr, R. House and A. Filley, "Relation of Leader Consideration and Initiating Structure to Research and Development Subordinate Satisfaction," *Administrative Science Quarterly*, Vol. 16, No. 1, 1971, pp. 19–30.

[24]For an essay on transformational leadership see Noel M. Tichy and David O. Ulrich, "The Leadership Challenge. A Call for the Transformation Leader," *Sloan Management Review*, Fall 1984, pp. 59–68.

fulfillment of their needs. People comply with the requests of others if they perceive them to be able to influence specific outcomes. These outcomes could be desirable, such as a salary increase, or undesirable, such as a reprimand or demotion. The influence over others is referred to as *managerial power.* Managers, as well as anyone else, use this power to achieve interpersonal influence, which is leadership in its applied form. Therefore power is the force that when successfully activated motivates others toward desired results. In the next section, we will look specifically into the power spectrum that is available to leaders in a technology-oriented environment. Further, we will discuss the situational effectiveness of various leadership styles.

Power-Sharing and Dual Accountability

Engineering managers must often cross functional lines to get required support. This is especially true for managers who operate within a matrix structure. Almost invariably, the manager must build multidisciplinary teams into cohesive work groups and successfully deal with a variety of interfaces such as functional departments, staff groups, team members, clients, and senior management. This is a work environment in which managerial power is shared by many individuals. In contrast to the traditional organization, which provides position power largely in the form of legitimate authority, the power of engineering managers needs to be supported extensively with knowledge that comes from expertise and credibility that comes from having an image as a sound decision-maker.

Like many other components of the management system, leadership style has also undergone changes over time. With increasing task complexity, increasing dynamics of the organizational environment, and the evolution of new organizational systems, such as the matrix, a more adaptive and skill-oriented management style evolved. This style complements the organizationally derived power bases such as authority, reward, and punishment with bases developed by the individual manager—examples are technical and managerial expertise, friendship, work challenge, promotional ability, fund allocations, charisma, personal favor, project goal identification, recognition, and visibility. This so-called System II management style evolved particularly with the matrix. A descriptive summary of both styles is presented in Figure 10.7. Effective engineering management combines both styles.

Various research studies by Gemmill, Thamhain, and Wilemon provide an insight into the power spectrum available to project managers.[25] Figure 10.8 indicates the relative importance of nine influence bases in gaining support from engineering subordinates and assigned personnel. Technical and managerial expertise, work challenge and influence over salary are the most important influences that engineering

[25]The original study was published by G. Gemmill and H. Thamhain in "Influence Styles of Project Managers: Some Project Performance Correlates," *Academy of Management Journal,* June 1974. This study used nonparametric Kendall's rank-order correlation techniques to measure the association between the strength of a power base on which project managers seemed to rely and certain performance measures, such as project support, communications effectiveness, personal involvement and commitment, and overall management performance. The findings were later validated for engineering work environments in general.

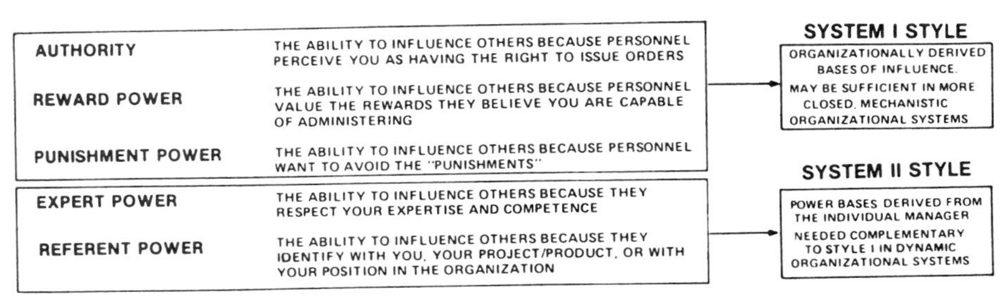

FIGURE 10.7 Commonly recognized bases of managerial influence.

managers seem to have, while authority, fund allocations, and penalties appear least important in gaining support from engineering subordinates.

Role of Salary

Salary plays a very special role in the power spectrum. It is interesting to note that engineering managers did not realize the relatively high importance of salary as an influence base. In fact, in the aggregate engineering managers ranked salary second to last in importance while their subordinates ranked it third highest, as shown in Figure 10.8. In analyzing this difference between managerial and subordinates' perception, it was found that professionals indeed make extra efforts for many other reasons than an increase in salary. Frequently engineering personnel were asked why they stayed after hours to fix a problem or why they worked casual overtime during weekends. The answers are invariably related to what Maslow would call the

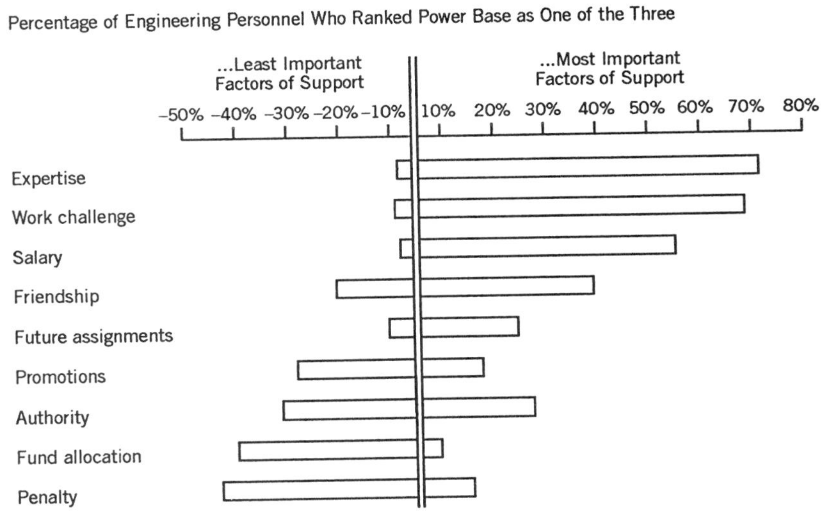

FIGURE 10.8 Power style profile: Engineering Personnel's ranking of the most and least important factors in a manager gaining their support.

fulfillment of growth needs, such as work challenge, recognition, project goal identification, or professional pride. Few people would make these extra efforts because they got paid last Friday or because they try to position themselves for an 8% raise to be budgeted for next year.

However, the above argument holds only if the personnel perceive a fair and adequate compensation. Otherwise salary becomes a barrier to effective teamwork, a handicap for attracting and holding quality people, and a source of steady conflict. To illustrate, a person who is motivated to make an extra effort might indeed enjoy the praise and recognition that comes with the well-done job. The person may further infer from the manager's action that the job was important to the company and that if the employee continues to receive praise, recognition, and visibility for high performance during the year, he or she may certainly feel in line for a raise or promotion. Now let's suppose that instead, the employee receives a zero increase. The normal response of the employee would be disbelief in any sincerity or value to praise, recognition, or other intrinsic rewards received in the past or anticipated for the future. The employee would most likely be angry and frustrated over being manipulated. He or she might also feel cheated and confused. Overall, this is *not* a situation that leads to long-range motivation, sustained personal drive, and high morale. To the contrary, one should expect resignation, game-playing, mistrust, and conflict, the very ingredients of low-performance engineering personnel. Salary is a very important power base, which must be used judiciously, but which also must be in line with the employee's output, efforts, and the salary expectations built by management over time.

Correlations to Engineering Management Performance

The perceived importance of certain power bases, such as expertise, authority, and work challenge, does not permit by itself any conclusion regarding their effectiveness. This was originally investigated by Gemmill and Thamhain.[26]

Expanded into a more general engineering work environment, the effectiveness of each managerial power base, as seen by subordinates, has been correlated to four performance indicators: (1) degree of support received by managers, (2) communication effectiveness as indicated by subordinates; (3) the degree of personal drive, involvement, and commitment of engineering personnel, and (4) the engineering managers' overall performance as seen by their superiors. Table 10.4 summarizes the performance correlates, which are based on Kendall tau rank-order techniques.[27] Table 10.4 indicates that two influence methods are particularly favorably associated with management performance: expertise and work challenge. The more expertise, both managerially and technically, engineering managers are perceived as having: (1) the more support they seem to get from their personnel, (2) the better communications seem to be, (3) the more commitment and involvement is generated, and, in the end,

[26]See note 25.

[27]For methods of computing Kendall's tau rank-order coefficients see S. Siegel, *Nonparametric Statistics for the Behavioral Sciences*, New York: McGraw–Hill, 1956.

TABLE 10.4 Managerial Style versus Effectiveness in Engineering and Technology Work Environments, Correlations

Engineering Manager's Influence Method as Perceived by Project Personnel	Engineering Personnel's			Engineering Manager's Effectiveness Rating
	Degree of Support	Willingness to Disagree	Project Involvement	
Expertise	0.15	0.30*	0	0.40*
Work Challenge	0.10	0.25*	0.45* 2	0.25
Salary	−0.20	−0.10	−0.15	−0.15
Friendship	0	0	0	0.17
Future Assignment	0.25	0	0	0.10
Promotion	0	0	0	0.08
Authority	−0.10	−0.20	−0.35**	−0.30
Coercive Power	−0.45**	−0.10	−0.20	−0.25*

Kendall's tau: *$p < .05$; **$p < .01$.

(4) the higher is their performance rating by their superiors. The same favorable relationship exists for work challenge. Conversely, engineering managers who are perceived as relying strongly on authority, emphasizing salary, or relying on coercive measures seem to get lower support, less open communications, and lower performance ratings.

Management Implications. The popularity of a particular influence base is not necessarily an indication of its effectiveness. Expertise, work challenge, and salary were cited as the three most important reasons for compliance. The performance correlates are somewhat different, however. They indicate that engineering managers who were perceived by their personnel as emphasizing work challenge and expertise not only achieved higher effectiveness ratings on overall project performance, but also tended to foster a climate of better communications and higher involvement among their project personnel. Conversely, the findings suggest that the use of authority, salary, and coercion as an influence method has a negative effect, resulting not only in a lower level of performance but also in less communication and involvement among project personnel. Therefore the more engineering managers rely on expertise and work challenge and the less they emphasize organizationally derived influence bases, such as authority, salary, and penalty, the higher their ability to manage in engineering. One of the most interesting findings is the importance of work challenge as an influence method. Work challenge appears to encompass integrating the personal goals and needs of project personnel with project goals, more than with any other influence methods; work challenge is primarily oriented toward the intrinsic motivation of engineering personnel, while other methods are oriented more toward extrinsic rewards without regard to people's preferences and needs. To make the assignments of engineering personnel enriched in a professionally challenging way may indeed have a beneficial effect on project performance. In addition the assignment of challenging work is a variable over which engineering managers may have a great deal of control. Even if the total task structure is fixed, the method by which work is assigned and distributed is discretionary in most cases.

10.6 HOW TO MAKE IT WORK: SUGGESTIONS FOR INCREASING EFFECTIVENESS

The nature of engineering management, the need to elicit support from various organizational units and personnel, the frequently ambiguous authority definition, and the often temporary nature of multidisciplinary engineering activities all contribute to the complex operating environment that engineering managers experience in the performance of their roles. A number of suggestions may be helpful to increase the engineering manager's effectiveness.

1. Understand Motivational Needs. Engineering managers need to understand the interaction of organizational and behavioral elements in order to build an environment conducive to their personnel's motivational needs. This will enhance active participation and minimize dysfunctional conflict. The effective flow of communication is one of the major factors determining the quality of the organizational environment. Since the manager must build task and project teams at various organizational levels, it is important that key decisions are communicated properly to all task-related personnel. Regularly scheduled status review meetings can be an important vehicle for communicating and tracking project-related issues.

2. Adapt Leadership to the Situation. Because their environment is temporary and often untested, engineering managers should seek a leadership style that allows them to adapt to the often conflicting demands existing within their organization, support departments, customers, and senior management. They must learn to "test" the expectations of others by observation and experimentation. Although it is difficult, they must be ready to alter their leadership style as demanded both by the specific tasks and by their participants.

3. Accommodate Professional Interests. Engineering managers should try to accommodate the professional interests and desires of supporting personnel when negotiating their tasks. Task effectiveness depends on how well the manager provides work challenges to motivate those individuals who provide support. Work challenge further helps to unify personal goals with the goals and objectives of the organization. Although the total work of an engineering department may be fixed, the manager has the flexibility of allocating task assignments among various contributors.

4. Build Technical Expertise. Engineering managers should develop or maintain technical expertise in their fields. Without an understanding of the technology to be managed, they are unable to win the confidence of their team members, to build credibility with the customer community, to participate in search for solutions, or to lead a unified engineering effort.

5. *Plan Ahead.* Effective planning early in the life cycle of a new engineering program is highly recommended. Planning is a pervasive activity that leads to personnel involvement, understanding, and commitment. It helps to unify the task team, provides visibility, and minimizes future dysfunctional conflict.

6. *Provide Role Model.* Finally, engineering managers can influence the work climate by their own actions. Their concern for project team members, ability to integrate personal goals and needs of personnel with organizational goals, and their ability to create personal enthusiasm for the work itself can foster a climate that is high in motivation, work involvement, open communication, creativity, and engineering performance.

A situation approach to engineering management is presented in Figure 10.9, which indicates that the intrinsic motivation of engineering personnel increases with the manager's emphasis on work challenge, their own expertise, and their ability to provide professional growth opportunities. On the other hand, emphasis on penalty measures, authority, and manager's inability to manage conflict lowers personnel motivation.

Managerial position power is further determined by such variables as formal position within the organization, the scope and nature of work, earned authority, and ability to influence promotion and future work assignments. Engineering managers who have strong position power and can foster a climate of highly motivated personnel not only obtain higher support from their personnel, but also achieve high overall performance ratings from their superiors.

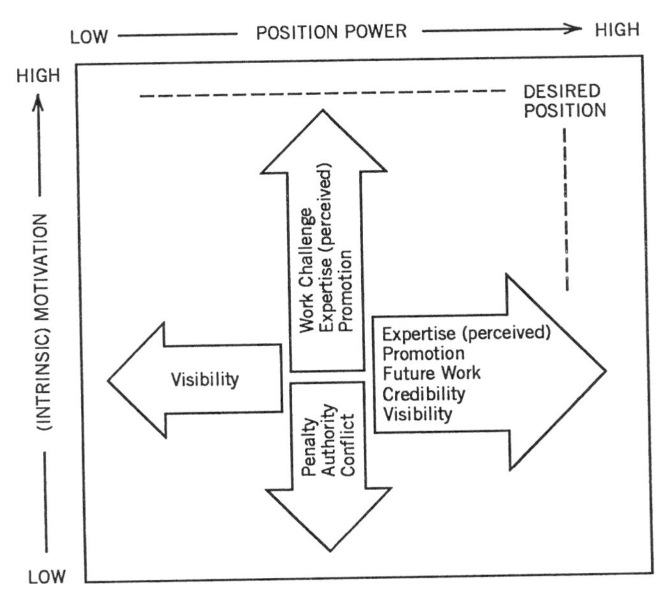

FIGURE 10.9 Situational effectiveness of environmental factors and influences on motivation and managerial power.

BIBLIOGRAPHY

Abatti, Pier A., "Technology: A Key Strategic Resource," *Management Review,* Volume 78, Issue 2, (Feb 1989), pp. 37–41.

Aikens, C. Harold, "How IE Training Can Facilitate Implementation of Deming's 14 Points," *Industrial Engineering,* Volume 23, Issue 8 (Aug 1991), pp. 33–36.

Allen, T. J., D. Lee, and M. L. Tushman, "R&D Performance as a Function of Internal Communication, Project Management, and the Nature of Work," *IEEE Transactions on Engineering Management,* Volume 27, Number 1 (Feb 1980).

Aryee, Samual, "Combating Obsolescence: Predictors of Technical Updating Among Engineers," *Journal of Engineering & Technology Management,* Volume 8, Issue 2 (Aug 1991), pp. 103–119.

Autry, James, *Love and Profits: The Art of Caring Leadership,* Hawaii: William Morrow, 1991.

Avots, Ivars, "Why Does Project Management Fail?" *California Management Review,* (Fall 1979).

Babcock, Daniel L., *Managing Engineering and Technology,* Englewood Cliffs, NJ: Prentice-Hall, 1991.

Badaway, Michael K., "What We've Learned Managing Human Resources," *Research-Technology Management,* Volume 31, Issue 5 (Sept/Oct 1988), pp. 19–35.

Baker, B. N., D. C. Murphy, and D. Fisher, "Factors Affecting Project Success," in D. I. Cleland and W. R. King, eds., *Project Management Handbook,* New York: Van Nostrand Reinhold, 1983.

Barcey, Hyler, *Managing from the Heart,* Pittsburgh Enterprise Press, 1990.

Barczak, Gloria and David Wilemon, "Successful New Product Team Leaders," *Industrial Marketing Management,* Volume 21, Issue 1 (Feb 1992), pp. 61–68.

Blake, Robert R. and Jane S. Mouton, *The Managerial Grid,* Houston: Gulf Publishing, 1964.

Blanchard, Frederick L., *Engineering Project Management,* New York: Marcel Dekker, 1990.

Boyer, Marcel, "Leadership, Flexibility, and Growth," *Canadian Journal of Economics,* Volume 24, Issue 4 (Nov 1991), pp. 751–773.

Braham, James, "Where are the Leaders?" *Machine Design,* Volume 63, Issue 20 (Oct 10, 1991), pp. 58–62.

Brown, Warren B. and Necmi Karagozoglu, "A Systems Model of Technological Innovation," *IEEE Transactions on Engineering Management,* Volume 36, Issue 1 (Feb 1989).

Butler, Arthur G., "Project Management: A Study in Organizational Conflict," *Academy of Management Journal,* (Fall 1970).

Cleland, David I., *Project Management: Strategic Design and Implementation,* New York: McGraw-Hill, 1990.

Corman, Joel, Benjamin Perles, and Paula Yancini, "Motivational Factors Influencing High-Technology Entrepreneurship," *Journal of Small Business Management,* Volume 26, Issue 1 (Jan 1988), pp. 36–42.

Davis, S. M. and P. R. Lawrence, "Problems of Matrix Organizations," *Harvard Business Review,* (May/June 1978).

Denton, D. Keith, "Effective Appraisals: Key to Employee Motivation, *Industrial Engineering,* Volume 19, Issue 12 (Dec 1987), pp. 24–25, 46.

Eckerson, Wayne, "Challenging Environment Keeps Workers Motivated," *Network World,* Volume 8, Issue 2 (Jan 14, 1991), pp. 25–26.

Einsiedel, Albert A., Jr., "Profile of Effective Project Managers," *Project Management Journal,* Volume 18, Issue (Dec 1987), pp. 51–56.

Emshoff, James R., *Analysis of Behavioral Systems,* New York: Macmillan, 1971.

Fiedler, Fred, *A Theory of Leadership Effectiveness,* New York: McGraw-Hill, 1967.

Fleishman, Edwin A., "Twenty Years of Consideration and Structure," in E. A. Fleishman and J. G. Hunt, *Current Developments in the Study of Leadership,* Carbondale, IL: Southern Illinois University Press, 1973.

Fox, Douglas A., "Employee Motivation: Technological Twists," *Incentive,* Volume 165, Issue 4 (Apr 1991), pp. 63–64.

Freedman, David, "The Call for Leadership," *CIO,* Volume 5, Issue 6 (Jan 1992), pp. 22–28.

Fukuda, Ryuji, *Managing Engineering,* Cambridge, MA: Productivity Press, 1986.

Gellerman, Saul W., and William G. Hodgson, "Cyanamid's New Take on Performance Appraisal," *Harvard Business Review,* Volume 66, Issue 3 (May/June 1988), pp. 36–41.

Gemmill, Gary R. and Hans J. Thamhain, "Influence Styles of Project Managers: Some Project Performance Correlates," *Academy of Management Journal,* Volume 17, Number 2, (June 1974).

Gemmill, Gary R. and Hans J. Thamhain, "The Effectiveness of Different Power Styles of Project Managers in Gaining Project Support," *IEEE Transactions on Engineering Management,* Volume 20, (May 1973), and *Project Management Quarterly,* Volume 5, Number 1 (Spring 1974).

Gemmill, Gary R. and David L. Wilemon, "The Power Spectrum in Project Management," *Sloan Management Review,* (Fall 1970).

Groves, Ray, "Leadership in Tomorrow's Global Marketplace: Quality Practices," *Vital Speeches,* Volume 58, Issue 5 (Dec 15, 1991), pp. 144–146.

Gunneson, Alvin O., "Communicating Up and Down the Ranks," *Chemical Engineering,* Volume 98, Issue 6 (June 1991), pp. 135–140.

Harmon, Frederick, G., *The Executive Odyssey,* New York: Wiley, 1989.

Hawk, Stephen R. and Brian L. Dos Santos, "Successful System Development: The Effect of Situational Factors on Alternate User Roles," *IEEE Transactions on Engineering Management,* Volume 38, Issue 4 (Nov 1991), pp. 316–327.

Hersey, P. and K. Blanchard, *Management of Organization Behavior and Utilizing Human Resources,* 4th ed., Englewood Cliffs, NJ: Prentice-Hall, 1982.

Herzberg, Frederick, "One More Time, How Do You Motivate Workers," *Harvard Business Review,* (Jan/Feb 1968), pp. 53–62.

Hodgetts, Richard M. and Fred Luthans, "Japanese HR Management Practices: Separating Fact from Fiction," *Personnel,* Volume 66, Issue 4 (Apr 1989), pp. 42–45.

Horton, Thomas R. and Peter C. Reid, *Beyond the Trust Gap,* Homewood, IL: Irwin, 1990.

Horwitt, Elisabeth, "Management by Motivation," *Computerworld,* Volume 21, Issue 52 (Dec 28, 1987/Jan 4, 1988), pp. 105, 107.

Huseman, R. C., J. D. Hatfield, and E. W. Miles, "A New Perspective on Equity Theory," *Academy of Management Review,* (Nov 12, 1987), pp. 232–234.

Jameson, Brenda and Elaine Soule, "Leadership Development: Meeting the Demands of Innovation," *Quality Progress,* Volume 24, Issue 9 (Sept 1991), pp. 84–87.

Japan Human Relations Association (ed.), *Kaizen Teian-1,* Cambridge, MA: Productivity Press, 1992.

Johnston, Neil M., "How to Create a Competitive Workforce," *Industrial and Commercial Training* (UK), Volume 23, Issue 2 (1991), pp. 4–6.

Kaplan, Robert E., *Beyond Ambition: How Driven Managers Can Lead Better and Live Better,* Jossey-Bass, 1991.

Katz, F. E., "Explaining Informal Work Groups in Complex Organizations," *Administrative Science Quarterly,* No. 10, 1965.

Kerr, S. R. House and A. Filley, "Relation of Leader Consideration and Initiating Structure to Research and Development Subordinate Satisfaction," *Administrative Science Quarterly,* Volume 16, Number 1 (1971), pp. 19–30.

Kerzner, Harold, *Project Management for Executives,* New York: Van Nostrand Reinhold, 1982.

Kexsbon, Deborah S., "Leadership and Influence: The Challenge of Project Management," *AACE Transactions,* (1988), pp. I.2.1–I.2.4.

Kirby, Tess, "Delegating: How to Let Go and Keep Control," *Working Woman,* Volume 15, Issue 2 (Feb 1990), pp. 32, 37.

Kirk, Roger E., *Experimental Design & Procedures for the Behavioral Sciences,* Belmont, CA: Wadsworth, 1968.

Kouzes, James M. and Barry Z. Posner, *The Leadership Challenge: How to Get Extraordinary Things Done in Organizations,* Jossey-Bass, 1991.

Kulik, Carol T., "The Effects of Job Categorization on Judgments of the Motivating Potential of Jobs," *Administrative Science Quarterly,* Volume 34, Issue 1 (Mar 1989), pp. 68–90.

Kupfer, Andrew, "How American Industry Stacks Up," *Fortune,* Volume 125, Issue 5 (Mar 9, 1992), pp. 30–46.

Langer W. C., *Psychology of Human Living,* New York: Century, Crofts, 1937.

Lawler, Edward E. III, "Substitutes for Hierarchy," *Incentive,* Volume 163, Issue 3 (Mar 1989), pp. 39–45.

Lawrence, Paul R., and Jay W. Lorsch, "New Management Job: The Integrator," *Harvard Business Review,* (Nov/Dec 1967).

Lea, Dixie and Richard Brostrom, "Managing the High-Tech Professional," *Personnel,* Volume 65, Issue 6 (June 1988), pp. 12–22.

Likert, Rewsis, *New Patterns of Management,* New York: McGraw-Hill, 1967.

Livingston, J. Sterling, "Pygmalion in Management," *Harvard Business Review,* (July/Aug 1966, pp. 81–89.

Ludeman, Kate, "Developing You Own Olympic Champions," *Executive Excellence,* Volume 5, Issue 10 (Oct 1988), pp. 13–14.

Maccoby, Michael, "Closing the Motivation Gap," *Research-Technology Management,* Volume 34, Issue 1 (Jan/Feb 1991), pp. 50–51.

Main, Jeremy, "The Winning Organization," *Fortune,* Volume 118, Issue 7 (Sept 26, 1988), pp. 50–60.

Manz, Charles C., *Mastering Self-Leadership: Empowering Yourself for Personal Excellence,* Englewood Cliffs, NJ: Prentice-Hall, 1991.

Maslow, Abraham, "A Theory of Human Motivation," *Psychology Review,* Volume 50, (1943), pp. 370–396.

Mayo, Elton, *The Human Problems of an Industrial Civilization,* New York: Macmillan, 1933.

McDonough, Edward F., III and Gloria Barczak, "Speeding Up New Product Development: The Effects of Leadership Style and Source of Technology," *Journal of Product Innovation Management,* Volume 8, Issue 3 (Sept 1991), pp. 203–211.

McGregor, Douglas, *The Human Side of the Enterprise,* New York: McGraw-Hill, 1960.

Miller, Donald B., "Challenges in Leading Professionals," *Research-Technology Management,* Volume 31, Issue 1 (Jan/Feb 1988), pp. 42–46.

Miner, John B., Norman R. Smith, and Jeffrey S. Bracker, "Role of Entrepreneurial Task Motivation in the Growth of Technologically Innovative Firms," *Journal of Applied Psychology,* Volume 74, Issue 4 (Aug 1989), pp. 554–560.

Nash, Laura I., *Good Intentions Aside,* Boston: Harvard Business School Press, 1990.

Ouchi, William, *Theory Z,* Reading, MA: Addison-Wesley, 1981.

Pettersen, Normand, "Selecting Project Managers: An Integrated List of Predictors," *Project Management Journal,* Volume 22, Issue 2 (June 1991), pp. 21– 26.

Pitman, Ben, "Technical and People Sides of Systems Project Start-Up," *Journal of System Management,* Volume 42, Issue 4 (Apr 1991), pp. 6–8.

Pryor, Fred, "Energetically Managing Your Organization," *Manage,* Volume 40, Issue 1 (Apr 1988), pp. 23–25, 34.

Raudsepp, Eugene, "How to Delegate Effectively," *Machine Design,* Volume 61, Issue 8 (Apr 20, 1989), pp. 117–120.

Rayner, Bruce C. P., "The Rising Price of Technological Leadership," *Electronic Business* (Mar 18, 1991), pp. 52–56.

Roberts, Edward B., "The Personality and Motivations of Technological Entrepreneurs," *Journal of Engineering & Technology Management* (Netherlands), Volume 6, Issue 1 (Sept 1989), pp. 5–23.

Rosenbaum, Bernard L., "Leading Today's Technical Professional," *Training and Development,* Volume 45, Issue 10 (Oct 1991), pp. 84–87.

Roussel, Philip A., Kamal N. Saad, and Tamara J. Erickson, *Third Generation R&D,* Boston: Harvard Business School Press, 1991.

Sanno Management Development Research Center, *Vision Management: Translating Strategy into Action,* Cambridge, MA: Productivity Press, 1992.

Shatzner, Linda, and Linda Schwartz, "Managing Intrapreneurship," *Management Decision,* Volume 29, Issue 8 (1991), pp. 15–18.

Siegel, Sydney, *Nonparametric Statistics for the Behavioral Sciences,* New York: McGraw-Hill, 1956.

Silverman, Melvin, *The Art of Managing Technical Projects,* Englewood Cliffs, NJ: Prentice-Hall, 1987.

Sizemore House, Ruth, *The Human Side of Project Management,* Reading, MA: Addison-Wesley, 1988.

Skinner, B. F., *Science and Human Behavior,* New York: Macmillian, 1953.

Smilor, Raymond W. and David V. Gibson, "Accelerating Technology Transfer in R&D Consortia," *Research-Technology Management,* Volume 34, Issue 1 (Jan/Feb 1991), pp. 44–49.

Smilor, Raymond W. and David V. Gibson,"Technology Transfer in Multi-Organizational

Environments: The Case of R&D Consortia," *IEEE Transactions on Engineering Management,* Volume 38, Issue 1 (Feb 1991), pp. 3–13.

Tang, Thomas Li-Ping, Peggy Smith Tollison, and Harold D. Whiteside, "The Effect of Quality Circle Initiation on Motivation to Attend Quality Circle Meetings and on Task Performance," *Personnel Psychology,* Volume 40, Issue 4 (Winter 1987), pp. 799–814.

Tannenbaum, Robert and Warren H. Schmidt, "How to Choose a Leadership Pattern," *HBR Classic,* (May/June 1973).

Taylor, Frederick, *Principles of Scientific Management,* New York: Harper, 1911.

Thamhain, Hans J. and David L. Wilemon, "Leadership, Conflict, and Project Management Effectiveness," *Sloan Management Review,* Volume 19, Number 1, (Fall 1977).

Thamhain, Hans J., "Managing Engineers Effectively," *IEEE Transactions on Engineering Management,* Volume 30, Number 6 (Aug 1983), pp. 231–237.

Thamhain, Hans J. and David L. Wilemon, "Leadership Effectiveness in Project Management," in Donald J. Reifer (ed.), *Tutorial: Software Management,* Los Angeles: 1983, IEEE Computer Society, pp. 235–241.

Thamhain, Hans J., "Managing Technologically Innovative Team Efforts Towards New Product Success," *Journal of Product Innovation Management,* Volume 7, Number 1 (Mar 1990).

Thamhain, Hans J., "Managing Technology: The People Factor," *Technical and Skill Training Journal,* Volume 1, Number 2 (Aug/Sept 1990).

Thamhain, Hans J., "Developing Project Management Skills," *Project Management Journal,* Volume 22, Issue 3 (Sept 1991), pp. 39–44, 53.

Tichy, Noel M. and David O. Ulrich, "The Leadership Challenge. A Call for the Transformation Leader," *Sloan Management Review,* (Fall 1984), pp. 59–68.

van de Meer, Jacques B. H., and Roland Calori, "Strategic Management in Technology-Intensive Industries," *International Journal of Technology Management* (Switzerland), Volume 4, Issue 2 (1989), pp. 127–139.

von Braun, Christoph-Friedrich, "The Acceleration Trap," *Sloan Management Review,* Volume 32, Issue 1 (Fall 1990), pp. 49–58.

Vroom, Victor, *Work and Motivation,* New York: Wiley, 1964.

Weatherall, David, "New Technology and Motivation," *Management Services* (UK), Volume 32, Issue 6 (June 1988), pp. 28–30.

Westwood, Albert R. C. and Yukiko Sekine, "Fostering Creativity and Innovation in an Industrial R&D Laboratory, *Research-Technology Management,* Volume 31, Issue 4 (Jul/Aug 1988), pp. 16–20.

Zawacki, Robert A., "Motivating the IS People of the Future," *Information Systems Management,* Volume 9, Issue 2 (Spring 1992), pp. 73–75.

11

DEVELOPING THE TECHNOLOGY-BASED ENGINEERING ORGANIZATION

11.1 CHANGING VIEWS ON ACHIEVING ORGANIZATIONAL EFFECTIVENESS

To survive and prosper in the changing world of technology-oriented business, we must respond to market needs with high-quality, cost-effective, and timely products and services. Many new techniques evolved within the company with the objective of making the various components of product design, development, production, and field support more effective. From concurrent engineering to all-inclusive total quality management concepts, these techniques were designed to increase organizational effectiveness, leading to better management of risks, technological integration, customer involvement, product reliability, and many other facets critical to linking today's markets, products, and technology. However, as we already have learned from Sloan, organizations do not just "evolve automatically" into an effective management system; they need to be systematically steered toward self-assessment and renewal.[1] This involves task redesigns, structural changes, skill developments, information handling, and many other aspects of the sociotechnical system of the organization. While management consultants forty and fifty years ago concentrated their efforts on improving efficiency in selected functional areas such as production or sales, today's organizational studies focus on developing the total enterprise. This new awareness is certainly present among businesses, but also is present in govern-

[1]Alfred P. Sloan introduced decentralized management at General Motors in the 1920s to cope with the company's business challenges and opportunities. However it took more than a policy directive to change to this new form of management. Extensive efforts at all levels and organizational functions, plus a sweeping array of new management principles and controls, were necessary to implement the planned change successfully.

ment organizations, hospitals, universities, and religious organizations. Even the Bank of England, which shielded itself from outsiders for several hundred years, solicited help from an American consulting firm for assessing its needs and for reorganizing to better respond to today's challenges.

The process of systematically improving organizational effectiveness, as measured against its objectives, is defined as organizational development (OD). It focuses on planned change, which usually involves all social, technical, behavioral, and administrative subsystems. Approaches to OD during the 1950s focused on the behavioral subsystems, at which time laboratory sensitivity training groups (t-groups) were the best-known vehicle to induce change. More recent developments in OD include a much broader approach to organizational improvement. This more modern approach to OD includes all subsystems of the organization. It tries to achieve improved organizational performance by diagnosing potential problems via feedback from organizational members, problem-solving, and follow-up. Since feedback is the essential ingredient for designing the planned change, this type of OD process is referred to as *survey research and feedback.* Today's efforts at renewing and developing the organization, as practiced by quality circles and total quality management initiatives, combine both laboratory training and survey research and feedback techniques into one process, often referred to as the *diagnostic-prescription cycle.*

Recent Changes Affecting Technological Organizations

The 1980s were the start of a period of rapid change. The primary factors that influence our competitive position today include the following:

- *Shifts in the Global Scientific and Technology Base.* The broad-based technological advantages that United States companies enjoyed up to the 1970s are rapidly diminishing. In many product areas there are Pacific Rim and European companies that outspent the United States in technology or engineering education and original R&D efforts and produce technologically superior results.
- *Shifts in Global Competition.* Increased levels of education, skills, government support, increased access to scientific and technological information, as well as some cost advantages have helped to position companies around the globe very favorably in international as well as in our own domestic markets.
- *Shifts in Customer Awareness, Needs, and Options.* Customers today are better skilled to compare the value of products and services. The industrial customer in particular, has broad access to a global market and can shop for specific features, quality, cost, delivery, reliability, technical excellence and product support.
- *Shifts in Information Access.* Knowledge of processes, materials, components, and general scientific discoveries relevant to today's product development are available with little delay or barriers to virtually all people around the globe.
- *Shifts in Time to Market.* Driven by both increased market demand and advanced technology, new product development cycles continuously are being

reduced. Companies must manage their new product development time effectively to compete favorably.

- *Shifts in Environmental and Social Awareness.* Increasing concern about environmental issues worldwide and about human factors—the desire for meaningful work, safety, and health—have lead to new business policies and management philosophies.

Areas of Organizational Development Focus

In response to the changing business environment, organizations have begun to develop process-oriented people skills together with technical expert knowledge. Particular emphasis is being placed on the development of those skills that enhance cross-functional teamwork. When describing their organizational development efforts, engineering managers and consultants often speak of four, often interrelated, areas of OD objectives and focus:

1. *Technological Readiness.* Technological readiness focuses on the development of the people, facilities and processes necessary to meet the technological challenges of managing the organization effectively in today's competitive world. This focus usually ranges across all organizational areas from product development to production and field services. The impact is on cost, product performance, quality, and timeliness.

2. *Productivity Improvements.* Productivity improvements focus on the effective utilization of resources. They include on the people side motivational issues, skill-building, and teamwork. But they also include processes, facilities, and organizational structures. Perhaps more than any other area, productivity issues are highly interfunctional and cannot be treated in isolation, as we have learned from factory and office automation. The desired impact areas of productivity improvements are primarily cost and response time.

3. *Innovation and Creativity.* Innovation and creativity focus on people and their creative skills in solving challenging problems in the technical as well as administrative areas. The winning edge of an engineering organization is its innovative capacity, the special magic of the people as individuals as well as a group. Although the primary focus is on the development of the people, the support systems, facilities, organizational environment, and leadership styles also play an important role. The impact of improved innovation and creativity is widespread. They affect product features, quality, cost, and timing.

4. *Minimizing Risk and Uncertainty.* This OD concern focuses largely on organizational process and structure. Examples are management procedures regarding new product identification, bid proposals, and project management. The business impact is on resource effectiveness and cash flow.

As competitive pressures mount and organizational environments continue to change, pressures become more intense on managers to utilize their resources in the most optimal way. All this does not mean, however, that engineering managers should become OD specialists and spend most of their prime time analyzing and training. Lead engineering personnel can, however, facilitate many of the organizational development processes as an integrated part of good management practice. The OD methodology is straightforward. The following discussion focuses on the pragmatics of diagnosing and managing change.

11.2 PHASES OF THE ORGANIZATIONAL DEVELOPMENT PROCESS

Modern OD practices have many facets. They range from individual skill development, through team-building, up to grid approaches for entire organizations. However, regardless of the massive applications, strategies, and approaches, the OD process usually includes four primary phases, which are schematically shown in Figure 11.1. In principle, this process was originally recommended by Kurt Lewin[2] for managing social change:

Phase 1: Initiating. Establishes the principal objectives, scope, needs, desires, and responsibilities for the organizational change to be considered. This also includes identifying, at least in principle, the perceived problems, threats, and

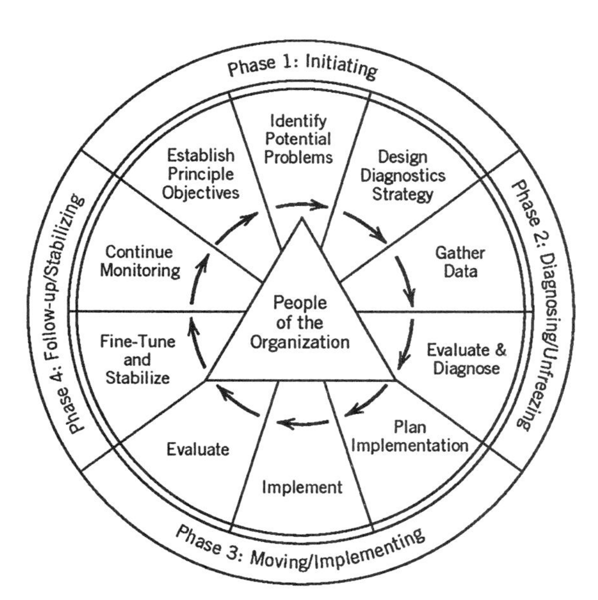

FIGURE 11.1 Organization development using the diagnostic prescription cycle.

[2]See Kurt Lewin, "Frontiers in Group Dynamics: Concept, Model, and Reality in Social Science," *Human Relations,* Vol. 1, No. 1, 1947, pp. 5–41.

opportunities. It also includes establishing a strategy for the intervention process.

Phase 2: Unfreezing (Diagnosing). Assesses the situation, defines the problems and challenges, prescribes the changes, and establishes the criteria for evaluating the organizational performance. The unfreezing phase usually involves the members of the organization in an effort to gain realistic situational insight, to neutralize resistance to change, to establish a willingness to participate, and to support the change-implementation later on.

Phase 3: Moving. Phase 3 refers to the intervention or implementation of the change. It may involve individuals, groups, or the whole organization. The methods include skill training, sensitivity training, conflict-resolution meetings, team-building, and communication workshops, to name a few of the most popular interventions.

Phase 4: Refreezing (Follow-Up). Refreezing involves the fine-tuning of the system after the change has been introduced. It includes follow-up on problems and complaints, assistance in coping with the newly changed system, and maintenance and stabilization of the new system. Progress is assessed against objectives and used as feedback for fine-tuning of the organizational system.

Organizational development should not end abruptly. The evaluation and fine-tuning of the intervention, together with possible external changes that may have occurred during the OD process, often make it necessary to continue the OD cycle. That is, the process iterates through the first three phases until the system is stabilized and the desired results have been achieved, which includes the acceptance of the new system and the commitment to it by the people of the organization.

11.3 SELF-ASSESSMENT: THE KEY TO ORGANIZATIONAL RENEWAL

Peter Drucker says managers are paid for getting the right things done.[3] In today's changing environment, few managers can hope to focus their efforts and resources on the right thing simply as a by-product of hard work. Avoiding the "activity trap" requires constant reassessment of goals, objectives, structures, and operations. Being a manager means taking responsibility for your own performance. Tom Peters and Bob Waterman suggest that once in awhile we should look at ourselves in the mirror.[4] Self-assessment is the starting point of any organizational development effort. It also is the most crucial step in facilitating success or failure of the OD effort.

Self-assessment refers to the organization's process of dealing with the threats and opportunities present in its environment. Self-assessment is a prerequisite for organizational renewal; it is part of the "diagnosis" and "unfreezing" phase. Designing a

[3]Peter Drucker, *The Effective Executive,* New York: Harper & Row, 1967.

[4]Thomas J. Peters and Robert H. Waterman, Jr., *In Search of Excellence,* New York: Harper & Row, 1982.

diagnostic strategy is one of the critical activities to be undertaken by management at the outset of the OD effort. First, management must decide who is responsible for the diagnostics and what areas to probe. Second, the organization must be properly prepared for such a self-assessment and the forthcoming interventions. Many of today's initiatives for improving organizational effectiveness, such as concurrent engineering, benchmarking, statistical process control, computer-integrated manufacturing, computer-aided engineering, and total quality management, need a high level of organizational readiness, which comes via broad-based involvement and education of people from all organizational levels. An organizational development program should not come as a surprise to people. The program must be properly introduced by a well-orchestrated set of meetings, newsletters, bulletin boards, and management actions to assure that people understand the what, why, and how of the upcoming OD effort. This will minimize anxiety and lead to personal involvement, interest, and commitment to the effort and its objectives. The actual situational assessment must be designed and conducted. The interpretation of diagnostic data completes the self-assessment phase.

Problems of Assessment

Designing proper diagnostics is difficult. Designing the diagnostic strategy is a true challenge to management and to the OD specialist (discussed further in Section 11.4), because it requires probing the various issues that may be associated with global OD objectives. For example, a typical organizational objective could be to improve the productivity of engineering personnel, as measured in reduced time to market of a new product. The diagnosis will most likely include an assessment and quantification of (1) the objectives, (2) technology, (3) engineering tools and techniques employed, (4) resources available, (5) organizational structure, (6) policies and procedures, and (7) management style.

In addition, the diagnosis must also probe the informal aspects of the organization, which often contain hidden, less obvious factors, such as (1) attitudes, (2) opinions, (3) feelings, (4) group norms, (5) power and politics, and (6) interpersonal dynamics. These "cultural" factors can be very important aspects of organizational life and may be strong drivers or barriers to organizational development objectives. These covert factors are often difficult to diagnose and to measure, because they are hidden from the view of the observer. People are reluctant to disclose them and sometimes are not even aware of their existence. A useful model is provided by Stanley Herman of the TRW Systems Group,[5] who categorizes the various factors that may influence organizational effectiveness into two groups, formal and informal, as shown in Figure 11.2. Herman suggests viewing the typical organization as an iceberg, with most of the organization's information and activities lying beneath the surface, hidden from the view of those not immediately connected with these aspects. The model shown

[5]Stanley N. Herman, "TRW Systems Group," in Wendell L. French and Cecil H. Bell, Jr., *Organization Development: Behavioral Science Interventions for Organizational Improvement*, Englewood Cliffs, NJ: Prentice-Hall, 1978, p. 16.

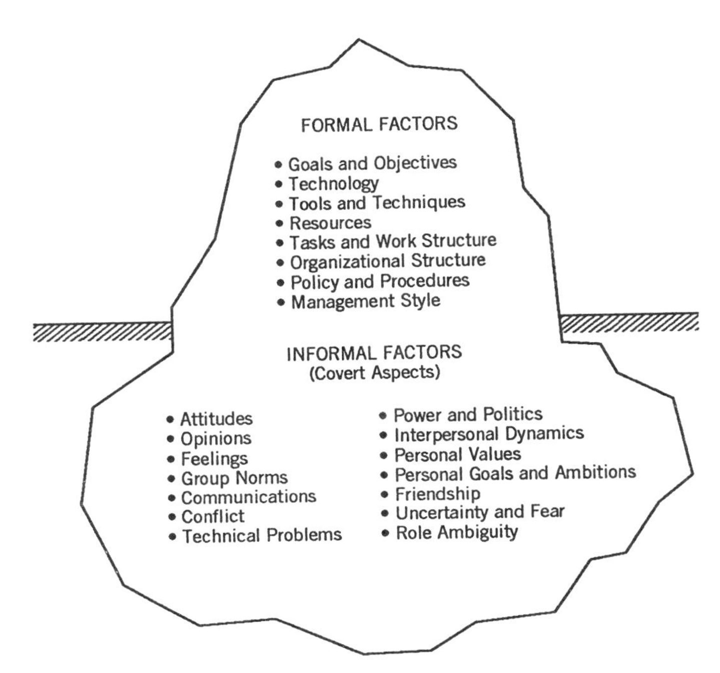

FIGURE 11.2 The organizational iceberg. Formal and informal factors that influence organizational effectiveness. Adapted from Stanley N. Herman, "TRW Systems Group," in Wendell L. French and Celil H. Bell, Jr., *Organization Development: Behavioral Science Interventions for Organization Improvement,* Englewood Cliffs, NJ: Prentice-Hall, 4e, 1990, p. 19, with permission.

in Figure 11.2 can be helpful in designing the diagnostic strategy as it helps to identify, categorize, and organize the various areas of organizational life for subsequent investigation.

Diagnostic Designs

Diagnostic designs are often complex and bear the risk of yielding costly and erroneous results. Depending on the situation and specific aspects of the organizational system to be investigated, a combination of the four diagnostic approaches listed below may be considered.

1. Interviews. One-on-one or small-group interviews can be conducted in either a structured or unstructured form; questions can be asked according to a specific predesigned questionnaire or in a more open-ended fashion that requires more explanation. While the structured format leads to more uniformity, comparison, and quantification, the unstructured format allows probing into unknown areas and facilitates a more spontaneous, candid response. Therefore, both forms of interview techniques are often employed side by side. Given a skilled interviewer, interviews can discover a great deal about the people and systems in an organization. If designed correctly, interviews also can produce core information from a selected sample in a relatively short period of time. On the other hand, interviews become very costly and time-consuming for larger samples. In addition, interviews contain many biases and often need careful interpretation and skillful translation of the information into the common framework of data to be investigated. Yet few organizational diagnoses can be completed without interviews.

2. Survey Questionnaires. Because of their cost- and time-effectiveness for large sample sizes, questionnaires are the most widely used instruments for data collection in situational analysis. However, to be effective as a survey instrument, questionnaires must be carefully designed and pretested by specialists. Many survey instruments for measuring common variables such as personal attitudes, values, fears, likes, and needs can be purchased in the open market at a fraction of the cost of custom design and with the built-in effectiveness of a tested design.[6] Some of the more sophisticated survey packages also include data scoring, analysis, and statistical summaries as a purchase option.

3. Observation. On-site observation by the line manager or an independent observer can be a very powerful technique for diagnosing process-related problems, conflict, or systemic interrelations. Observations, in combination with other survey techniques, can often validate findings, or can detect previously unrecognized problems. On the negative side, observation techniques are costly and time-consuming. Furthermore, the presence of the observer may make people uncomfortable and cause them to behave in an unnatural work mode. Because of these potential problems, observations are often modified or augmented using group discussions and focus on group formats. Observations must be carefully designed to minimize these biases and to make them economically affordable.

4. Records and Documents. Today's organizations capture a wealth of information that is available not only for review, but also can be readily accessed for statistical evaluation and summary. Examples range from personnel records on turnover, sick leave, awards, safety, and skill levels, to data on reliability, machine downtime, material flow, design cycle time, and budgets. Modern techniques such as statistical process control (SPC) are typical examples of utilizing available operating data or setting up measurement systems for capturing data for diagnosing potential operating problems and development needs. Records often present factual information of organizational health. Like a barometer, they can measure an increase or decrease of a standardized organizational variable as a precursor to a potential problem or improvement. Records are also very effectively used to compare the outcome of an organizational development intervention. For example, in Company Y, remodeling the cafeteria was supposed to improve morale and employee spirit and ultimately help reduce costly personnel turnover. Comparison of personnel records before and after this intervention will clearly verify the degree of its effectiveness.

[6]One of the publishers of OD instruments is University Associates, Inc., 8517 Production Avenue, San Diego, California 92126. University Associates published *Instrumentation in Human Relations Training* (1976) by J. William Pfeiffer, Richard Heslin, and John E. Jones, a collection of 92 survey instruments.

11.4 ORGANIZATIONAL DEVELOPMENT: OBJECTIVES AND CHALLENGES

Organization development efforts come in many forms and levels of intensity. However their aim is usually the systematic improvement of organizational performance. One of the original OD definitions that is often quoted dates back to the Western Organization Development Conference in San Francisco in 1970: "OD is a systematic effort to improve organization performance by changing the way organizations operate through modifying the structure and process of the management system that guides their behavior."

Other definitions stress the psychological subsystem, with the objective of developing social processes such as communications, problem-solving, mutual trust, and cooperation. Taken together, ten specific objectives can be listed that are common for a typical OD program:

1. To foster a sense of organizational purpose and ownership in individuals
2. To encourage openness and feedback among people
3. To encourage risk-taking and double-loop learning[7]
4. To strengthen trust, cooperation, and support
5. To facilitate problem-solving via true problem-finding and correction
6. To enhance the sense of accomplishment and pride in workmanship and motivation among employees
7. To develop communication and feedback channels
8. To increase personal accountability, responsibility, and commitment
9. To facilitate willingness to change
10. To create participation and involvement (people support what they create) and help to unify people at all levels of the organization behind established business objectives, values, and strategies.

When to Use Organizational Development

Organizational development focuses on the improvement of overall organization effectiveness and performance rather than on the improvement of an individual manager's performance. In an engineering environment, typical situations appropriate for OD may include those listed below:

1. A Potentially Ineffective Organizational Structure. An organizational unit (e.g., R&D, manufacturing, or marketing) does not interface and communicate properly with other units. OD can help the people in the various organizational units to think through the issues and challenges involved in this interface. It can further help in identifying potential solutions in a participative way, which may in turn facilitate the staff's willingness and commitment to the desired change.

2. Intergroup Conflict. Dysfunctional conflict may exist between two or more work groups or departments. The OD process may involve an open exchange of information between the conflicting groups to determine the true barriers to effective work interface and cooperation. This may lead to a systematic attempt to correct these deficiencies through change of process, structure, and personal attitudes.

3. Formation of a New Project Team. In order to form a new team of people from different functional areas or to integrate newcomers quickly into a project team, some group interventions and one-on-one sign-ons of new members to the team can enhance the sense of belonging and commitment to the team.

4. Project Recovery After a Setback. Recovering from a project setback usually is associated with a series of morale, trust, and motivational problems, which affect the productivity of the project team. OD can help to clarify the new mission and unify the team behind it.

5. Reorganization. Any reorganization, whether associated with personnel changes or not, usually has an effect on people's productivity. OD techniques such as working with groups may help in the understanding of the dynamics and organizational relations of the new work environment and may turn potential anxieties into productive energy.

6. Introducing New Technology. Introducing a new technology or implementing a new technology-oriented policy requires the cooperation and support of many people. OD techniques combined with participative planning and involvement can enhance cooperation in such a venture and the personal commitment to it.

7. Training. Building specialty skills for your employees or retraining part of the work force could utilize a combination of on-the-job training and classroom education, both very popular and effective OD modes of intervention.

8. Organizing Quality Circles. Continual organizational self-improvement and self-renewal are necessary to stay at the cutting edge of today's competitive environment. The formation of special professional interest groups and task forces can be an effective OD technique to fine-tune organizational performance.

9. Improving General Morale and Attitudes. Survey–feedback techniques are often used as an OD tool to determine the sources and causes of employee dissatisfaction and grievances. Bringing the true problems out in the open often enables management to deal with them more effectively and participatively.

10. Troubleshooting. One of the most challenging assignments of a manager is to improve organizational performance or staff performance on a particular project that seems to be below par for no obvious reason. OD can help with a variety of techniques, which include task forces, survey–feedback, and team-building.

Challenges to Organizational Development's Adoption

Despite all of the advantages and benefits of OD, managers—especially those in technical organizations—are often skeptical that OD interventions will actually produce the desired results. The quality circle is a typical example of an OD approach that has received a mixed reception regarding its effectiveness. Most skeptical are those managers who tried some of the interventions and got disappointing results. Many of the problems and challenges are associated with the risk of applying the wrong intervention or solving the wrong problem, especially when using technique-oriented tools. Not only is there danger in wasting substantial time and energy but, even more seriously, the process can produce highly undesirable consequences, such as increased conflict, power struggles, confusion, apathy, lower motivation, and risk-avoidance. Take for example two engineering departments, electrical and mechanical engineering, which do not work well together because tension and interpersonal problems exist at the upper management level. Unless this problem is properly diagnosed and dealt with, no team-building, task force or survey–feedback technique at the department level will improve the situation. To the contrary, people might get frustrated and angry over their inability to resolve the problem. Possible effects of survey feedback are shown in Figure 11.3.

Other challenges are associated with the different cultures and value systems that exist within an organization. Not all OD processes are appropriate for integrated problem-solving that reaches across functional lines or geographic borders. For example, a joint problem between engineering and manufacturing might need a totally different approach than if the situation were confined to engineering only. Even more challenging are multinational ventures involving people from different countries. A self-awareness or team-building session that is jointly conducted with people from the regions involved—Middle East, Europe, Japan, etc.—may serve as a vehicle to better international understanding. Yet the exercise might not lead to productivity improvements. To the contrary, these interventions can sometimes lead to great confusion and even resentment by people who do not buy into the change objectives or who feel manipulated or threatened. Thus great challenges exist to properly facilitating these interventions.

Yet another concern expressed by managers is over the amount of time and effort usually required to conduct OD, which often produce little short-term benefits. Typically, training and self-awareness programs take people away from the workplace for days and sometimes for a full week. The benefits are long-range, hard to

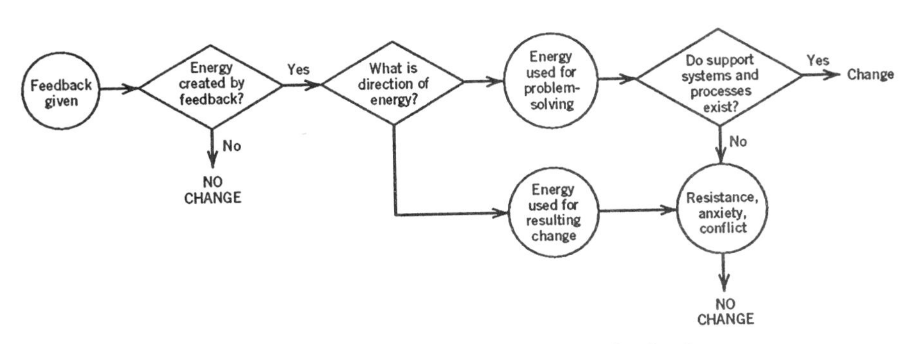

FIGURE 11.3 Energy flow and results of survey feedback process.

measure, and primarily for the individual. Therefore, it is not surprising that many managers are skeptical about the value of OD. They often see only the immediate costs and work interruptions, which are painful in an environment that is driven by quarterly production quotas and profit measures.

Every manager has a different set of challenges and problems that he or she is concerned about in applying OD techniques effectively. However, by and large these challenges are as noted below:

1. OD may concentrate on the wrong problem. Proper diagnostics is critically important.
2. Many OD interventions are psychologically complex and require personnel with specialty skills to be effective.
3. Many of the technique-oriented approaches don't work well in multidisciplinary or multicultural environments.
4. Benefits are mostly long-range and are difficult to measure and to quantify.
5. OD efforts can be very time-consuming and costly.
6. Results are not always predictable and can have undesirable side effects, such as people becoming more cooperative but less aggressive and less willing to take risks.
7. People may be skeptical and don't participate in OD ventures because they perceive it as being threatening and manipulative.
8. To be effective, most OD efforts must be supported by upper management.
9. OD efforts tend to equalize organizational power while weakening traditional line authority.
10. OD efforts often shift the management style toward more democratic, team-oriented management, which weakens central leadership and decision-making.

From these challenges it is clear that OD approaches require more than casually applied management skills or the use of canned techniques. It usually takes very careful planning and often the assistance of an OD specialist. The sequence of the process is critical. Problem diagnostics must come first before a particular OD intervention can be designed. Many of the methods used by OD practitioners today have been tested only in a very limited way. Every day new methods and refinements of existing methods emerge. Examples range from concurrent engineering to risk management, statistical process control, work flow analysis, robust design, and the all-encompassing concept of total quality management. However, managers in modern engineering organizations cannot wait for the "ultimate" knowledge to evaluate their situations. Practicing engineering managers must apply whatever theory is available in seeking improved organizational performance. After all, management is, at least in part, an art, and managerial performance is judged by a reasonable success rate in a probabilistic environment. The OD concepts discussed here can be powerful tools in the hands of a skilled practitioner.

11.5 ORGANIZATIONAL DEVELOPMENT TOOLS AND TECHNIQUES

The Organizational Development Specialist

Organizational development efforts are often too complex and intricate to be managed as a sideline by functional managers. Further, line managers frequently underestimate the psychological skills required to conduct some of these OD interventions which, at the outset, often look simple, behaviorally safe and straightforward. Many managers have found out too late that they were inadequately prepared and skilled to handle a particular intervention. The cost of such a misjudgment is not only borne in the failed OD venture, as measured in time, money, and effort, but also in lost credibility, trust, and position power, plus undesirable organizational side effects such as additional conflict, lower morale, confusion, and lower work effort.

The role of the OD specialist is usually as consultant to management. The specialist should become part of the team and its OD effort, while reporting to the manager who is directly responsible for the specific organizational development or implementation of change. The advantages of bringing an OD specialist on board are in several areas. First, the specialist brings considerable amount of experience, skills, and relevant OD tools into the situation. If we have a serious plumbing problem we might be better off with a professional plumber, even though it might be tempting to try fixing it ourselves. Secondly, the specialist might be able to communicate with the team more candidly and cut through some of the existing biases better than someone within the organization. The specialist seems to be more impartial to the situation than a line manager. Third, the specialist gets the job done with a minimum amount of interference with managerial duties, but for the line manager the OD effort becomes an additional job, for which the manager is usually not well trained, and which drains a lot of his or her prime time. To summarize, the specialist is a facilitator who assesses the situation, helps to diagnose the problem, designs the proper interventions together with management, and then plans, conducts, and follows up on the actual OD program. The final responsibility for solving the problem or obtaining specific results rests, however, with management.

Typical Organizational Development Steps

The following steps are typical for an organizational development effort:

Step 1: Global Objectives. Management states its broad objectives, potential problems, and desires and obtains agreement among the management group. Management communicates information to an OD specialist, who clarifies it.

Step 2: Preliminary Assessment. The OD specialist conducts a preliminary assessment of the situation by on-site observations, analysis of existing data, and brief interviews.

Step 3: Process Definition. OD specialist defines the process or method of diagnosing the situation, evaluating the findings, and prescribing a particular solution (change strategy).

Step 4: Management Review. OD specialist discusses his/her findings from the preliminary assessment, validates the earlier assumptions and verifies the feasibility of the global OD effort, or makes alternate suggestions. The specialist also presents the suggested method of approach. Discussion follows.

Step 5: Management Approval. Management either approves the suggested approach, modifies it, or sends the specialist back to step 2 with new directions.

Step 6: Kickoff I. Once an agreement has been reached on the OD approach, the total work group affected by the upcoming OD effort, or, at least the key personnel, are called for a kickoff meeting with the following agenda:

- Statement of objectives by senior management
- Introduction and endorsement of the OD specialist
- Broad overviews of situation and forthcoming efforts
- Description of principal steps in the OD effort
- Schedule and key assignments.

Step 7: Situational Analysis. The OD specialist conducts a detailed assessment of the situation and tries to diagnose specific problems, challenges, and limitations pertaining to the improvement objectives by way of interviews, on-site observations, historic data, brainstorming sessions, group discussions. Data analysis, integration of findings, report, and offering potential solutions follow.

Step 8: Management Review. This is a presentation of the findings from the situational analysis to management, followed by discussion and integration of any additional management perspective. To be effective, these reviews require strong management direction and leadership.

Step 9: Prescription. The OD specialist designs specific OD interventions appropriate for improving organizational performance in moving toward desired goals.

Step 10: Management Review, Fine-Tuning, and Approval. The OD specialist presents the suggested OD plan (intervention) to management for discussion, fine-tuning, and approval. This step requires intense management involvement and often several iterations.

Step 11: Status Report. A summary report of findings is given to personnel involved in the diagnosis effort.

Step 12: Final Intervention Plan. The specialist performs fine-tuning of the approach and detailed planning of the OD effort (intervention). There is management review and approval of the plan.

Step 13: Kick-Off II. There is a group meeting with all personnel to be involved in the upcoming OD effort; it has the following agenda:

- Statement of objectives by senior management
- Introduction of key individuals

- Broad overview of upcoming effort
- Description of principal steps of OD effort and expected contributions and outcomes
- Schedule and key assignments.

Step 14: OD Intervention. The OD specialist facilitates the implementation of the OD plan. This requires strong involvement and participation of management and the continuous fine-tuning of the change strategy and plan according to the dynamics of the evolving situation. Regular status assessments, tracking against established performance measures, and strategy reviews are needed. An important component of the intervention is effective feedback[8] to the participants, as summarized in Table 11.1.

Step 15: Monitoring, Evaluation and Follow-up. OD programs do not end abruptly at the end of the intervention. Management and OD specialist assess the benefits, the potential for further benefits, and potential destabilizing factors. It is necessary to design, review, approve, and implement the follow-up program. Many organizations must be "refueled," for instance via team-building interventions, for the organizational change to become stable and permanent.

TABLE 11.1 Criteria for Effective Survey Feedback

Feedback given to survey participants must be:

1. *Relevant.* Only information meaningful to the recipients should be fed back.
2. *Understandable.* Communication should be clear and simple. Language and symbols should be familiar to the recipients.
3. *Descriptive.* Data should be in the form of real-life examples, with which the recipients can identify.
4. *Verifiable.* Recipients should be able to test the validity and accuracy of the data fed back to them.
5. *Limited.* Only significant highlights should be presented, to avoid an information overload.
6. *Impactable.* Recipients should be given information on situations that they can directly control.
7. *Comparative.* Comparative data should be provided to let recipients know where they stand in relation to others.
8. *Unfinalized.* Feedback of information must be presented in such a way that participants see them as a beginning and a stimulus rather than as a final document.

[8]The process of using feedback for OD plan implementation is discussed by David A. Nadler in *Feedback and Organization Development*, Reading, MA: Addison-Wesley, 1977.

Applying Organizational Development Tools and Techniques to Target Groups

A large number of OD tools and techniques have been developed for correcting organizational deficiencies. These tools are by and large designed to increase effectiveness via organizational change. That is, they are administered after the target group has been unfrozen and the diagnosis is complete. One useful way of classifying these OD techniques is in terms of target groups, as shown for twelve popular interventions in Figure 11.4 and discussed below.

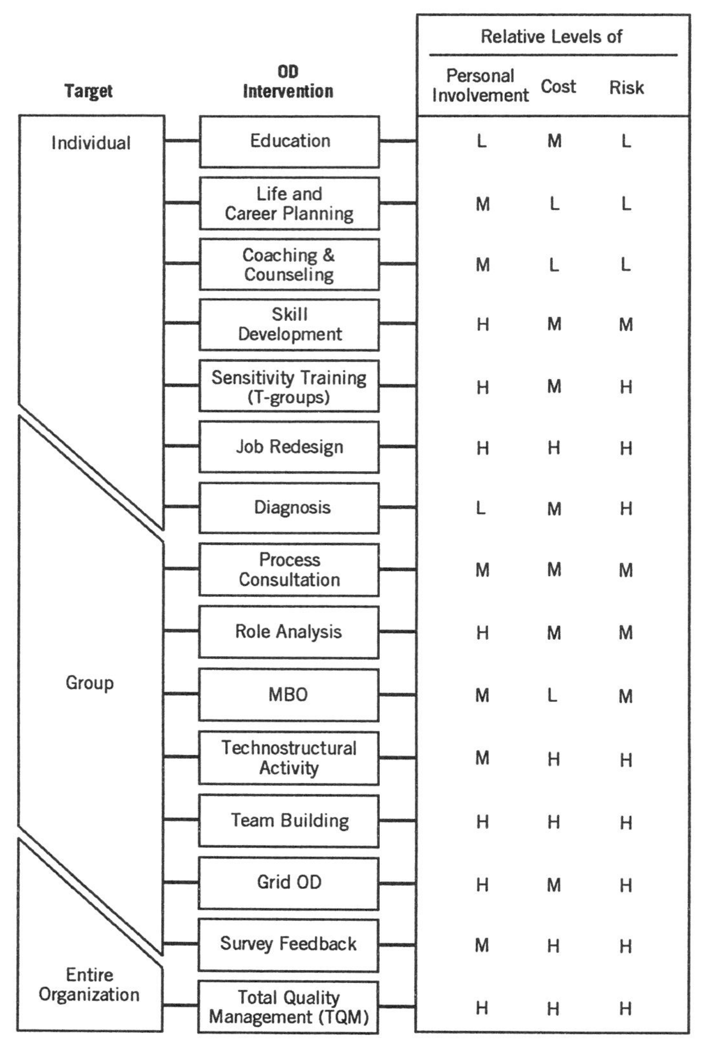

FIGURE 11.4 Various OD techniques and their typical target groups, and levels of personal involvement, cost, and risk.

OD techniques can be directed toward various organizational entities, ranging from a single individual to an entire organization. Theoretically any OD technique can be useful for a variety of organizational structures; in actuality, certain techniques are more likely to be used with certain groups. The following target group classification, adopted from James Stoner,[9] puts some order into the large number of techniques and applications that are summarized in Figure 11.4. Figure 11.4 also indicates the relative levels of involvement, cost, and risk for each. A brief description of the available development activities is provided in Table 11.2.

1. The Individual. Based on the pioneering work by Kurt Levin and by the National Training Laboratory (NTL), sensitivity training became a widespread technique for encouraging individuals to develop changed attitudes and behavior, and ultimately, improved job performance.[10] The training process involves 10–15 people, usually from different organizational units or companies. The "T" (training) groups, are guided by a trained leader to increase the participants' sensitivity to and skills in handling interpersonal relationships. The emphasis is on self-awareness of one's behavior and its impact on others. Many of those who participate in T groups find the experience intense but also rewarding. However, others may feel hurt and have difficulties coping with the anxieties generated within the T group. Sensitivity training is now less frequently used by organizations, and participants are usually screened to make certain that they can withstand group pressures.

Today, a broad range of activities is available to managers to further individual development. In addition to sensitivity training, these activities include life and career planning, skill development and training, education, and coaching and counseling. In contrast to traditional development approaches, which focused on improving work-related skills via "classroom" learning, today's developments combine classroom sessions with experiential learning via role-playing, analysis of videotapes, and the application of newly acquired skills to actual work situations. The feedback obtained from the application reinforces the skill-building and desired behavior. Ultimately, transfer of knowledge into job skills is realized through continued practice, involvement, and reinforcement of the newly learned behavior.

2. Teams or Groups. Today, organization development efforts most frequently concentrate on the effective functioning of work groups or teams. Team development efforts typically focus on member relations, team processes such as communications and decision-making, and task integration. The specific interventions include diagnostic meetings, role analysis, process consultation, technostructural analysis, and traditional transactional analysis. In addition, many organizations have established an environment for ongoing team development via quality circles within the framework

[9]James A. Stoner, *Management* Englewood Cliffs, NJ: Prentice-Hall, 1982, Chapter 14.

[10]Sensitivity training dates back to 1946, when the National Training Laboratory asked Kurt Levin to develop and conduct a training program for community leaders. In group sessions, the community leaders discussed various social problems. Independent observations were made, analyzed, and fed back to the participants. This feedback appeared to be well received and helped the leaders to increase awareness about their behavior and its impact on others. From this beginning, sensitivity training has become a widely used technique for individual change efforts and developments.

TABLE 11.2 Some Intervention Activities Available for Organization Development

1. Life- and Career-Planning Activities. Designed to help individuals focus on their life and career objectives and develop plans for achievement. The emphasis is on diagnosing personal strengths, weaknesses, and objectives and determining what is needed to strengthen deficiencies. This intervention is of low intensity and focuses on the long-range benefit to both the individual and the organization.

2. Training and Education Activities. Designed to improve the knowledge, skills, and abilities of individuals regarding job performance and interpersonal competence. Intervention is of low intensity but high effort and focuses on both short and medium-range benefits.

3. Sensitivity Training (T Groups). Group sessions where participants learn interpersonal skills via increased awareness about their behavior and its impact on others. Intervention is of high intensity and focuses on immediate needs.

4. Job Redesign. Realigns task components to fit the needs and capabilities of the individual and the organization better. Short- or medium-term focus.

5. Coaching and Counseling. Directed toward the individual or a small group with the objective of working more effectively within the existing organizational system. Counseling is often used, in conjunction with career planning and skill development. Medium intensity; short-range needs focus.

6. Role Analysis. Especially in contemporary organizations such as the matrix, a systematic analysis and discussion of individual roles can help to clarify roles, increase cooperation, and reduce dysfunctional conflict and tension. Low intensity, short-range needs focus.

7. Diagnostic Activities. Fact-finding activities that examine operational conditions such as attitudes, processes, and interpersonal dynamics. The objective is to diagnose potential problems or deficiencies and provide a basis for correction. Interventions range from attitude surveys to informal meetings and brain-storming sessions. Usually of low to medium intensity, with short-range needs focus.

8. Process Consultation. Third-party observation of critical group processes (e.g., communication, conflict, and decision-making). The impartial observer then analyzes and integrates the findings and provides feedback to key individuals for potential improvement of the organizational process. Diagnostics and process consultation activities often overlap. Medium intensity; short-range needs focus.

9. Team-Building Activities. Designed to improve the effectiveness of work groups or teams. Especially in engineering, team-building has become a very popular OD intervention. It can take many forms, ranging from simple group meetings, such as a project kick-off, to more intense laboratory training. Medium to high intensity; short- to medium-range needs focus.

10. Technostructural Activities. Interventions focus on the technological subsystem of the organization; its human, structural, and procedural components, as well as its tools and techniques. Examples: determining the optimum levels of automation, job design, centralization versus decentralization of resources, work flow, and computer-aided operations. Especially in engineering environments, technostructural interventions become highly complex, reaching across almost every organizational subsystem and activity, with interrelated and often opposing objectives, high risks, and uncertainties. Medium intensity; long-range focus on organizational needs.

11. Management by Objective. Designed to improve the focus of organizational efforts or needs. Activities formalize an MBO framework throughout the organization, so that organizational objectives are clearly linked to one another and tie into the individual efforts. Medium intensity; short- to medium-range focus on organizational needs.

12. Survey–Feedback Activities. Comprehensive and systematic data collection through interviews and questionnaires to identify attitudes and needs. The data are analyzed and presented in an integrated, understandable form to the people who supplied them. This feedback helps people to understand organizational issues and potential problems as well as suggest possible changes and solutions. Survey feedback is most likely the starting point for various OD interventions, such as skill development, role analysis, and team-building. Medium to high intensity, medium- to long-range focus on organizational needs.

13. Organizational Grid Activities. Based on the managerial grid developed by Blake and Mouton,[1] grid activities are designed to evaluate the total organization and its resources regarding overall effectiveness. The focus is on upgrading managers, skills and leadership abilities, teamwork, planning, goal setting, and monitoring of events within the organization. The grid approach is considered by many managers as the most comprehensive kind of OD, because it moves the organization through the unfreezing–moving–refreezing phases in a systematic and continuous way as summarized in Table 12.3. Medium to high intensity; focus is on long-range organizational needs.

14. Total Quality Management (TQM). Utilizes the whole spectrum of OD intervention activities for continuous improvement of products, processes, and services. Medium intensity; focus is on medium- to long-range organizational needs.

[1]See Robert R. Blake and Jane S. Mouton, *The Managerial Grid*, Houston: Gulf Publishing Co., 1978.

of total quality management, which will be discussed in more detail in the next section of this chapter.

3. Intergroup Relations. Most organizations are structured as a set of relatively autonomous work groups and departments with specific charters, and they focus on group-internal activities and performance. This is especially true for technology-oriented firms, in which R&D, engineering, production, and marketing, are usually relatively autonomous entities. Each group usually performs its specialty function well. Yet, when it comes to transferring technology, time to market, or overall product quality, the aggregated performance of a firm is often marginal; which may be attributed to poor intergroup cooperation, integration, and involvement. Intergroup conflict is often inevitable and frequently becomes a strong barrier to intergroup relations. Confronting the reasons for this dysfunctional conflict and minimizing it becomes the primary objective for most intergroup interventions. Frequently, a confrontation meeting is used to assess intergroup performance and to set up action plans for improvement. Typically, this is a one-day meeting with the key personnel of all affected organizations. The objective is to discuss problems, analyze their underlying causes, and plan remedial actions. Such a confrontation meeting is often used after an organization has undergone a major change caused by restructuring, merger, or introduction of new technology.

4. The Total Organization. Many organizational objectives require highly orchestrated efforts of the whole organization to respond to broad-based organizational issues such as productivity improvement, restructuring, mergers, or technology shifts. Interventions with organization-wide impact are complex and often risky. They include quality of worklife programs, productivity improvement and gain-sharing, the managerial grid, and corporate venturing, just to name a few of the more popular ones. The survey–feedback technique is the principal methodology for these interventions, aimed toward improving total organizational operations. It involves conducting attitude and other surveys, whose results are systematically reported to organizational members. Members then determine what actions need to be taken to solve the problems and to exploit the opportunities uncovered in the surveys.

Today, development of the total organization is conducted under the increasingly popular and powerful approach of total quality management (TQM). TQM is an integrated operational strategy for continuous improvement of products, processes, and services regarding their overall quality and cost. TQM utilizes the whole spectrum of traditional OD tools and techniques in a highly integrated fashion, with very strong involvement of all the people from participating organizations. Gradually, the labels traditionally used for total organizational development, such as Grid OD[11] and Survey–Feedback Techniques, are being replaced by the total quality management notion, which will be discussed in more detail later in this chapter.

The Organizational Development Toolkit

Out of the large spectrum of OD techniques, a selected number of interventions is briefly described in Table 11.2 and shown in Figure 11.4 relative to their target groups. While many of these techniques overlap in their objectives and activities, each of these interventions focuses on a particular subset of the population, either the individual, group, or entire organization as shown in Figure 11.4. However, this does not exclude interventions from being designed for applications to several target groups. In addition, each technique requires a different level of personal involvement and is associated with various costs and risks as shown in Figure 11.4. Although the cost of a given intervention is usually predictable within reasonable boundaries, the risk of failure and unintended consequences, such as conflict, collusion, and sabotage, is often difficult to assess at the outset. Their effective application requires careful planning, organizational involvement, and managerial leadership. Many of these interventions also require the skillful guidance of an experienced OD specialist, who may be an employee of the organization or an outside change agent.

[11]Grid OD refers to a comprehensive organization development program that uses a combination of tools and techniques for developing organizational effectiveness based on both, concern for production and concern for people. Grid OD has its roots in the managerial grid originally developed by Blake and Mouton (see Figure 10.4).

11.6 TOTAL QUALITY MANAGEMENT: AN INTEGRATED APPROACH TO DEVELOPING THE TECHNOLOGICAL ORGANIZATION

Background of Total Quality Management

Total quality management (TQM) is a management philosophy that evolved in Japan over the last 40 years. Its objective is the continued improvement of products, processes, and services with focus on quality and cost. TQM also has its roots in the sociotechnical and behavioral sciences as pioneered by the Tavistock Institute in Great Britain. It is the sociotechnical integration of all systems and processes of an organization toward total quality improvement that is often identified as the primary criterion for successful implementation of TQM. As part of this tight integration of organizational systems, TQM requires intense involvement and participation of people from all functions and levels of the organization.

In spite of the visibility of TQM and its apparent success in Japan and other Pacific Rim countries, United States businesses have been slow in adopting the TQM philosophy for several reasons. One primary concern centered around the change of management style and power distribution required by the TQM philosophy. TQM focuses on participative management and strong operational accountability at the individual contributor level. This means that managers have to share power and trade managerial control for higher levels of individual accountability, pride, and workmanship. Its implications range from organizational command and control to structure, risk management, work integration, technology transfer, and personnel training. In essence, TQM affects virtually all areas of organization and management, requiring often radical changes in traditional operating methods. With the lack of examples of successful TQM operation until the mid 1980s, and the lack of available management education in this area, most American industries were suspicious that the potential benefits of TQM might not justify the associated costs and risks. Also, managers had, and in many organizations still have, a vested interest in maintaining the power structure associated with traditional methods of management, and thus resist change. The combination of all of these factors led to a wait-and-see attitude of U.S. companies toward TQM, and hence a propagation of the status quo.

Pressures Toward Change

The 1980s brought enormous pressure on United States companies to produce their goods and services more cost effectively and at a higher quality. Shrinking market shares and profits were clear signs that foreign competition, especially noticeable from Pacific Rim countries, penetrated our business here and abroad. It became clear that price–performance of many foreign goods and services was superior. In addition to a higher level of performance regarding features, reliability, service, and overall quality, these companies responded to market needs more accurately and faster.

Faced with these challenges, American industry and business was searching for new methods to operate more effectively and in a manner that was more market-oriented. This new awareness is also reflected by several national initiatives, such as the Malcolm Baldridge National Quality Award created by Public Law in 1987 and the *Guidelines for Quality and Productivity Improvement* developed by the U.S. Office of Management and Budget under Executive Order 12637. W. E. Deming, the great visionary of quality-based productivity improvement summarized the situation, perhaps better than anyone else[12]:

> U.S. industry needs to do better than just compete. That's nonsense. We need to get ahead of the competition. We do everything except apply knowledge. We have the knowledge but don't know how to use it. Everyone doing his best is not the answer. It is necessary that people know what to do. The first step is to learn how to change. A long term commitment to new learning and new philosophy is required of any management that seeks to improve quality and productivity. The timid and the fainthearted, and people that expect quick results, are doomed to disappointment.

Gradually the process and operating philosophy of total quality management became appealing and acceptable to American management during the 1980s. At first specific quality systems and methods such as statistical process control (SPC) emerged for production-oriented operations, where the benefits of TQM were perceived greatest. However, as we entered the 1990s, TQM had been extended into many other operational areas, especially within the high-tech industries. Most noticeably, these additional areas included R&D, engineering, marketing, and field services. Concurrent engineering, benchmarking, and design/build became well-known processes for proactively achieving superior performance in a changing business environment. In addition, organizations as diverse as banking, health care, and many U.S. federal government departments and agencies recognized the value of TQM and started implementation initiatives.

Total Quality Management Overview

Total quality management is an operational philosophy for continued organizational development toward quality improvement of products, processes and services. It is also a proactive process to change the organization, in a structured fashion, toward superior performance. Many companies have translated the principles of TQM into their own operating environment. One such graphical description, developed by the McDonnell Douglas Space Systems Company is shown in Figure 11.5. It summarizes the components of TQM and its supporting functions very effectively in three principal areas: people, discipline, and support environment. Greater detail and depth is shown in Figure 11.6. Yet, inspite of well-established definitions and practices, to many managers, TQM is mystifying. It is still evolving and the hard requirements and techniques are still emerging in U.S. industries. Yet, TQM is an organizational development that relies on established processes for introducing organizational change, as was shown in Figure 11.1 and described earlier. Applied to TQM, the four phases of the OD-oriented diagnostics–prescription cycle are as described below.

[12]W. Edwards Deming, *Out of Crisis,* Cambridge, MA: MIT Press, 1986.

FIGURE 11.5 The elements of total quality management and the five keys to self-renewal lead to continuous improvement. Adapted with permission from McDonnell Douglas Space Systems Company, *Principles for Total Quality Management,* McDonnell Douglas Co., Huntington Beach, California, 1990.

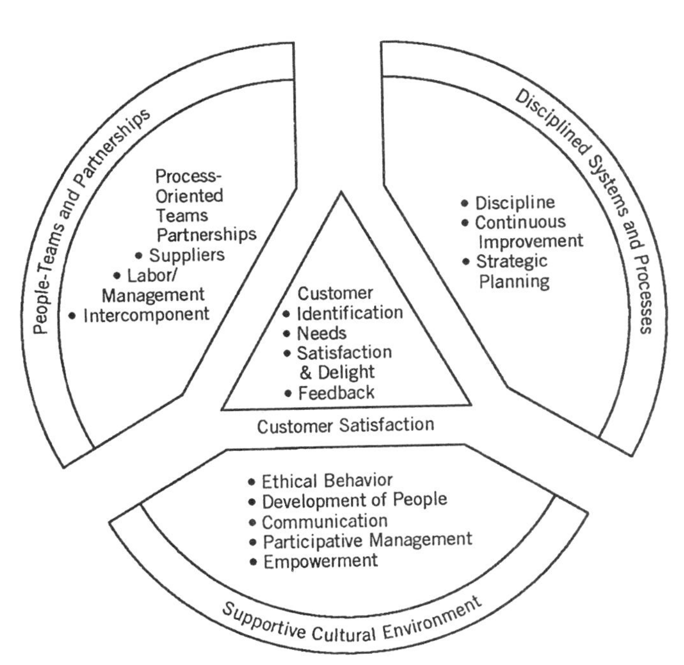

FIGURE 11.6 Total quality management principles established for the McDonnell Douglas Space Systems Company. Adapted with permission from McDonnell Douglas Space Systems Company, *Principles for Total Quality Management,* McDonnell Douglas Co., Huntington Beach, California, 1990.

Phase 1: Initiating a TQM Cycle. Before embarking on a TQM effort, the organization must be clear in its purpose. Management should establish the principal objectives of the development effort as integrated with the long-range strategic business goals. The initiation phase also includes creating a climate conducive to change and establishing a plan for organizational development.

Phase 2: Diagnostics and Action Plan. This phase includes a self-study of the various processes within the organization, such as product design, manufacturing, and field services. It also includes an integrated study of the company's products and services, their markets, competition, technologies, supplies and support functions. The objectives of this diagnostics phase are (1) to identify benchmarks of high quality and performance, (2) to identify organizational strengths and weaknesses, (3) to identify drivers and barriers toward performance improvements, (4) to establish principal objectives and action plans for improvement, (5) to define specific methods, (6) to explore alternatives for reaching desired objectives, (7) to develop a resource plan, and (8) to obtain organizational support and commitment to the organizational development plan. The diagnostics phase should examine all organizational systems, such as human resources, technology, communications, decision-making, and leadership. The final output of this phase is the management-endorsed TQM action plan, accompanied by resource allocations, policy directives, and benchmark measures.

Phase 3: Implementation. This phase implements the TQM action plan while concurrently evaluating progress toward the desired objectives. Similar to other organization development efforts, the implementation phase relies on strong involvement of the people throughout the organization and at all levels. It also relies on strong, continuous top-management involvement, support, and encouragement.

Phase 4: Stabilizing. Once certain organizational changes and objectives have been achieved, efforts must be made to sustain these accomplishments by firmly establishing the new operational norms in the organizational value system. Continued monitoring of organizational processes, training, and positive reinforcement of desired performance are necessary to stabilize the new modus operandi. As a by-product of this stabilization effort, the organization can—if directed properly—engage in a continuous self-study leading to an ongoing self-improvement and self-renewal of the organization and its performance.

Management Guidelines for Successful Total Quality Management

Implementing a total quality management (TQM) philosophy depends on the ability to manage change. Many models have been suggested in recent years for planning and implementing organizational changes in moving toward quality and productivity improvement. Some of the best-known models and techniques are summarized in Table 11.3.They include the seven process steps of Kaizen; the six-sigma process; the

TABLE 11.3 Concept Models for Planning and Implementing Quality and Productivity Improvements

Benchmarking: Collection and analysis of data critical for competitive performance and comparison to other companies in the industry (e.g., time to market)

Computer-Integrated Manufacturing (CIM): Organizational concept that uses computer/information technology to integrate all engineering and manufacturing support functions

Concurrent Engineering (CE): Process of planning and executing product design and manufacturing simultaneously, with strong cross-functional participation. Emphasis is on increased quality, speed, and cost-effectiveness

Design/Build Team: Similar in concept to concurrent engineering, but with additional emphasis on multifunctional team integration

Design of Experiments: Statistical modeling for tracing quality problems in the production process

Failure Mode and Effects Analysis (FMEA): Procedure for tracing the source or reason of a component failure

Kaizen: Japanese-originated management philosophy and practice, which aims at continuous improvement of quality and efficiency through the direct on-the-job involvement of all people

Life Cycle Cost Management (LCCM): Product planning and design that focuses on the total life cycle cost, including acquisition, maintenance, repair, down-time, and replacement

Process Action Team (PAT): Multifunctional task force appointed to analyze an internal process, such as design, vendor selection, merit increase, or training for potential improvement. Requires strong involvement of host function(s)

Quality Circles (QC): Special work interest group, organized on ad hoc basis, to study particular areas for potential improvement: efficiency, quality, work life, etc.

Quality Management Board (QMB): Multifunctional steering committee that develops operational guidelines and policies for process and quality improvements

Robust Design: Practice of product design for effective manufacturability and reliable field operation

Six-Sigma Process: Concept of designing for highly uniform production so that only a six-sigma deviation for the norm is acceptable. Thus the probability of a part's or product's rejection because of defective quality is less than one in a billion

Shewhart Cycle: A phased procedure for continuous improvement: plan, do, execute, check and compare results, and take action for improvement; then continue the cycle

Statistical Process Control (SPC): Method for improving product quality via qualifying, measuring, and diagnosing quality problems throughout the production cycle, using statistical models

Taguchi Method: Method for product and manufacturing process engineering, that relies on experimental design and other statistical techniques to improve product quality

Total Quality Management (TQM): Philosophy for continuous improvement of products, processes, and services to enhance quality. Uses quantitative methods and human resource techniques to manage all functions and processes of an organization, to optimize customer value and satisfaction

Value Engineering: Method for product improvement via systematic multifunctional analysis and fine-tuning of its design

benchmarking technique, the Shewhart cycle (PDCA, Plan–Do–Check–Act),[13] and the process action team, just to name a few of the more popular techniques. Many of these techniques rely on statistical tools and methods. Most noticeably, the techniques originally developed and promoted by Deming, Juran and Taguchi focus on quality improvement in the area of manufacturing and are based primarily on well-established statistical concepts.[14] However, during the 1980s more sophisticated tools were developed to aid in the data analysis and diagnostics of potential quality problems. Many of them incorporate graphical techniques for better diagnostics and communication effectiveness. Examples include Ishikawa diagrams, Lambda plots, and Pareto charts, just to name a few. A detailed listing of the more popular diagnostic tools used today in production engineering for quality control and improvement is shown in Table 11.4.

Change management, especially when it is broad-based and organizationally intricate, obviously requires more than a simple process. However, the problem is not with the process, which actually works beautifully in many companies, but with the way we try to implement it. That is, in order to describe or teach change management

TABLE 11.4 Diagnostic Tools and Statistical Methods Used in Production Engineering for Quality Control and Improvement

- Accumulated Analysis
- Analysis of Transformation
- Analysis of Variance
- Bayes Plots
- Design of Experiments
- Ishikawa Diagram
- Kempner–Tragoe Method
- Lambda Plot
- Minute Analysis
- Orthogonal Arrays
- Pareto Chart
- Sensitivity Analysis
- Signal-to-Noise Ratio
- Tests of Significance

[13]Named after Walter Shewhart, who first devised and introduced the PDCAQ procedure. It is described in greater detail later in this chapter.

[14]For more detailed discussion and an overview of the techniques for quality engineering, see Khosrow Dehnad (ed.), *Quality Control, Robust Design, and the Taguchi Method*, Pacific Grove, CA: Wadsworth & Brooks/Cole, 1989.

and to put some logic and credibility into it, these processes have been sanitized to a point where they have lost their identity as part of a live organization. In addition, any major organizational change takes between five and seven years to be fully implemented and institutionalized. It must be visibly supported and led by upper management and a significant effort is required to build the infrastructure necessary to support and sustain the new TQM operating philosophy. Employee education plays an important role in establishing such an operating philosophy and the corresponding TQM processes. This education effort also helps to develop the necessary communication system among interdependent organizations. It further stimulates participation in the change and development process, as well as a sense of ownership and commitment to successful implementation.

Understanding and Implementing Total Quality Management

Understanding and implementing TQM is so difficult because it requires a fundamental change in the way we organize and manage. The key components of successful change toward a TQM operating philosophy in a technology-focused organization are described below.

1. Collective Vision. Effective change management begins with identifying the company's future needs and wants. This requires clear corporate vision. Management, together with the people at all levels, must formulate a long-term central focus for the organization. This vision must eventually be articulated in a mission statement and broad-based operating policy. Technical professionals can play a significant part in the development of these documents, which should focus on the customer, business environment, and company's internal strength and processes.

2. Understand the Company's Unique Operation. Companies must recognize their uniqueness before they copy someone else's success strategy. Initiating change for the sake of change is not only useless, but can be devastating. Companies must analyze and clearly understand their strengths, weaknesses, opportunities and threats, and the trends of their dynamic business environment. Using benchmarks might help in comparing current organizational performance with that of competitors. It also can help in identifying trends and future needs. Once a company has determined its course, and its most likely versus desired future position, it can assess the gap and develop change strategies for bridging the gap. Following this methodology promotes long-range planning with a "consistency of purpose," which is a key point in W. E. Deming's theory of total quality management.[15]

3. Ensure Management Commitment. If TQM is to succeed, it must be managed as a continuous process, rather than an "on-again off-again" scenario. This requires a solid top-down management commitment, backed with active management involvement, policies, incentives, training, and resources.

[15]See W. Edwards Deming. *Quality, Productivity, and Competitive Position,* Cambridge, MA: MIT Press, 1982.

4. Translate TQM Mission Into Engineering. TQM, its mission statements, and company policies focus on the customer; therefore, the engineering organization must pay close attention to the process of developing products and services and their life cycles. If new developments are planned and managed according to the "consistency of purpose of the company," total quality has to start with the product concept. Economic manufacturability, product reliability, functionality, and serviceability all must be designed into the product. Engineering influences the ultimate product "quality" via the design process; technology selection; composition, material, and vendor selection; technical competency of the personnel; system integration; and technology transfer. It is further important to involve the internal and external customers of the engineering organization in defining the process requirements and value expectations as a prerequisite for engineering policies and TQM operating guidelines.

5. Provide and Improve Support Systems. TQM support systems include four basic components: (1) training and education, (2) design process support (drafting, prototyping, configuration, CAD/CAE), (3) information systems, and (4) integration and project management. All four support systems are usually considerably interrelated. The constant improvement of every support system and process associated with product planning, design, and development is the challenge to management.[16] It is crucial however, that any new or improved system is in turn supported by the existing processes and systems. Deming refers to the Shewhart cycle as a procedure for managing the never-ending process and system improvement. As a continuous loop, the *Shewhart cycle* includes the following steps: (1) prepare a plan for learning, experimentation, and improvement; design experiments and collect data; (2) carry out the plan; (3) study the results, diagnose possible problems, check against the original plan, improve the plan and reiterate if necessary; and (4) act on the lessons learned from the effort to improve the process. The Shewhart cycle is a simple but powerful method for implementing continuous improvement opportunities. It reinforces the importance of planning and provides a checking mechanism against the plan. It also minimizes the barriers to implementation and change. Learning results from iteration through the cycle in a continuous fashion, which reduces the temptation to force-fit some solutions to a problem.

6. Take a Team Approach. Process improvements do not occur from a few good ideas alone, but are the result of collaborative efforts of many people across all the functional areas involved in the work process. Each function "owns" part of the knowledge necessary for system improvement. A great deal of leadership is required at all levels to foster a collaborative environment, which unifies a multifunctional team in working toward the established quality objectives.

Another important aspect of implementing change is that people must be willing to participate. This is more likely to happen when people are involved with each other, participate in quality improvement planning, understand the benefits and risks,

[16]This challenge has been discussed by many, including W. E. Deming, who deals with this issue in point number five of his fourteen points for top management in "Improvement of Quality and Productivity Through Actions by Management," *National Productivity Review*, Winter 1982, pp. 12–22.

and are willing to share them. The more people of a work group share the same values and norms and buy into the overall TQM concept, the stronger is the desire of all the people in the organization to participate and to contribute to the ongoing quality-improvement efforts. Change is considered more positive and desirable if we are participating and contributing to it. This active participation and team enforcement is also professionally stimulating and rewarding. It further minimizes anxieties over uncertainty and over the possible negative impacts that the changes may cause in participants' work lives. When people will be affected by the change and they are needed to help implement it, it is especially important to make them coarchitects of the change. This is one reason why quality circles have been so successful in continuous quality improvement efforts.

7. Change Incrementally. Rosabeth Moss Kanter, in her book *The Change Masters,*[17] makes an important point for making changes in small steps: "Change brings pain when it comes as a jolt; when it is seemingly abrupt and shocking." The more radical the change, the less likely it is that the organization will adapt to it. Changes are viewed more evolutionarily, as a natural part of corporate life, if they come gradually and as a result of the continuous involvement of the people at all organizational levels.

Several other reasons speak favorably for incremental changes. Incremental change allows management to test the results early, with opportunities for fine-tuning and celebrations for accomplishments; it minimizes the risk of wholesale failure; it eliminates the need for perfectionism, since we will learn incrementally from both success and failure; it minimizes the impact on ongoing operations; and, finally, incremental implementation of change is an excellent way to gain visibility, promote participation, and refuel commitment to the established TQM program and its objectives.

8. Celebrate Success. Innovation, experimentation, and risk-taking must be encouraged and rewarded. People put their efforts in the direction of greatest rewards. Effective managers create an environment where people are proud and excited to participate in a change effort. The manager is a cheerleader who creates a can-do attitude and stimulates confidence in people at all levels that they are capable of making a difference. Celebrating success can come in many forms. The General Electric Company awards nearly $2 million each year to their employees for improvement ideas. As CEO of Texas Instruments, Mark Shepard constantly reminded his people that their rewards and professional growth within the organization will largely be based on their ideas toward organizational improvement. Westinghouse conducts Total Quality Fitness Reviews as a basis for some resource allocations, as well as for team and individual rewards. There are many forms for celebrating success. They do not have to be only money. Well earned recognition in forms of public praise, a plaque, a testimonial, a complementary letter, a formal announcement by upper management, or just a recognition luncheon with your team for a job well done can be very effective in promoting the sense of ownership and commitment to

[17]New York: Simon & Schuster, 1983.

TQM. It also helps to unify the team behind the established objectives and strengthens communication links and cooperation.

9. Use Change Agents. Of course, the manager usually is the primary change agent. However promoting change is a monumental task, and managers need help. For one thing, change is disruptive to the organization, even if it comes with a TQM label. It often causes conflict, tension, and increases barriers between functional groups. Change agents minimize the dysfunctional components associated with the change process and help to drive the organization toward the desired change. Committees and task teams are classical examples of change agents. These are groups of people organized on an ad hoc or permanent basis to examine the organizational system for potential improvement. Since the system, its processes, problems, and changes are multifunctional, these committees should have a similar cross-functional membership. These task teams can not only help identify areas for improvement, but also can screen, evaluate, and select ideas for implementation; identify most valuable contributions; recommend rewards; provide recognition and acknowledgment. These task forces have a more democratic characteristic than a management directive. Their members have organizationally distributed respect, trust, and credibility, although they are perceived as part of the "worker bees" who will be affected by changes in the same way as anyone else. Examples of these task teams and committees are: process actions teams, quality management board, total quality fitness review committees, quality circles, new product review boards, corporate quality councils, ABC study teams and award selection committee. In addition, an overlying process such as benchmarking, concurrent engineering, life cycle cost management, or statistical process control can serve as an effective change agent. Finally, many organizations have used a self-study format for promoting change. An example is the extensive self-studies conducted by many companies in preparation for the Malcolm Baldridge National Quality Award. Although only a few companies receive the award each year, all self-studies lead to substantial quality-improvement initiatives, driven by self-assessment.

10. Assure Leadership. Leading organizational change is tough. It's similar to *Star Trek's* mission, "to boldly go where no person has gone before." It's risky. Therefore, it is even more important to have proactive managers who can create a vision and inspire the organization to move toward change. Proactive managers are facilitators who can influence the organizational environment to see a need for change; they can also communicate opportunities for improvement and their payoffs to the participants. Some of the desirable behaviors of leadership and proactive management in relation to TQM are: (1) active management involvement and encouragement for self-studies to move the organization toward continuous improvement; (2) issuing clear directives and policies communicating broad-based TQM objectives; (3) support in establishing a TQM infrastructure of policies, procedures, committees, training, and planning; (4) allocating adequate resources; (5) encouraging cross-functional cooperation; (6) motivating people to commit to quality improvement; (7) willingness to trade managerial control for individual accountability, innovation, and risk-taking; (8) recognizing and rewarding accomplishments, (9) making TQM visible throughout the organization and unifying people behind the TQM objectives, and (10) managing the inevitable conflicts and anxieties associated with the organizational development process.

11.7 HOW TO MAKE ORGANIZATIONAL DEVELOPMENT WORK FOR YOU

Each Manager Is an Organizational Development Practitioner

Developing the organization should be an ongoing process in any well-managed company or institution. OD is synonymous with fine-tuning the organization to get the best from its people and supporting resources. Such a continuous OD effort is the job of managing by keeping in touch with people, the organizational systems, and technology. This very pragmatic approach to OD is summarized today under the umbrella of total quality management (TQM). It has several advantages over a major intervention that involves consultants and results in radical changes. First, the ongoing approach is not disruptive. Feedback is given to people, suggestions are made, and changes are introduced incrementally as new situations unfold. Further, the ongoing approach is adaptable to dynamic situations that require fast reaction time and the integration of many variables. These situations often require a great deal of intuitive thinking and experiential decision-making. Yet another advantage of an ongoing approach to organizational development is that the process is less threatening to the people as situations and changes unfold in a natural way, often with the help and approval of the people involved. In addition, managers who are in close touch with their organizations are more likely to develop their own management skills and the skills needed to conduct any of the OD interventions. They are also using heuristic processes, that is, learning by doing what techniques work in what situation with whom. Finally, there are some side benefits for managers, who stay involved and in charge of their organization. These managers will most likely create involvement among their people and foster a work environment that is high on participative problem-solving and rich in committed people who are willing to work toward a necessary change. This is usually also an environment of mutual trust, low conflict, and high productivity.

A More Complex Situation Requires More Formality

Of course, not all organizational problems can be solved by continuous fine-tuning as part of day-to-day management. Major opportunities or threats, long-term internal or external changes, a new business mission or corporate objective, may require "major surgery." Such a situation calls for a highly systematic effort to diagnose, prescribe, and implement organizational developments that change the way the organization operates. In the process, the organizational structure, its management system, and its people may undergo significant changes. This is the type of OD program that requires the full commitment of top management and of the key people in the organization. In addition, the effort must be well planned and executed with the help of a competent OD practitioner. Finally, the organizational environment must be conducive to the OD activities and the changes required.

Criteria for Organizational Development Success

A practical set of conditions necessary for an organizational development program to succeed has been identified by French and Bell;[18] it is used in a similar format by many OD practitioners and change management consultants today.[19] Its main conditions are listed below:

1. Recognition and Endorsement. Recognition by top management that the organization has problems and full management endorsement of OD activities are crucial. Without these, it is highly unlikely that the necessary time, effort, and money will be invested in OD.

2. Use of an Outside Consultant Experienced in OD. Internal change agents are unlikely to have, or to be seen as having, the experience, objectivity, expertise, or freedom required to implement a major change program.

3. Support and Involvement of Top-Level Managers. Top managers play a key role in overcoming initial resistance to change. Lack of involvement and support from them would signal to lower-level managers that the activity is not considered important.

4. Involvement of Work Group Leaders. Activities to improve the effectiveness of existing work groups are frequently an important part of OD programs. To be successful, such programs require active support and involvement by the manager of the work group.

5. Making Early Successes with the OD Effort Visible. When the first OD changes are made and prove successful, organization members are motivated to continue the process and to attempt larger-scale changes. Early failures may destroy the credibility of the change agents and may weaken the commitment of top-level supporters.

6. Educating Organization Members About OD. People are likely to feel manipulated if they do not understand the reasons for changes and, to some extent, the theories on which OD change programs are based. Making behavioral science knowledge widely available will minimize these barriers to change.

7. Acknowledgment of Managers' Strengths. Managers who are already using good management techniques are likely to resent an OD change agent who overplays the "expert" or "teacher" role. The things managers do well need to be acknowledged and reinforced.

[18]Wendell L. French and Cecil H. Bell, Jr., *Organization Development,* Englewood Cliffs, NJ: Prentice-Hall, 1978, pp. 177–187.

[19]For a typical application of these criteria and their framework to managing change in IBM's production planning function, see Johnson A. Edosomwan, "A Framework for Managing Change," *Industrial Management,* Vol. 31, No. 5 (September/October 1989), pp. 12–15.

8. Involvement with Managers of Personnel Departments. Personnel managers have as their primary responsibility the management of the organization's human resources. Their expertise and support are thus essential in designing and implementing changes in such areas as employee evaluation, development, and reward policies.

9. Development of Internal OD Resources. Outside change agents cannot assume indefinite or total responsibility for the organization's change efforts. Internal change agent expertise should be developed; the organization's line managers must acquire many of the change agent skills.

10. Effective Management of the OD Program. Change agents and their clients must work together to coordinate and control the OD program. Otherwise, the organization's change effort may lose its impetus and become isolated from the needs that the organization members seek.

11. Measurements of Results. The success of the organization in meeting its organizational and human goals must be monitored. Success in a particular subunit will suggest that the OD program is working; failure will suggest that the program needs to be modified. Results provide feedback on the organization's change efforts compared to the objectives it has established.

A Final Note

From its beginnings in the areas of laboratory training and survey research and feedback, organizational development has evolved into a systematic approach to planned change, better known today under the aggregated notion of total quality management (TQM). OD techniques focus on the people of an organization to achieve higher performance via better communications, mutual trust, improved cooperation, and more effective problem-solving and decision-making. The OD objectives must include the total organization. OD is a continuous process of developing social conditions conducive to improved organizational performance by developing the people, but also by changing the way the organization operates and by modifying the structure and process of the management system that guides human behavior.

During the 1980s, organizational development efforts emphasized quality improvements, with a focus on manufacturing; today companies take a much more comprehensive approach and look to operational improvement of the total organization. Changes in the market and technology have blurred the distinctions of functional responsibility for product success. Reliability, time to market, and cost-effectiveness are by all criteria that affect the product development–production–marketing process. Today, organization development is usually a highly coordinated joint effort among all functions.

For technology-intensive companies, the key management issues confronting executives today are improved quality, speed, and creativity. R&D, engineering, and manufacturing are major stakeholders in fulfilling these goals. The greatest challenge for managers in the years to come will not come from a single threat, but from a multitude of factors and forces that continuously change our business environment. The companies that succeed and prosper will be those that understand the changing environment and can adapt to it effectively. This requires a careful integration of all company functions with a unified commitment to excellence. Managing the organization in its movement toward change must also be focused. Improved time to market, yield, cost, and design to manufacturing interfaces are specific measures of organizational performance. Our competitive, market-driven business environment requires change managers and change leaders who recognize the unique position of their organizations and can apply the available techniques effectively to improve operations.

In this chapter, we discussed a number of organizational development techniques, with emphasis on performance improvement via total quality management. Although there is a great need for additional empirical research and refinement of the concepts, managers in business, government, and institutional organizations cannot wait until the ultimate knowledge emerges. Often what distinguishes the successful companies from others is their ability to use the best available practices—such as competitive benchmarks, employee involvement, team-based approaches to product and process development, and training—and adapt them to their unique situations. This will improve the collective knowledge of these concepts and develop new insight into the effective real-world applications.

Managing organizational improvements via change requires leadership, vision, courage, wisdom, people skills, and the willingness to take risks. To quote Thomas Watson, Jr., the former chair of IBM[20]:

> An organization will stand out only if it is willing to take on seemingly impossible tasks. The people who set out to do what others say cannot be done are the ones who make the discoveries, produce the inventions, and move the world ahead.

BIBLIOGRAPHY

Altany, David, "Benchmarkers Unite: Clearinghouse Provides Needed Networking Opportunities," *Industry Week,* Volume 241, Number 3, (Feb 3, 1992), pp. 25.

Aly, Nael A., Venetta J. Maytubby, and Ahmad K. Elshennawy, "Total Quality Management: An Approach & Case Study," *Computers and Industrial Engineering,* Volume 19, Number 1–4, (1990), pp: 111–116.

Autry, James, *Love and Profits: The Art of Caring Leadership,* Hawaii: William Morrow, 1991.

Axland, Suzanne, "Two Awarded NASA's Prize Trophy," *Quality Progress,* Volume 24, Number 12, (Dec 1991), pp. 51–52.

Babcock, Daniel L., "Some Implications of Team Management," *Engineering Management Journal,* Volume 2, Number 4 (Dec 1990), pp. 29–32.

Badiru, Adedeji B., "A Systems Approach to Total Quality Management," *Industrial Engineering,* Volume 22, Number 3, (Mar 1990), pp. 33–36.

[20]Thomas J. Watson, Jr., *A Business and Its Beliefs: The Ideas That Helped Build IBM,* New York: McGraw-Hill, 1963.

Barcey, Hyler, *Managing from the Heart,* Pittsburgh, Enterprise Press, 1990.

Brown, Mark Graham, "Developing a Plan to Win the Malcolm Baldridge National Quality Award," *Journal for Quality and Participation,* Volume 14, Issue 6 (Dec 1991), pp. 16–23.

Butman, John, "Quality Comes Full Circle," *Management Review,* Volume 81, Number 2, (Feb 1992), pp. 49–51.

Byrne, Patrick M., "Global Leaders Take a Broad View," *Transportation and Distribution,* Volume 33, Issue 2 (Feb 1992), pp. 61–62.

Chapman, Ross L., Paul Clarke, and Terry Sloan, "TQM in Continuous-Process Manufacturing: Dow-Corning (Australia) Pty Ltd," *International Journal of Quality and Reliability Management,* Volume 8, Number 5, (1991), pp. 77–90.

Cole, Raymond C., Jr. and H. Lee Hales, "How Monsanto Justified Automation", *Management Accounting,* Volume 73, Number 7, (January 1992), pp. 39–43.

Cole, Raymond C., Jr. and Paul F. Murphy, "Six Steps to Quality Improvement," *Journal for Quality and Participation,* Volume 14, Issue 6 (Dec 1991), pp. 24–26.

Coombes, Peter, "Implementing the Program: Olin," *Chemical Week,* Volume 149, Number 20, (Dec 11, 1991), p. 42.

Crosby, Philip B., "The Next Effort," *Management Review,* Volume 81, Number 2, (Feb 1992), p. 64.

De Ceiri, Helen, Danny A. Samson, and Amrik S. Sohal, "Implementation of TQM in an Australian Manufacturing Company," *International Journal of Quality and Reliability Management,* Volume 8, Number 5, (1991), pp. 55–65.

Dertouzos, M. L., R. K. Lester, R. M. Solow, and the MIT Commission on Industrial Productivity, *Made in America: Regaining the Productive Edge,* Cambridge, MA: MIT Press, 1989.

Edosomwan, Johnson A. and Wanda Savage-Moore, "Assess Your Organization's TQM Posture and Readiness to Successfully Compete for the Malcolm Baldridge Award," *Industrial Engineering,* Volume 23, Number 2, (Feb 1991), pp. 22–24.

Eldred, William J., "A Proposed Approach to Computer-Supported TQM in Maintainance Work Induction and Accomplishment," *Computer and Industrial Engineering,* Volume 21, Number 1–4, (1991) pp. 123–127.

Fenwick, Alan C., "Five Easy Lessons: A Primer for Starting a Total Quality Management Program," *Quality Progress,* Volume 24, Number 12, (Dec 1991), pp. 63–66.

Higgins, Ronald C. and Michael L. Johnson, "Total Quality Enhances Education of U. S. Army Engineers," *National Productivity Review,* Volume 11, Number 1, (Winter 1991/1992), pp. 41–49.

Hoffmann, Gerard C., "Prescription for Transitioning Engineers into Managers," *Engineering Management Journal,* Volume 1, Number 3 (Sept 1989), pp. 3–7.

Horton, Thomas R. and Peter C. Reid, *Beyond the Trust Gap,* Homewood, IL: Irwin, 1990.

Hunt, Daniel V., *Quality in America,* Homewood, IL: Irwin, 1992.

Irving-Monshaw, Susan, "Quality Advice: Listen to Your Customers," *Chemical Engineering,* Volume 97, Number 6, (June 1990), pp. 27–30.

Japan Human Relations Association (ed.), *Kaizen Teian-1,* Cambridge, MA: Productivity Press, 1992.

Kern, Jill Phelps, "The QIS/TQM Balancing Act," *Quality,* Volume 30, Number 12, (Dec 1991), pp. 18–41.

Maddux, Gary A., Richard W. Amos, and Alan R. Wyskida, "Organizations Can Apply Quality Function Deployment as Strategic Planning Tool," *Industrial Engineering,* Volume 23, Number 9, (Sept 1991), pp. 33–37.

McCarthy, Kimberly and Ahmad K. Eishennaway, "Implementing Total Quality Management at the U. S. Department of Defense," *Computers and Industrial Engineering,* Volume 21, Number 1–4, (1991), pp. 153–157.

Murphy, Paul F., "Six Steps to Quality Improvement," *Journal for Quality & Participation,* Volume 14, Number 6 (Dec 1991), pp. 24–26.

Nash, Kim S., "Ernst & Young CASE Quality Service Bows," *Computerworld,* Volume 26, Number 4, (Jan 27, 1992), p. 20.

Nelson, Gregory V., "Managing Change: Quality in R&D at R. L. Mitchell Technical Center," *Engineering Management Journal,* Volume 2, Number 3 (Sept 1990), pp. 39–44.

Noori, Hamid, "TQM and Its Building Blocks: Learning from World-Class Organizations," *Optimum,* Volume 22, Issue 3 (1991/1992), pp. 31–38.

O'Lone, Richard G., Michael A. Dornheim, William A. Scott, and Philip J. Klass, "Boeing 777 Transport: A Risky Development Approach," *Aviation Week and Space Technology,* Volume 134, Number 22, (June 3, 1991), pp. 34–61.

Panchak, Patricia L., "How to Implement a Quality Management Intitiative," *Modern Office Technology,* Volume 37, Issue 2 (Feb 1992), pp. 27–31.

Patten, Thomas H., Jr. "Beyond Systems—The Politics of Managing in a TQM Environment," *National Productivity Review,* Volume 11, Number 1, (Winter 1991/1992), pp. 9–19.

Pfau, Loren D., "Total Quality Management Gives Companies a Way to Enhance Position in Global Market," *Industrial Engineering,* Volume 21, Number 4, (Apr 1989), pp. 17–21.

Postula, Frank D., "Cost Engineering's Role in Total Quality Management," *AACE Transactions,* (1990), pp. Q.5.1–Q.5.8.

Rosenblatt, Alfred, "Success Stories in Instrumentation, Communications—Case History 4: ITEK Optical Systems," *IEEE Spectrum,* Volume 28, Number 7, (July 1991), pp. 36–37.

Rossler, Paul E. and Scott Sink, "A Roadmap for Quality and Productivity Improvement," *Engineering Management Journal,* Volume 2, Number 3 (Sept 1990), pp. 17–23.

Sarchet, Bernard R., "Engineering Management—Key to the Future," *Engineering Management Journal,* Volume 1, Number 2 (Mar 1989), pp. 4–7.

Smith, Bruce A. and William B. Scott, "Douglas Tightens Controls to Improve Performance," *Aviation Week and Space Technology,* Volume 132, Number 23, (June 4, 1990), pp. 16–20.

Sproles, Gary W., "Total Quality Management Applied in Engineering and Construction," *Engineering Management Journal,* Volume 2, Number 4 (Dec 1990), pp. 33–38.

Stocker, Gregg D., "Reducing Variability—Key to Continuous Quality Improvement," *Manufacturing Systems,* Volume 8, Number 3, (Mar 1990), pp. 32–36.

Sunno University (ed.), *Vision Management,* Japan: Sunno University, 1990.

Tribus, Myron, "Applying Quality Management Principles to R&D," *Engineering Management Journal,* Volume 2, Number 3 (Sept 1990), pp. 29–38.

Tribus, Myron, "My CEO Does Not Understand Quality: So What Can I Do to Save the Company?" *National Productivity Review,* Volume 11, Number 1, (Winter 1991/1992), pp. 41–49.

Turney, R. D., "The Application of Total Quality Management to Hazard Studies and Their

Recording," *International Journal of Quality and Reliability Management,* Volume 8, Number 6, (1991), pp. 47–53.

Velocci, Anthony L., Jr., "TQM Makes Rockwell Tougher Competitor," *Aviation Week & Space Technology,* Volume 135, Number 23, (Dec 9, 1991), pp. 68–69.

Whetsell, George W., "Total Quality Management," *Topics in Health Care Financing,* Volume 18, Number 2, (Winter 1991), pp. 12–20.

Whiting, Rick, "Engineers Say 'Prove It!'" *Electronic Business,* Volume 17, Number 19, (Oct 7, 1991), pp. 89–92.

12 TEAM-BUILDING IN ENGINEERING

12.1 THE CHALLENGES: THE SCIROCCO PROJECT EXAMPLE

Scenario: Five months after the Scirocco project was formally kicked off, Brian Perry is still experiencing serious problems with his team of 18 people, who were assigned to him for the ten months' duration of the project. Although a dedicated office space was furnished so that all Scirocco team members could be located in one area, only 9 people moved to the new area, in spite of pressures from upper management. The others work out of their regular offices. Easy access to functional support, data files, and technical facilities seems to be more important to them than physical proximity to the Scirocco group.

Real problems surfaced two months ago when the concept developments of the various functions did not integrate into an acceptable systems solution. Many team members expressed frustration and confusion over unclear technical requirements and objectives. A great deal of conflict and mistrust has developed among the various functional subgroups over the inability to work out suitable interface requirements. There have been rumors that the project is in serious trouble, several key people asked for transfers, and management has become more involved in the technical aspects of Scirocco. Meanwhile, Brian Perry seems to be losing credibility and trust with the team and has difficulties in controlling the project according to the established plan.

What went wrong on the Scirocco program? The problems and challenges faced by Brian Perry are quite common in today's engineering organization. To a large degree, they are also predictable. Especially in complex task environments, as are found in engineering, these challenges center around the points made in Table 12.1, which will be discussed in this chapter and further expressed in Exercise 12.1 at the end of the chapter.

TABLE 12.1 Challenges of Building Engineering Task Teams

- Defining and negotiating the appropriate human resources for the project team
- Bringing together the right mix of competent people, who will develop into a team
- Integrating individuals with diverse skills and attitudes into a unified work group with a unified focus
- Dealing with support department values, managers, and priorities
- Directing multifunctional work teams across organizational lines where one may have little formal authority
- Maintaining project direction and control without stifling innovation and creativity
- Coordinating and integrating the various task group activities into a complete system
- Fostering a professionally stimulating work environment where people are motivated to work effectively toward established project objectives
- Maintaining leadership at each task group in spite of often informal organizational structures and control systems
- Coping with changing technologies, requirements, and priorities while maintaining project focus and team unity
- Dealing with peer struggles and conflicts
- Dealing with technical complexities in an integrated multidisciplinary framework
- Building lines of communication among task teams as well as to upper management and to the project sponsor or customer
- Keeping upper management involved, interested, and supportive
- Sustaining high individual efforts and commitment to established objectives
- Encouraging innovative risk-taking without jeopardizing fundamental project goals
- Providing or promoting equitable and fair rewards for individual team members
- Building the specific skills needed in the particular task team
- Facilitating team decision-making
- Providing an organizational framework to unify the team
- Providing overall project leadership in an often loosely structured, temporary team environment.

Framework for Analyzing Challenges

The framework for analyzing these challenges in project team situations focuses on the following questions:

1. What challenges and problems does the situation present?
2. What characteristics do we expect in a high-performing project team?
3. What are the early warning signals of poor project performance?
4. What are the drivers and barriers toward high team performance?
5. What can we do to improve team performance?

These questions are being asked by project leaders and senior engineering managers, whether they work on new product development, a defense contract, or on any other technical project. However, the intensity of their concern is increasing with project size, complexity, and the development/implementation cycle. Yet team-building is not a new idea. Its basic concept can be traced back to periods of early civilization. For an example, in 4000 B.C., Egyptian pyramid-builders demonstrated the ability to organize and control large work groups to reach specific results. Modern team-building concepts and practices evolved with the multidisciplinary task complexities of the 1960s, which also led to new management styles and techniques and contemporary organization forms, such as the matrix. Traditional bureaucratic hierarchies have declined and horizontally oriented teams and work units have become increasingly important. The principal challenge is to transform an ad hoc collection of people assigned to a particular task into a coherent, integrated work group.

The Characteristics of an Integrated Project Team

The unique characteristics of an integrated project team are the interdependence of its members and the satisfaction and pleasure that team members derive from their association with the group. Specifically, some of the more important characteristics of a fully integrated engineering team (into which the Scirocco group apparently never developed) are:

- Being part of a team satisfies members individual needs.
- Members have shared interests.
- There is a strong sense of belonging.
- Members have pride and enjoyment in group activity.
- There is a commitment to team objectives.
- Members treat each other with high trust, and low conflict.
- Members are comfortable with interdependence.
- There is a high degree of intragroup interaction.
- There are strong performance norms and result-orientation.

12.2 TEAM BUILDING FOR TODAY'S ENGINEERING ACTIVITIES

Team-building is important for many activities. It is especially crucial in a technology-oriented work environment, where complex multidisciplinary activities require the integration of many functional specialties and support groups. To manage these multifunctional activities, it is necessary for managers and their task leaders to cross organizational lines and deal with resource personnel over whom they have little or no formal authority. Yet another set of challenges is the contemporary nature of project organizations, with their horizontal and vertical lines of communication and control, their resource-sharing among projects and task teams, the multiple reporting relationships to several bosses, and their dual accountability.

To manage projects effectively in such a dynamic environment, task leaders must understand the interaction of organizational and behavioral variables in order to foster a climate conducive to multidisciplinary team-building. Such a team must have the capacity to innovatively transform a set of technical objectives and requirements into specific products or services that compete favorably against other available alternatives in the marketplace.

A New Management Focus

Building effective task teams is one of the prime responsibilities of engineering managers. Team-building involves a whole spectrum of management skills, which are required to identify, commit, and integrate various task groups from traditional functional organizations into a multidisciplinary task-management system. This process has been known for centuries. However, as we could see in the Scirocco project, this process becomes more complex and requires more specialized management skills as bureaucratic hierarchies decline and horizontally oriented teams and work units evolve. Starting with the evolution of formal project organizations in the 1960s, managers in various organizational settings expressed increasing concern and interest on the concepts and practices of multidisciplinary team-building. Responding to this interest, many field studies were conducted, investigating work-group dynamics and criteria for building effective, high-performing project teams.[1] These studies contributed to the theoretical and practical understanding of team-building and form the basis for the discussion of the fundamental concepts in this chapter.

When is Teamwork Especially Critical?

Team-building is an ongoing process that requires leadership skills and an understanding of the organization, its interfaces, authority and power structures, and motivational factors. This process is particularly critical in certain project situations, such as:

- Establishing a new program
- Transferring technology
- Improving project–client relationships
- Organizing for a bid proposal
- Integrating new project personnel
- Resolving interfunctional problems
- Working toward a major milestone
- Reorganizing a company
- Moving a project into a new activity phase.

[1]For specific field studies on team work and its effective management see articles by Altier (1986), Benningson (1972), Rauftl (1978), Thamhain and Wilemon (1987), and Ward (1987) cited in the bibliography at the end of this chapter.

Today, team-building is considered by many management practitioners and researchers as one of the most critical leadership qualities determining the performance and success of multidisciplinary efforts. The outcome of these projects critically depends on carefully orchestrated group efforts, requiring the coordination and integration of many task specialists in a dynamic work environment with complex organizational interfaces. Therefore, it is not surprising to find a strong emphasis on teamwork and team-building among today's managers, a trend that is expected to continue and most likely intensify for years to come.

A Simple Model for Analyzing Team Performance

Team-building is defined as the process of taking a collection of individuals with different needs, backgrounds, and expertise and transforming them into an integrated, effective work unit. In this transformation process, the goals and energies of individual contributors merge and support the objectives of the team.

The characteristics of a project team and its ultimate performance depend on many factors. Using a systems approach, Figure 12.1 provides a simple model for organizing and analyzing these factors. It defines three sets of variables which influence the team characteristics and its ultimate performance: (1) environmental factors, such as working conditions, job content, resources, and organizational support factors; (2) leadership style, and (3) specific drivers and barriers toward desirable team characteristics and performance. All of these variables are likely to be interrelated in a complex, intricate form. However, using the systems approach allows researchers and management practitioners to break down the complexity of the process and to analyze its components. It can further help in identifying the drivers and barriers to transforming resources into specific results under the influence of managerial, organizational, and other environmental factors.

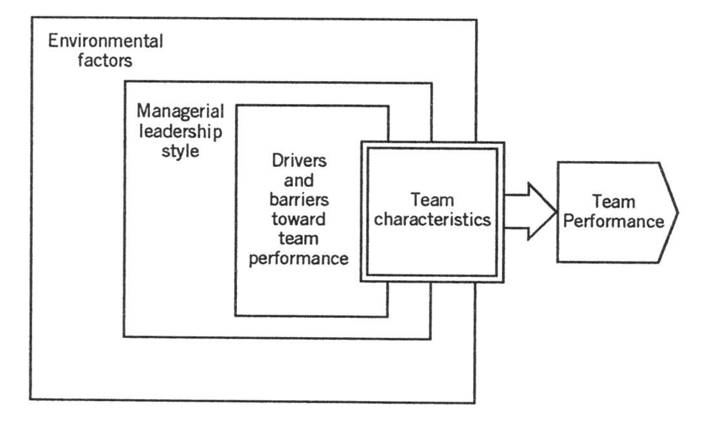

FIGURE 12.1 A simple model for analyzing project team performance.

Characteristics of an Effective Project Team

Obviously, each organization has its own way to measure and express performance of a project team. However, in spite of the existing cultural and philosophical differences, there seems to be a general agreement among engineering managers on certain factors that are included in the characteristics of a successful project team:

1. Technical project success, measured according to agreed-upon objectives.
2. On-time performance.
3. On-budget performance (staying within resource limitations).[2]

Over 60% of those managers who were interviewed ranked these three measures in the above order.

When describing the characteristics of an effective, high-performing project team, engineering managers point to the factors summarized in Figure 12.2. These managers stress consistently that a high-performing team not only produces technical results on time and on budget, but is also characterized by specific job- and people-related qualities.

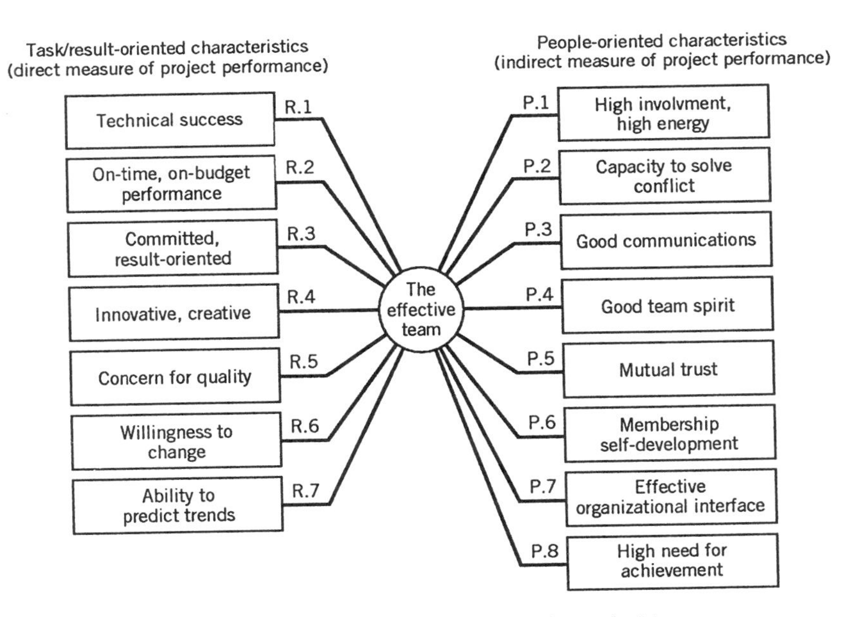

FIGURE 12.2 Characteristics of an effective project team.

[2]Over 90% of the project managers interviewed during a survey by Thamhain and Wilemon (1987) mentioned these measures among the most important criteria of team performance.

In fact, field research shows a statistically strong association between the team qualities listed in Figure 12.2 and team performance at a confidence level of $p = 95\%$ or better.[3]

Determining team performance characteristics offers some clues as to what an effective team environment looks like. This can stimulate management's thoughts and activities for effective team-building. It also allows us to define measures and characteristics of an effective team environment for further research on organization development efforts, such as defining drivers and barriers to team performance.

12.3 DRIVERS AND BARRIERS TO HIGH TEAM PERFORMANCE

Studies by Gemmill, Thamhain, and Wilemon on work-group dynamics clearly show significant correlations and interdependencies among work factors and environmental factors and team performance.[4] These studies indicate that high team performance involves four primary factors: (1) managerial leadership, (2) job content, (3) personal goals and objectives, and (4) work environment and organizational support. The actual correlation of 60 influence factors to the project team characteristics and performance provided some interesting insight into the strength and effects of these factors. One of the important findings was that only 12 of the 60 influence factors were found to be statistically significant.[5] All other factors seem to be much less important to high team performance. The six drivers that had the strongest positive association to project team performance were:

1. Professionally interesting and stimulating work
2. Recognition of accomplishment
3. Experienced engineering management personnel
4. Proper technical direction and leadership
5. Qualified project team personnel
6. Professional growth potential.

The strongest barriers to project team performance were:

1. Unclear project objectives and directions
2. Insufficient resources

[3]Specifically see H. Thamhain and D. Wilemon (1987), who used a Kendall's-tau rank-order correlation model to measure an average association of $\tau = .37$. Moreover, there appears to be a strong agreement between managers and project team members on the importance of these characteristics, as measured via a Kruskal–Wallis analysis of variance at a confidence level of $p = 95\%$.

[4]For detailed methods and results of these field studies see articles by G. Gemmill and H. Thamhaim (1974), H. Thamhain and D. Wilemon (1978), and Wilemon (1991), cited in the bibliography at the end of this chapter.

[5]Kendall's tau rank-order correlation was used to measure the association between these variables. Statistical significance was defined at a confidence level of 95% or better. For specific results and discussion see H. Thamhaim and D. Wilemon, "Building High Performing Engineering Project Teams," *IEEE Transactions on Engineering Management,* Vol. 34, No. 3, August 1987, pp. 130–137.

3. Power struggle and conflict
4. Uninvolved, disintegrated upper management
5. Poor job security
6. Shifting goals and priorities.

It is interesting to note that the six drivers not only correlated favorably to the direct measures of high project team performance, such as the technical success and on-time/on-budget performance, but also were positively associated with the 16 indirect measures of team performance, ranging from commitment to creativity, quality, change-orientation, and need for achievement. The six barriers had exactly the opposite effect. The complete listing of the 16 performance measures is shown in Figure 12.3. In addition, the actual correlation table developed during the field study is shown and discussed in the Appendix of this chapter (Table 12.A.1). The results provide some quantitative support to previous field studies by Thamhain and Wilemon.[6]

What we find consistently is that successful organizations pay attention to the human side. They seem to be effective in fostering a work environment conducive to innovative, creative work, where people find the assignments challenging, leading to recognition and professional growth. Such a professionally stimulating environment also seems to lower communication barriers and conflict and enhances the personal desire to succeed. This seems to increase organizational awareness and the ability to respond to changing requirements. The second finding is that a winning team appears to have good leadership. Management understands the factors crucial to success. Managers are action-oriented, provide the needed resources, properly direct the im-

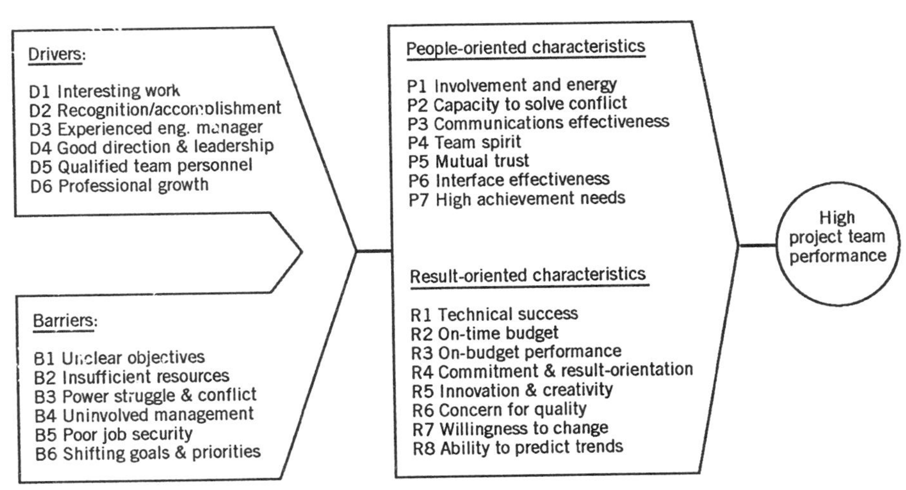

FIGURE 12.3 Major drivers and barriers to project team performance.

[6]The research on team characteristics and drivers versus barriers is based on a field study by H. Thamhain and D. Wilemon, "Building High Performing Engineering Project Teams," *IEEE Transactions on Engineering Management*, Vol. 34, No. 3, August 1987, pp. 130–137.

plementation of the project plan, and help in the identification and resolution of problems in their early stages. Taken together, the findings support three propositions:

Proposition 1. The degree of project success seems to be primarily determined by the strength of six driving forces and six barriers, which are related to (a) leadership, (b) job content, (c) personal needs, and (d) the general work environment.

Proposition 2. The strongest driver toward project success is a professionally stimulating team environment, characterized by (a) interesting, challenging work; (b) visibility and recognition for achievements; (c) growth potential; and (d) good project leadership.

Proposition 3. A professionally stimulating team environment also leads to low perceived conflict, high commitment, highly involved personnel, good communications, change orientation, innovation, and on-time/on-budget performance.

To be effective in organizing and directing an engineering project team, the leader must not only recognize the potential drivers and barriers, but also must know when in the life cycle of the project they are most likely to occur. The effective project leader takes preventive actions early in the project life cycle and fosters a work environment that is conducive to team-building as an ongoing process.

The effective team-builder is usually a social architect who understands the interaction of organizational and behavioral variables and can foster a climate of active participation and minimal dysfunctional conflict. This requires carefully developed skills in leadership, administration, organization, and technical expertise. It further requires the project leader's ability to involve top management and to assure organizational visibility, resource availability, and overall support for the project activities throughout its life cycle.

It is this organizational culture that adds yet another challenge to project team-building. The new team members are usually selected from hierarchically organized support departments, which are led by strong individuals who often foster internal competition rather than cooperation. In fact, even at the contributor level, many of the highly innovative and creative people are strongly individualistically oriented and often admit their aversion to cooperation. The challenge to the engineering manager or project leader is to integrate these individuals into a team that can produce innovative results in a systematic, coordinated, and integrated way that is consistent with the overall project plan. Many of the problems that occur during the formation of the new project team or during its life cycle are normal and often predictable. However, they present barriers to effective team performance. They must be quickly identified and dealt with.

12.4 ORGANIZING THE NEW ENGINEERING PROJECT TEAM

Too often the engineering manager or assigned project manager, under pressure to start producing, rushes into organizing the project team without establishing the proper organizational framework. While initially the prime focus is on staffing, the project manager cannot effectively attract and hold quality people until certain organizational pillars are in place. At a minimum, the basic project organization and various tasks must be defined before the recruiting effort can start.

These pillars are not only necessary to communicate the project requirements, responsibilities, and relationships to team members, but also to manage the anxiety that usually develops during team formation. This anxiety is normal and predictable. It is a barrier, however, to getting the team quickly focused on the task.

The anxiety may come from several sources. For example, if the team members have never worked with the project leader, they may be concerned with his or her leadership style and its effect on them. In a different vein, team members may be concerned about the nature of the project and whether it will match their professional interests and capabilities. Other team members may be concerned about whether the project will be helpful to their career aspirations. Team members may also be anxious about life style or work style disruptions. As one project manager remarked: "Moving a team member's desk from one side of the room to the other can sometimes be just as traumatic as moving someone from Chicago to Los Angeles to build a power plant[7]." As the quote suggests, seemingly minor changes can result in sudden anxiety among team members.

Another common concern among newly formed teams is whether or not there will be an equitable distribution of workload among team members and whether each member is capable of pulling his or her own weight. In some newly formed teams, members not only have to do their own work, but they must also train others. Within reason this is bearable, necessary, and often expected. However, when it becomes excessive, anxiety increases and morale can fall.

A number of procedural guidelines are provided below to help engineering personnel in organizing their multidisciplinary task teams for high efficiency and performance. These guidelines are presented in four sections: (1) make functional ties work; (2) structure your team organization; (3) define the project; and (4) recruit your team members.

Make Functional Ties Work For You

The multidisciplinary nature of engineering task teams usually requires dual accountability of team personnel to both the project leader and the functional resource manager. It is a mistaken belief that strong ties of team members to their functional organizations are bad for effective multidisciplinary team work and should be eliminated. On the contrary, loyalty to both the project and the functional organization is a natural, desirable, and often very necessary condition for project success. For example, in the most common of all engineering organizations, the matrix, the project leader gives operational directions to the functional support personnel. He or she is responsible for the budget and schedule, while the functional organization provides

[7]D. Wilemon and H. Thamhain, "Team Building in Project Management," *Project Management Quarterly*, July 1983.

technical guidance and personnel administration. Both the program manager and the functional managers must understand this process and perform accordingly or severe jurisdictional conflicts can develop.

Structure Your Team Organization

The keys to successfully building a new team organization are to have clearly defined and clearly communicated responsibilities and organizational relationships. The tools for systematically describing the project organization were discussed in Chapter 2. They come, in fact, from conventional management practices, and include the following: (1) the charter of the program, project, or team organization; (2) the project organization chart; (3) the responsibility matrix; and (4) the job description.

Define the Project. It is seldom a problem to define the technical components of the project. Project team members are usually very competent in their technical areas. They also enjoy the technical content of the project. Yet, this is only one of the four parameters of the project that must be defined, at least in principle, before staffing can begin. They are listed in Table 12.2. Regardless of how vaguely and preliminarily these project parameters are defined at the beginning, the initial description will help in recruiting the appropriate personnel and eliciting commitment to the preestablished parameters of technical performance, schedule, and budget. The core team should be formed prior to finalizing the project plan and contractual arrangements. This will provide the project management team with the opportunity to participate in tradeoff discussions and customer negotiations that will lead to technical confidence and commitment of all parties involved.

TABLE 12.2 Project Parameters That Need Definition

1. The Work	2. Timing
Overall specifications	Master schedule
Requirements document	Milestone chart
Statement of work	Network
System block diagram	Critical path analysis
Work breakdown structure	
List of deliverables	
3. Resources	**4. Responsibilities**
Budget	Task matrix
Resource plan	Project/task roster
	Project charter
	Work packages

Recruit Your Team Members

Staffing the project organization is the first major milestone during the project formation phase. Because of the pressures on the project manager to produce, staffing is often done hastily and without properly defining the basic project work to be performed. As a result, team members are often poorly matched to the job requirements, resulting in conflict, low morale, suboptimum decision-making, and in the end poor project and team performance.

The comment of a team leader who was pressed into quick staffing actions is indicative of these potential problems: "How can you interview task managers when you cannot show them what the job involves and how their responsibilities tie in with the rest of the project?"

Therefore, only after the project organization and the tasks are defined, in principle, can project leaders start to interview candidates.

All project assignments should be negotiated individually with each prospective team member. Each task leader should be responsible for staffing his or her own task team. Where dual-reporting relationships are involved, staffing should be conducted jointly by the two managers. The assignment interview should include a clear discussion of the specific task, the outcome, timing, responsibilities, reporting relation, potential rewards, and importance of the project to the company. Task assignments should be made only if the candidate's ability is a reasonable match to the position requirements and the candidate shows a healthy degree of interest in the project.

The Interview Process. The interview process normally has five facets, which are often interrelated.

1. *Informing the Candidate About the Assignments*

- What the objectives are for the project.
- Who will be involved and why.
- What the structure of the project/team organization is and what its interfaces are.
- Importance of the project to the overall organization or work unit; short- and long-range impact.
- Why the team member was selected and assigned to the project. What role he or she will perform.
- Specific responsibilities and expectations.
- What rewards might be forthcoming if the project is completed successfully.
- A candid appraisal of the problems and constraints that are likely to be encountered.
- What the rules of the road are that will be followed in managing the project, such as regular status review meetings, controls, and rewards.
- The challenges and recognition the project is likely to provide.
- Why the team concept is important to success and how it should work.

2. Determining Skills and Expertise

- Probe related experience; expand from resumé.
- Probe candidate's aptitude relevant to your project environment: technology involved, engineering tools and techniques, markets and customer involvement, and product applications.
- Probe into the program management skills needed. Use current project examples: "How would you handle this situation . . . ?" Probe leadership, technical expertise, planning and control, administrative skills, and so on.

3. Determining Interests and Team Compatibility

- What are the professional interests and objectives of this candidate?
- How does the candidate manage and work with others?
- How does the candidate feel about sharing authority, working for two bosses, or dealing with personnel across functional lines with little or no formal authority?
- What suggestions does the candidate have for achieving success?

4. Persuading the Candidate to Join the Project Team

- Explain specific rewards for joining the team, such as financial rewards, professional growth, recognition, visibility, work challenge, and potential for advancement.

5. Negotiating Terms and Commitments

- Check candidate's willingness to join team.
- Negotiate conditions for joining: salary; hired, assigned, or transferred; performance reviews and other criteria.
- Elicit candidate's commitment to established project objectives and modus operandi.
- Secure final agreement.

In essence, this interview process with potential team members should be similar in nature to the hiring of a new employee from outside the company. It is a process that involves two-way selling. The candidate presents his or her qualities, skills, and interests in the task to be performed and the manager tries to sell the desirability of the new assignment. In this dialogue, both parties can assess their mutual compatibility, or lack of compatibility, for the job. The discussion not only clarifies on the requirements and potential rewards, but also may stimulate interest in the assignment, lead to a better understanding of the needs, and ultimately result in *commitment* of the team member to the project and its objectives. This "sign-on process" has been used in various degrees by many companies for establishing and unifying their multidisciplinary task teams. However, it was not until Tracy Kidder's book, *The Soul of a New Machine*, described the process of forming the design team for Data General's new Eclipse minicomputer system in 1972, that this sign-on process was formally recognized as an effective management tool for team-building.

12.5 SUGGESTIONS FOR HANDLING THE NEWLY FORMED ENGINEERING TEAM

During its formation stage, the project group represents just a collection of individuals who have been selected for their skills and capabilities as collectively needed to perform the upcoming project task. However, to be successful, the individual efforts must be integrated. Even more demanding, these individuals have to work together as a team to produce innovative results that fit together to form an integrated new system, as conceptualized in the project plan.

Initially, there are many problems that prevent the project group from performing as a team. Although these problems are normal and often predictable, they present barriers to effective team performance, the problems therefore must be quickly identified and dealt with. The following list presents typical problems that occur during a project team formation:

- Confusion
- Responsibilities unclear
- Channels of authority unclear
- Work distribution load uneven
- Assignment unclear
- Communication channels unclear
- Overall project goals unclear
- Mistrust
- Personal objectives unrelated to project
- Measures of personal performance unclear
- Commitment to project plan lacking
- Team spirit lacking
- Project direction and leadership insufficient
- Power struggle and conflict.

Certain steps taken early in the life of the team can help the project leader in identifying specific problems and dealing with them effectively. These steps may also provide preventive measures, which will eliminate the potential for these problems to develop in the first place. Specific suggestions are made below.

1. The Assignment Should Be Clear. Although the overall task assignment, its scope, and objectives might have been discussed during the initial sign-on of the person to the project, it takes additional effort and involvement for new team members to feel comfortable with the assignment. The thorough understanding of the task requirements usually comes with the intense personal involvement of the new members with the project team. Such involvement can be enhanced by assigning the new member to an action-oriented task that requires team involvement and creates visibility, such as a requirements analysis, an interface specification, or a producibility study. In addition, any committee-type activity, presentation, or data-gathering will help to involve the new team member. It also will enable that person to better understand the specific task and his or her role in the overall team effort.

2. *New Team Members Must Feel Professionally Comfortable.* The initial anxieties and lack of trust and confidence that new members may feel are serious barriers to team performance. New team members should be properly introduced to the group, and their roles, strengths, and importance to the project should be explained. Providing opportunities for early results allows the leader to give early recognition for professional accomplishments, which will stimulate the individual's desire for the project work and build confidence, trust, and credibility within the group.

3. *Team Organization Should Be Clear.* Project team structures are often considered very "organic" and inconsistent with formal chain-of-command principles. However, individual task responsibility, accountability, and organizational interface relations should be clearly defined to all team members. A simple work breakdown structure (WBS) or task matrix, together with some discussion, can facilitate a clear understanding of the team structure, even with a highly unconventional format.

4. *Locate Team Members in One Place.* Members of the newly formed team should be closely located to facilitate communications and the development of a team spirit. Locating the project team in one office area is the ideal situation. However, this may be impractical, especially if team members share their time with other projects or if the assignment is only for a short period of time. Regularly scheduled meetings are recommended as soon as the new project team is formed. These meetings are particularly important where team members are geographically separated and do not see each other on a day-to-day basis.

5. *Provide a Proper Team Environment.* It is critical for management to provide the proper environment for the project to function effectively. Here the project leader needs to tell the management at the onset of the program what resources are needed. The project manager's relationship with senior management support is critically affected by his or her credibility and the visibility and priority of the project.

6. *Manage.* Especially during the initial stages of team formation, it is important for the project leader to keep a close eye on the team and its activities to detect problems early and to correct them. It is often the engineering manager's responsibility to influence the climate of the work environment by his or her own actions. The manager's concern for project team members, ability to integrate personal goals and needs of project personnel with project objectives, and ability to create personal enthusiasm for the work itself can foster a climate that is high on motivation and work involvement, and as a result, high on project performance.

7. Team Commitment. Engineering managers and their project leaders should determine lack of team member commitment early in the life of the project and attempt to change possible negative views toward the project. Since insecurity is often a major reason for lacking commitment, managers should try to determine why insecurity exists, then work on reducing the team members' fears. Conflict with other team members may be another reason for lack of commitment. It is important for the project leader to intervene and mediate the conflict quickly. Finally, a team member's professional interests may lie elsewhere. The project leader should examine ways to satisfy part of the team member's interests by bringing personnel and project goals into perspective.

12.6 TEAM-BUILDING AS AN ONGOING PROCESS

Although proper attention to team-building is crucial during the early phases of a project, it is also a never-ending process. It is a shared responsibility between the project leader and the engineering resource managers who normally control the assignment of the work. These managers continually monitor team functioning and performance to see what corrective action may be needed to prevent or correct various team problems.

Several barometers provide good clues of potential team dysfunctioning. First, noticeable changes in performance levels for the team and/or for individual team members should always be followed up. Such changes can be symptomatic of more serious problems, such as conflict, lack of work integration, communication problems, and unclear objectives. Second, the project leader and team members want to be aware of the changing energy level in various team members. This, too, may signal more serious problems, or indicate that the team is tired and stressed. Sometimes changing the work pace, taking time off, or setting short-term targets can serve as a means to reenergize team members. More serious cases, however, can call for more drastic measures, such as reappraising project objectives and/or the means to achieve them. Third, verbal and nonverbal clues from team members may be a source of information and team functioning. It is important to hear the needs and concerns (verbal clues) and to observe how team members act in carrying out their responsibilities (nonverbal clues). Finally, detrimental behavior of one team member toward another can be a signal that a problem within the team warrants action.

It is highly recommended that project leaders hold regular meetings to evaluate overall team performance and deal with team functioning problems. The focus of these meetings can be directed toward "What are we doing well as a team?" and "What areas need our team's attention?" This approach often brings positive surprises in that the total team will be informed on progress in diverse project areas, such as a breakthrough in technology, a subsystem schedule met ahead of the original target date, or a positive change in a client's behavior toward the project. After the positive issues have been discussed, attention should be focused on areas that need improvement or help. The purpose of this part of the review session is to focus on real or potential problem areas. The meeting leader should ask each team member for his or her observations on these issues. Then an open discussion should be held to ascertain how significant the problems really are. Assumptions should, of course, be separated from the facts of each situation. Next, assignments should be agreed upon regarding how to best handle these problems. Finally, a plan for follow-up should be developed. The process should result in a better overall performance and promote a feeling of team participation and high morale.

Over the life of a project, the problems encountered by the project team are likely to change. It is therefore quite normal and predictable that as the old problems are identified and solved, new ones will emerge; they should be dealt with in an ongoing fashion.

In summary, effective team-building is a critical determinant of success in the multidisciplinary work environment of the engineering manager. While the process of team-building can entail frustrations and require the use of energy on the part of all concerned, the rewards can be great.

Social scientists generally agree that there are several indicators of effective and ineffective teams. At any point in the life of the team, the project manager should be aware of certain effectiveness–ineffectiveness indicators, which are summarized in Table 12.3.

As we go through this decade, we anticipate important developments in team-building. As shown in Figure 12.4, these developments will lead to higher performance levels, increased morale, and a pervasive commitment to final results that can withstand almost any kind of adversity.

12.7 MANAGING TECHNOLOGICALLY INNOVATIVE TEAM EFFORTS TOWARD DESIRED RESULTS

The important role of technological innovation for successful project performance has long been recognized. Innovation can generate a competitive advantage for one firm, while eroding the market position of another.[8] However, deriving competitive advantages from innovation is a highly complex process. Research on the determinants of product innovation, for example, has traditionally focused on the qualities of the individuals who are crucial elements in the innovation process. In recent years, an increasing number of studies have broadened the investigations into other areas, which include planning, entrepreneurship, product champions, top management involvement, and marketing factors. Researchers such as Edward Roberts have attempted to model the macroprocess of innovation, integrating people, organizational processes, and plans.[9]

[8]For discussion see William J. Altier, "A Perspective on Creativity," *Journal of Product Innovation Management*, Vol. 5, No. 2, June 1988, pp. 154–161 and Peter F. Drucker, *Innovation and Entrepreneurship: Practice and Principles*, New York: Harper & Row, 1985.

[9]Edward Roberts, "Managing Inventions and Innovations," *Technology Management*, Volume 31, Number 1 (Jan/Feb 1988), pp. 11–30; and *Generating Technological Innovation*, New York: Oxford University Press, 1987.

TABLE 12.3 Characteristics of Effective and Ineffective Project Teams

Characteristics of Effective Team	Characteristics of Ineffective Team
• High performance and task efficiency	• Low performance
• Result-oriented	• Activity-oriented
• Innovative/creative behavior	• Low level of involvement and enthusiasm
• Committed, result-oriented. High achievement needs. Change-oriented	• Low commitment to project objectives
• Professional objectives of team members coincide with project requirements. Interested in membership self-development	• Unclear project objectives and fluid commitment levels from key participants
• Technically successful	• Quality problems
• On-time/On-budget performance	• Schedule and budget slips. Frequent surprises. Image problems (credibility)
• Team members highly interdependent, interface effectively	• Uninvolved management
• Capacity for conflict resolution, but conflict encouraged when it can lead to beneficial results	• Unproductive gamesmanship, manipulation of others, hidden feelings, conflict avoided at all costs
• Communicates effectively	• Confusion, conflict, inefficiency
• High trust levels	• Subtle sabotage, fear, disinterest, or foot-dragging. Cliques, collusion, isolating members
• High energy levels and enthusiasm. High morale	• Lethargic/unresponsive. Anxieties and insecurities

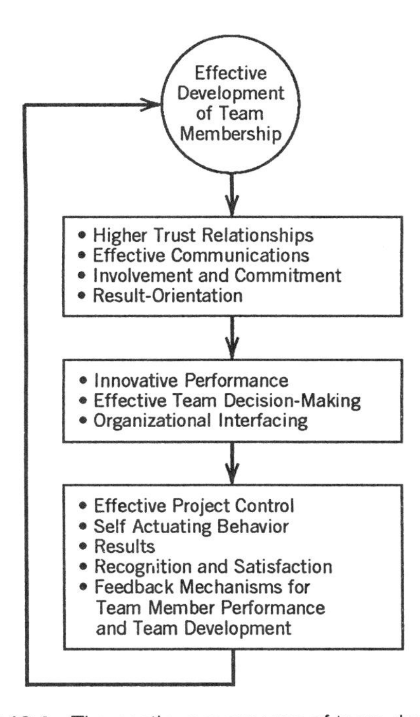

FIGURE 12.4 The continuous process of team development.

Technology-based innovation needs strong integration and orchestration of cross-functional activities. Engineering management in particular involves a sequence of interrelated multifunctional efforts. Examples include new product design, product assurance, licensing, production systems, materials and materials handling, communications, customer applicants, deliveries, and field service. Management of innovation spans the complete project life cycle: from recognition of an opportunity and creation of new knowledge and concepts, to product research, development and engineering, transferring technology into manufacturing and the market, product distribution, upgrading and service. Thus, innovation involves a complex array of interrelated processes, as summarized in the partial listing in Table 12.4.

Consequently, many business leaders and product managers are greatly concerned about their organization's ability to manage innovative efforts effectively. This concern is magnified by the realities of our competitive business environment, that is, end-date driven schedules, limited resources, organizational uncertainties, and multidisciplinary team efforts. In addition, many engineering managers operate in a matrix environment with power- and resource-sharing, multiple accountability, and limited control over support personnel. Any and all of these factors may serve to complicate a firm's effort toward product innovation.

Focus on Team-Building

One of the managerial challenges of successful engineering work often consists of building a unified multifunctional team committed to the innovative implementation of the established project or work plan. Because these team members come from different organizations with different needs, backgrounds, interests and expertise,

TABLE 12.4 Activities and Processes Involved in Innovation

Invention and creativity
Idea generation
Strategic business planning
Managerial decision-making
Needs assessment
Feasibility studies
Research and development
Product development and engineering
Prototyping, testing, and integration
Product improvement and upgrading
Production
Field engineering and service
Marketing and marketing communication
Bid proposals
Technology transfer to other areas

they must be transformed into an integrated work group that is unified toward the project objectives.

Characteristics of Innovative Team Performance

There are obviously cultural and philosophical differences that influence the way companies define innovative performance. Even within one company or department, performance is seen as a complex set of interrelated variables. However, there seems to be general agreement among engineering managers on factors inherent in the success formula for the majority of innovations. Specific data obtained in a field study of 270 engineering managers and their supervisors[10] show that five measures are particularly important for successful innovative performance: (1) number of innovative ideas commercialized, adopted, or recognized by the organization; (2) established organizational objectives met; (3) adaptability to changing requirements and conditions; (4) commitment; and (5) ability to meet schedule and budget objectives.

The study also shows that team leaders and first-line managers rate their group's innovative performance highly (an average of "good"), while their superior's perception is considerably less favorable, ranging from "marginal" to "satisfactory."

The significance of determining innovative performance characteristics is in two areas. First, it shows some commonality of concern and measures used among managers. Second, it establishes the basis for standards of innovative output measurement. In addition, these characteristics can be useful in establishing performance criteria for employee appraisals.

Characteristics of an Innovative Work Environment

Technical professionals perceive many factors influencing innovative performance. However, the field study test referred to earlier showed that only a relatively small number of factors correlate at a statistically significant level and thus affect innovative performance in a significant way.[11] As shown in Table 12.5, the factors with the strongest association to innovative performance relate to three primary issues: (1) task definition, (2) people management and (3) organizational support. It is further interesting to note that these drivers not only correlate favorably to the direct measure of innovative performance, the upper management perception, but also correlate positively with four other, less-direct measures: the number of innovative ideas generated, meeting established goals, adaptability, and commitment.

[10] H. Thamhain, "Managing Technologically Innovative Team Efforts Toward New Product Success," *Journal of Product Innovation Management*, Vol. 7, March 1990, pp. 5–18.

[11] H. Thamhain, "Managing Technologically Innovative Team Efforts Toward New Product Success," *Journal of Product Innovation Management*, Vol. 7, March 1990, pp. 5–18. Kendall's tau rank-order correlation was used to measure the association between perceived conducive factors in the work environment and actual innovative performance. Only those factors that correlated at a confidence level of 90% or better were deemed "significant" drivers toward innovation.

TABLE 12.5 Performance Correlates: Work Environment versus Innovative Performance

Drivers and Barriers to Innovation	Characteristics of Innovative Performance				
	No. of Ideas P_1	Meeting Goals P_2	Change Orient. P_3	Commitment P_4	Senior Management Perception of Innov. Performance P_5
TASK-RELATED FACTORS					
Clear objectives and plans	.38	.45	.15	.35	.45
Technology direction and leadership	.50	.30	.10	.45	.38
Autonomy and challenge	.25	.30	0	.20	.30
Experienced personnel	.10	.15	.15	.25	.25
Project involvement and visibility	.32	.45	.25	.37	.25
Aggregated task variables	.38	.35	.13	.33	.38
PEOPLE RELATED FACTORS					
Personal work satisfaction	.40	.15	.20	.35	.50
Mutual trust, team spirit	.15	.15	.15	.20	.30
Good communication	.20	.35	.32	.28	.35
Low conflict	.35	.35	.15	.45	.15
Low threat/fail-safe	.25	.25	.30	.30	.15
Aggregated people variables	.27	.33	.28	.34	.31
ORGANIZATION-RELATED FACTORS					
Organizational stability	.15	.28	.20	.30	.32
Sufficient resources	.05	.30	0	.15	.30
Involved management	.20	.45	.15	.38	.30
Rewards and recognition	.25	.45	.05	.20	.25
Stable goals and priorities	.35	.20	.10	.38	.20
Aggregated organization variables	.20	.38	.12	.30	.28

Significance levels: $\tau \geq .15$. . . . $\alpha \geq 90\%$; $\tau \geq .25$. . . . $\alpha \geq 95\%$; $\tau \geq .35$. . . . $\alpha \geq 99\%$.
Sample size: $N(P_1, P_2, P_3, P_4) = 270$; $N(P_5) = 90$.

A Framework for Innovative Team Performance

The characteristics of an innovative engineering team depend on many factors. While this chapter study looks only at a limited set of variables, it suggests many areas for benchmarking and further investigation by practitioners and scholars. The system approach offers a simple and effective format for summarizing the current findings and organizing additional investigations. As shown in Figure 12.5, a simple input–output diagram can help in analyzing innovative team performance. The diagram shows the three types of input variables from the team environment grouped into task-, people- and organization-oriented sets, that is, those that produce the strongest correlation to innovative performance. In other situations, innovative team performance may be associated with an even larger set of variables, all likely to be interrelated in a complex, intricate form. However, using the system approach allows researchers and management practitioners to break down the complexity of variables and transfer processes involved in the study of innovative team performance. Further, the framework can provide a starting point for studying innovative team performance at multiple phases of a project or new product development cycle, such as (1) product idea generation, research and advanced development, (2) concept development and feasibility, (3) product design and engineering, (4) manufacturing, (5) marketing and sales and (6) field service.

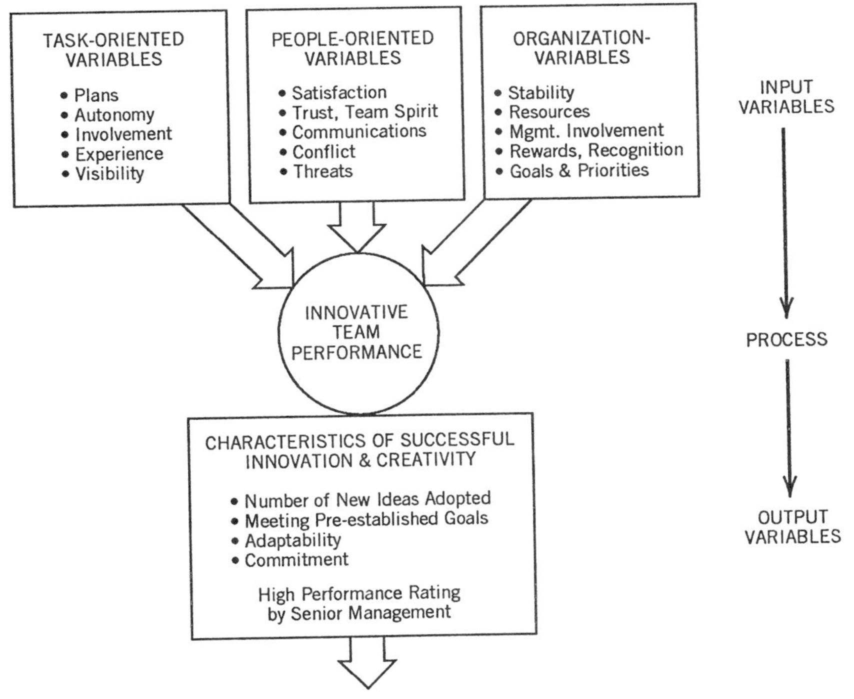

FIGURE 12.5 Input–output diagram of innovative team performance.

The findings demonstrate consistently that successful organizations and their managers pay attention to human factors. They appear effective in fostering a work environment conducive to innovative work, one where people find challenging assignments that lead to recognition and professional growth. Such a professionally stimulating environment seems to lower communication barriers, reduce conflict, and enhance the desire of personnel to succeed. Also, it seems to enhance organizational awareness of greater environmental trends and enhance the ability to prepare and respond to these challenges effectively.

Successful innovative work groups have good leadership. Management understands the factors crucial to success and makes proper provisions for them. Leadership is action-oriented, provides the needed resources, properly plans and directs the implementation of its programs, and helps in identification and resolution of problems in their early stages. Many early warning signs of low innovative team performance can be identified, as summarized in Table 12.6. Effective team leaders monitor such feedback and focus their efforts on problem avoidance. The effective team leader recognizes the potential interrelationship among drivers and barriers to innovative performance. It may be difficult to assess the exact impact of a "small" problem and its nominal effects. The effective manager keeps an eye on all potential and actual problems and minimizes them wherever possible.

A Final Note

The successful management of technological innovation involves a complex set of variables. There is, of course, no single set of broad guidelines that guarantees instant success. The role of the team leader and engineering manager is a difficult one. However, by understanding the variables, and their interrelationships, that drive technological innovation and product success, managers can develop a better-integrated insight into the factors that contribute to success. This may help in fine-tuning styles, actions, and strategies of managing resources toward successful technology-based innovation.

12.8 ADDITIONAL RECOMMENDATIONS FOR EFFECTIVE TEAM MANAGEMENT

The engineering manager must foster an environment where team members are professionally satisfied, are involved, and have mutual trust. As shown in Figure 12.3, the more effective the project leader is in stimulating the drivers and minimizing the barriers, the more effective the manager can be in developing team membership and the higher the quality of information contributed by team members, including their willingness and candor in sharing ideas and approaches. By contrast, when a team member does not feel part of the team and does not trust others, information will not be shared willingly or openly. One project leader emphasized the point: "There's nothing worse than being on a team where no one trusts anyone else. Such situations

TABLE 12.6 Early Warning Signs of Innovative Performance Problems

- Project perceived as unimportant
- Unclear task/project goals and objectives
- Excessive conflict among team members
- Unclear mission and business objectives
- Unclear requirements
- Perceived technical uncertainty and risks
- Low motivation, apathy, low team spirit
- Little team involvement during project planning
- Disinterested, uninvolved management
- Poor communications among team members
- Poor communications with support groups
- Problems in attracting and holding team members
- Unclear role definition, role conflict, power struggle
- No agreement on project plans
- Lack of performance feedback
- Professional skill obsolescence
- Perception of inadequate rewards and incentives
- Poor recognition and visibility of accomplishments
- Little work challenge (professionally not stimulating)
- Fear of failure, potential penalty
- Fear of evaluation
- Mistrust, collusion, protectionism
- Excessive documentation
- Excessive requests for directions
- Complaints about insufficient resources
- Strong resistance to change

lead to gamesmanship and a lot of watching what you say because you don't want your own words to bounce back in your own face . . . "

A number of specific recommendations have been provided above for project leaders and engineering managers responsible for the integration of multidisciplinary tasks to help in their complex efforts of building high-performing project teams. A few more are mentioned below:

1. *Management Commitment.* Project leaders must continuously update and involve their managements to refuel their interests and commitments to the new project. Breaking the project into smaller phases and being able to produce short-range results frequently, seem to be important to this refueling process.

2. *Image-Building.* Building a favorable image for the project or team activities, in terms of high-priority, interesting work, importance to the organization, high visibility, and potential for professional rewards is crucial to the ability to attract and hold high-quality people. It is also a pervasive process, which fosters a climate of active participation at all levels; it helps to unify the new project team and minimize dysfunctional conflict.

3. *Effective Planning Early in the Project Life Cycle.* This will have a favorable impact on the work environment and team effectiveness. This is especially so because project managers have to integrate various tasks across many functional lines. Proper planning means more than just generating the required pieces of paper. It requires the participation of the entire project team, including support departments, subcontractors, and management. These planning activities, which can be performed in a special project phase such as requirements analysis, product feasibility assessment, or product/project definition, usually have a number of side benefits, besides generating a comprehensive roadmap for the upcoming program. They provide insight and perspective to the upcoming tasks, create visibility and interest, minimize confusion, conflict, and anxiety, and help to unify the task team toward the overall project objectives.

4. *Involvement of all Personnel.* One of the side benefits of proper project planning is the involvement of personnel at all organizational levels. Project leaders and engineering managers should drive such an involvement, at least with their key personnel, especially during the project-definition phases. This involvement will lead to a better understanding of the task requirements, stimulate interest, help unify the team, and ultimately lead to commitment to the project plan regarding technical performance, timing, and budgets.

5. *Organization Development Specialists.* Team leaders should watch for changes in performance on an ongoing basis. If performance problems are observed, they should be dealt with quickly. If the project manager has access to internal or external organization development specialists, they can help diagnose team problems and assist the team in dealing with the identified problems. These specialists can also bring fresh ideas and perspectives to difficult and sometimes emotionally complex situations.

6. *Early Problem Recognition.* Team leaders should focus their efforts on troubleshooting. That is, the project leader, through experience, should recognize potential problems and conflicts at their onset and deal with them before they become big and their resolution consumes a large amount of time and effort.

A Final Note

The potential gains from increased productivity are great for individuals, organizations, and society as a whole. Such gains are possible only if we can utilize project resources effectively. One such improvement is through effective team-building.

Over the next decade, we anticipate important developments in team-building, which will lead to higher performance levels, increased morale, and a pervasive commitment to final results. This chapter should help both the professional in the field of engineering management as well as the scholar who studies contemporary organizational concepts to understand the intricate relationships between organizational and behavioral elements by providing a conceptual framework for specific situational analysis of engineering team-building practices. The case description of the MK-2000 New Product Development, in the Appendix at the end of this book, provides an opportunity for stimulating and analyzing a technical project situation involving many facets of project team management. It is a vehicle for applying the various concepts and techniques of team-building via conceptual thinking, group discussion, and role-playing.

APPENDIX RANK-ORDER CORRELATION OF DRIVERS AND BARRIERS TO TEAM PERFORMANCE

Association Between Team Characteristics and Team Performance

The association was measured by utilizing Kendall's tau rank-order correlation and partial rank-order correlation. First, projects were rank-ordered by senior managers according to their perception of overall project performance. Then the various factors describing the team characteristics were each rank-ordered by both project leaders and team members, according to their strength. Project team members were asked to rate each of the influence factors shown as drivers and barriers in Fig. 12.3. The rating measured the presence of each of these factors in the team environment, using a five-point scale ranging from "strongly agree" to "strongly disagree." The team rankings were based on performance as perceived as senior managers (R scores) and project managers (P scores). Finally the tau coefficients and their significances were calculated for each association, as shown in Table 12.A.1.[12] Those influences which correlated predominantly positive were characterized as drivers, those that correlated predominantly negatively were characterized as barriers. The variables in Table 12.A.1 are identical to those of Figure 12.3; the statistical significance is indicated as follows:$\tau \geq 0.25$ indicates a 95% confidence level ($p \leq 0.05$), and $\tau \geq 0.35$ indicates a 99% level ($p \leq 0.01$).

The Kruskal–Wallis One-Way Analysis of Variance by Ranks

The Kruskal-Wallis analysis is a test for deciding whether K independent samples are from different populations. In our study the test verifies that both managers and other project team members believe in essentially the same qualities that should be present within an effective, high-performing project team. For mathematical procedure see S. Siegel.[12]

[12]For mathematical procedure, see S. Siegel, *Nonparametric Statistics for the Behavioral Sciences*, New York: McGraw-Hill, 1956.

TABLE 12.A.1 Drivers and Barriers To Technical Team Performance: Kendall's Tau Correlation of Team Characteristics and Team Performance

	PEOPLE-ORIENTED CHARACTERISTICS								RESULT-ORIENTED CHARASTERISTICS								
Drivers and Barriers	P1	P2	P3	P4	P5	P6	P7	P8	R1	R2	R3	R4	R5	R6	R7	R8	AVGE. $\overline{PR}$
DRIVERS (+τ):																	
D1 Interesting Work	+.45	.55	.35	.40	.30	.10	.20	.55	.30	.30	.20	.50	.25	.25	.25	.10	.32
D2 Recognition/Accomplishment	+.10	.35	.20	.25	.30	.30	.15	.60	.25	.25	.15	.35	.15	.40	.10	.15	.27
D3 Experienced Eng. Manager	+.20	.10	.25	.20	.20	.25	.30	.25	.35	.30	.30	.30	.25	.30	.30	.35	.26
D4 Proper Direction & Leadership	+.10	.12	.35	.20	.05	.10	.20	.30	.55	.35	.30	.30	.25	.30	.25	.30	.25
D5 Qualified Team Personnel	+.12	.20	.30	.25	.10	.30	.20	.25	.25	.35	. 30	.10	.35	.45	.10	.30	.24
D6 Professional Growth Potential	+.15	.10	.10	.15	.10	.10	.05	.25	.10	.15	. 10	.25	.10	.30	.10	.20	.14
BARRIERS (−τ):																	
B1 Unclear Objectives	−.45	.45	.20	.35	.40	.20	.35	.25	−.40	.20	.20	.55	.25	.15	.30	.35	.31
B2 Insufficient Resources	−.30	.35	.05	.35	.25	.05	.10	.20	−.35	.40	.55	.40	.20	.35	.10	.35	.26
B3 Power Struggle & Conflict	−.25	.60	.10	.40	.45	.30	.25	.25	−.20	.15	.20	.35	.20	.30	.20	.10	.26
B4 Uninvolved Management	−.35	.25	.25	.45	.30	.05	.10	.05	−.35	.10	.15	.35	.20	.30	.15	.35	.23
B5 Poor Job Security	−.10	.30	.20	.40	.40	.10	.15	.10	−.30	.20	.15	.35	.15	.35	.20	.30	.23
B6 Shifting Goals & Priorities	−.30	.25	.15	.20	.15	.05	.25	.15	−.20	.35	.35	.15	.15	.40	.25	.10	.22

Significance Levels:
$\tau \geq 0.25 \ldots p \leq .05$
$\tau \geq 0.35 \ldots \leq p\ .01$.

P1: Involvement and energy
P2: Capacity to solve conflict
P3: Communications effectiveness
P4: Team spirit
P5: Mutual trust
P6: Membership self-development
P7: Interface effectiveness
P8: High achievement needs

R1: Technical success
R2: On-time performance
R3: On-budget performance
R4: Commitment and result orientation
R5: Innovation and creativity
R6: Concern for quality
R7: Willingness to change
R8: Ability to predict trends

EXERCISE 12.1: PROJECT TEAM DEVELOPMENT

The lead-in scenario presented at the beginning of this chapter is repeated with specific instructions for analysis and delivery of results. The exercise can be done either by an individual or by a group of people. The exercise should stimulate critical thinking of the issues involved in organizing and building a high-performing project team and formulating an actual team development plan.

Scenario

Five months after the Scirocco project was formally kicked off, Brian Perry is still experiencing serious problems with his team of 18 people who were assigned to him for the 10 months' duration of the project. Although a dedicated office space was furnished so that all Scirocco team members could be located in one area, only 9 people moved to the new area in spite of pressures from upper management. The others work out of their regular offices. Easy access to functional support, data files, and technical facilities seems to be more important to them than physical proximity to the Scirocco group.

Real problems surfaced two months ago when the concept developments of the various functions did not integrate into an acceptable systems solution. Many team members expressed frustration and confusion over conflict and mistrust has developed among the various functional subgroups over the inability to work out suitable interface requirements. There have been rumors that the project is in serious trouble, several key people asked for transfers, and management has become more involved in the technical aspects of Scirocco. Meanwhile, Brian Perry seems to be losing credibility and trust with the team and has difficulties in controlling the project according to the established plan.

Analysis and Preparation

Analyze the scenario and prepare five lists:

1. Challenges and problems that the Scirocco team faces
2. Characteristics of a high-performing project team
3. Early warning signals of poor project team performance
4. Drivers and barriers to a high-performing project team
5. Recommendations to management and to the project manager for improving the current Scirocco team's effectiveness.

Present your findings to management: 10 minutes for each of the above categories.

Delivery

The exercise can be completed in several modes: (1) individually as an exercise to test the understanding of project team building; (2) individually or in groups, as a homework assignment, take-home exam, or group exercise in a professional development seminar; and (3) in a multigroup workshop mode, with specific roles assigned to various discussion groups, such as project leader, functional manager, sponsor/customer, and upper management.

BIBLIOGRAPHY

Altier, William J., "A Perspective on Creativity," *Journal of Product Innovation Management,* Volume 5, Number 2 (June 1988), pp. 154–161.

Aquilino, J. J., "Multi-Skilled Work Teams: Productivity Benefits," *California Management Review,* (Summer 1977).

Aram, J. D. and C. P. Morgan, "Role of Project Team Collaboration in R&D Performance," *Management Science,* (June 1976).

Barczak, Gloria and David Wilemon, "Communications Patterns of New Project Development Team Leaders," *IEEE Transactions on Engineering Management.* Volume 38, Number 2 (May 1991), pp. 101–109.

Beckert, Beverly A., "Concurrent Engineering: Changing the Culture," *CAE,* Volume 98, Number 98 (Sept 1991), pp. 195–200.

Belzer, Ellen J., "Twelve Ways to Better Team Building," *Working Women,* Volume 14, Number 8 (Aug 1989), pp. 12, 14.

Benningson, Lawrence, "The Team Approach to Project Management," *Management Review,* Volume 61, (Jan 1972) pp. 48–52.

Brendan, W. and Kaleel Jamison (eds.), *Team Building Blueprints for Productivity and Satisfaction,* Alexandria, VA: National Training Laboratory, 1988.

Cleland, David I., "The Cultural Ambience of Project Management—Another Look," *Project Management Journal,* Volume 19, Number 3 (June 1988), pp. 49–56

Cleland, David I. and William R. King (eds.) *Project Management Handbook,* New York: Van Nostrand Reinhold, 1988.

Cleland, David I., *Project Management: Strategic Design and Implementation,* New York: McGraw-Hill, 1990.

DiMarco, Nicholas, Jane R. Goodson, and Henry F. Houser, "Situational Leadership in a Project/Matrix Environment," *Project Management Journal,* Volume 20, Number 1 (Mar 1989), pp. 11–18.

Drucker, Peter F., *Innovation and Entrepreneurship: Practice and Principles,* New York: Harper, 1985.

Dyer, William, *Team Building: Issues and Alternatives,* Reading, MA: Addison-Wesley, 1987.

Eckerson, Wayne, "Training in Team Building Critical to Complex Projects," *Network World,* Volume 7, Number 40 (Oct 1, 1990), pp. 23–34.

Editorial, "CAD/CAM Planning: Building an Engineering Team," *CAE,* Volume 10, Number 7 (July 1991), pp. CC4–CC6.

Fischer, Robert L., "HRIS Quality Depends On Teamwork," *Personnel Journal,* Volume 70, Number 4 (Apr 1991), pp. 47–51.

Gatto, Rex P., *Team Building and Communication: Unifying People Power,* Pittsburgh, PA: GTA Press, 1987.

Gemmill, Gary and Hans J. Thamhain, "Influence Styles of Project Managers: Same Project Performance Correlates," *Academy of Management Journal,* Volume 17, Number 2 (June 1974).

Gersick, Connie J. G., "Time and Transition in Work Teams: Toward a New Model of Group Development," *Academy of Management Journal,* Volume 31, Number 1 (Mar 1988), pp. 9–41.

Gupta, Ashok K. and David L. Wilemon, "Accelerating the Development of Technology-Based New Product," *California Management Review,* Volume 32, Number 2 (Winter 1990), pp. 24–44.

Hahn, Udo, Matthias Jarke, and Thomas Rose, "Teamwork Support in a Knowledge-Based Information Systems Environment," *IEEE Transactions on Software Engineering,* Volume 17, Number 5 (May 1991), pp. 467–482.

Hall, Stephen M., "Building a Team for Design Projects," *Chemical Engineering,* Volume 97, Number 9 (Sept 1990), pp. 189–196.

Harris, Philip R. and Dorothy L. Harris, "High Performance Team Management," *Leadership and Organization Development Journal* (UK), Volume 10, Number 4 (1989), pp. 28–32.

Hensey, Mel, "Making Teamwork Work," *Civil Engineering,* Volume 62, Number 2 (Feb 1992), pp. 68–69.

Hull, Frank M. and Koya Azumi, "Teamwork in Japanese and U.S. Labs," *Research Technology Management,* Volume 32, Number 6 (Nov/Dec 1989), pp. 21–26.

Jacobs, R. C. and J. G. Everett, "The Importance of Team Building in a High-Tech Environment," *Journal of European Industrial Training* (UK), Volume 12, Number 4 (1988), pp. 10–16.

Japan Human Relations Association (ed.), *Kaizen Teian-1,* Cambridge, MA: Productivity Press, 1992.

Katz, F. E., "Explaining Informal Work Groups in Complex Organizations," *Administrative Science Quarterly,* Number 10 (1965).

Kernaghan, John A. and Robert A. Cooke, "Teamwork in Planning Innovative Projects: Improving Group Performance by Rational and Interpersonal Interventions in Group Process," *IEEE Transactions on Engineering Management,* Volume 37, Number 2 (May 1990), pp. 109–116.

Kidder, John Tracy, *The Soul of a New Machine,* New York: Avon Books (1982).

Kliem, Ralph L., "Project Proficiency," *Computerworld,* Volume 25, Number 29 (July 22, 1991), pp. 77–78.

Kozar, Kenneth A., "Team Product Reviews: A Means of Improving Product Quality and Acceptance," *Journal of Product Innovation Management,* Volume 4, Number 3 (Sept 1987), pp. 201–216.

Krouse, John, Robert Mills, Beverly Beckert, Laura Carrabine, and Lawrence Berardinis, "Building an Engineering Team," *Industry Week,* Volume 240, Number 14, (July 15, 1991), pp. CC4–CC6.

Melymuka, Kathleen, "Teamwork Tools," *CIO,* Volume 5, Number 3 (Nov 1, 1991), pp. 52–55.

Mille, Robert, "Linking Design and Manufacturing," *CAE,* Volume 10, Number 3 (Mar 1991), pp. 42–48.

Monti, Gary, "Project Organization and Fast-Tracking: Communications Considerations," *Project Management Journal,* Volume 19, Number 2 (Apr 1988), pp. 15–16.

Mower, Judith C. and David Wilemon, "Rewarding Technical Teamwork," *Research-Technology Management,* Volume 19, Number 2 (Apr 1988), pp. 24–29.

Murphy, Ken, "Venture Teams Help Companies Create New Products," *Personnel Journal,* Volume 71, Number 3 (Mar 1992), pp. 60–67.

Perry, Tekla S., "Teamwork Plus Technology Cuts Development Time," *IEEE Spectrum,* Volume 27, Number 10 (Oct 1990), pp. 61–67.

Phillips, Steven L. and Robin L. Ellidge, *Team Building Source Book,* San Diego, CA: Pfeifer & Co., 1989.

Powell, Mary, "Long Term Commitment to Teamwork," *Journal for Quality and Participation,* (July/Aug 1990), pp. 86–89.

Ranftl, Robert M., *R&D Productivity (A Productivity Improvement Study Report),* Los Angeles: Hughes Aircraft Company, 1978.

Rich, Ben R., "The Skunk Works Management Style: It's No Secret," *Vital Speeches,* Volume 55, Number 3 (Nov 15, 1988), pp. 87–93.

Robert, Edward B., "Managing Inventions and Innovations," *Technology Management,* Volume 55, Number 3 (Nov 15, 1988), pp. 87–93.

Sprague, David A. and Randy Greenwell, "Project Management: Are Employees Trained to Work in Project Teams?" *Project Management Journal,* Volume 23, Number 1 (Mar 1992), pp. 22–26.

Stinson, Terry, "Teamwork in Real Engineering," *Machine Design,* Volume 62, Number 6 (Mar 22, 1990), pp. 99–104.

Stokes, Stewart L., Jr., "Building Effective Project Teams," *Journal of Information Systems Management,* Volume 7, Number 3 (Summer 1990), pp. 38–45.

Thamhain, Hans J., "Managing Technologically Innovative Team Efforts Toward New Product Success," Journal of Product Innovation Management, Volume 7 (Mar 1990), pp. 5–18.

Thamhain, Hans J., "Managing Technology: The People Factor," *Technical and Skill Training,* Volume 1, Number 2 (Aug/Sept 1990).

Thamhain, Hans J. and David L. Wilemon, "Leadership, Conflict, and Project Management Effectiveness," *Executive Bookshelf on Generating Technological Innovation Sloan Management Review,* Edward Roberts, ed., (Fall 1987).

Thamhain, Hans J. and D. Wilemon, "Building High Performing Engineering Project Teams," *IEEE Transactions on Engineering Management,* Volume 34, Number 3 (Aug 1987), pp. 130–137.

Thamhain, Hans J., "Managing Engineers Effectively," *IEEE Transactions on Engineering Management,* (Aug 1983).

Todryk, Lawrence, "The Project Manager as Team Builder: Creating an Effective Team," *Project Management Journal,* Volume 21, Number 4, pp. 17–22.

Tolle, Ernest F., " Management Team Building: Yes But!, "Engineering Management International (Netherlands), Volume 4, Number 4 (Jan 1988) pp. 277–285.

Turino, Jon, "Making It Work Calls for Input From Everyone," *IEEE Spectrum,* Volume 28, Number 7 (July 1991), pp. 30–32.

Ward, Kim and T. Hillman Willis, "Managing Change as an Operations System," *Management Decisions (UK),* Volume 28, Number 7 (1990), pp. 17–21.

Wiegner, Kathleen K., "Teamwork," *Forbes,* Volume 148, Number 4 (Aug 19, 1991), p. 106.

Wilemon, David, and Hans J. Thamhain, "Team Building in Project Management, *Project Management Quarterly,* (July 1983).

Wolff, Michael F., "Building Teams—What Works (Sometimes)," *Research-Technology Management,* Volume 32, Number 6 (Nov/Dec 1989), pp. 9–10.

Wolff, Michael F., "Before You Try Team Building," *Research-Technology Management,* Volume 31, Number 1 (Jan/Feb 1988), pp. 6–8.

Wolff, Michael F., "Teams Speed Commercialization of R&D Projects," *Research-Technology Management,* Volume 31, Number 5 (Sept/Oct 1988), pp. 8–10.

Wolff, Michael F., "Building High-Performing R&D Teams: Collaborate and Interact—Profitably," *Research-Technology Management,* Volume 34, Number 5 (Sept/Oct 1991), pp. 11–15.

13
MANAGING CONFLICT AND CHANGE

Conflict is fundamental to complex task management and is often determined by the interplay of the engineering organization and its support functions. Complex organizational relationships, dual accountability, and shared managerial powers are factors that contribute to a new-frontier environment, where conflict is inevitable. Understanding the determinants of conflict is important to an engineering manager's ability to deal with conflict effectively. When conflict becomes dysfunctional, it often results in poor technical and managerial decision-making, lengthy delays over operational issues, low innovation and creativity, and a disruption of the team's efforts, all negative influences on engineering performance. However, contrary to conventional wisdom, conflict can be beneficial when it produces involvement, new information, and competitive spirit. This is clearly a new perspective, which traditional management concepts fail to recognize. This emerging new view on conflict, summarized in Table 13.1, recognizes conflict as a potentially creative force in today's engineering organizations.

Managers must be able to deal with inevitable conflict effectively. Research studies by Thamhain and Wilemon[1] have found that accomplished engineering managers have developed a "sixth sense" to indicate when conflict is desirable, what kind of conflict will be useful, and how much conflict is optimal for a given situation.

[1]H. Thamhain and D. L. Wilemon, "Leadership, Conflict, and Project Management Effectiveness," *Executive Bookshelf on Generating Technological Innovation, Sloan Management Review*, Fall 1987, and "The Effective Management of Conflict in Project-Oriented Work Environments," *Defense Management Journal*, July 1975.

TABLE 13.1 Old and New Assessments of Conflict

Traditional View

1. Conflict should be avoided
2. Conflict is caused by troublemakers and prima donnas
3. Conflict is bad
4. Managers must eliminate conflict

New View

1. Conflict is inevitable
2. Conflict is part of change
3. Conflict is determined by the structure of the system and the interplay of its components
4. Conflict *may be* beneficial; it may:
   - Enhance communications
   - Stimulate innovation
   - Unify the team
   - Provide early warning signs

13.1 HOW TO ANTICIPATE TYPICAL SOURCES OF CONFLICT

Engineering managers frequently admit that they are unprepared to deal effectively with conflict. Yet understanding conflict and its determinants is a prerequisite for effectively dealing with the multitude of work and personality problems that come into play during the execution of modern engineering programs.

Analyzing conflict provides insight into the engineering management environment and its dynamics, thus enabling managers to choose appropriate resolution modes and thereby manage disagreements more effectively. Engineering managers can often prepare for and deal more effectively with operational problems if they can anticipate these problems and understand their specific sources.

Conflict in engineering and project organizations has been investigated by Thamhain and Wilemon,[2] who delineated typical sources of conflict in seven propositions, which were tested against expert opinions. Table 13.2 shows these propositions and the reactions to the propositions, gathered from project managers, regarding their agreement or disagreement with each one. Propositions are listed in decreasing order of acceptance of each statement in Table 13.2. Figure 13.1 shows the relative intensity of conflict perceived by these managers from each of the seven sources.

[2]For details of this research study, see H. J. Thamhain and D. L. Wilemon, "Diagnosing Conflict Determinants in Project Management," *IEEE Transactions on Engineering Management*, February 1975.

TABLE 13.2 Beneficial and Detrimental Consequences of Selected Conflicts: Propositions and Opinions

Propositions and Opinions[a]	Potential Benefits	Potential Detriments
1. The less the specific objectives of a project are understood by project team members, the more likely that conflict will develop. (Agree: 90%; Disagree: 10%)	Open exchange may develop, which clarifies objectives and details of the project	Conflict over project goals and priorities; wasted motion by project team; inability to measure project performance
2. The more members of a functional area perceive that the implementation of project management will adversely affect their traditional organizational roles, the greater the potential for conflict. (Agree: 85%; Disagree: 15%)	May cause functional area to relate more effectively to overall goals of organization	Lack of support; withdrawal; sabotage; delay in project accomplishment
3. The greater the ambiguity of roles among participants of a project team, the more likely that conflict will develop. (Agree: 85%; Disagree: 15%)	May ensure that difficult project issues are responsibly covered; may encourage constructive competition; may foster exchange of ideas in early project phases; project team members assist in own role definition	Lack of project focus; conflict over various "turf issues"; confusion; avoidance of responsibility
4. The greater the agreement on top management goals, the lower the potential for detrimental conflict at project level. (Agree: 80%; Disagree: 20%)	Lowers potential for parochial conflict among departments; goal congruency	Top management goals may not be good for organization or project; discourages constructive dialogue of key project issues
5. The lower the project manager's formal authority over supporting organizational units, the more likely conflict will occur. (Agree: 70%; Disagree: 30%)	Encourages open exchange of ideas; constructive criticisms and analysis of key project issues	Slows down decision-making process; uncertainty and disagreement over priorities, work, force, and resource allocations
6. The lower the project manager's power of reward and punishment, the greater the potential for conflict to develop. (Agree: 65%; Disagree: 35%)	Encourages open exchange between project manager and supportive groups	May delay decision-making process; may not be able to reward key contributors satisfactorily
7. The greater the diversity of expertise among the participants of a project team, the greater the potential for conflict. (Agree: 50%; Disagree: 50%)	Enhances the decision-making process by providing high-quality informational inputs	May slow project decision-making process due to alternative problem-solving approaches suggested

[a]Results are simplified in this table for clarity. Actual expert opinions were measured from 100 project managers on a five-point scale: (1) strongly agree, (2) agree, (3) neutral, (4) disagree, and (5) strongly disagree. For details see H. Thamhain and D. Wilemon, "Conflict Determinants in Project Management," *IEEE Transactions on Engineering Management*, Vol. 22, No. 1 (February 1975), pp. 31–40.

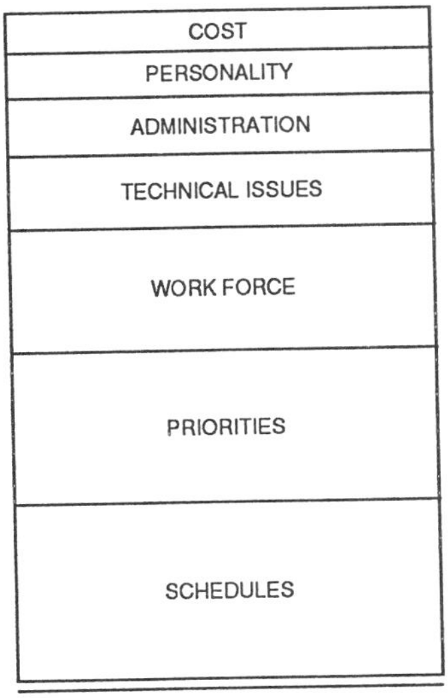

COST
PERSONALITY
ADMINISTRATION
TECHNICAL ISSUES
WORK FORCE
PRIORITIES
SCHEDULES

FIGURE 13.1 Relative intensity of conflict from seven sources, as perceived by engineering managers. Larger size indicates higher intensity.

Schedules. As indicated in Figure 13.1, disagreement over schedules is the most intense source of conflict in the engineering project life cycle. Scheduling conflicts often occur with other support departments, over which the engineering or project manager may have limited authority and control. Scheduling problems also often involve disagreements and differing perceptions of organizational departmental priorities, the second most frequent and most intense conflict source. For example, an issue that is urgent to the engineering manager may receive a low-priority treatment from support groups and/or staff personnel because of a different priority structure in the support organization. Conflicts over schedules frequently result from the cumulative effects of other factors involved in technical performance, namely, technical difficulties, changing scope, and excessive skill requirements.

Conflicts Over Priorities. Conflicts over priorities rank second-highest over the project life cycle. In discussions with engineering managers, many indicated that this type of conflict frequently developed because the organization did not have prior experience with a project or task; consequently, the pattern of project priorities changed from the original forecast, necessitating the reallocation of crucial resources and schedules, a process that was often susceptible to intense disagreements and conflicts. Similarly, priority issues often developed into conflict with other support departments, whose established schedules and work patterns were disturbed by the changed requirements.

Conflict Over Work-Force Resources. These were ranked as the third most important source of conflict. Engineering managers frequently complain that there is little "organizational slack" in terms of work-force resources, a situation in which they often experience intense conflicts. They note that most conflicts over personnel resources occur with those departments who either assign personnel to a project or support a project internally.

Disagreements Over Technical Issues. These were the fourth-strongest source of conflict. Often, engineering departments support a major project; they are primarily responsible for technical input and performance standards, but do not have a broad management overview of the total project. The project manager, on the other hand, is accountable for costs, schedules, and performance objectives. The project manager, for example, may be presented with a technical issue from a support group, involving alternative ways of solving a technical problem. Often he or she must reject the technical alternative because of cost or schedule restraints. In other cases, the project manager may find that he or she disagrees with the opinions of engineering managers on strictly technical grounds. Or, the situation could be reversed; that is, the technical requirements of the project organization may be unacceptable to the engineering manager. In either case conflict may develop.

Conflict Over Administrative Procedures. This ranks fifth in the list of the seven conflict sources. It is interesting to note that conflicts over administrative procedures are distributed almost uniformly throughout functional departments and across all levels. Conflicts may involve disagreements over the engineering manager's authority and responsibilities, reporting relationships, administrative support, and status reviews, especially when the manager's activities involve crossing of organizational boundaries. For the most part, disagreements over administrative procedures involve issues of how the engineering manager will function and how he or she relates to the organization's top management.

Personality Conflicts. Personality conflicts ranked sixth in intensity. Discussions with engineering managers indicate that although the intensity of personality conflicts may not be rated as high as that of some of the other sources of conflicts, they are the most difficult to deal with effectively. Personality issues also may be obscured by communications problems and technical issues. A support person, for example, may stress the technical aspect of a disagreement with a project manager when, in fact, the real issue is a personality conflict.

Conflicts Over Costs. Like schedules, costs are often a basic performance measure in engineering management. Relative to other sources, cost ranks lowest on the list of conflict areas. Disagreements over cost frequently develop when engineering managers, under budget constraints, negotiate with support groups who want to maximize their part of the budget. In addition, conflict may occur as a result of technical problems or schedule slippages, which increase costs.

13.2 CONFLICT IN THE PROJECT LIFE CYCLE

It is important to examine some of the principal determinants of conflict from an aggregate perspective; however, engineering activities are executed in the project life cycle. Specific and useful insights can be gained, therefore, by exploring various conflict sources in each project life-cycle stage: project formation, project build-up, main program phase, and phaseout. Figure 13.2 summarizes these profiles graphically; they are discussed below.

Project Formation. A great deal of conflict from all sources is experienced by project managers during project formation. The project formation phase has some unique characteristics. The project manager must launch the project within the larger host organization. Frequently conflict develops among people during the project start-up over how to define, organize, and prioritize the project, including balancing prior commitments made to other line and staff groups. To eliminate or minimize the detrimental consequences that could result, project managers need to evaluate and plan the support interfaces carefully. This should be accomplished as early as possible in the program life cycle.

Administrative procedures ranked highest in the project-formation stage as a source of conflict. The conflicts center around critically important management issues such as: How will the project organization be designed? To whom will the project manager report? What is the authority of the project manager? Does the project manager have control over work-force and material resources? What reporting and communication channels will be used? Who established schedules and performance specifications? Most of these areas are negotiated by the project manager, and conflict frequently occurs during the process. To avoid prolonged problems over these issues, it is important to establish procedures clearly as early as possible.

Technical opinions, schedules, costs, and manpower issues are other intense conflict areas during start-up. These conflicts are predictable because the members of the newly formed project group see the evolving technical, organizational, and business challenges from their own local functional perspective, often highly differentiated from each other. Further established groups may have to accommodate to the newly formed project organization by adjusting their own operations. Most project managers attest that this adjustment process is highly susceptible to conflict, even under ideal conditions, since it may involve a reorientation of present operating patterns and "local" priorities in the support departments. These same departments may be fully committed to other projects. For similar reasons, negotiations over support personnel and other resources can be an important source of conflict during the project formation stage. Schedule challenges are often an integral part of project definition activities. While by themselves they may not cause a great deal of conflicts, they contribute significantly to the pressures, anxieties, and challenges of technical and organizational issues and contribute considerably to overall conflict during the project definition phase. Thus, effective resource planning and negotiating at the beginning of a project are important.

Project Build-Up. The conflict aggregated from all sources seems to reach the highest level during project build-up phase. Disagreements over technical opinions, priorities, schedules, and administrative procedures continue as important determinants of conflict. Some of these sources of conflict appear as an extension from the previous program phase. Additional conflicts surface during negotiations with other

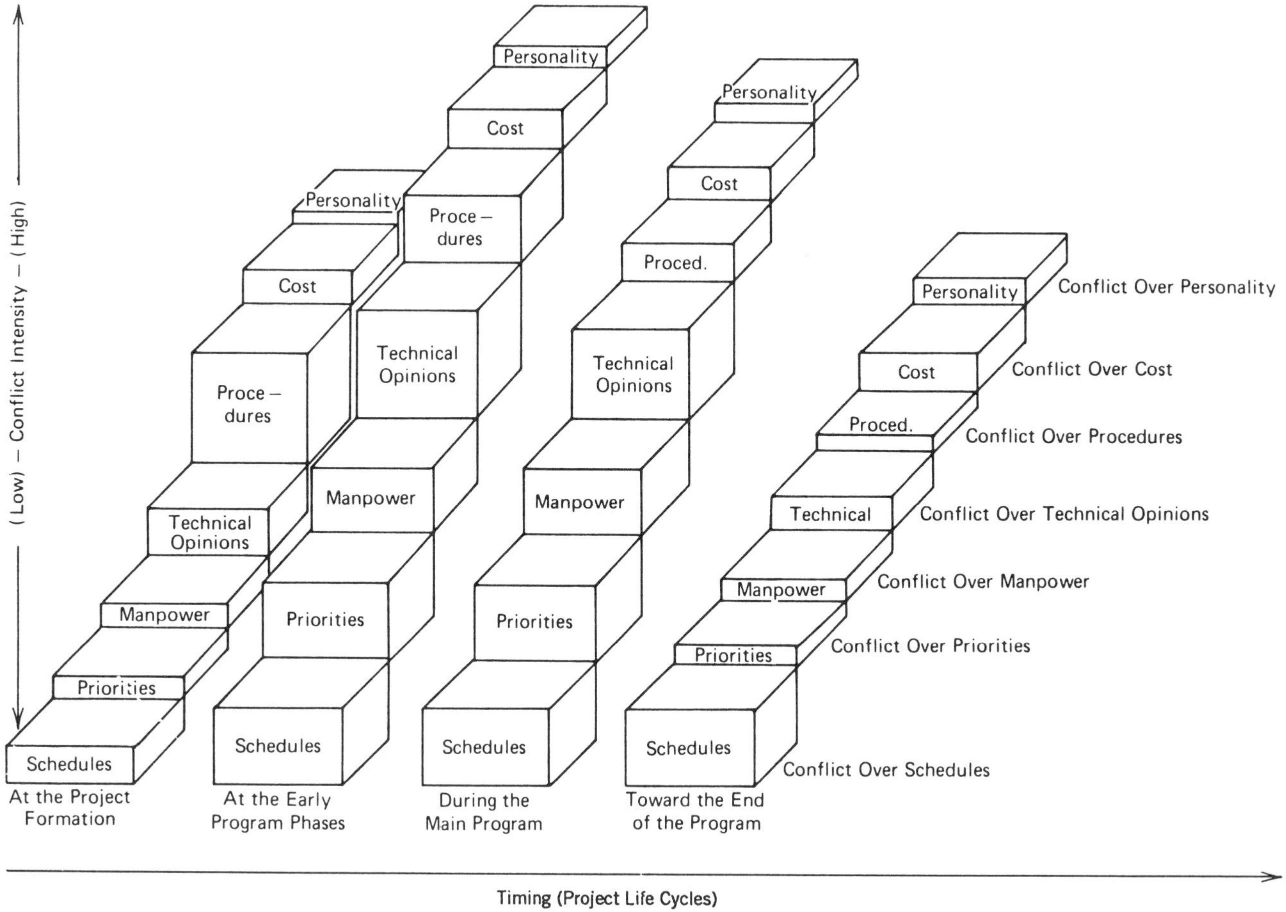

FIGURE 13.2 Conflict throughout the project life cycle.

groups in the build-up phase. During project build-up, schedules are again a great source of conflict. Many of the schedule conflicts arise in the first phase over the establishment of schedules and their effect on other issues. By contrast, in the build-up phase, conflict develops over the enforcement of schedules and necessary schedule changes.

Conflict over technical issues becomes especially pronounced in the build-up phase, doubling in its intensity. Conflicts over technical issues often result from disagreements with support groups who are unable to meet technical requirements, or who want to design for unnecessarily high performance. Such action usually adversely affects the project manager's cost and schedule objectives.

Project managers further emphasize that personality conflicts start flaring up during project build-up. While relatively low in measurable intensity, personality conflicts are more disruptive and detrimental to overall program effectiveness than intense conflicts over nonpersonal issues, which can often be handled on a more rational basis.

Main Program Phase. The main program phase reveals a conflict pattern similar to that of project build-up, with a particularly high conflict over schedules and technical opinions. The explanation for this finding might be that in the main program phase the meeting of schedule commitments by various support groups becomes critical to effective project performance. In complex task management, the interdependence of various support groups dealing with complex technology frequently gives rise to slippages in schedules. When several groups or organizations are involved, this in turn can cause a "whiplash" effect throughout the project. In other words, a slippage in schedule by one group may affect other groups if they are in the critical path of the project.

Technical conflicts are important sources of conflict in the main program phase. There appear to be two principal reasons for the rather high level of conflict. First, the main program phase is often characterized by the integration of various project subsystems. Due to the complexities involved in this integration process, conflicts frequently develop over a lack of subsystem integration and poor technical performance of one subsystem which may, in turn, affect other components. Second, simply because a component can be designed in prototype, does not always ensure that all the technical anomalies have been eliminated. In extreme cases, the subsystem may not even be producible in the main program phase. Such problems can severely impact the program and generate intense conflicts. Disagreements also may arise in the main program phase over reliability and quality control issues, various design problems, and testing procedures. All these problems cause conflicts for the project manager and impact on overall project performance.

Conflict over work-force resources also ranks high in the main program phase. The need for personnel usually reaches the highest levels in the main program phase. Since support groups provide personnel with various projects, severe strains over work-force availability and project requirements can develop.

Phaseout. The final project phase has an interesting shift in the principal causes of conflict, resulting in the lowest level of overall conflict.

Schedules are the most likely form of conflict to develop during the project phaseout. Project managers frequently indicate that many of the schedule slippages that developed in the main program phase tend to carry over to project phaseout and become cumulative.

Personality conflict becomes somewhat more pronounced during project phaseout. This can be explained in two ways. First, it is not uncommon for project participants to be tense and concerned with future assignments. Second, interpersonal relationships may be quite strained during this period because of the pressure on project participants to meet stringent schedules, budgets, and performance specifications and objectives.

Conflict over work-force resources in the phaseout period seem to be related to the personality issue. Disagreements over personnel resources may develop because of new projects phasing in and competing for personnel that are critically needed during the final project phase. Project managers also may experience conflicts over the absorption of personnel back into the functional areas, impacting the budgets and organizational variables.

Conflict over priorities in the phaseout stage appears to be directly related to the competition with other project start-ups. The combined pressure on schedules, the work force, and on personnel creates a climate that is highly vulnerable to conflicts over priorities.

Costs, technical problems, and administrative issues rank lowest as conflict sources during the final phase. Costs, somewhat surprisingly, are not a major determinant of conflict. Discussions with project personnel suggest that while cost control can be troublesome in this phase, intense conflicts usually do not develop, because cost problems develop gradually and provide little ground for arguments. There are various other reasons why conflicts over cost are low. First, some of the project components may have been purchased externally on a fixed-fee basis. In such cases, the contractor bears the burden of the costs. Second, while cost is difficult to control at an established level, project budgets are often adjusted for an increase in material and work-force costs. This incremental cost growth frequently eliminates the "sting" from cost overruns. Moreover, some projects in the high-technology area are managed on a cost-plus basis. Precise cost estimates cannot always be rigidly adhered to in some of these projects. The reader should be cautioned, however, that the low level of conflict is by no means indicative of the importance of cost performance to the overall rating of a project manager. During discussions with top management, they repeatedly emphasized that cost performance is one of the key evaluation measures in judging the performance of project managers.

Technical and administrative procedures rank lowest as a source of conflict during project phaseout. When a project reaches this stage, most of the technical issues are usually resolved. A similar situation holds for administrative procedures.

It is important to note that although a determinant of conflict may be ranked relatively low in its frequency, this does not describe its weight regarding its significance and severity. A single conflict over a technical issue can be as detrimental and jeopardizing to project performance as a series of schedule slippages. This point should be kept in mind in any discussion on project-management conflict. Moreover, problems may develop that appear virtually "conflict-free," such as technological anomalies or problems with suppliers, but they may be just as detrimental to project performance as any of the conflict issues discussed.

13.3 WHAT TO DO ABOUT CONFLICT

An important first step in dealing effectively with the inevitable conflict situations is to recognize the potential causes and intensities of conflicts. This section provides the basis for selecting conflict-resolution style most effective for a given situation. The next section will provide some insight into the complex topic of effective conflict management.

It is equally important to realize the specific sources of conflict throughout the project life cycle. Preventive actions are often simple and represent good management practices. Table 13.3 provides a summary of specific recommendations for minimizing unproductive conflict.

Conflict-Resolution Approaches

When conflict situations arise, effective project managers often attempt to resolve the conflict at the project level rather than resorting to arbitration by higher management. Yet a formal hierarchical system is established in many organizations to arbitrate conflicts that cannot be resolved at the project level. More typically, project personnel have straightforward meetings with other team members to address task-oriented conflicts. Such direct confrontation methods often can dispose of many problems before they become detrimental to the overall project.[3]

Five methods, defined by Blake and Mouton, are frequently used to describe modes for handling conflict.[4] These approaches are listed here in increasing order of interpersonal involvement. Table 13.4 characterizes each approach.

What Method of Conflict-Resolution Works Best?

Project managers are quick to point out the highly situational effectiveness of each conflict-resolution mode. A field study by Thamhain and Wilemon shows that, in spite of their situational nature, certain modes are more important to project managers, while others are less favored.[5] As indicated in Figure 13.3, the problem-solving approach of confrontation and compromise seem to be the two most important to project managers and the most frequently used; use of force and withdrawal seem to

[3]See P. R. Lawrence and J. W. Lorsch, "New Management Job: The Integrator," *Harvard Business Review*, November–December 1967, pp. 142–152.

[4]R. R. Blake and J. S. Mouton, *The Managerial Grid*, Gulf Publishing Company, 1964; see also R. J. Burke, "Methods of Resolving Interpersonal Conflict," *Personnel Administration*, July–August 1969 and R. Likert and J. Likert, *New Ways of Managing Conflict*, New York: McGraw-Hill, 1976.

[5]For details see H. J. Thamhain and D. L. Wilemon, "Conflict Management in Project Life Cycles," *Sloan Management Review*, Spring 1975.

TABLE 13.3 Recommendations for Minimizing Unproductive Conflict, by Project Life-Cycle Phase

Conflict Source and Recommendations
Project Formation Phase
PRIORITIES
Clearly define plans. Joint decision-making and/or consultation with affected parties. Stress importance of project to goals of the organization.
PROCEDURES
Develop detailed administrative operating procedures to be followed in conduct of project. Secure approval from key administrators. Develop statement of understanding or charter.
SCHEDULES
Develop schedule commitments in advance of actual project commencement. Forecast other departmental priorities and possible impact on project.
Build-up Phase
PRIORITIES
Provide effective feedback to support areas on forecasted project plans and needs via status review sessions.
SCHEDULES
Carefully schedule work breakdown packages (project subunits) in cooperation with functional groups.
PROCEDURES
Contingency planning on key administrative issues.
Main Program Phase
SCHEDULES
Continually monitor work in progress. Communicate results to affected parties. Forecast potential problems and consider alternatives. Identify potential trouble spots needing closer surveillance.
TECHNICAL
Early resolution of technical problems. Communication of schedule and budget restraints to technical personnel. Emphasize adequate early technical testing. Facilitate early agreement on final designs.
WORK FORCE
Forecast and communicate work-force requirements early. Establish work-force requirements and priorities with functional and staff groups.
SCHEDULES
Close monitoring of schedules throughout project life cycle. Consider reallocation of available work force to critical project areas prone to schedule slippages. Attain prompt resolution of technical issues which may impact schedules.
PERSONALITY AND WORK FORCE
Develop plans for reallocation of work force upon project completion. Maintain harmonious working relationships with project team and support groups. Try to loosen up high-stress environment.

TABLE 13.4 Five General Modes for Handling Conflict

Conflict-Handling Modes and Their Characteristics
Withdrawing: Retreating from a conflict issue. Here the engineering project manager does not deal with the disagreement. He or she may ignore it entirely, may withdraw out of fear, may feel inadequate to bring about an effective resolution, or may want to avoid "rocking the boat." Withdrawing may intensify the conflict situation. On the other hand, withdrawing can be beneficial either as a temporary strategy to allow the other party to cool off or as a strategy to buy time so that the manager can study the issue further.
Smoothing: Emphasizes common areas of agreement and deemphasizes areas of difference. Like withdrawing, smoothing may not address the real issues in a disagreement. Smoothing can be more effective, however, as it may set a more cooperative stage for seeking solutions. Further, project work can often continue in areas where there is agreement by the parties.
Compromising: Bargaining and searching for solutions which bring some degree of satisfaction to the parties involved in conflict. Since compromise yields less-than-optimum results, the project manager must weigh such actions against program goals. Compromise is always the outcome of a negotiation.
Forcing: Exerting one's viewpoint at the expense of another, characterized by competitiveness and win/lose behavior. Forcing is often used as a last resort by project managers, since it may cause resentment and deterioration of the work climate. However, many organizational or final technical decisions are most effectively made via forcing. Forcing requires that the leader has the proper position power.
Confronting or problem-solving: Involves a rational problem-solving approach. Disputing parties solve differences by focusing on the issues, looking at alternative approaches, and selecting the best alternative. Confronting may contain elements of other modes, such as compromising and smoothing.

be the two methods least favored. Even more important is the finding that confrontation and compromise were found the most effective methods of dealing with conflict in most project situations,[6] while forcing and withdrawal seem to be the least effective methods in the management of multifunctional activities.

The effectiveness of conflict-resolution approaches is highly situational and depends upon the type of conflict to be solved, the personnel and the organization involved, and the power relationship that exists among the parties engaged over a particular problem. It appears that confrontation, also called the problem-solving approach, not only is the most frequently used method, but also seems to result in a higher project performance, as measured by general management personnel. This holds particularly in situations of complex, unstructured decision-making, which do not follow traditional lines of authority or preestablished rationales.

[6]H. J. Thamhain and D. L. Wilemon, "Leadership, Conflict and Program Management Effectiveness," *Sloan Management Review*, Vol. 19, No. 1, Fall 1977, pp. 68–89.

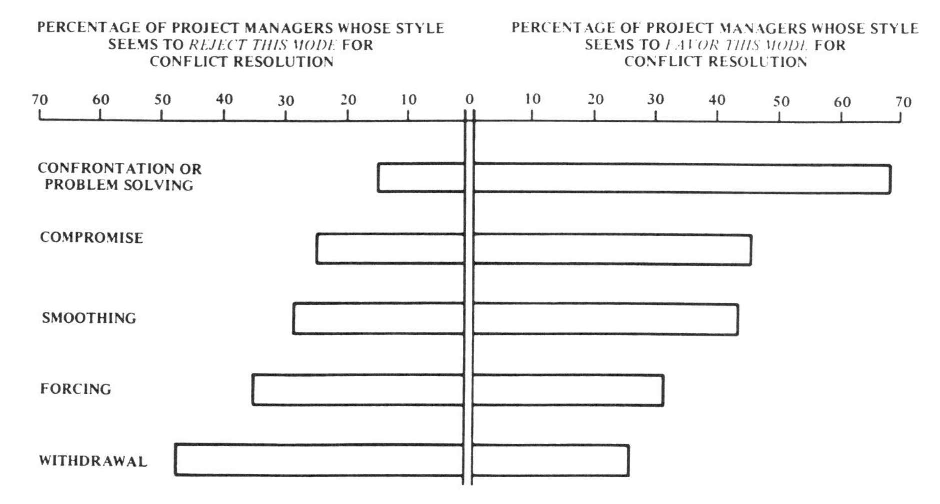

FIGURE 13.3 How frequently managers use various modes of conflict-resolution. Reprinted with permission from Hans J. Thamhain, *Engineering Program Management*, New York: John Wiley & Sons, 1986, p. 252.

In contrast to studies of general management, research in project-oriented environments suggests that it is less important to search for a best mode of effective conflict management. It appears to be more significant that project managers and engineering managers, in their capacity as integrators of diverse organizational resources, employ the full range of conflict-resolution modes. While confrontation was found as the ideal approach under most circumstances, other approaches may be equally effective, depending upon the situational content of the disagreement. Withdrawal, for example, may be used effectively as a temporary measure until new information can be sought, or to "cool off" a hostile reaction from a colleague. As basic long-term strategy, however, withdrawal may actually escalate a disagreement if no resolution is sought eventually.

In other cases, compromise or smoothing might be considered an effective strategy by the project manager if it does not severely affect the overall project objectives. Forcing, on the other hand, often proves to be a win/lose mode. Even though the project manager may "win" on the specific issue, an effective working relationship with the "forced" party may be jeopardized in the future. Nevertheless, some project managers find that forcing is the only viable mode in some situations. Confrontation, or the problem-solving mode, may actually encompass all conflict-handling modes to some extent. For example, in solving a conflict, a project manager may use confrontation in combination with withdrawal, compromise, forcing, and smoothing to eventually get an effective resolution to the issue in question, one in which all affected parties can live with the eventual outcome.

Like any other method of organizational development, conflict resolution must be managed in a disciplined behavioral framework. The integrity of such a procedure becomes even more crucial for larger groups in conflict or for solving conflicts among groups. A typical procedure for solving such multi-group conflict is suggested in Table 13.5

Conflict is fundamental to complete task management. It is not only important for project managers to be cognizant of the potential sources of conflict, but also to know when in the life cycle of a project they are most likely to occur. Such knowledge can help the project manager avoid the detrimental aspects of conflict and maximize its beneficial aspects. Conflict can be beneficial when disagreements result in the development of new information, which can enhance the decision-making process. Finally, when conflict does develop, the manager needs to know the advantages and disadvantages of each resolution method in order to optimize the effectiveness of his or her approach.

Skills for Effective Management of Conflict

The work environment is changing rapidly and dramatically. With increasing technology and changing organizational structures, we expect people to be more accountable, self-directed, and team-oriented. These expectations lead to new performance norms and leadership styles. Especially for dealing with conflict, we have learned that autocratic styles are ineffective. While autocratic or forcing methods of handling conflict can often stop the apparent conflict, they do not solve the underlying problem. Conflict may continue below the surface unchecked; it might even get stronger, less diagnosable and manageable. At the same time, potential benefits are being lost.

In today's environment, effective management of conflict requires proper attitudes and skills. There are as many approaches to conflict management as there are managers. Anyone who has handled a conflict situation has quickly learned some lessons on what to do or not to do in a similar situation the next time around. However, what we found systematically through formal research[7] is that managers who could deal effectively with conflict had certain skills and attitudes in common, which are summarized below:

- Before trying to mediate, managers tried to understand the nature and implications of the conflict, not just the symptoms. The effective manager can diagnose the conflict and determine why it occurred, who is involved, and where it leads.
- The team or its leader could successfully isolate the problem areas and eventually deal with them successfully.
- They built trust with their people and earned credibility to be qualified to help in the search for solutions.
- They were perceived as sincere in their attempt to help and find acceptable solutions to all parties.
- They facilitated communications with all parties involved in the conflict and built coalitions among competing groups. Steering committees, quality circles,

[7]See H. J. Thamhain and D. L. Wilemon, "Leadership, Conflict, and Program Management Effectiveness," *Sloan Management Review*, Vol. 19, No. 1, Fall 1977, pp. 68–69.

TABLE 13.5 A Multiphased Approach to Solving Organizational Conflict Among Groups

Process/Activity
Phase 1: Preliminaries
• The leader prepares for the conflict-resolution meeting by collecting and analyzing relevant information and defining the approach, objectives, and timing
• The parties to the conflict state their willingness to participate
• The effort is legitimized and sanctioned
Phase 2: Perspective
• Each group identifies the issues involved in the conflict, their role in the conflict, how they see the other parties involved in the dispute, and how they think the other parties see them
Phase 3: Image Exchange
• All conflicting parties exchange with other parties their own perception of the conflict:
a. How do you see yourself
b. How do you see the other groups
c. How do you think the others see you
• Exchange information, interpret, clarify
Phase 4: Problem Identification
• All parties identify the existing problems and causes of conflict as they see them
• Present these problems to other parties. Discuss, clarify
• Consolidate problems
Phase 5: Organize for Problem-Solving
• Select most important problems. Form cross-functional problem-solving groups
• Establish process for problem-solving
• Obtain endorsement and commitment from all parties towards resolution
Phase 6: Problem-Solving
• Cross-functional task team works out solutions. Obtain endorsement from group members and management
• Define specific implementation plan with priorities, schedules, and resource requirements
• Obtain commitment from all parties
Phase 7: Implementation
• Execute the solution plan. Fine-tune and stabilize
• Management facilitates and supports implementation and provides visibility and recognition for successful organizational development effort
Phase 8: Continuing Improvement
• A cross-functional team should remain in place to monitor post-transition developments and ensure continuous organizational improvement
• This cross-functional team can also be used to monitor, diagnose, and solve related problems in a Kaizen fashion

design reviews, and technology transfer teams are examples of organizational vehicles for facilitating cross-functional communications and support.

- They could stimulate professional excitement, work challenge, and a clear business direction. When people see the need for their contributions and the potential for professional recognition, they are more likely to cooperate toward the organizational mission and its objectives. Such a professionally stimulating work environment is one of the strongest catalysts toward effective conflict-resolution while preserving the benefits of enhanced communications, creativity, and team unification.
- They minimized power struggles and polarizations of groups along functional, professional, or ethnic lines. They ensured a collegial work environment, where people from all parts of the organization were proud of their contributions and had a strong sense of ownership.
- They could sense personality conflicts and handle them "off line."
- They recognized a deadlocked situation and its dysfunctional consequences and could mediate, intervene or, if necessary, force a conflict resolution.
- They could lead a group toward resolving their internal conflict or a conflict with other groups. Managing such organizational conflict usually required a multiphase approach.
- They helped their people to implement solutions, which were worked out and agreed on.
- Finally, effective conflict managers monitored the organization carefully after a conflict had apparently been resolved to ensure that the organization stabilized completely and absorbed the new solution as permanent.

Managing conflict, now and in the years to come, requires a skillful assessment of the situation, a good understanding of the organization (its people and the interaction of all business components, including the market) and great leadership. As a change agent and facilitator, the manager provides the organizational framework and human interaction to continuously deal with the inevitable conflicts. This requires vision, empathy, wisdom, and leadership, and the courage to experiment beyond the known parameters of conventional methods.

13.4 MANAGING CHANGE

Engineering Management Is Change Management

The ability to manage change will be one of the hallmarks of the organization of the future. In fact, many senior managers see change as the only constant in the organizational equation.[8] Changes in technology such as computers and semiconductor large-

[8]This was already one of the central themes advanced by the well-known futurologist Alvin Toffler in *Future Shock* (New York: Random House, 1970). More contemporary writings in engineering management stress that good management practices must incorporate in their planning both assumptions and uncertainties and must allow for flexibility and change in their plans and management practices. See P. A. Roussel, K. N. Saad, and T. J. Erickson, *Third Generation R&D*, Boston: Harvard Business School Press, 1991.

scale integration (LSI) are constant drivers to business strategy and organizational design changes. Similarly, the changing energy situation, the emergence of new economic powers, Europe 1992, the new developments in Eastern bloc, and social currents in the U.S. all represent strong sources of change that affect managerial and organizational practices. In addition, engineering managers have to cope with yet another layer of changes based on their contemporary work and its turbulent environment. Organizational support may wane and need reenergizing; conflict with key interfaces may develop and need prompt resolution; clients may change the scope of contract work, which poses both business challenges and opportunities; and unforeseen technology developments may require sudden redirection of entire engineering programs. Such developments can severely tax even the most mature engineering organizations and their leaders.

Managing change also is required for building multidisciplinary teams into cohesive groups and successfully dealing with a variety of interfaces. For example, functional departments, staff groups, team members, clients, and senior management all need to be unified behind common goals and project plans. Each interface group may have different expectations. To be effective, engineering managers must be able to introduce changes in strategy, technology, and organization, enabling their firms to anticipate and manage changes in their business environment.

The Process of Change Management

Current thinking on the process of managing change in modern organizations was influenced by the pioneering research of Kurt Lewin.[9] His process of inducing change also applies to changing attitudes and behavior in individuals and small groups. Lewin's theory is based on the notion of countervailing forces. In the management of engineering activities there will be certain driving forces propelling the developments and projects toward success and certain restraining forces that may work against them. In a steady state there is a balance between the driving forces pushing for change (or success) and the restraining forces that resist change. Lewin believed that change is the result of a shifting of counterbalanced forces. In other words, if the driving forces can be increased or if the restraining forces can be minimized, change is likely to occur. These processes may occur separately or together.

The formal study of these forces is known as *forcefield analysis*. It is a simple yet powerful technique that can help engineering managers to identify those forces that "drive" the engineering organization toward success and the barriers or restraining forces that may keep it from attaining its goal.

To understand where these driving and restraining forces originate, we need to be more precise about the meaning and origin of change in engineering management. Change can be introduced to the engineering organization via many subsystems. At least four global categories can be defined:

1. *Changes Introduced by the Organization.* Examples of organization-introduced changes are: reorganizations, expansions, personnel turnover, layoffs,

[9]Kurt Lewin, "Frontiers in Group Dynamics," *Human Relations*, Vol. 1, No. 1, 1947. Also see Lewin's *Field Theory in Social Science*, New York: Harper & Row, 1951.

new policies and procedures, management directives, business plans, changing management support, and changed working conditions.

2. *Changes Introduced by Technical Requirements.* Examples of technically introduced changes are those related to project specifications and requirements, schedules, budgets, teaming, subcontracting, contingencies, technical risks, innovative requirements, and technological challenges.
3. *Changes Originated by the Engineering/Project Team.* The professional interests, abilities, motivation, and team spirit of project personnel may change over the life cycle of an engineering program, affecting team performance in terms of ineffective communications, increased conflict, lower quality decisions, less innovative thinking, and in the end, lower productivity.
4. *Changes Introduced by the External Environment.* Changes introduced external to the project team and its parent organization can originate with the customer community, regulatory agencies, or suppliers, or they may be part of general socioeconomic trends. These trends include technological changes such as those in computer development or factory automation. Demographic changes, the recognition of minority rights, and the interconnection of world markets also are sources of constant changes from the external environment.

Recognizing Resistance to Change

In order to function effectively, engineering managers must lead in situations requiring change-agent skills. Engineering managers must first be able to diagnose situations where change is needed. Second, they must be able to design strategies for accomplishing the desired change. And third, they must be able to successfully implement these changes in their organizations and programs.

The way engineering personnel respond to implementing change seems to depend on the perceived risk–benefit factor. Each engineering/project team member has developed a psychological contract with either the engineering supervisor in the functional organization or the project leader, or with both. If the change is perceived as a violation of this contract, the individual is likely to resist its implementation.[10] However, if the change is consistent with prior agreements, or even is perceived as a potential benefit or opportunity, the individual is often quite cooperative. Other reasons for resisting change are related to the risks involved in having to perform unfamiliar tasks, learn new skills, and deal with new situations. These types of change are often resisted by engineering personnel, because they fear to be asked to act beyond their capabilities. Other reasons for resisting change are related to uncertainty or rumors.

Minimizing Resistance to Change

Engineering managers have to operate in an environment that is dynamic and constantly changing. The ability to successfully deal with change regarding organiza-

[10]For further discussion of how these understandings, shared by a group of people, develop and become part of an operational norm, see L. Schein, *A Manager's Guide to Corporate Culture*, New York: The Conference Board, 1989.

tional, human, and technical factors determines whether a manager is successful or unsuccessful. A field study by Dugan, Thamhain, and Wilemon defines specific driving forces that propel technical project activities toward success because they foster an environment conducive to innovative work, involvement, trust, and a low resistance to change.[11] The principal driving and restraining forces to successful change management are summarized in Table 13.6. Until recently, it was a commonly held view that resistance to change must be minimized or, better yet, eliminated. Research shows, however, that resistance to change is not only natural and predictable, but can also be beneficial.[12] It might bring out new information, reveal some legitimate concerns, validate assumptions, establish communications links, get people involved, and represent a barometer of organizational vitality and morale. Of course, when the resistance to change becomes too intense, it becomes a restraining factor which impairs the manager's ability to implement the change. Such resistance can lead to very dysfunctional conflict, power struggles, and overall operational deficiencies, including collisions and outright sabotage.

Table 13.6 can be used as a checklist for monitoring and managing specific organizational factors. Using the forcefield analysis technique, engineering managers can identify forces for and against change. To move the organization toward the desired state requires increasing the driving forces or decreasing the restraining forces, or, most effectively, doing both. While the main topics of Table 13.6 follow a well-known format established by Kotter and Schlesinger,[13] the specific driving and restraining forces listed as subcategories represent the actual factors that are found specifically in *engineering* organizations.[14]

Rather than trying to isolate a single driver or barrier, engineering managers must consider the whole spectrum of forces when mapping their change strategies. Further, the traditional prescription of participation and involvement addresses only a very limited range of change-management problems. Imagine the futility of trying to gain the support and commitment of a design group for utilizing more large-scale integrated components in favor of detailed design work, thus eventually eliminating 30% of their work force through layoffs. Or requesting a 20% cut in schedule and budget for a new product development project currently in progress. Obviously, participation is not the panacea for all change-management problems. Managers need a much broader array of methods for dealing with changes in today's complex

[11]H. S. Dugan, H. J. Thamhain, and D. L. Wilemon, "Managing Change in Project Management," *Proceedings of the Annual Symposium of the Project Management Institute*, Chicago, October 23–26, 1977. The study uses the concept of forcefield analysis, defining the driving forces that push the project environment toward change and the restraining forces that resist it.

[12]P. E. Conner and L. K. Lake, *Managing Organizational Change*, New York: Praeger, 1988.

[13]John P. Kotter and Leonard S. Schlesinger, "Choosing Strategies for Change," *Harvard Business Review*, March–April 1979.

[14]See research paper by H. S. Dugan, H. J. Thamhain, and D. L. Wilemon, "Managing Change in Project Management," in *Proceedings of the Annual Symposium of the Project Management Institute*, Chicago, October 23–26, 1977. For a more detailed discussion of the original work, see Harold Kerzner and Hans Thamhain, Chapter 25, "How to Manage Change," in *Project Management for Small and Medium Size Businesses*, New York: Van Nostrand Reinhold, 1984.

TABLE 13.6 Driving and Restraining Forces to Successful Change Management in Engineering

Driving Forces Toward Successful Change	Forces Restraining Successful Change
EDUCATION AND COMMUNICATION • Open communication throughout organization • Clear organizational goals • Technical colloquia • Training programs to increase awareness • Gatekeepers • Assistance to problem-solving • Regular project reviews and sponsor involvement	SURPRISE • Poor communication • Rumor • Lack of involvement • Lack of understanding and awareness • Unclear plans and objectives
PARTICIPATION AND INVOLVEMENT • Broad team involvement in project planning and tracking • Individual involvement in commitment • Organizational goal-setting and diagnostics for change such as quality circles • Project visibility and recognition • Good interpersonal relations and team spirit • Professionally stimulating work • Management interest and involvement	FEAR OF FAILURE • Unclear expectations, plans • Excessive expectations • Deficient skills • Deficient resources • Lack of organizational support • Lack of management • Team problems • Risks and contingencies • Boss lacks credibility • Uncontrollable factors • Work interruptions • Technical difficulties and problems • Changing technology • Changing skill requirements
FACILITATION AND SUPPORT • Direction and leadership • Assistance to problem-solving and professional development • Team unification behind project/organizational goals • Defense against outside pressures • Minimum interpersonal conflict • Protection from in-fighting and power struggles • Recognition of achievements • Job security and continuity • Sense of belonging • Mutual trust • Resource provisions • Functional department support • Autonomy and freedom	LACK OF TRUST • Lack of management credibility • Gamesmanship, politics • Power struggle • Organizational conflict • Changing requirements • Poor direction and leadership • Unfair treatment • Apathy
	HABIT AND INERTIA • Comfortable with status quo • Lack of professional challenges • Security concern • Laziness • Activity trap • Rigid plans • Fear of change • No incentive or benefit

(continued)

TABLE 13.6 *(continued)*

Driving Forces Toward Successful Change	Forces Restraining Successful Change
NEGOTIATION AND AGREEMENT	LACK OF SECURITY
• Project planning and team involvement	• Punishment of failure
• Clear project objectives	• Next assignment unclear
• Project importance recognized by team members	• Uncertainties
• Clear interface relationships	• Change leads to reduced status or power
• Clear role and authority definition	• Rumors
• Workable project/task plan	• Management insecurity
• Sufficient expertise	• Poor or improper communications
• Management commitment	SOCIAL AND ETHICAL FACTORS
• Qualified personnel	• Undesirable side effects
• Available resources	• Team pressure, norms
• Desirable work, benefits	• Distasteful results
MANIPULATION AND COOPERATION	• Changing reporting relations
• Co-sponsored activities, consortia	• Morale problems
• Committee actions	• Emotional side effects
• Employee involvement in management function	• Lack of sensitivity and tact
• Cross-functional involvement	• Breakup of work group
• Group spokesperson	• Hostility and conflict
• Incentive	
• Friendship	
COERCION	
• Clear direction and leadership	
• Power	
• Clearly defined need	
• Importance and urgency of tasks	
• Respect and credibility of leaders	

engineering environment. Six categories of *driving forces* listed in Table 13.6 provide some options for our consideration below.

1. Education and Communication. This option is especially effective during the early phases of the change process for keeping people informed about the need for change and its potential challenges and benefits. It advocates prevention rather than cure. If meetings are conducted on an ongoing basis, such as regular project reviews or staff meetings, the need for change becomes apparent more gradually and can often be dealt with in small increments. Proper communication also helps to unify the work groups behind the need for change and toward acceptable solutions.

2. *Participation and Involvement.* Participation, usually an integral part of many other driving forces, helps to get people interested in the diagnostics of problems and the search for solutions. It helps to defuse fear and uncertainty and builds the desire to work toward the change. It also helps to unify the team. Team involvement in project planning, organizing, and review activities are practical examples of stimulating interest in a new venture via participation. Such participation can be further enhanced and transferred into commitment via proper project visibility, recognition of efforts, and active management support.

3. *Facilitation and Support.* This driver has many facets. It is especially effective to overcome fear of failure and lack of trust and security. Managers can assist in problem-solving, provide training and resources, and build a "can-do" image. Managers can also facilitate a vibrant team environment through recognition of achievements, visibility, minimal conflict and power struggle, and some protection from pressures from outside the team.

4. *Negotiation and Agreement.* Often a by-product of participation in activity planning, negotiations over schedules, budgets, and deliverables lead to a commitment to the project objectives or desired changes. Team members become personal stakeholders and are more likely to make commitments toward success. In order to come to an agreement, however, the assignments must be clear, workable, and desirable.

5. *Manipulation and Cooptation.* This is just another strategy for enhancing involvement and communication while building trust and minimizing fears. Many committees, cross-functional task forces, and worker–management interactions represent effective drivers toward effective change management.

6. *Coercion.* Fear can be an effective driver toward change. However, it is usually considered as a last resort. Its potential for undesirable side-effects include noncommitment, reprisals, legal actions, collusion, and sabotage. Yet, when properly directed with position power and clearly defined needs, it may be accepted and even be supported under the notion of effective leadership. Obviously, for many change actions, such as disciplinary measures, firing, or project termination, managers have few or no other options available.

Most important, successful change implementation requires an effective management style and leadership. Whether the need for change involves the work, its process, or its administration, people are more likely to accept and support the change, and even recognize need for change, suggest solutions, and work toward its successful implementation, the more comfortable they feel about the situation. The above drivers, if appropriately employed, have a stabilizing effect on the work environment. When in place, people feel more assured about their work and the viability of their organization. They see the necessity for the requested change as more likely, or may even see it as an opportunity rather than a risk or threat that violates their psychological contract and reaches beyond their capabilities. The barriers have exactly the opposite effect. Both drivers and barriers and their underlying causes, can be seen as change agents which, when optimized, help to manage the successful implementation of desired changes. Taken together, an effective change management process includes four phases: (1) diagnosing the situation; (2) defining the type of change that deals with the change situation most effectively; (3) designing a strategy for achieving the desired change; and (4) implementing and managing the change to achieve the desired results. All four phases of the change process can benefit from the cooperation and positive attitudes of the people in the organization. Understanding and optimizing the driving and restraining forces are absolute prerequisites for effective change management.

Guidelines for Successful Management of Change

A number of suggestions may be beneficial for developing the change-agent skills needed by engineering managers to perform effectively in today's dynamic environment. Many of the driving and restraining forces noted in Table 13.6 can be used to monitor the state of a particular engineering team as it progresses through the phases of a project's life cycle. For example, in the project formation phase, the need for accomplishment, motivational attitudes, clearly defined objectives, and senior management commitment are important ingredients for effectively launching a project. An uncooperative and unmotivated team, poor project leadership, technological limitations, and severe financial constraints may be early warning signals for project failure. Thus, the data presented in Table 13.6 may be used as a comparative barometer of anticipated engineering performance over the life cycle of a given program.

Managers can use a forcefield analysis framework (cf. footnote 9) for auditing engineering activities, developments, and projects on an ongoing basis. An engineering manager or project leader may, for example, periodically assess the overall strengths of the team and the project performance, then define the existing driving forces and their potential for improvement so that the team can focus on the major areas of concern—for example, the barriers. This is very similar to using an Ishikawa diagram[15] for cause-and-effect analysis, a very effective technique for organizational improvement, originally developed in Japan. Increasing a driving force does not always assure an increase in benefits. In some instances, the increase of a driving force may cause the opposite of the reaction intended. For example, a project team may feel it has the support and visibility of senior management. A further increase in this driving force (participation of senior management) may be viewed as unnecessary and unwarranted meddling and interference by senior management, or jealousy might arise with functional support departments since they may feel that too many resources and too much attention are already being devoted to the project.

To minimize the identified restraining forces or barriers, the manager may want to assess their potential for change, that is, to find ways to minimize and neutralize

[15]Although referred to as the "Fishbone Diagram," the Ishikawa diagram can be used to graphically show the cause-and-effect sequence of events and incidents that contribute to a problem. For detailed application, see K. Ishikawa, *What Is Total Quality Control*, Englewood Cliffs, NJ: Prentice-Hall, 1985.

them, or even turn the restraint into a driving force. For example, the negative attitude of a manager in a functional support group may be identified as a restraining force early in the life cycle of an engineering program. Unless the support manager's attitude and behavior toward the program and its host organization are changed, project performance may suffer. Knowing this possibility, the responsible engineering manager should attempt to change the support manager's resistance to project support through rational problem-solving approaches. The conflict issues should be discussed in detail and the importance of the project to the overall organization should be emphasized.

Involving the engineering/project team in a forcefield exercise can be conducive to team-building. Such an approach involves the team members in an audit of, "What's going right, and what do we need in the management of our project?" When forcefield analysis is used as an integral part of team-building, the project leader and the responsible engineering manager should actively seek the team's advice on dealing with the identified issues.

The findings have important implications for senior management. The role of senior management in each life cycle phase should be clearly identified for each engineering program, whether it concerns a new development product, process, or service.The support of senior management is an important driving force necessary for success at the project level, and the lack of it is viewed as a detriment. Senior managers should continually reevaluate their roles and relationships to their functional organizations, and their overlying project activities. They must establish the necessary organizational climate for the engineering activities to flourish as an integrated and often overlying part of the functional and project organizations. Top management must facilitate the necessary power-sharing and resource-sharing, and promote the delicate balance necessary for the engineering organization to function properly within their multilayered structure. This includes maintaining a balance between necessary direction, advice, and feedback without preempting the engineering managers' and project leaders' responsibility. Finally, senior management should be selective in staffing key engineering positions. The deficiencies in engineering manager leadership are an important restraining force. A more careful selection and training of engineering managers may significantly reduce leadership problems, lack of team motivation, and change-management problems.

13.5 RECOMMENDATIONS FOR IMPROVING ENGINEERING MANAGEMENT EFFECTIVENESS

The findings presented in this chapter should help both the professionals who operate in an engineering-oriented environment and the scholars who study and research contemporary organizational concepts to understand the complex interrelationships among managerial influence, conflict-resolution approaches, and engineering management effectiveness.

A number of suggestions can potentially increase the manager's effectiveness in resolving conflict and may ultimately improve overall engineering performance.

1. Understand the Culture and Value System. Project managers need to understand the interaction of organizational and behavioral elements in order to build an environment conducive to their team's motivational needs. This will enhance active participation and minimize dysfunctional conflict. The effective flow of communication is one of the major factors determining the quality of the organizational environment. Since both engineering and project managers must build support teams at various organizational layers, it is important that key decisions be communicated properly to all project-related personnel. By openly communicating the project objectives and the subtasks, unproductive conflict can be minimized. Regularly scheduled status review meetings can be an important vehicle for communicating project-related issues.

2. Have Flexible Leadership Style. Because their environment is temporary and often untested, engineering project managers should seek a leadership style that allows them to adapt to the often conflicting demands of their functional units, project organization, parent organization, or client requirements. They must learn to "test" the expectations of others by observation and experimentation. They must be ready to alter their leadership style as demanded by both the status of the project and its participants, although it is difficult to do.

3. Respond to Situational Variables. Since the ability to manage conflict is affected by many situational variables, a manager should (1) recognize the primary determinants of conflict in his or her environment and when they are most likely to occur in the life of the engineering activity or project, (2) consider the effectiveness of the conflict-handling approach he or she has used in the past to manage these conflicts, and (3) consider experimenting with alternative conflict-handling modes if better performance is warranted.

4. Provide Interesting Work. The manager should try to accommodate the professional interests and desires of supporting personnel when negotiating their tasks. Project effectiveness depends upon work challenge as a motivator. If the manager can match personal goals with objectives of the project and of the overall organization, he or she will lead a more motivated, committed, and result-oriented work force. Although the scope of the work may be fixed, the manager usually has a degree of flexibility in allocating task assignments among various contributors.

5. Have Technical Expertise. Engineering project managers should develop or maintain technical expertise in their fields. Without an understanding of the technology they are managing, they are unable to win the confidence of team members or to build credibility within the customer community or with top management.

6. Do Effective Planning. Effective planning early in the life cycle of an engineering project is another factor that may have a favorable impact on the organizational climate. This is particularly true as project managers have to integrate various disciplines across functional lines. Insufficient planning may eventually lead to interdepartmental conflict and discontinuity in the work flow; it also invites often unnecessary changes and work interruptions.

7. Display Personal Drive. Managers can influence the climate of the work environment by their own actions. Concern for their team members, ability to integrate personal goals and needs of their engineering personnel with project and organizational goals, and ability to create personal enthusiasm for the work itself can foster a climate high in motivation, work involvement, open communication, and resulting engineering project performance.

Probably the most important research finding in the area of change and conflict management is that engineering managers who foster a climate of highly motivated personnel not only obtain higher support from their personnel, but also achieve high overall performance ratings from their superiors. Another factor is *position power*, a perception of personnel based on the manager's formal position within the organization. It includes managerial influence bases such as earned authority, scope and nature of the work, and the ability to influence rewards, salary, promotions. The higher the engineering manager's perceived position power, as influenced by expertise, credibility, trust, and position, the better is the potential for effective management performance.

A situational approach to engineering management effectiveness is presented in Figure 13.4. It summarizes the effects of managerial influence style on two variables: (1) organizational effectiveness, which is measured in short-range results, and (2) willingness to change and adapt to new situations, which is more long-range-oriented. Figure 13.4 indicates that organizational effectiveness is primarily determined by managerial leadership and motivational power, which includes such variables as formal position within the organization, the scope and nature of the work, earned authority via expertise, credibility and trust, and ability to influence promotions and

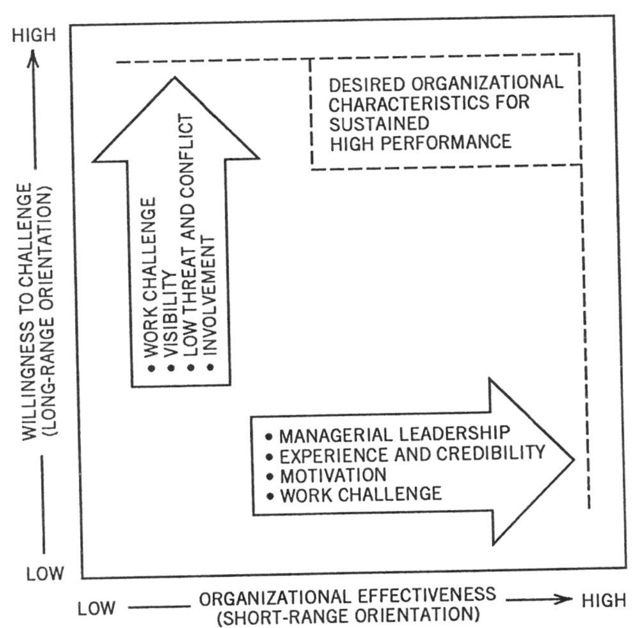

FIGURE 13.4 Key variables for sustaining high performance of a technology/engineering-focused organization.

future work assignments. Figure 13.4 further illustrates that organizational effectiveness also increases as work challenge increases. That is, professionally stimulating and interesting work is a strong intrinsic motivator, which has a positive effect on communications, commitment, personal drive, innovation, and teamwork and ultimately leads to high organizational performance. Further, it is interesting to note that organizational effectiveness is measured on a short-range basis, while the effects of organizational change are long-range. At first glance this appears inconsistent, however, it illustrates the reality of organizational life. That is, while the organization must constantly reposition itself to improve its long-range competitive posture, it must also produce tangible business results on a daily basis. These results are important for sustaining continuity, resources, and management commitment for any engineering activity, organizational change, or improvement.

Finally Figure 13.4 indicates the factors that contribute to long-range organizational performance via the people's willingness to change and to adapt to changing situations. Work challenge appears again as a major driver toward favorable conditions. It seems to promote cooperation and organizational unity. It further lowers conflict and perceived threats associated with change. Other major factors that promote willingness to change include visibility of the activities and the level of involvement with them.

Managing change and adapting to it is one of the most difficult tasks facing managers today. This is especially true in technology-oriented environments. In contrast to the manager who works in a conventional functional environment, engineering managers must live with constant change. In their efforts to integrate various disciplines and projects across functional lines, they must learn to cope with the pressures of the changing work environment.

The nature of engineering management, the need to elicit support from various organizational units and personnel, the frequently ambiguous authority definition of the matrix, and the temporary nature of engineering operations all contribute to the complex operating environment that engineering managers experience in the performance of their roles and the often conflicting pressures for change exist from top management, project leaders, and functional support groups. Demanding compliance to rigid rules, principles, and techniques is often counterproductive. Only by understanding those variables that contribute to more effective role performance can one develop meaningful insight into engineering management effectiveness. This chapter was intended to contribute to the building blocks for a theory of engineering management effectiveness and to add understanding for project leaders, engineering managers, and scholars who study complex organizational systems.

BIBLIOGRAPHY

Barcey, Hyler, *Managing from the Heart,* Pittsburgh, PA: Enterprise Press, 1990.

Barker, Jeffrey, Dean Tjosvold, and Robert I. Andrews, "Conflict Approaches of Effective and Ineffective Project Managers: A Field Study in a Matrix Organization," *Journal of Management Studies* (UK), Volume 25, Number 2 (Mar 1988), pp. 167–178.

Barratt, Alan, "Doing Business in a Different Culture: The Implications for Management Development," *Journal of European Industrial Training* (UK), Volume 13, Number 4 (1989), pp. 28–31.

Bartolome, Fernando, "When You Think the Boss Is Wrong," *Personnel Journal,* Volume 69, Number 8 (Aug 1990), pp. 66–73.

Blake Robert R. and Jane S. Mouton, *The Managerial Grid,* Houston: Gulf Publishing, 1964.

Borys, Walter, Jr., "Are Differing Values Causing Your Management Difficulties?" *Supervision,* Volume 50, Number 4, (April 1989), pp. 9–12.

Burke, R. J., "Methods of Resolving Interpersonal Conflict," *Personnel Administration,* (July/Aug 1969).

Butera, Ann M., "Conflict Management," *Retail Control,* Volume 55, Number 10 (Dec 1987), pp. 37–38.

Butler, Arthur G., Jr., "Project Management: A Study in Organizational Conflict," *Academy of Management Journal,* (Mar 1973).

Chan, Marjorie, "Intergroup Conflict and Conflict Management in the R&D Divisions of Four Aerospace Companies," *IEEE Transactions on Engineering Management,* Volume 36, Number 2 (May 1989), pp. 95–104.

Cleland, David I., "The Deliberate Conflict," *Business Horizons,* (Feb 1968).

Collyer, Margaret E., "Resolving Conflicts: Leadership Style Sets the Strategy," *Nursing Management,* Volume 20, Number 9 (Sept 1989), pp. 77–80.

Conner, P. E. and L. K. Lake, *Managing Organizational Change,* New York: Praeger, 1988.

DuBose, Philip B. and Charles D. Pringle, "Choosing a Conflict Management Technique," *Supervision,* Volume 50, Number 6, (June 1989), pp. 10–12.

Dugan, H. S., H. J. Thamhain, and D. L. Wilemon, "Managing Change in Project Management," *Proceedings of the Annual Symposium of the Project Management Institute,* Chicago, October 23–26.

Ertel, Danny, "How to Design a Conflict Management Procedure That Fits Your Dispute," *Sloan Management Review,* Volume 32, Number 4 (Summer 1991), pp. 29–42.

Evan, W. M., "Superior-Subordinate Conflict in Research Organizations," *Administrative Science Quarterly,* Volume 10, (1965), pp. 52–64.

Ferris, Frank D., "Labor Relations: A Conflict Gap," *Bureaucrat,* Volume 20, Number 1 (Spring 1991), pp. 47–51.

Firth, Jane, "A Proactive Approach to Conflict Resolution," *Supervisory Management,* Volume 36, Number 11 (Nov 1991), pp. 3–4.

Fishman, Robert, "Administration of Diversity," *Administration in Social Work,* Volume 12, Number 2 (1988), pp. 83–94.

Gemmill, Gary R. and Hans J. Thamhain, "Influence Styles of Project Managers: Some Project Performance Correlates," *Academy of Management Journal,* Volume 17 (1964), pp. 216–224.

Horton, Thomas R. and Peter C. Reid, *Beyond the Trust Gap,* Homewood, IL: Irwin, 1990.

Japan Human Relations Association (ed.), *Kaizen Teian-1,* Cambridge, MA: Productivity Press, 1992.

Kelly, J., "Make Conflict Work for You," *Harvard Business Review,* (July/Aug 1970).

Kerzner, Harold and Hans J. Thamhain, Chapter 25, "How to Manage Change," in *Project*

Management for Small and Medium Size Businesses, New York: Van Nostrand Reinhold, 1984.

Kotter, John P. and Leonard S. Schlesinger, "Choosing Strategies for Change," *Harvard Business Review,* (Mar/Apr 1979).

Kozan, Kamil M., "Cultural Influences on Styles of Handling Interpersonal Conflicts: Comparisons Among Jordanian, Turkish, and U.S. Managers," *Human Relations,* Volume 42, Number 9 (Sept 1989), pp. 787–799.

Lewin, Kurt, "Frontiers in Group Dynamics," *Human Relations,* Volume 1, Number 1 (1947).

Likert, R. and J. Likert, *New Ways of Managing Conflict,* New York: McGraw-Hill, 1976.

Link, Patricia B., "How to Cope with Conflict Between the People Who Work for You," *Supervision,* Volume 51, Number 1 (January 1990), pp. 7–9.

Mallak, Larry A., Gerald R. Patzak, and Harold A. Kurstedt, Jr., "Satisfying Stakeholders for Successful Project Management," *Computers and Industrial Engineering,* Volume 21, Number 1–4 (1991), pp. 429–433.

Margerison, Charles, "Introducing Change: Advisors We Consult and the Methods They Use," *Management Decision* (UK), Volume 27, Number 6 (1989), pp. 22–26.

Marsh, Cynthia E. and Val J. Arnold, "Address the Cause—Not the Symptoms—Of Behavior Problems," *Personnel Journal,* Volume 67, Number 2 (1988), pp. 92–98.

Matejke, Kenneth, Diane Dodd-McCue, and Neil D. Ashworth, "Managing the Difficult Boss," *Journal of Managerial Psychology* (UK), Volume 3, Number 1 (1988), pp. 3–7.

McLaurin, Donald L. and Shareen Bell, "Open Communication Lines Before Attempting Total Quality," *Quality Progress,* Volume 24, Number 6 (June 1991), pp. 25–28.

Nelson, Gregory V., "Managing Change: Quality in R&D at Mitchell Technical Center," *Engineering Management Journal,* Volume 2, Number 3 (Sept 1990), pp. 39–44.

Pascae, Richard Tanner, "On Using Conflict Management," *Modern Office Technology,* Volume 36, Number 2 (Feb 1991), pp. 12,14.

Pondy, L. R., "Organizational Conflict: Concepts and Models," *Administrative Science Quarterly,* Volume 12 (1967), pp. 296–320.

Poole, Marshall Scott, Michael Holmes, and Geraldine DeSanctis, "Conflict Management in a Computer-Supported Meeting Environment," *Management Science,* Volume 37, Number 8 (Aug 1991), pp. 926–953.

Rhinesmith, Stephen H., "Going Global from the Inside Out," *Training & Development,* Volume 45, Number 11, (Nov 1991), pp. 42–47.

Scheid-Cook, Teresa L, "Ritual Conformity and Organizational Control: Loose Coupling or Professionalization?" *Journal of Applied Behavioral Science,* Volume 26, Number 2 (1990), pp. 183–199.

Schien, Edward H., *A Manager's Guide to Corporate Culture,* New York: The Conference Board, 1989.

Thamhain, Hans J. and David L. Wilemon, "Diagnosing Conflict Determinants in Project Management," *IEEE Transactions of Engineering Management* (Feb 1975).

Thamhain, Hans J. and David L. Wilemon, "The Effective Management in Project Life Cycles," *Sloan Management Review,* (Spring 1985).

Thamhain, Hans J. and David L. Wilemon, "Leadership, Conflict, and Program Management Effectiveness," *Executive Bookshelf on Generating Technological Innovation, Sloan Management Review,* (Fall 1987).

Thamhain, Hans J. and David L. Wilemon, "Developing Project/Program Manager," *Proceedings of the Annual Symposium of the Project Management Institute,* Toronto, (October 1982).

Toffler, Alvin, *Future Shock,* New York: Random House, 1970.

Roussel, P. A., K. N. Saad, and T. J. Erickson, *Third Generation R&D,* Boston: Harvard Business School Press, 1991.

Van de Vliert, Evert and Boris Kabanoff, "Toward Theory-Based Measures of Conflict Management," *Academy of Management,* Volume 33, Number 1 (Mar 1990), pp. 199–209.

Victor, Bart, "Coordinating Work in Complex Organizations," *Journal of Organizational Behavior* (UK), Volume 11, Number 3 (May 1990), pp. 187–199.

Walton, R. E. and J. M. Dutton, "The Management of Interdepartmental Conflict: A Model and Review," *Administrative Science Quarterly,* Volume 14 (Mar 1969).

Wilemon, David L., "Managing Conflict on Project Teams," *Management Journal,* (Summer 1974).

Zinober, Joan Wagner, "Resolving Conflict in the Firm," *Law Practice Management,* Volume 16, Number 6 (Sept 1990), pp. 20–26.

14 CAREER DEVELOPMENT IN ENGINEERING

14.1 INTRODUCTION

The need for identifying, selecting, and developing engineering personnel is very clear to the majority of senior managers. The issue discussed here is not so much the need but the process and its effectiveness. There are two major special factors that affect technical personnel development that differ from personnel in other professions: (1) the special skill requirements needed by engineering personnel, including their managers, to fulfill their roles effectively and (2) the high mobility of engineering personnel needed for the variety of programs of functional assignments, which often results in a high turnover rate.

Because of the dynamic environment and experiential nature of skill requirements, many senior managers feel that engineering personnel, and specifically the new managers, must be trained and developed *on the job*. However, different organizations take sharply different approaches regarding their methods and practices of training. This chapter provides an insight into human resource planning and development practices, with a focus on engineering-oriented environments. The findings may help senior managers to define the tools and methods for selecting and developing the personnel needed in today's complex engineering organizations.

14.2 CAREER LADDERS IN ENGINEERING

Engineering offers many career avenues. It provides careers for individual contributors as well as opportunities for administrative, technical, and business management positions. Typical positions and their responsibilities were discussed in Chapters 1 and 2. Career ladders in engineering management usually parallel those in other functional areas with many crossovers into other operations, such as manufacturing and marketing. For example, an engineering professional might start his career as a designer, then take a field engineering project assignment, later work as a project manager, and then return to a functional position as product manager. Project assignments often provide excellent opportunities for the individual to gain a better understanding of the organization and its interfaces. The project activities cut across various functional lines, which requires dealing with a broad variety of personnel. Moreover, project personnel receive management recognition and company-wide visibility for their work and accomplishments.

The Incremental Nature of Career Advancement

Career advances in engineering, and especially into engineering management, usually are made incrementally. Assignments for a particular task or project responsibility are made for a fixed time period, namely for the duration of the project. Thereafter, the individual returns to his or her previous position. This gives both management and the individual a chance to evaluate the past assignment regarding job performance, as well as likes and dislikes for the new assignment. This process has a built-in fail-safe mechanism. In contrast to traditional management appointments, the assignment is not permanent, and reassignment to similar or higher-level responsibilities depends on a mutually satisfactory assessment of the individual's performance. Moreover, career growth in project management can be effectively supported and enhanced by on-the-job training. A person interested in an engineering management career can be assigned a job as an assistant to a task manager, administrator or project manager. At the more junior level these "assistant-to" positions might have titles such as analyst, project administrator, or technical assistant. At more advanced career stages, actual management responsibilities are assigned. Further, assignments can vary in size, complexity, and duration to reflect the individual's experience level and career ambitions.

14.3 A PLANNED APPROACH TO ENGINEERING PERSONNEL DEVELOPMENT

Successful personnel development programs usually consider the total job cycle from need-identification to placement. Four specific phases should be considered: (1) identifying staffing needs; (2) defining the work environment; (3) designing a personnel development plan; and (4) implementing the plan. Each phase should be planned, organized, and managed via regular reviews and updates.

Identify Staffing Needs

An important first step is to identify the number and types of engineering personnel needed over the next few years, as both individual contributors and managers. Three years may be a convenient planning horizon. This phase needs the involvement of all senior management personnel. Specific *management tools* to aid this phase are business plans, organizational charts, job descriptions, and manpower plans.

Define the Work Environment

Many executives feel that their engineering organizations evolved in a natural process, as a by-product of getting the required work done. Therefore, it is often argued, the need for formal policies, procedures, and directives is minimal. It should be realized, however, that especially in a less structured environment as is found in engineering, which is often organized along matrix lines, the various management processes must be clearly spelled out and integrated into the organization. These processes become a crucial input to the design of any personnel development program. Three types of processes are used to define the work environment:

1. Role specifications, which define the authority and reporting relations. They include job descriptions, policies, procedures, directives, and organizational charts.
2. The project management support system, which provides the basis for integrated decision-making. It includes planning, bidding, reports and reviews, and controls.
3. The reward system, which establishes accountability and assesses performance as a basis for rewards. It includes the appraisal system, the salary and bonus structure, promotional policies, and career development.

Each of these processes is operational in every organization, at least to some degree. However, it often helps to stabilize the organization and makes management's job easier if an effort is made to formalize and to document the work environment.

Design the Personnel Development Plan

The personnel development plan should include all aspects, from identifying candidates to placement and training. Companies have learned to pay more attention to the team approach to personnel development, in which people become intrinsically interested, involved, and motivated to take on project responsibilities and grow with the project operations. Employees also become involved in the staffing and training process of others. The major facets of the development plan include articulating the following:

- Means of identifying and attracting candidates for various levels of project management positions
- Specific training and development methods
- Appraisal and assessment of training effectiveness; reviews
- Personnel placement and advancement.

Implement the Plan

To be successful, the personnel development plan must have an approved budget and the total commitment of the organization at all levels of management. Such a commitment is more likely to evolve if management was actively involved in the generation of the plan. Further, it is important to assess the effectiveness of personnel development from both sides, that of the management and that of the people in training. If people are not attracted to the development opportunity, it is doomed to fail. Regular interviews should be held with all participants involved with the program. The specific management tools to aid this phase are: the personnel development plan, formal and informal feedback from participants, review meetings, and independent audits.

14.4 TECHNIQUES FOR DEVELOPING NEW ENGINEERING MANAGERS

Companies that believe in developing their managers and help senior engineering personnel in preparing for management assignments have a wide array of methods and techniques to choose from. Table 14.1 provides an overview of these methods, which may be divided into three categories: (1) on-the-job training, (2) schooling (which focuses on the individual), and (3) organizational system developments, which focuses on the supporting infrastructure of the company. However, when actual management practices were analyzed in a field study, it was found that only 30% of the companies investigated had an engineering management development function formally established.[1] Most companies build their management-development efforts around experiential, on-the-job training methods, supplemented by special courses, workshops, and professional activities. On-the-job training involves application-oriented skill-building using hands-on methods, while the employees continue to contribute to the organization, often on a full-time basis, thus minimizing any training cost to the company. Table 14.2 indicates the relative time and effort that was spent on each training method. On the average, managers estimate that 60% of the training time and effort should be spent in on-the-job training. Given the complexities and challenges of managing in engineering, it is not surprising that experiential learning is the primary method of gaining the needed skills and qualifications.

How Learnable Are Engineering Management Skills?

Can we actually develop technical managers by channeling them through some training program? Or do people have to meet certain preconditions for qualifying for management training? Is the ability to lead and manage a personal characteristic that

[1]Details of this field study are discussed by H. Thamhain in "Skill Developments for Project Managers," *Project Management Journal*, Vol. 22, No. 3 (September 1991), and "Managing Engineers Effectively," *IEEE Transactions on Engineering Management*, Vol. 30, No. 4 (November 1983).

TABLE 14.1 Methods and Techniques for Developing Engineering Managers

EXPERIENTIAL ON-THE-JOB TRAINING (FOCUS ON INDIVIDUAL)

- Working with experienced professional leader
- Working with project team member
- Assigned a variety of project management responsibilities, consecutively
- Job rotation
- Formal on-the-job training
- Supporting multifunctional activities
- Customer-liaison activities

CONCEPTUAL TRAINING AND SCHOOLING (FOCUS ON INDIVIDUAL)

- Attending courses, seminars, workshops
- Participating in simulations, games, cases
- Participating in group exercises
- Doing hands-on exercises in using project management techniques
- Attending professional meetings
- Attending conventions, symposia
- Reading books, trade journals, professional magazines

ORGANIZATIONAL DEVELOPMENT (FOCUS ON SUPPORT SYSTEM)

- Formally established and recognized project management function
- Proper project organization
- Project support systems
- Project charter
- Project management directives, policies, and procedures

develops with the individual early in life? Is management too complex and multifaceted to be taught in a simple training program? Organizations have spent much time addressing these questions. Every company has its own reasons and justifications for its management training approaches. However, in recent years the list of premanagement qualifiers seems to have gotten shorter. More and more executives believe that managers are made rather than born. This is further supported by several field studies.[2] Engineering managers sampled across the industrialized world felt that although many skills were very important to effective role performance in engineering management, as indicated in Figure 14.1, these skills were learnable by the methods shown in Table 14.2 and Figure 14.2. Managers indicated that they believed all of the seven skill categories surveyed were learnable, on the average, up to a 95% proficiency level. This included (rank-ordered by learnability) (1) technical

[2]Details of these field studies are discussed in articles by H. Thamhain, "Skill Developments for Project Managers," *Project Management Journal*, Vol. 22, No. 3 (September 1991) and "Developing Technology Management Skills," *Research and Technology Management Journal*, March 1992.

TABLE 14.2 Percent of Time and Effort Spent on Four Training Methods

Experiential learning on the job	60%
Formal education and special courses	20%
Professional activities, seminars	10%
Readings	10%

Source: H. Thamhain "Skill Development for Project Managers," *Project Management Journal*, Vol. 22, No. 3 (September 1991).

expertise, (2) organizational skills, (3) administrative skills, (4) leadership, (5) team-building, (6) interpersonal skills, and (7) conflict-resolution skills.[3]

This is not to say that every person can be trained to become an effective manager. Nor does it say that all people benefit equally from a management-development program. But the personal traits and qualities that are helpful for managerial effectiveness seem to vary considerably among work situations. It is therefore often difficult to define a set of rigid conditions for managerial success or for the selection of prospective candidates. Further, executives are worried that formal selection of

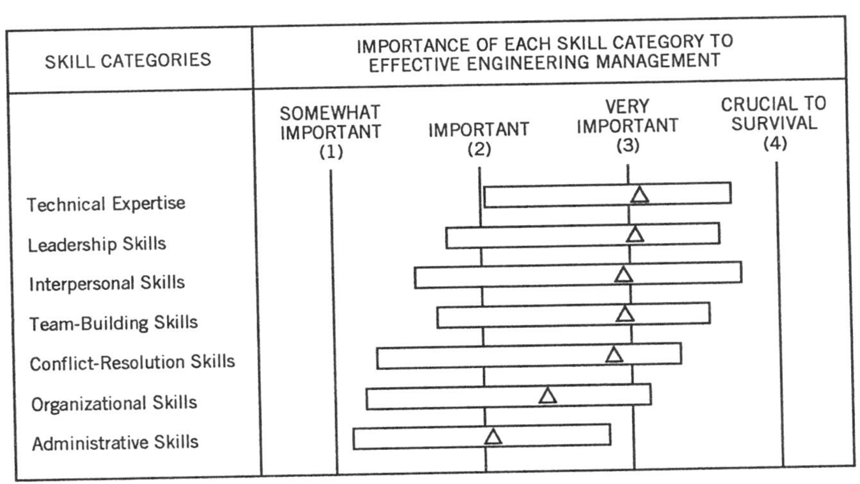

FIGURE 14.1 How important are specific management skills for effective engineering management performance? For graphical display purposes, the length of each bar was calculated as follows: the low endpoint represents the weighted average of all score 1 and score 2 observations; the high endpoint of each bar represents the weighted average of all 3 and 4 observations. 1, 2, 3, and 4 are the score/scale points in the figure as well as the weights. Δ indicates mean value of all managerial perceptions.

[3]Data from H. Thamhain, "Skill Developments for Project Managers," *Project Management Journal*, Vol. 22, No. 3 (September 1991), and "Managing Technology: The People Factor," *Technical & Skill Training Journal*, Vol. 1, No. 2 (August/September 1990).

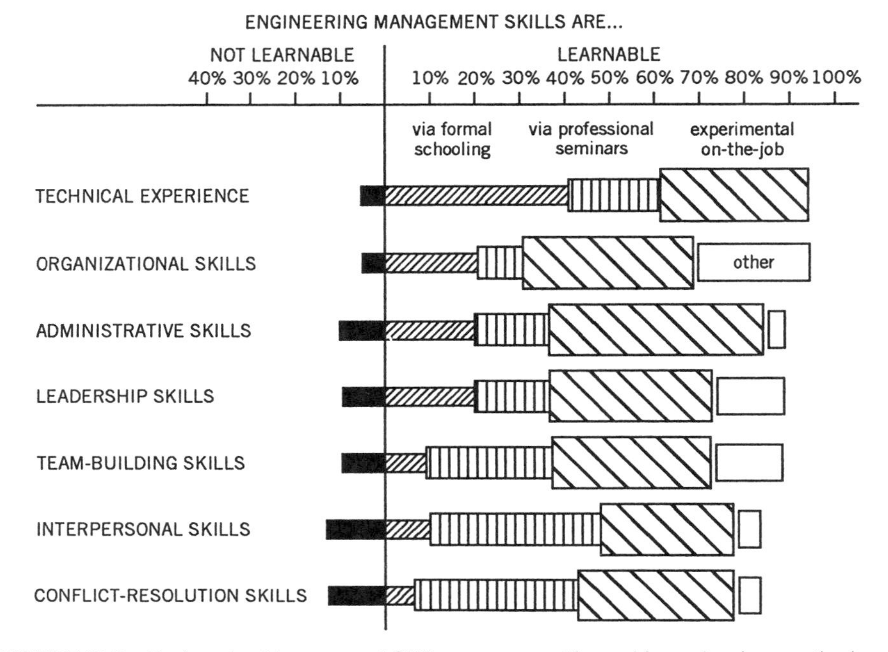

FIGURE 14.2 Engineering Management Skills are or are not learnable, and various methods as to how they may be learned.

management trainees might lead to a promotional expectation, which may not materialize. In that case, the candidates are likely to perceive the training and their own development efforts as a failure, with all the usual undesirable consequences of such perception. Because of all of these complexities, concerns, and risks, executives often choose an incremental approach to management selection and development, which does not single out people for either success or failure but provides a more gradual process; we will discuss it next.

The Making of an Engineering Manager

Most of today's technical managers evolve with their jobs and within their organizations. Certainly, formal training, including professional courses, seminars, and MBA programs, enhance this evolution and accelerate skill development. But few companies select individual contributors without prior team-leadership experience and train them to become engineering managers.

The following process, which is graphically illustrated in Figure 14.3, is typical for the preparation and development of today's new engineering managers. The process includes four basic phases as part of an interactive cycle: observing the potential manager's performance, determining career ambitions, providing training and development, and evaluating feedback. During this interactive process, with its

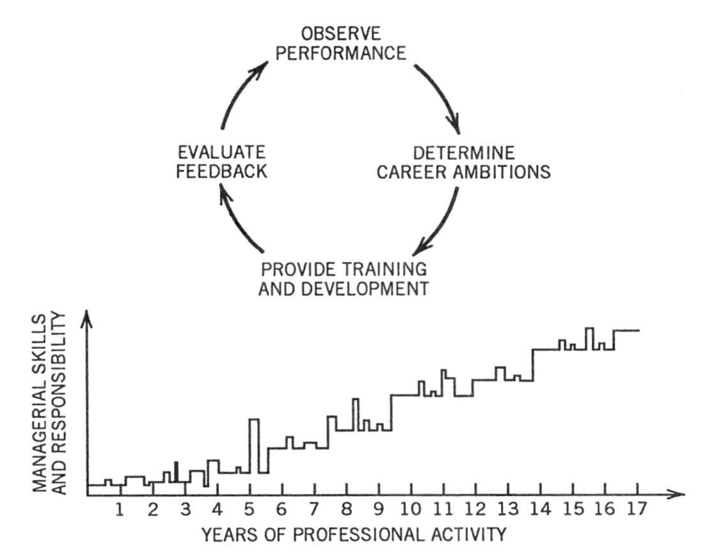

FIGURE 14.3 The iterative process of professional development.

often overlapping phases and cycles, people assume various levels of responsibility and incrementally develop the necessary skills that are needed to manage in a technical environment, as elaborated below:

1. Observing the Potential Managers Professional Performance. This is part of every supervisor's responsibility anyway, aside from any management-development activity. Management should especially try to assess each individual's ability to perform technical tasks as an integrated part of his or her team. Management should judge the person's ability to work with others, to lead task teams, to plan and organize part of a program, to tend to administrative details, and to resolve conflicts.

2. Determining Career Ambitions. Usually as part of the personnel appraisal and guidance process, supervisors should meet at least once a year with their personnel to discuss their professional growth needs and career ambitions. Especially after an employee has had the opportunity to work as a task leader, a more realistic assessment can be made, from both sides, of career ambitions compared to the employee's ability. At that time, it might be possible to make a specific assessment of professional activities that may facilitate career development. The employee's desire for a specific career track, such as engineering management, must come first, but to be realistic, it must be in accord with the individual's potential.

3. Provide Training and Development Opportunities. At the individual contributor level, any training and personal development activity that enhances teamwork, communications, and technical skills should be promoted, regardless of an individual's ambitions toward management. Two basic types of professional development activities are available:

I. Development activities that can be pursued by the employee without management approval, such as readings, professional meetings, involvement with certain assignments, and continued education within policy guidelines.

II. Development activities that need management approval, such as professional seminars and workshops, any training requiring time off from work, special assignments, and job rotations. Employees are often selected by management to take part in such an activity.

In reality, Type I and II activities often overlap. An example might be a supervisor directing an employee to do certain readings or to go to a professional meeting; or an employee might request a particular activity, such as job rotation, and might obtain management approval for it.

It is management's responsibility to assure that proper professional development opportunities are available and to encourage people to keep abreast and advance in the most appropriate way consistent with long-range organizational needs.

4. Evaluate Feedback. Feedback from the various development activities should be studied regarding both the employee's reaction and his or her job performance. In essence, we go back to Step 1 and iterate.

In such an ongoing process of continuous experiential development and training, nobody is singled out for management training, but everyone participates in professional development activities, working selectively and incrementally toward career ambitions, as well as toward filling immediate organizational needs. Invariably this process leads to a steady improvement of organizational productivity, work quality, and the quality of work life.

Developing Senior Engineering Management Personnel

The methods described so far are relevant mainly for developing first-level engineering management personnel. This is a very critical area. Management often sees the identification and development of new management personnel as more important than development of their senior managers. The reason is that new managers, once successful, often move on to new jobs. They leave a void that must be filled. Without a development program, the needed entry-level management personnel are often not readily available, or are poorly qualified, and in the end are often highly disillusioned and frustrated about management.

With respect to developing senior management personnel, the same methods and techniques apply as discussed for the new engineering manager. However, the process is usually less formal than that used for cultivating new managers. It relies more on the individual's desire, interest, and drive to advance than on formally established training and development programs. In recent years, many companies have learned to pay more attention to career-pathing as a way of making their organizations more effective. Career-pathing is a dynamic process of matching existing organizational needs with individual interests and capabilities, while developing people's skills. Job-rotation programs have become especially prevalent in recent years to support such organizational development programs. Depending on organizational needs, individuals may move from a functional management assignment to a project task responsibility or a program office, then to a corporate staff position and back to functional management, perhaps with a different functional charter and higher responsibilities than their previous niche. It is common, for example, to put a functional manager in charge of a new product's development, with responsibilities for the total product cycle through engineering, production, and market development. In this process, the manager gains additional multifunctional skills, while making unique contributions by applying his or her current senior-level knowledge. After the program is completed, the manager often has a choice of returning to his or her former position, enriched by the experience, or taking on a higher level of responsibility. Making career-pathing work usually requires a combination of personal drive and desire as well as mentoring and a certain degree of organizational dynamics, in which new assignments emerge and the need for higher-level responsibilities evolves periodically.

14.5 THE ROLE OF PERFORMANCE APPRAISALS AND FINANCIAL REWARDS

Why do we need yet another administrative system? Research has suggested that goal-setting improves performance at both the managerial and nonmanagerial level.[4] It also indicates that to be realistically evaluated, improved performance can only be measured against preestablished goals. The level of performance has been shown to improve with the level of goal acceptance.[5] Yet another aspect is the positive relationship among perceived task performance, rewards, and long-range motivation and drive.

Performance Appraisal

Engineering executives tend to agree that the performance-appraisal system ties together the management functions of goal-setting, performance-monitoring and control, rewards, and career development. Senior managers do not treat the appraisal system in isolation, but make it part of an integrated management system. A typical example is Texas Instrument's OST System (*O*bjectives, *S*trategies, and *T*actics), which includes the personnel function as part of the total management approach, consistent with the overall company operating needs and management philosophy.

Let's take a look at how the performance appraisal system is supposed to work in engineering. Traditionally, the purposes of the performance appraisal are:

[4] H. John Bernadin and Richard W. Beatty, *Performance Appraisal: Assessing Human Behavior at Work*, Boston: Kent, 1984.

[5] A study of 140 engineers and other technical personnel shows that the level of task performance is directly related to the level of goal acceptance (M. Erez, P. C. Early, and C. I. Hulin, "The Impact of Participation on Goal Acceptance and Performance," *Academy of Management Journal*, Vol. 28, No. 2 (May 1985, pp. 50–66).

- To assess the employee's work performance, preferably against preestablished objectives
- To provide a justification for salary actions
- To establish new goals and objectives for the next review period
- To identify and deal with work-related problems
- To serve as a basis for career discussions.

The realities are, however, that the first two objectives are in conflict with the other three. As a result, the traditional performance appraisal essentially becomes a salary discussion, the objective of which is to justify subsequent managerial actions.[6] Furthermore, a discussion dominated by salary actions usually is not conducive to future goal-setting, problem-solving, or career-planning.

Managers recognize the complex nature of engineering jobs, in which specific goal-setting along the traditional management by objective (MBO) process lines may be impractical. In fact, an inappropriate goal-setting process, which attempts to be too specific, narrow, and inflexible, may lead to bureaucratic behavior in which goal-setting becomes a goal in itself. People focus on the process, they confuse activity with achievement, and they may achieve the original goal but the output may be useless because of the dynamics in the environment. To avoid this classical case of an "activity trap," managers try to define goals together with their subordinates in a dynamically qualified manner, where the objectives depend on certain environmental assumptions and the objectives themselves must be reassessed and changed if the evolving engineering project or organizational environment necessitates such changes. Using task complexities as an excuse for not defining goals is just as bad, however; therefore managers and their reporting personnel must be innovative in defining goals and assessing performance.

The first challenge of assessing work performance is on content, that is, to decide what to review and how to measure performance. Modern management practices try to individualize accountability as much as possible. Furthermore, subsequent incentive or merit increases are tied to business performance. Although most companies apply these principles to their engineering organizations, they do it with a great deal of skepticism. Practices are often modified to assure balance and equity for jointly performed responsibilities. A similar dilemma exists in the area of profit accountability. The comment of an engineering manager at the General Electric Company is typical of the situation faced by business managers: "Although I am responsible for business results of a large program, I really can't control more than 20 percent of its cost." Acknowledging the realities, organizations are measuring performance of their engineering managers in at least two areas:

[6]For detailed discussions, see Robert I. Lazer and Walter S. Wikstrom, "Appraising Managerial Performance: Current Practices and Future Directions," *Report 723*, New York: The Conference Board, 1977. Excellent discussions of the challenges of performance appraisal practices and suggestions for improvements are found in Ed Yager, "A Critique of Performance Appraisal Systems," *Personnel Journal*, Vol 60, February 1981, pp. 129–133; and H. Kent Baker and Philip I. Morgan, "Two Goals in Every Performance Appraisal," *Personnel Journal*, Vol. 63, September 1984, pp. 74–78.

1. Contribution to business results, as measured by profits, margins, return on investment, new business, and revenues; also, on-time delivery, meeting of contractual requirements, and within-budget performance
2. Managerial performance, as measured by overall engineering effectiveness, organization, direction and leadership, and team performance.

The first area applies only if the engineering manager is indeed responsible for business results, such as contractual performance or new-business acquisitions. Many engineering managers work with company-internal sponsors, as in the case of new product development. In these situations, producing the results within the agreed-on scheduling and budget constraints becomes the primary measure of performance.

The second area of managerial performance (#2 above) is more directly related to the individual manager's performance and control. It is also used as the basis for salary actions. Nevertheless, because of its normative nature, managerial performance is difficult to quantify. Therefore, it is not surprising that performance appraisals, at any level, are associated with considerable tension and some mistrust and conflict. Moreover, if handled improperly, performance appraisals will lead to manipulation and game-playing. It also will make it difficult, if not impossible, to set dynamic objectives in a changing environment. Table 14.3 provides some guidelines for defining and evaluating specific engineering management performance.

For nonmanagerial engineering professionals, performance appraisals should focus primarily on the ability to direct implementation of a specific engineering program:

- Technical implementation measured against requirements, quality, schedules, and cost targets
- Team performance as measured by ability to staff, build an effective task group, interface with other groups, and integrate among various functions.

Specific performance measures for nonmanagerial personnel are shown in Table 14.4. In addition, the performances of engineering managers and their resource personnel should be assessed taking into consideration the conditions under which they were achieved: The degree of task difficulty, complexity, size, changes, and general business conditions.

Finally, one needs to decide who is to conduct the performance appraisal and to make the salary adjustment. Where dual accountabilities are involved, good practices call for inputs from both bosses. Such a situation could exist for project managers who report functionally to one superior but are also accountable for specific business results to another person. While dual accountability of project managers is an exception for most organizations, it is common for resource personnel who are responsible to their functional superior for the quality of the work and to their project manager for meeting the requirements within budget and schedule. Moreover, resource personnel may be shared among many projects. Only the functional or resource manager can judge overall performance of resource personnel, but should

TABLE 14.3 Performance Measures for Engineering Managers

Who Performs Appraisal: Functional superior of engineering manager
Source of Performance Data: Functional superior, resource managers, project leaders, general managers

PRIMARY MEASURES

1. Engineering manager's success in leading the engineering organization toward preestablished global objectives
   - Target costs
   - Key milestones
   - Profit, net income, return on investment, contribution margin
   - Quality
   - Technical accomplishments
   - Market measures, new business, follow-on contract
2. Project manager's effectiveness in overall project direction and leadership during all phases, including establishing
   - Objectives and customer requirements
   - Budgets and schedules
   - Policies
   - Performance measures and controls
   - Reporting and review system

SECONDARY MEASURES

1. Ability to utilize organizational resources
   - Overhead cost reduction
   - Working with existing personnel
   - Cost-effective make–buy decisions
2. Ability to build effective project team
   - Project staffing
   - Interfunctional communications
   - Low team conflict, complaints, and hassles
   - Professionally satisfied team members
   - Work with support groups
3. Effective project planning and plan implementation
   - Plan detail and measurability
   - Commitment by key personnel and management
   - Management involvement
   - Contingency provisions
   - Reports and reviews
4. Customer/client satisfaction
   - Perception of overall project performance by sponsor
   - Communications, liaison
   - Responsiveness to changes
5. Participation in business management
   - Keeping management informed of new project/product/business opportunities
   - Bid proposal work
   - Business planning, policy development

ADDITIONAL CONSIDERATIONS

1. Difficulty of tasks involved
   - Technical tasks
   - Administrative and organizational complexity
   - Multidisciplinary nature
   - Staffing and start-up
2. Scope of the project
   - Total project budget
   - Number of personnel involved
   - Number of organizations and subcontractors involved
3. Changing work environment
   - Nature and degree of customer changes and redirections
   - Contingencies

solicit feedback from the project manager who directs the integration of the efforts and their personnel.

Merit Increases and Bonuses

Professionals have to come to expect merit increases as a reward for a job well done. However, under inflationary conditions, which we have experienced for many years, pay adjustments often lag behind cost-of-living increases. To deal with this salary compression and to give incentive for management performance, companies have introduced bonuses uniformly to all components of their organizations. The problem is that these standard plans for merit increases and bonuses are based on individual accountability, but engineering personnel often work in teams with shared accountabilities, responsibilities, and controls. It is usually very difficult to credit success or failure of a particular engineering program or mission to a single individual or a small group.

Most managers that found themselves with these quandaries turned to the traditional reward system of performance appraisals and merit increases. If done well, the appraisal should provide particular measures of job performance that assesses the level and magnitude at which the individual contributed to the success of the organization, including the managerial performance and team performance components. Therefore, a properly designed and executed performance appraisal, which includes the inputs from all matrix areas and the basic agreement of the employee to the conclusions, is a sound basis for future salary reviews. Often more important than the actual increase is the salary adjustment relative to other employees. Paying equitably for performance and position is crucial to employee morale and satisfactory productivity, a very important area that deserves careful management attention.

14.6 RECOMMENDATIONS FOR DEVELOPING ENGINEERING MANAGERS

Making the engineering management development system work requires more than just another plan. It requires the total commitment of management at all levels. To be successful, companies must not only consider the training of their personnel but also devise systems for identifying prospective candidates, develop their management support systems, assure equitable and attractive rewards consistent with the challenges and responsibilities involved, and have the role of engineering management personnel delineated through charters, job descriptions, policies, procedures, and directives. Most importantly, management must foster a work environment that is professionally stimulating and conducive to teamwork. In such an environment the engineering management function will be self-developing to a large extent. That is, personnel from all levels and disciplines will see the professional and personal incentives of participating in and seeking out engineering challenges. There will be many volunteers competing for job openings in management.

A number of suggestions are advanced below that can potentially increase the effectiveness of management training and development and ultimately enhance operational efficiency of engineering:

1. Carefully consider and analyze the specific needs of your engineering function, including the development of managers, project personnel, support systems, and organizations.

2. Define your work environment. Charters, job descriptions, and policies are helpful in all organizational endeavors, but they are absolutely crucial in defining and communicating the work environment, a prerequisite for training and developing management personnel.

3. Understand your engineering organization, its interfaces, cultures, and value system. This will provide the basis for building a professionally stimulating work environment, the prerequisite for a self-developing management system.

4. Plan your management needs and developments in sufficient detail to make it operational. The plan should include key results, measurable milestones, budgets, and responsible individuals for each major activity.

5. Involve your management and human resource people in this organizational development, starting at the initial planning. This will develop the needed support, endorsement, and commitment.

6. Use the experiential, on-the-job training method as the cornerstone of your personnel development system.

7. Assure systematic tracking and follow-through of your development plan. Many management development programs do not achieve their anticipated results because of poor tracking and control. The initial enthusiasm and interest may evaporate quickly if progress is slow. The development program should be tracked and controlled like any other project, including status reviews and progress reports.

8. Management should tie the specific results of the development program into the performance reviews of the responsible individuals. Rewards can come in many forms, ranging from recognition for accomplishments, to incentives for higher operating efficiencies, to monetary rewards.

9. Take a multidisciplinary approach to organizational development. Usually the problems contain a mixture of human, technical, management, and organizational facets.

10. Avoid the label of a staff function for your organization development. Involve all line and senior management.

11. Be aware of resistance to change. Your development efforts may not be in tune with deeply rooted cultures and values of the organization. Your success will depend on cooperation, participation, trust, and team spirit. Early involvement and participation of key personnel throughout the organization may help to ease this resistance to change.

12. Take a long-range approach. Organizational developments are long-range rather than short-term fixes.

13. Assure that incentives and rewards for personnel are consistent with responsibilities and challenges, to induce people to move in and up the organization.

14. Do not panic if results are initially slow in coming. Complex multidisciplinary developments are time-consuming. Moreover, the results are perceived on the basis of potential benefits. If management and personnel in training see benefits in terms of their own career advancement or operating efficiencies or recognition by their peers and bosses, then the development program will be labeled successful. Continuous involvement of personnel at all levels, review meetings, short reports, and proper follow-through of action items will help to build the image of a credible and useful organization development program.

15. Assure that tangible results are produced early in the program. These results could include actual staffing of engineering or project management positions, promotions, provision of an additional service/support function, or reduction of some backlog or schedule.

16. Conduct audits of your development activity regularly, for instance semiannually. This will expose problems at their early stages where you can deal with them more readily. It will also help to build confidence and credibility for the program and your efforts.

14.7 HOW TO ADVANCE YOUR ENGINEERING MANAGEMENT CAREER

Technical management provides an environment for fast-track career growth. Engineering organizations are dynamic regarding size, charter, and staffing levels. Workloads and manpower requirements fluctuate. No one has tenure in a project-oriented engineering environment. People are recognized for their skills relevant to the immediate organizational needs and their real-time contributions. Responsibility and authority seem to gravitate toward those who are best suited to handle a particular job rather than being vested in a title or position. This is a work environment that recognizes job performance. High achievers usually receive additional responsibility, recognition and financial rewards. They usually also have high upward mobility.

People who are considering moving into a career in engineering management or advancing in one should first find out what specific project or functional leadership opportunities exist in their company, then assess the requirements, challenges, and benefits against their individual needs, wants, and career goals. Once an individual concludes that he or she wants to pursue this career path, a specific development plan should be worked out by the individual and discussed with the superior. Although this discussion may not lead to an immediate change in responsibility and job content, it communicates the employee's career objectives to the boss and provides an opportunity for discussing the realism of the employee's ambitions. The boss may also offer helpful advice to the employee regarding specific experiences and skill developments needed. Even if the superior does not support the employee's career plan, the discussion may prepare him or her for seeking out opportunities with other organizations. Advancing a career requires a great deal of strategic vision plus skills in communication, information-gathering and -processing as a basis for organizing the array of career objectives and opportunities into a cohesive career development plan. Regardless of the level of career objective sought and the person's point in the career path, some specific suggestions may help in preparing for the next career move and in attaining continuous professional growth:

1. Assess your own career needs in terms of job content responsibilities, recognition, and financial needs.
2. Assess the realities of your work environment by talking to people who already hold the positions you aspire to. This will give you an insight into the type of work available, its challenges, and its skill requirements.
3. Develop the skills needed in the new job. This can be done in part through courses, seminars, readings, and professional meetings, and experientially through your current position.
4. Use your current position as a springboard. Apply the new knowledge you have gained in your current job. Seek out additional responsibilities that may prepare you for your next career move.
5. Seek out specific opportunities for advancement, such as participation in a bid proposal effort or the formation of a new program.
6. Make your management skills visible. Project activities, management reviews and reports are good vehicles for showing to your management and your peers that you have the needed action-orientation and possess the needed skills for planning and control. You can demonstrate that you clearly understand the management tools, such as scheduling and budgeting, and that you can apply them at an appropriate level.

7. Give credit for accomplishment to other individuals and organizations who participated in and contributed to your efforts. This builds a network of supporters.
8. Grow with your present job. Find out about the broader business goals and objectives of your department. Seek out additional responsibilities that would support your career ambitions.
9. Delegate. Utilize organizational resources and support as much as possible. You cannot take an additional responsibility unless you delegate the work to others. This type of delegation does not require that support personnel be actually reporting to you, but simply means that you find out what technical and administrative support is available, company-internally or externally, and that you involve these support groups, plan for their participation, agree on the required resources, and subcontract with them.
10. Keep abreast and stay current in your professional field. There is a great variety of activities and resources to choose from. Some involve major commitment while others can be very pleasantly integrated into the daily routines of a busy engineering manager. Some of the resources to consider include:

    - Trade shows and fairs
    - Seminars and short courses
    - Audio cassettes and videotapes
    - Trade magazines and journals
    - Professional books and journals
    - Professional meetings
    - New assignments and job rotation
    - Team interactions and observations
    - Management meetings and briefings
    - Ad-hoc assignments
    - Committees and task forces
    - Formal courses and training programs.

APPENDIX: TEST YOUR APTITUDE FOR ENGINEERING MANAGEMENT*

High-technology businesses have long recognized the importance of developing new engineering talent. But despite enormous advances in evaluating, educating, and training technical professionals for managerial responsibilities, new engineering managers often find themselves ill-prepared for their new assignments.

*This test was developed by Hans Thamhain. It originally appeared in *Training & Development*, September 1991 (pp. 66–70) and is reprinted by permission.

Even if they wanted managerial duties in the first place, many first-time supervisors are surprised at the scope and content of their new jobs. They may feel unsure about their career development.

In a survey I conducted recently (and reported on in "Managing Technology: The People Factor," in *Technical & Skills Training,* August/September 1990), 85% of engineering managers considered the development of new engineering management talent crucial to the survival and growth of their businesses.

Many studies have defined the type and extent of skills and training an effective manager needs. But most have focused on the continuing development of managers, not on the development of professionals in technology-oriented environments who have moved from being individual contributors to managing.

In addition, the leaders I surveyed considered engineering management more complex and multifunctional than other types of management.

Engineering managers must operate in multidisciplinary environments within their companies. They work with many support groups over whom they have little or no formal authority. Outside their companies, they have to cope with constant changes in technology, markets, regulations, and socioeconomic factors. To achieve productivity, they have to motivate and lead their work forces toward innovative results in work environments that are often unstructured.

Yet the formal education system that creates engineers does little to prepare them for advancement into management.

There is strong interest among managers and engineers in tools to help assess technical-management potential. Responding to that interest, I organized a study to investigate engineering-management aptitudes and define a simple aptitude test.

My findings and the resulting instruments can help engineering professionals and managers to determine engineers' potential for advancement and for effective performance as managers. The instruments can help in the management selection, transition, and development processes as well.

Research Methodology

The data for this study, collected during a series of management seminars and in-house consulting assignments, come from some 800 engineering professionals and their superiors. The sample contains personnel from 55 technology-oriented companies. They manufacture such products as computers, electronic equipment, integrated circuits, photographic equipment, tooling machinery, and bioengineering and pharmaceutical products.

The investigation involved the following four steps:

Step 1: Pilot Study Before the principal investigation, I used questionnaires and personal interviews to obtain data from a sample of 450 research, development, and engineering managers. The data identified specific personal characteristics that seemed to be critical for a successful transition into engineering management and for subsequent effective performance.

Step 2: Detailed Research Design From the data, I defined five global aptitude categories, based on the frequency of the personal characteristics that were mentioned:

- Personal desire
- People skills
- Technical knowledge
- Administrative skills
- Business acumen.

Secondary aptitudes were grouped under each category, resulting in the aptitude test instruments shown in Table A.1. Each section of the questionnaire measures the aptitude for a category by asking the respondent to rate 10 statements on a modified, 10-point Likert scale, according to his or her level of agreement or disagreement. The goal was to have a quantified measure of engineering management aptitude.

Step 3: Data Collection. A large group—210 managers and 640 of their subordinates—responded to the questionnaires. The subordinates included 155 people recently promoted from engineers to managers. The sample also included 90 senior managers, such as directors of R&D and vice-presidents of engineering.

Individual contributors and the recently promoted managers assessed their own managerial aptitudes. Supervisors rated their subordinates.

Step 4: Data Evaluation. Table A.2 shows a summary of the aptitude measures. Measured also is the association between the individuals' self-assessments and their superiors' assessments, using Kendall's tau rank-order correlation. Kendall's tau was also used to measure the association between perceived engineering management aptitude and actual performance.

Preparing for Engineering Management

"Engineering managers can be developed." This is the strong message from 300 managers interviewed in technology-oriented companies.

Exhibits 14.A1 and 14.A2 briefly describe the ways to prepare for and make a successful transition. They show that a person's preparation, his or her supervisor's assistance, and organizational support have a significant impact on his or her ability to become a candidate for a management position and ultimately to succeed in it.

One of the key criteria for success, stressed by newly promoted managers and management veterans alike, is a person's desire to become a manager. This desire seems to have a positive effect on many of the components needed to prepare for a career in engineering management.

This association was verified statistically using Kendall's correlation: the association between personal desire to become a manager and actual promotion was tau =

TABLE 14.A.1 Sample Questions from the Aptitude Tests

Use a 10-point scale to indicate your agreement with each of the statements (1 = strong disagreement; 10 = strong agreement).

PEOPLE SKILLS

1. I feel at ease communicating with people from other technical and administrative departments
2. I can effectively solve conflict over technical and personal issues, and don't get involved
3. I can work with all levels of the organization
4. I am a good liaison person to other departments and outside organizations
5. I enjoy socializing with people
6. I can persuade people to do things that they normally don't want to do
7. I can get commitment from people, even if they don't report directly to me
8. People enjoy working with me and follow my suggestions
9. I am frequently asked by my colleagues for my opinion and to present ideas to upper management
10. I think the majority of my department would select me as their team leader (vote for me)

TECHNICAL KNOWLEDGE

1. I understand the technological trends in my area of responsibility as well as those for the business environment of my company
2. I understand the product applications, markets, and economic conditions for my business area
3. I can effectively communicate with my technical colleagues from other disciplines
4. I can unify a technical team toward project objectives and can facilitate group decision-making
5. I have a system perspective in my area of technical work
6. I have technical credibility with my colleagues
7. I can use the latest design techniques and engineering tools
8. I can recognize work with potential for technological breakthrough early in its development
9. I can measure work/project status and technical performance of other people on my team
10. I can integrate the technical work of my team members

ADMINISTRATIVE SKILLS

1. I don't mind administrative duties
2. I am familiar with techniques for planning, scheduling, budgeting, organizing, and personal administration, and can perform them well
3. I can estimate and negotiate resources effectively
4. I can measure and report work status and performance
5. I find policies and procedures useful as guidelines of my activities
6. I have no problem delegating work even though I could do it myself, probably quicker than someone else
7. I don't mind writing reports and preparing for meetings, and I do it well
8. I can handle changed requirements and work interruptions effectively
9. I am good at organizing a party
10. I can work effectively with administrative support groups throughout the company

PERSONAL DESIRE TO BE A MANAGER

1. Managing people is professionally more interesting and stimulating to me than solving technical problems
2. I am interested and willing to assume new and greater responsibilities
3. I am willing to invest considerable time and effort into developing managerial skills
4. I have an MBA (or am working intensely on it)
5. I am prepared to update my management knowledge and skills via continuing education
6. I have discussed the specific responsibilities, challenges, and skills requirements with managers who hold similar positions to the one that I would like to grow into
7. I have defined my specific career goals and mapped out a plan for achieving them
8. I would be willing to change my professional area of engineering activity if an advanced (managerial) opportunity would occur
9. Managerial and business challenges are more interesting and stimulating to me than technical challenges
10. Achieving a managerial promotion within the next few years is a top priority, very important to the satisfaction of my personal and professional needs

BUSINESS ACUMEN

1. I would be good at directing the activities of my department toward the overall business objectives of my company
2. I am productive
3. I enjoy long-range planning and find the time to do it
4. I am willing to take risks to explore opportunities
5. I feel comfortable working in dynamic environments associated with uncertainty and change
6. I would enjoy running my own company
7. I consider myself more of an entrepreneur than an innovator
8. In social functions, I tend to get involved more in business discussions rather than technical discussions
9. I enjoy being evaluated, in part, on my contributions to my company's business environment
10. I have been more right than wrong in predicting the business environment

MANAGERIAL APTITUDE AND DEVELOPMENT PLAN

Aptitude measure	Individual's score	Supervisor's score	Action plan	Potential score
Personal desire to be a manager	______	______	______	______
People skills	______	______	______	______
Technical knowledge and competence	______	______	______	______
Administrative skills	______	______	______	______
Business acumen	______	______	______	______
Composite measure	______	______	______	______

TABLE 14.A.2 Distribution of Aptitude Test Scores, as Assessed by Engineers and their Supervisors

Aptitude Measure	Assessed by	Percentage of People Rated Below the Score								
		10%	20%	30%	40%	50%	60%	70%	80%	90%
Personal Desire	Engineer:	30	40	50	58	65	71	76	80	83
	Supervisor:	25	34	43	51	57	60	62	64	66
People Skills	Engineer:	33	45	57	66	74	78	81	83	85
	Supervisor:	32	42	52	60	67	72	74	75	77
Technical Knowledge	Engineer:	40	47	56	66	77	85	89	91	92
	Supervisor:	38	46	52	58	64	71	75	78	80
Administrative Skills	Engineer:	15	26	37	46	55	63	70	75	79
	Supervisor:	16	26	36	45	53	60	66	70	72
Business Acumen	Engineer:	30	35	40	45	49	54	58	62	65
	Supervisor:	25	29	33	37	41	44	47	49	50
Overall composite measure	Engineer:	30	38	46	52	58	65	72	77	80
	Supervisor:	27	35	45	50	56	61	65	67	69

3.5; between desire and subsequent managerial performance, tau = .30. Both statistics are significant at a confidence level of 92% or better.

Despite this favorable association, personal desire alone is insufficient to gain a promotion. In the final analysis, personal competence and organizational needs are the deciding factors. People who receive promotions usually meet five key requirements:

1. They are competent in their current assignments. An engineer must master the duties and responsibilities of the current position, have the respect of his or her colleagues, and receive favorable recommendations from the supervisor.
2. They have the capacity to take on greater responsibility. The person must demonstrate the ability to handle larger assignments that have new and more challenging responsibilities. Good time management, the willingness to take on extra assignments, and the expressed desire to advance toward a management assignment are usually good indicators that a person is ready for advancement.
3. They have prepared for the new assignments. A new management assignment requires new skills and knowledge. Candidates who have prepared themselves through courses, seminars, on-the-job training, professional activities, and special assignments will have the edge. Managers perceive such initiatives as evidence that a candidate is committed to the new career path, is willing to develop new skills, and wants to go the extra distance.

PREPARING FOR AN ENGINEERING MANAGEMENT CAREER

The first list gives specific steps that a technical professional can take in order to prepare for a management career. The other two sections outline ways in which the employee's supervisor and organization can support the employee in preparing for such a transition.

Individual Preparation

- Define specific objectives and plans.
- Gain experiential learning.
- Take on administrative assignments.
- Practice team motivation and leadership.
- Participate in task forces.
- Seek our multifunctional assignments.
- Participate actively in professional organizations.
- Publish in professional journals and speak at conferences
- Maintain technical expertise.
- Take courses and seminars.
- Read management literature.
- Complete an MBA.
- Talk to managers.

Supervisor's Assistance

- Help the employee in assessing career ambitions.
- Facilitate assignments that may provide management learning experiences.
- Encourage leadership and assistance to management.
- Facilitate dual-career ladders.
- Encourage and support management training.
- Use project management as a training ground.
- Use temporary assignments and titles for incremental skill development (such positions include task manager, lead engineer, proposal manager, and team leader).
- Recognize the value of managerial skills for engineering assignments.

Organizational Support

- Establish policy guidelines for managerial development.
- Develop managerial staffing plans.
- Provide resources.
- Establish dual-career ladders.

EXHIBIT 14.A1

4. They are good matches with organizational needs. The candidate's ambitions, desires, and capabilities have to match both the current and long-range needs of the firm.

   Obviously, a new job opening can create an immediate opportunity for advancement, but many companies make detailed, long-range plans for their managerial staffing needs. They also encourage managers to identify and develop future managers.

   In such environments, engineers with management ambitions must be recognized early as qualified candidates. Their companies then may help them develop management skills with training and special job assignments. When the need for a new manager arises, the company can select from the pool of prequalified candidates.
5. They have an aptitude for management. If the previous requirements are met, higher-level managers will have a lot of confidence in the candidate, but there is no guarantee that the candidate will perform well in the new assignment.

The other four qualifications are based on behavior in a known environment and on the assumption that the candidate will adapt to the new management situation. But in a nonmanagement situation, it is difficult to prepare for the challenges in leadership, power, personnel administration, and the management of change and conflict. Therefore, engineers who want to be considered for management must show strong aptitudes for demanding leadership positions.

Determining Aptitudes

The test instruments in Table 14A.1 are helpful in many ways. For engineers, they provide a set of criteria for self-assessment, self-study, and potential career development actions. They also allow engineers to compare their own aptitudes with the general engineering population (see Table 14A.1).

For managers, these instruments provide a set of criteria for initial assessment of an engineer's potential for management, for defining career objectives and development plans, and for comparing candidates and attitude changes over time. They are also helpful to the management researcher, who can use them for comparative studies and refinement of criteria, as well as for validation of the aptitude scores regarding the probability of management success.

Here are some suggestions for using the aptitude tests properly and facilitating successful career development.

1. Realize the criteria for becoming a manager. The instruments should serve as guidelines rather than absolute measures for promotion. Personal judgment should remain an important factor in the final decision.
2. By the same token, don't use the numbers in an absolute sense. Realize that the scores, whether they come from you or others, are subjective and should serve as a basis for comparison and critical thinking, not as a mechanical selection tool.

MAKING THE TRANSITION TO AN ENGINEERING MANAGEMENT CAREER

Here are some tips for making the transition from a technical professional to a manager of technical professionals.

The first list gives strategies for an engineer who is making such a transition. The other lists give suggestions for how his or her supervisor and firm can help.

Individual Actions

- Behave like a manager.
- Develop more of a business perspective.
- Build motivation and leadership skills.
- Build credibility.
- Practice delegation.
- Develop a managerial style.

Supervisor's Assistance

- Facilitate the transition with personal support.
- Provide a charter of job responsibilities.
- Help to establish communication channels.
- Help to establish interaction with other departments.
- Establish key performance objectives.

Organizational Support

- Provide management development and training.
- Provide networks for new managers.

EXHIBIT 14.A2

3. Use the instruments for formulating career development plans. When engineers use these instruments to assess their own managerial aptitudes, they can get a better understanding of personal strengths, weaknesses, and desires, and of areas in which they need to increase their efforts. In other words, the test instruments may help in formulating career plans and specific action items.
4. Use the instruments to assist in management development planning. In a conceptual manner, supervisors can use both the criteria and the instruments themselves to identify subordinates with management interest and potential.

5. Remember to obtain data from multiple sources. In addition to your own evaluations of a candidate, solicit evaluations from other sources. Your peers, project leaders, customers, and support personnel who know the candidate may be good possibilities. You can solicit their input during informal discussions or by using formal questionnaires or letters.
6. Focus on managerial strength. If the aptitude scores indicate a strength in a specific area, the management candidate should seek opportunities for additional responsibilities in that area. That will build skills in areas of most likely success and provide favorable learning experiences. Management should encourage and support such learning experiences.
7. Use incremental skill-building. Participation in such managerial activities as proposal development, task management, feasibility studies, and technology assessments helps a candidate sharpen skills in cross-functional communication, planning, and organizing. The candidate also can develop new skills and test his or her desire for a career in management.
8. Recognize that self-assessment scores change with experience. The aptitude scores, especially the self-assessment, improve with actual management experience. Usually, the more favorable the experience, the more favorable the change in overall aptitude score.
9. With that in mind, consider also the confidence level and bias of the scoring: aptitude measures should be more realistic after a candidate has gained some management-related experience.
10. Encourage discussions with managers. People who seek careers in engineering management should discuss their ambitions with managers and get insight into related tasks, responsibilities, and skills. Attending such functions as management conferences and business luncheons also can be beneficial in measuring management abilities and providing a realistic background against which to take formal aptitude tests.
11. Base any decision regarding career paths, promotions, or training methods on personal judgment rather than on mechanical scores. As mentioned before, the aptitude test may help to provide insight into a person's values and attitudes toward engineering management. In the final analysis, though, you must make an integrated judgment that considers all the factors against the person's professional background and the specific managerial challenges at hand.

Making Managers of Engineers

Career decisions and selections for managerial training are complex. To succeed in a challenging business environment, organizations must be able to select and prepare future engineering managers effectively.

The instruments for determining managerial aptitudes can be helpful in supporting professional development decisions, but we lack a method for accurately determining management potential. The challenge is for managers and engineers to develop enough understanding of the dynamics of their organizations and their managerial demands and to relate them to candidates' personal strengths and desires. This should provide the basis for integrating aptitude scores with a person's individual background and the needs of the organization.

BIBLIOGRAPHY

Badiru, Adedeji B., "Training the IE for a Management Role," *Industrial Engineering,* Volume 19, Issue 12 (Dec 1987), pp. 18–23.

Bailyn, Lotte, "The Hybrid Career: An Exploratory Study of Career Routes in R&D," *Journal of Engineering & Technology Management,* Volume 8, Issue 1 (June 1991), pp. 1–14.

Baker, H. Kent and Philip I. Morgan, "Two Goals in Every Performance Appraisal," *Personnel Journal,* Volume 63, (Sept 1984), pp. 74–78.

Bernadin, H. John and Richard W. Beatty, *Performance Appraisal: Assessing Human Behavior at Work,* Boston: Kent, 1984.

Braham, James, "Engineering Your Way to the Top," *Machine Design,* Volume 63, Issue 17 (Aug 22, 1991), pp. 65–68.

Brandt, Ellen, "Building a Career in Engineering," *Chemical Engineering,* Volume 98, Issue 5 (May 1991), pp. 83–86.

Cavazos, Lauro F., "The Role of Technical Education," *Occupational Outlook Quarterly,* Volume 35, Issue 1 (Spring 1991), pp. 23–25.

Cherney, Steven D., "Career Management: Are You in Control?" *Chemical Engineering,* Volume 95, Issue 5 (Apr 11, 1988), pp. 75–78.

Edosomwan, Johnson A., "Professionals Must Train for Factory of Future's Integrated Work Environment," *Industrial Engineering,* Volume 21, Issue 10 (Oct 1989), pp. 20–23.

Ercz, M., P. C. Early, and C. I. Hulin, "The Impact of Participation on Goal Acceptance and Performance," *Academy of Management Journal,* Volume 28, Number 2 (May 1985), pp. 50–66.

Ferguson, Gary, "Developing a Curriculum for IE Graduates of Today and Tommorrow," *Industrial Engineering,* Volume 23, Issue 11 (Nov 1991), pp. 46–50.

Franklin, Jerry, "For Technical Professionals: Pay for Skills and Pay for Performance, *Personnel,* Volume 65, Issue (May 1988), pp. 20–28.

Goldstein, Mark L. "Dual-Career Ladders: Still Shakey but Getting Better," *Industry Week,* Volume 236, Issue 1 (Jan 4, 1988), pp. 57–60.

Hawkins, Peter and Ian Barclay, "The Engineering and Manufacturing Managers of the 21st Century: Part II: Career Development and Progression," *Management Decision* (UK), Volume 28, Issue 6 (1990), pp. 48–54.

LaPlante, Alice, "Survival Tips for the Coming Year," *Computer World,* Volume 24, Issue 52, 53 (Dec 24, 1990/Jan 1, 1991), pp. 51.

Lazer, Robert I. and Walter S. Wikstrom, *Appraising Managerial Performance: Current Practices and Future Directions, (Report 723),* New York: The Conference Board, 1977.

Leibowitz, Sandy B., Barbara H. Feldman, and Sherry H. Mosley, "Career Development Works Overtime at Corning, Inc.," *Personnel,* Volume 67, Issue 4 (Apr 1990), pp. 38–46.

Louchheim, Frank P. "Managing Your Career—For Better or Worse," *Chemical Engineering,* Volume 95, Issue 2 (Feb 15, 1988), pp. 93–94.

Murray, Margo, *Beyond the Myth and Magic of Mentoring,* Jossey-Bass, 1991.

Raudsepp, Eugene, "Hang Loose," *Machine Design,* Volume 63, Issue 23 (Nov 21, 1991), pp. 79–82.

Rimler, George W., "The Transition from Engineer to Manager," *Industrial Management,* Volume 33, Number 6 (Dec 1991), pp. 17–18.

Sheldon, Ronald and Brian Kleiner, "What Management Techniques can (or should) be applied by American Managers," *Industrial Management,* Volume 32, Number 3 (May/June 1990), pp. 17–19.

Thamhain, Hans J., "Managing Engineers Effectively," *IEEE Transactions on Engineering Management,* Volume 30, Number 4 (Nov 1983), pp. 231–237.

Thamhain, Hans J., "Managing Technology: The People Factor," *Technical and Skill Training Journal,* Volume 1, Number 2 (August/September 1990), pp. 23–30.

Thamhain, Hans J., "Skill Developments for Project Managers," *Project Management Journal,* Volume 22, Number 3 (Sept 1991), pp. 39–45.

Thamhain, Hans J., "Developing Technology Management Skills," *Research and Technology Management Journal,* Volume 35, Number 2 (March 1992), pp. 42–47.

Thompson, Nancy, "Balance Is Know-How with Professional Polish," *Computerworld,* Volume 26, Issue 7 (Feb 17, 1992), pp. 79.

Thornberry, Neal, "Pre-Appointment Programs Help Engineers Become Good Managers," *Industrial Engineering,* Volume 31, Issue 5 (Sept/Oct 1988), pp. 19–35.

Werskey, Gary, "Engineering People—How Japanese Electronics Firms Train Their Engineers," *Work and People* (Austrailia), Volume 12, Issue 3, (1986), pp. 14–20.

Yager, Ed, "A Critique of Performance Appraisal Systems," *Personnel Journal,* Volume 60, (February 1981), pp. 129–133.

APPENDIX 1
GLOSSARY OF TERMS

Acceptance. Process of delivery and takeover by the customer.

Acceptance tests. Formal customer-approved tests to demonstrate compliance with all specification requirements. These tests are limited to early manufacture or engineering models. They may be performed by engineering, factory, or the customer depending on the particular project. In any case, sufficient, component and system engineering time should be scheduled.

Account code structure. The framework of numbers, following the pattern of the work breakdown structure, which is used for charging (DSO and shop order charge numbers) and summarizing (summary numbers) the cost of program.

Accounting month. Period approximating a calendar month, but consisting of four or five whole weeks, for accounting purposes.

Actual completion date. Actual calendar date on which a work effort (activity, work package, or summary item), contract, or program/project is considered to be completed.

Actual cost variance. The difference between the budgeted cost for work performed and actual cost for work performed. An unfavorable variance exists when actual cost exceeds budgeted cost. If budgeted cost exceeds actual cost, a favorable variance exists.

Advance payments. Advances of money made by the government to a contractor prior to, in anticipation of, and for the purpose of complete performance under a contract or contracts; advance payments due to the contractor incident to performance of contracts. Since they are not measured by performance, they differ from partial, progress, or other payments that are made because of and on the basis of performance or partial performance of a contract. Advance payments may be made to prime contractors for the purpose of making subadvances to subcontractors.

Allowable costs. Costs incurred by a contractor, which the customer recognizes as reimbursable.

Analysis and Report. Process of review of information, measuring of performance, and reporting.

Anticipated business. Those programs, including extension of existing contracts, that the division marketing departments expect to book and those billings that are expected to materialize during the projected budget period. Anticipated business will be further categorized into anticipated/unfunded business. Unfunded business is that part of a definitized contract which has not been funded and on which bookings and billings are anticipated in the projected budget period.

Associate contractor. Each of two or more contractors who enter into a contract with a customer to provide specified items of materials or supplies and/or perform services in accordance with the contract requirements of the agency to meet the objectives of a single program project. Since no single contractor is a prime contractor, the government, in effect, becomes the prime contractor.

ASPR. Armed Services Procurement Regulations; regulations governing the award and administration of prime and subcontracts.

Billing. Contractor's submission of vouchers, etc., to customer for reimbursement for work performed.

Black box. Mechanical, electrical, or electromechanical assembly that is part of a system or subsystem.

Booked commitment. Contractual obligation to pay for goods or services to be received.

Budgeting, Budgets. The administrative process by which the contract price is allocated to planned, scheduled, and internally authorized subdivisions of the tasks required to be performed, by the terms of the contract. The estimated price of authorized but unpriced changes and unpriced work shall be included in budgeting. To the extent that detailed planning, subtask definition, and/or milestone-oriented scheduling may not have been accomplished at some discrete point in time (e.g., for future work tasks whose detailed definition and scheduling is dependent upon the outcome of nearer-term efforts), or to the extent that additional costs are originally anticipated but not allocated, as a means of holding tight controls, management reserve budgets are established. Operating budgets for DSOs or tasks are therefore the original planned costs as modified by changes. While control budgeting must be expressed in dollars, budgets may also be expressed in manpower units (e.g., man-hours or man-months) for convenience.

Change order. A written order, signed by the contracting officer, directing the contractor to make changes that the changes clause of the contract authorizes the contracting officer to order without the consent of the contractor.

Charge number. A number used for identifying the costs charged to a work package (e.g., a DSO or shop order).

Client. Sponsor.

Closure. Closing out of a contract on which performance has been completed.

Company funds. Use of its own funds by a contractor.

Constraints. Applicable restrictions that will affect the plan.

Consultants. Specialized members with professional status.

Contract. Any type of legal document of agreement or order for procurement of supplies and services, including awards and preliminary notices of award, contracts of a fixed price, cost, cost-plus-fixed-fee, or incentive type, also including task orders, task letters, letter contracts, modifications, and supplemental agreements with respect to the foregoing. The most usual types of contracts are listed below.

a. **Cost contract.** The cost (or cost-sharing) type of contract provides the payment to the contractor of allowable costs, to the extent prescribed in the contract, incurred in the performance of the contract.

b. **Cost-plus-fixed-fee contract.** The cost-plus-fixed-fee type of contract provides for payment to the contractor of all allowable costs as defined in the contract and establishes an estimate of the total cost. It differs from the cost contract in that it also provides for payment of a fixed fee based on the estimated cost of the contract. The fixed fee will not vary except as a result of a change in the scope of work under the contract.

c. **Fixed-price contract.** The fixed-price type of contract generally provides for a firm price or prices for supplies or services. It may include provision for price escalation or adjustment.

d. **Fixed-price contract with provision for redetermination of price.** This type of contract is a fixed-price contract with special provision for redetermining upward or downward the price or prices in the contract.

e. **Incentive-type contract.** The incentive-type contract may be either a fixed-price type or a cost-plus-a-fixed-fee type, with a special provision for redetermination of the fixed price or fixed fee. The incentive-type contract provides for a tentative base price or target price (called the "contract price") or a maximum price or maximum fee, with price redetermination after completion of the contract for the purpose of establishing a final price or fee, which varies inversely with the cost. In no event will the final price or fixed fee exceed the original amount stated in the contract.

Contract base. Contractual basis for project operations.

Contract number. The assigned government contract code.

Contractor. Company that engages in a contract.

Contractor-extended work breakdown structure (contractor extended WBS). The WBS that is evolved by the contractor as an extension of the contract summary WBS, representing the supplemental and traceable elements required by the contractor's management systems. The contractor-extended WBS is evolutionary in nature as information becomes available during the progressive definition of the work to be done during the phases of the contract.

Contract price. The total amount fixed by the contract (other than any portion of the contract specifically providing for cost reimbursement only), as amended, to be paid for complete performance of the contract. If the contract provides for escalation or for redetermination of price, this term means the initial price until changed and not the ceiling price. If the contract is of the incentive type, the term means the ceiling or maximum price. For letter contracts and similar preliminary contractual instruments, "contract price" means the maximum expenditure authorized by the contract, as amended.

Contract summary work breakdown structure (contract summary WBS). The WBS(s) that is prepared by the customer by selecting elements from the project summary WBS for application to and inclusion through negotiation in the individual contract or contracts, if more than one procurement is placed.

Contractual coverage. That which will permit the contractor to be reimbursed under the existing customer agreement for services performed and/or material delivered.

Control. Process of managing the project.

Cost accounts. The specific identifications of each WBS element at any level with a cost accounting number. Cost accounts may be chargeable as tasks or established as cost collection points. Task cost accounts are usually at the lowest level of the WBS and are referred to as "work packages." Cost accounts may be planned and controlled through the use of either work packages or time-phase budgets (level-of-effort), but normally will not be a combination of both. The DSO and shop order may be used for this purpose. Cost accounts must (a) summarize directly into the WBS without requiring allocations to two or more elements; (b) include all actual costs on the contract; (c) have separately assigned and identified budgets, which are obtained via task cost estimates; (d) be measurable in terms of specific start and completion points based upon network scheduling and definable work accomplishment.

Cost activity. An activity that employs resources, the costs of which are a direct charge to the program.

Cost category. The name and/or number of a functional, hardware, or other significant cost category for which costs are to be summarized.

Cost control. Control of expenditures and accounting for expenditures.

CFGG. Cost-plus-fixed-fee contract; the contractor's allowable costs are covered, profit is fixed.

CPIF. Cost-plus-incentive-fee contract; the contractor's allowable costs are covered, profit varies according to performance.

Critical path. The path(s) of activities and events on a PERT network that take the longest time to reach the objective event(s); the path(s) with the least amount of slack on which the slippage of any event will reflect as an equal slippage in the end objective.

Critical path method (CPM). Developed by Remington Rand and DuPont in 1957 as a networking technique employing arrows to show activities, but not including events on the network. It uses one-time estimate and employs a cost curve to determine minimum cost expediting of the project.

Customer. An organization, agency, or contractor who pays for work.

Definitized work. Described portions of an authorized contractual effort for which firm contract prices have been agreed to in writing by the parties to the contract.

Deliverable item. Hardware, software, services, reports, or other items deliverable to the customer in accordance with the terms of a contract.

Delivery schedule. Shipping dates of equipment to the customer.

Department. The next-lower level of formal organization within a function.

Design review. An administrative and technical control that is utilized to bring to each design the knowledge and experience of specialists who are intimately familiar with the total designs or portions of the total designs. Design reviews ensure consistency with other designs; application of advance techniques; increased reliability; reproducibility, and maintainability; and the reductions of costs of labor, parts, materials, and processes.

Development suborder (DSO). A document which initiates effort and states the definition of that portion of a task assigned to the lowest responsible organizational unit: The DSO also reflects a unique charge number, which will be utilized for accumulating costs expended on that portion of the task by that unit or at that unit's direction.

DoD. Department of Defense.

Earliest finish time (EF). Referring to PERT, the earliest time a project activity or project is expected to finish.

Earliest start time (ES). Referring to PERT, the earliest time an activity can begin according to schedule constraints and sequencing logic.

End item. An item complete in itself per the contract.

Engineering report. Document that clarifies the project on a detailed technical level.

Engineering task schedule. A schedule for each task, divided into a section for each lowest responsible organizational unit, and further divided into activities and events.

Estimate to complete. A time-phased detailed cost estimate from a discrete point in time through to completion of the DSO activities. It is the responsibility of the program office to provide adequate baseline (scope and schedule) information for each estimate to complete.

Event. A specific measurable point in time. This point must be selected and described so that there is no question as to whether or not it has been reached. This point may describe when a task has been initiated or completed, information has been generated, or a decision has been reached. An event takes no time, costs no money, and uses up no resources. Some of the descriptive terms that are associated with events are as follows: (a) shipped, (b) demonstrated, (c) reviewed, (d) announced, (e) released, (f) approved, (g) selected, decided, (h) received. On a PERT network, a geometric shape is used to signify an event.

Expenditure. Cost, funds expended for the project work.

Expenditure limitation. The amount of money that is authorized to be expended on a task or DSO, irrespective of the budget for that task or DSO. The expenditure may be limited in time as well as money.

Exposure. Amount of company funds risked, "exposed," in proceeding without customer funding coverage.

Feasibility. Assessment of probability of carrying out project plan.

Fee. Profit on a cost-type contract.

Fiscal management. Management of the flow of funds received from the customer and required in project operations.

Forecast. A timely estimate to indicate any changes from the latest approved estimate to complete. The forecast is prepared by the person responsible for the DSO and is submitted on a marked-up monthly project control report.

Forecasting. Estimating the schedule and funding.

FP. Fixed-price contract; one in which the contractor's total receipts are fixed at a certain dollar amount.

FPI. Fixed-price incentive contract; the contractor's reimbursable costs are fixed, but the profit varies according to how efficiently the contractor performs.

FPR. Fixed-price redeterminable contract; one in which the dollar amount is fixed, subject to relatively minor adjustments to be determined.

Free slack (FS). Referring to PERT, the time an activity can be delayed without increasing the actual total time to complete the project; calculation for activity x: $FS(x) = TS(x) - TS$ (most critical path) where TS = total slack.

Functional management. Process of planning, organizing, coordinating, controlling, and directing project efforts within a structure, which groups responsibilities according to the type of work to be performed so that functional activities are governed by consistent policies and objectives at all management levels.

Functional organization. Organization engaged in one general function, such as engineering, manufacturing, or marketing.

Funding, funds. Money for furthering a contract or project purpose.

Funding coverage. Customer funding available to cover or pay for contract work.

Funding limitation. The contractually imposed limitation of customer termination liability, irrespective of total contract price.

General and administrative (G&A) costs. That portion of indirect cost applicable to the general direction and control of the contractor's activities as a whole. Does not include those indirect costs that are applicable to a specific functional cost category such as engineering, tooling, or manufacturing. Usually includes such items as: salaries of executives and general office employees, legal and auditing, and corporate office expenses.

Government actual contract cost. Applied direct costs, indirect costs, and profit or fee earned for work performed or services rendered by a contractor.

Government-furnished property. Property in the possession of, or acquired directly by, the government and subsequently delivered or otherwise made available to the contractor.

Incentive contract. Contract in which the contractor's profit varies according to performance.

Incremental funding. Funding committed in time increments.

In-process commitment. Money set aside while a subcontract or other binding agreement is being processed.

Inspection. The examination (including testing) of supplies and services (including, when appropriate, raw materials, components, and intermediate assemblies) to determine whether the supplies and services conform to contract requirements as defined in all applicable drawings, specifications, and purchase descriptions.

Interaction. Cooperative action between member and other groups.

Interfaces. Organizational or personal points of interaction, cooperation, or commonality.

IR&D. Industrial research and development, a reimbursable effort for defense and aerospace contractors.

Kickoff meeting. A meeting between representatives of marketing and all activities involved in the cost estimating and tentative planning of a project, in response to a cost estimate request.

Labor. Direct man-hours and dollars expended by personnel involved in direct labor activities affecting the design, development, test, fabrication, and assembly of contract articles.

Latest allowable time. A term commonly used with PERT, it means the latest time that an event can be accomplished without causing slippage of the objective event.

Latest finish time (LF). Referring to PERT, the latest date or time a project activity or total project is permitted to finish if overall schedule constraints are to be met.

Level of effort. Application of resources (labor, material, funds) to an effort in a manner that is described in terms of time and general objectives rather than in terms of a specific end objective. For example, engineering man-hours applied over a month to achieve the "best results" possible as opposed to "result X."

Liability. The money a customer is obligated to pay a subcontractor or vendor in the event of a termination.

Line of balance. A time-scheduling control system for production-type projects. It utilizes graphical means to indicate the performance status of significant work elements of production.

Loading the contract. Charges of doubtful propriety against a contract.

Management control. The method by which management examines results and makes certain all operations are carried out in accordance with the adopted plan, the principles laid down, and the orders given. Control provides techniques for open discussion and constructive criticism. It stimulates planning, simplifies and strengthens organization, increases the efficiency of command, and facilitates coordination.

Management reserve. The algebraic difference between the contract price and the sum of all the budgeted costs.

Management system. A system composed of equipment, skills, procedures, and techniques, the composite of which forms an instrument of administration or control.

Manpower schedule. Manpower requirements established to meet the time schedule.

Manufacturing process. A written instruction by which a given part or assembly may be produced most economically, in the desired quantity, while adhering to quality, engineering, and customer specifications. A manufacturing process normally lists detailed operations in their proper sequence; the materials, tools, and equipment required to perform the operations; the rate for each operation type of labor used; and the activity performing the operation.

Manufacturing test specification information. Time required for engineering to advise and consult with manufacturing test process on production tests to be performed.

Material cost. Cost of material and purchased parts (including semifabricated) consumed in in-plant manufacturing processes and charged directly to the contract. Specifically excludes subcontract costs.

Measuring and testing equipment. All electrical and mechanical equipment used to determine the conformance of an equipment, an assembly, or components used therein, to specified requirements. This includes government- or customer-owned, contractor-owned, employee-owned, borrowed, and leased measuring and testing equipment.

Monitoring. Tracking project progress or results against cost and schedule.

NASA. National Aeronautics and Space Agency.

Operations plan. Plan for conducting project operations.

Organization. Main group, subgroups, and their functional structure.

Overexpenditure. Expenditure above the level planned or funded.

Overrun. Costs in excess of the contemplated or estimated contract cost.

Past performance. Analysis of historical data to determine any spending patterns or trends that can be of assistance to prepare a cost forecast.

PERT. Program evaluation and review technique; a management tool for planning, tracking, analyzing, and controlling a project.

Planning. Identification of activities, timing, and resources to be carried out to meet specified objectives; the establishment of logical sequences of activities or tasks.

PM. Project manager.

Policies. Basic set of project and/or corporate guidelines for managing organizational resources and obtaining objectives.

Postproject evaluation. At completion of project, comparison of final results to cost, efforts, and problems; lessons learned.

Problem statement. Documentation for defining and clarifying the problem.

Procedures. Systems and methods established and updated throughout the length of the project for the purpose of making decisions.

Productivity. Ability to yield or furnish results, benefits, profits.

Program. Related series of undertakings which continue over a period of time and which are designed to accomplish a specific goal.

Program directive. A document containing nonscope program information or instructions, pertaining to a particular program.

Program director. Position title of individual who manages a large program or a group of programs.

Program manager. Individual who is assigned the direct responsibility for program execution; if titles of both program and project manager are used, the program manager has the higher-level responsibility.

Program master schedule (PMS). The overall program schedule denoting major task phasing, milestones, and customer delivery rates relating to all engineering, manufacturing and subcontractor work.

Program office or program management office (PMO). That organizational unit within either the customer or contractor organization that has overall responsibility for the technical, schedule, and financial performance of the program.

Program plan package. The complete package of program documents (WBS/TFM, schedule, budgets, PA, PATD, etc.), which is reflected in the program plan summary.

Program plan summary. A program office document indicating the current revision status of the program plan package. By indicating the current revisions of task definitions, schedules, budgets, and DSOs, the program plan summary provides a method of displaying a consistent baseline against which to work, plan, report, and estimate.

Project. Undertaking with specific results, schedules, and resources; usually an integrated part of a larger program.

Project authorization and task definition (PATD). A document issued by the program office, which defines and authorizes the lowest discrete task identified on the project task summary. The PATD contains the dollars budgeted to perform the task, the funds presently available to perform the task (expenditure limit), and the task schedule.

Project base. Contractual basis for project operations; may involve more than one contract and more than one customer.

Project control report (PCR). A management report and an operations tool used to report financial/manpower information in a summarized manner both by organization and program.

Project funding. Process of making funds available.

Project management. Process of managing the project, or the technical and administrative lead personnel of a project team.

Project manager. Individual who is assigned the direct responsibility for project execution.

Project objectives. Goals, as measured by results, time parameters, constraints/targets, milestones, and control considerations, and general project sequencing.

Project office. Administrative project headquarters.

Project organization. Establishment of people, responsibilities, authority, and procedures.

Project parameters. Characteristics, factors, or specifications of the total project or of its subsystems.

Project proposal. Summary of a proposed project regarding its technical, managerial, and schedule content.

Project task summary (PTS). A document prepared by the program office which identifies the program and all the tasks to be performed by the major functional organization (e.g., engineering or manufacturing) to which it is issued. It includes that portion of the work breakdown structure appropriate to that organization.

Project team. All participants in the project, full- or part-time.

Quality assurance. Organizational system; begins with planning, proceeds to engineering, design, specifications, and materials selection along with scheduling.

Quality control. Management process that ensures the development of a quality product or service according to established standards.

Quality management. Planning, scoping, implementing, and monitoring of quality into all phases of the project from concept through the delivery aspects of the work.

Real time. Actual calendar time.

Records. Personnel- and project-related information.

Reporting. Presentation of results and analysis, usually to customer and management.

Rewards. Pay, salaries, commissions, bonuses, etc.

RFP. Request for proposal, or proposal request (PR).

Scheduling. Application of calendar time to project activities.

Scope statement. Document outlining the magnitude and depth of project, including global results.

Slippage. Failure to maintain the schedule due to slower progress than planned.

Solicitation. Process of requesting bids or proposals.

Specifications. Performance, environmental, size, weight, and other requirements that a deliverable item must meet.

Spending profile. Subdivision of project funds allocated over specific time periods.

Standards. Criteria against which a project is measured to determine the status of the project in relationship to its plan.

Statement of work. Part of a program plan which outlines the work to be performed.

Status index. Management tool for correlating planned and actual progress with planned and actual expenditures.

Stop work order. Order from the customer to stop work on a contract; work may later be resumed or the contract may be terminated.

Subcontractor. Contractor below the prime contractor.

Subsystem. Major part of a system, such as the propulsion subsystem of an ICBM system.

Subtask. One of several items of work required in accomplishing a task.

Succession. Strategy to afford continuity.

System. Operational capability, integrated from various subsystems, elements, and components.

System life cycle. Phases through which a system passes from conception to disposition.

System management. Process of planning, organizing, coordinating, controlling, and directing combined efforts to accomplish system program objectives.

System manager. Individual who is assigned responsibility for the execution of a major subdivision of project work.

Task. The lowest level of the work breakdown structure, other than organizational contributions, that denotes a single program objective (e.g., transmitter engineering model, study, test station manuals). This effort is spelled out on the PATD. A task is also the summary of a group of related DSOs commonly contributing to the single objective or purpose as defined in the PATD.

Team. All participants in the project.

Team management. Organizational and functional technique to achieve compatibility within an organization.

Termination. Directed termination of all or part of a contract.

Total slack (TS). Referring to PERT, the time an activity can be delayed without impacting the agreed-upon latest finishing time of the project (LF); calculation: TS – LF – EF = LS – ES.

Training. Process of developing an individual's ability to handle a particular job.

Trials. Processing of testing the completed project to assure specified performance.

Unallowable costs. Costs incurred by a contractor that are not reimbursable by the customer.

Users. Members of the customer community to whom the product will be delivered for end use.

Variance. Difference between planned and actual (or predicted) cost or performance.

Variance interpretation. Clarification of significance of variance with respect to overall objectives.

Verification. Assurance that the resource and time estimate matches the overall objectives.

Work breakdown structure (WBS). A product-oriented family tree division of hardware, software, services, functions, and other work tasks which organizes, defines, and may graphically display the products as well as the work to be accomplished in order to achieve the specified project objectives. This forms a common framework for financial and schedule definition and control.

Work-force planning. Determining the requirements and necessary procedures.

Work loads. Project effort per organizational unit or individual measured in man-hours, man-weeks, etc.

Workmanship. Individual craftsmanship and performance.

Work packages. Project subsystem grouped for effective execution.

CASE STUDIES IN ENGINEERING AND TECHNOLOGY MANAGEMENT

The first case study in Appendix 2, Managing the Development of the HP DeskJet Printer, is typical of the type of management challenges and organizational complexities that can arise in a matrix organization involved in a high-tech project. The second case, the PK 2000 New Product Team, describes and analyzes a real-world situation that was observed in an electronic equipment/computer systems company of the Fortune 500 category. It provides the basis for examining challenges, skill requirements, risks, and processes involved in building multifunctional project teams in a technology-oriented work environment. The third case study, Thermodyne Incorporated, presents an engineering management situation with facets of functional, task, and project management. The case follows an engineering contract from its acquisition to its development and troubled integration. Following each case are activities for students based on the cases.

APPENDIX 2.1

CASE STUDY: MANAGING THE DEVELOPMENT OF THE HP DESKJET PRINTER[1]

The HP DeskJet Printer case describes a high-technology product development from engineering design to its interface with manufacturing. The case focuses on the multidisciplinary challenges involved in the management of a complex engineering development. The critical role of marketing in defining the product plan and providing guidance for the engineering development effort is highlighted. The case provides the basis for discussing the challenges involved in organizing and managing a complex engineering development program.

CASE DESCRIPTION

The creation of a high-technology product is an enterprise that requires the contributions and skills of many people. The articles on the HP DeskJet printer in this issue[1] mainly explore the technical engineering problem solving that is essential to new-product development. But there is a bigger picture. There is much more to engineering than solving equations and setting up experiments.

In the case of the DeskJet printer, our organization and planning encompassed all of the functional departments of HP's Vancouver Division. Our development and management teams included members from R&D, manufacturing, marketing, and quality assurance, with special assistance from personnel and finance.

[1]This case study is based on an article by John D. Rhodes "Managing the Development of the HP DeskJet Printer," originally published in *The Hewlett-Packard Journal*, October 1988, and is reprinted by permission.

The core development teams, which worked on the product from its inception, consisted of about 25 engineers, split into three project teams—firmware, electronics, and mechanical—with a project manager leading each team. The three core project managers reported to a laboratory section manager who served to coordinate the entire project.

Within the lab, midway through the development process, additional teams of two to five engineers were formed. These teams developed character fonts, emulation software, and application drivers, and performed extensive verification and performance testing.

During the first year of development, the core management team's role was directed towards technical guidance, resource organization, planning, and progress tracking. In the second year of development, the management team's emphasis shifted to coordination and prioritization as the circle of people involved in the program grew larger, and as the date for product introduction grew closer.

Our project started with a loose collection of specifications—a list describing what features we felt customers needed—tempered by what we felt we could technically achieve given our time, people, and financial resource levels. In the broadest sense, the goal of the project teams was to transform that list into an engineering specification.

DESKJET PRINTER FEATURES

The HP DeskJet printer, is a personal-convenience printer that produces laser-quality output at a price comparable to other low-cost personal printers. Among its features are 300-dot-per-inch resolution, merged text and graphics, multiple fonts, two slots for font or personality cartridges, 120-character-per-second letter-quality speed, built-in cut-sheet feeder for common office paper, desktop design, and quiet operation.

The DeskJet printer comes with Centronics parallel and RS-232-D interfaces. It is supported on HP Vectra, Portable, and Touchscreen personal computers, HP terminals, the Apple II series, IBM PC/XT, PC/AT, and PS/2 computers, and compatibles. It is also supported on HP 3000, HP 1000 A Series, HP 9000, and HP 260 systems.

Many applications software packages, such as spreadsheet and word processing programs, support the DeskJet printer. For other packages, the HP LaserJet Series II printer driver will work well because the DeskJet printer uses the HP PCL (Printer Command Language) Level III command set. An Epson FX-80 driver will also work if used with the optional HP Epson DeskJet personality cartridge, which fits into one of the DeskJet printer's option slots.

TECHNICAL CHALLENGES

The development path is never smooth or level. There are steep hills and traverses across unexplored territory. The technical challenges on the DeskJet printer project were many:

- Extend the 96-dpi inkjet printing technology to 300 dpi, keeping the printing speed above 120 cps
- Ensure that the operation of the ink delivery system is totally transparent to the user, eliminating the messy image that inkjet printing had inherited from its early days
- Make this printing technology available on all standard office papers, eliminating the dependency of inkjet printing on special papers
- Provide these product features in a small-footprint package that does not dominate the desk on which it sits
- Make this product of traditional HP quality, with reliability unrivaled by any other printer
- Accomplish all of this within a tight development schedule of 22 months with a design that can be built, distributed, and profitably sold for a low target price.

DESIGN FOR RELIABILITY

The hallmark of a successful project is careful risk management, for touch or recalcitrant problems require intensive resources to solve. If the development team is considered as a problem-solving engine, then that engine has a specific capacity, and for a given complement of engineers, there is a limit to the number and difficulty of technical problems the engine can solve in a given time period.

With follow-on products, the design task is that of interpolating from a well-understood basis, peaking performance, adjusting features, or reducing costs.

In breakthrough products, the design task must include solutions that use unfamiliar technologies. It is these forays into unexplored regions that must be carefully limited to essential development, since there is little experience to guide progress or to gauge the potential difficulties. In other words, the design teams must carefully choose which problems they are going to solve.

An example will help to illustrate this point. Early in the DeskJet development project, the mechanical design team elected to use filled thermoset plastic for the major structural part (the chassis). This decision was based on experience with similar structural plastic parts in several HP Divisions. In fact, the material set chosen for the structure and gears was identical to that successfully used in the PaintJet printer. The DeskJet team sought to reduce its design load by using an existing and well-understood technology. Or so we thought! The first prototype printers assembled from the molded plastic parts showed rapid deterioration of bearing materials with resultant squeaking, galling, and seizing—often within a few tens of pages. It turned out that we had exceeded a critical PV (pressure–velocity) point in the bearing loads. Until solutions were found (it took several intensive weeks), all design teams were hampered in their development by a lack of working prototype printers.

In our laboratory, team techniques are an important contributor to rapid progress and reliable design. Development tasks always have a principal designer and a subsidiary designer. The principal designer has part responsibility, while the subsidiary designer is a valued consultant. This designer pairing is based on interacting part/subsystem functions. Thus, a web of pairings exists, connecting the designer teams.

This pairing has several advantages over solitary design. The synergism of two (or more) designers working on the same problem is remarkable, and the quality and quantity of potential solutions is superior. In addition, solutions always have two committed designers to argue their merits with the rest of the design team. Furthermore, the principal designer has a backup in the event of illness, a trip, or reassignment to another design task.

We also encourage informal and frequent design reviews. Typically, these occur when the designers have a concept worked out that is supported by preliminary analysis. The designers meet with those who have expertise in the area (in mechanical designs usually including a procurement engineer and a manufacturing engineer), and walk through the design with their peers. It is important that these reviews occur early in the design so ideas and suggestions can be incorporated easily.

These design reviews occur close in time to the development of the first crude prototypes. Initially, the early mockups examine a subsystem or specific function, such as picking paper with a platen roller. Later refinements are incorporated in a product breadboard—a working printer that demonstrates all of the critical subsystem functions.

TESTING THE DESIGN

Coincident with the emergence of the prototypes is testing. First the basic concept is examined to see if it addresses all of the design constraints. Next, normal operations are explored to find where the design is deficient. (It always is!) Finally, the design limits are probed through accelerated tests or abuse testing.

Throughout, the engineer repeatedly cycles through the test–analyze–fix process. Initial testing yields many easily discovered defects, which can be quickly resolved with engineering analysis. Subsequent testing is aimed at improving ruggedness and reliability; these test scenarios usually require many unit-hours of experience, and the conclusions must be reached by careful statistical inference.

Compounding the statistical problem is that of securing representative parts. Much of the initial testing is performed with parts from prototype tools or processes. The design limits are not well understood at this point, and the parts are varying because the process is still unstable and undeveloped. Tolerance analysis of the design is useful, but not sufficient, since the called-for tolerances must be satisfied by a production process. Much effort goes on at this time to allocate the tolerances and allowances between parts and process.

The final phases of testing involve tests under controlled conditions by impartial quality-assurance engineers. Here, testing to rigorous HP standards is completed, including temperature and humidity excursions, shock and vibration tests, and transportation and use/abuse tests that seek out the weak links in the design.

Life testing proceeds under accelerated and nonaccelerated conditions, probing the design for deterioration, wearout, and contamination.

In summary, reliable design requires more than theoretical design skills and analyses. Although good first-round designs are an essential foundation, the bulk of the engineer's efforts go into executing well-thought-out testing programs whose intent is to stress the design and uncover its limitations so that improvements in the subsystems and the integrated product can be made.

MARKET RESEARCH AS A DESIGN TOOL[2]

The DeskJet story is filled with thoughtful responses to design challenges, as are many product development histories. But in this case, the product is all the more successful because of the design team's speedy reaction to market research feedback, thus enabling the product to deliver a key benefit long sought by customers, but never before achieved.

For several years, low-end printer customers have been expressing the desire to have a "printer on my desk, comparable in cost to other low-cost personal printers, that produces output like a laser printer." Almost without exception, in qualitative research sessions (focus groups) conducted by the HP Vancouver Division, individual users expressed preference for the convenience of their own printer, on the desk, for low-volume printing to support their own work. Although inkjet printers were perceived by many as unreliable, it was discovered that customers really judge printers by the benefit they deliver—the quality of the output.

Thus was born the DeskJet concept. The challenge was to use thermal inkjet technology to satisfy that commonly heard wish for an inexpensive laser printer for the desktop, but to do so in a way that minimizes the inkjet technology issue and breaks through the clutter of the low-end printer market.

The challenge really boiled down to two components: (1) could the print quality be made good enough in a short enough time and at the right cost, and (2) how could we communicate the key benefit of the product, that is, what position should it occupy in the mind of the prospective buyer? To measure progress on the first point, early prototypes were taken to more focus groups, and print samples were taken to shopping mall test sites in late 1986 and early 1987. The results were discouraging. Even though we were proud of our achievements to date, printer users were not impressed. Specifically, the print was not black enough and not sharp enough. We learned exactly where we stood on a numerical scale with laser printers, daisy-wheel printers, and a major competitor: dead last. The product did not deliver the desired "laser printer on my desk."

After several months of intense work on ink formulation and font design, the DeskJet printer and print samples were again taken to shopping malls for testing. The print quality improvements were dramatic. Many respondents asked if the output came from a laser printer. We finally had what we felt customers had been asking for.

[2]"Market Research as a Design Tool," written by Alan Grube, Product Market Manager, Vancouver Division of Hewlett–Packard, and originally appeared in *The Hewlett-Packard Journal*, October 1988. It is reprinted by permission.

We finally had what we felt customers had been asking for. Now the task was positioning the product in a fashion to communicate the message, in everything from advertising to training to sales tools to public relations efforts.

A key market research effort consisted of extensive telephone interviews of over 800 printer users. It showed that there is a large segment of users who yearn to own a laser printer but never really intend to purchase one. They may drop back to a low-priced impact printer, usually one of the 24-wire models, because they can't afford or justify a laser printer. A marketer's dream was born: to position a low-cost product around the benefits of an upscale product. And since the design team had done an extraordinary job of controlling cost and keeping on schedule, the DeskJet printer was brought to market ahead of the competition at a price comparable to other personal desktop printers that do not deliver laser quality.

The idea of laser-quality output for a personal-printer price has been so effective in communicating the long-sought benefits of the product that the DeskJet printer received over 26,000 orders in its first month.

HP DESKJET PRINTER: RELATED ASSIGNMENTS

1. Analyze the case, read some articles on matrix management, and talk to some project managers.
2. Generate five lists:
   - 2.1 Challenges and problems that you expected the project manager had to face in planning, organizing, and integrating the project.
   - 2.2 Specific skills that you feel were crucial to the management of the DeskJet Printer.
   - 2.3 Specific job responsibilities of the project leader.
   - 2.4 Specific job responsibilities of the team leaders.
   - 2.5 Specific job responsibilities of the (a) design engineers and (b) financial specialists.
3. Who provided the technical direction for the mechanical design engineers?
4. Draw a flow chart of the major product development phases as you see them.
5. If during the early phases of the project definition, a cost and time estimate would be required for the total product development, how could one approach obtaining such an estimate?
6. (Preparation for group discussion.) Assume that you have been hired as a management consultant at the time of the formation of a new development project similar to the DeskJet Printer. You are being asked to advise HP management on the type of organizational and managerial provisions (changes, steps, developments, reorganization, procedures, etc.) they should make to assure that the project runs smoothly and is successful.
   - 6.1 What factors would you examine? What questions would you ask of management?

6.2 Using the attached article as a factual summary of what typically happens in an HP development project, what suggestions would you make to ensure effective management of the next project toward its successful completion?

7. Knowing that HP is matrix-organized for their engineering product developments, do you think this is a good structural choice? Why or why not? What alternative does HP have? What would you suggest?

8. Classroom Activities

8.1 Each student team displays their lists and the results of points 2–4.

8.2 Group discussion on point 5.

8.3 Group discussion on point 6. Some student teams will be assigned to present and discuss their suggestions to the other teams, who act as HP management and evaluate the recommendations.

8.4 Student teams are prepared to discuss point 7. Instructor will facilitate.

APPENDIX 2.2

CASE STUDY: THE MK 2000 NEW PRODUCT TEAM

The MK 2000 case study discusses a technical project that ran into difficulties during its transition from Engineering into Manufacturing. The two interfacing groups were not organized as one integrated project team, but operated as two separate work groups. A recently established central program office, responsible for multifunctional coordination, did not seem to work either. The case provides the basis for discussing the total spectrum of challenges and processes involved in building a multifunctional technical project team.*

CASE DESCRIPTION

Mark Jones, the program manager of the new MK 2000 computer product, came back from his monthly program review meeting highly frustrated and discouraged. Mark was responsible for the product development, including concept design, product engineering and the transitioning into volume production. Although not without problems, the program moved reasonably smoothly through the engineering design and prototyping phase. However, the transitioning of the work into production via manufacturing engineering seemed to face insurmountable problems. After a detailed review of design documents and the engineering prototype, the manufacturing engineering personnel concluded that the design was not conducive to low-cost volume production. In addition, the design specified several long-lead items. This came as a

*This case was developed to illustrate a technology-oriented management situation. Fictitious names have been adopted for people, products, and organizational units.

surprise to manufacturing, since such items were supposed to be identified two months earlier.

CORPORATE ORGANIZATION

The MK 2000 product development was one of many projects that was executed within the matrix organization of the Quantum Data Corporation (QDC), a volume producer of small computer systems for Original Equipment Manufacturers (OEM) and industrial end-user markets. The organization chart in Figure A2.1 shows a typical functional structure of the various resource departments at Quantum Data. A great deal of coordination is required among the resource departments to successfully integrate the various subsystems toward a new product introduction or product upgrade. The engineering function, managed by George Jackson, is primarily chartered with new product developments and product enhancements, plus some special projects involving customized systems. Because of the intense new product activity within engineering, a special department was established under Don Patterson, responsible for the integration of all product development projects throughout the company.

In addition to engineering, engineering services and manufacturing are primarily involved in bringing a new product on line. The engineering services department, managed by Beth Brady, includes testing facilities, model shop, and documentation. All projects use this resource. Advanced work planning is extremely difficult as many of the jobs surface spontaneously and priorities are often dictated by upper management, depending on the criticality of a particular project rather than on its original schedule. The manufacturing department, managed by Chuck Goldberg, includes some corporate-level manufacturing engineering, plant maintenance, purchasing, and material handling people, plus three major manufacturing plants, including one in Singapore. Each plant is responsible for the set-up and volume production of several product lines and some customized specialty systems. Each plant has its own manufacturing services, capable of handling the complete production cycle, from receiving to shipping.

PRODUCT DEVELOPMENT PROCESS

The new product development process begins with a formal request by the vice-president of marketing and sales to the vice-president of operations, after the feasibility and desirability of the new product has been established and the basic budget has been approved. At this point in time, the request includes a budget, a target date for the product introduction, and the basic product concept. The vice-president of operations then requests from the director of engineering a detailed project plan, which includes a verification of the operational feasibility of the new product according to the established time and budget constraints.

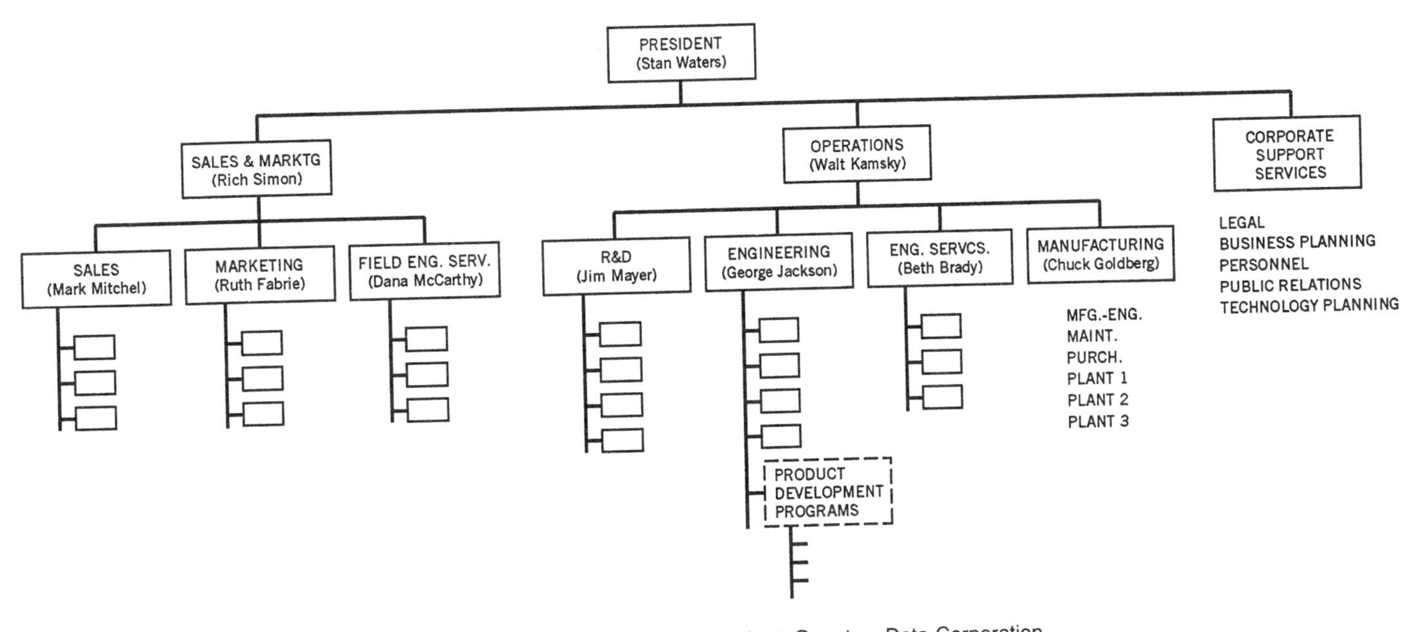

FIGURE A2.1 Organizational chart, Quantum Data Corporation.

Although the director of engineering is ultimately responsible for the new product's development and its transitioning into volume production, the project is managed by the newly established Product Development Programs (PDP) office, currently under Don Patterson. After a new product development request has been received, PDP appoints a technical program manager (TPM), who is responsible for all of the project definitions, including the verification of operational feasibility. Then the project plan is reviewed at the director's level, and if mutually agreed to, the plan is endorsed by all directors and vice-presidents who provide resources, including marketing/sales, who have to live with the final product and represent the actual sponsors of the program. After this director-level endorsement, the new product project plan becomes a quasi-contract for the technical program manager to implement. At this point there is strong upper-management support for the project plan; however, it is up to the TPM to negotiate and integrate operational resources from the various support groups.

SOME EARLY WARNING SIGNALS

Although the new product development process is well-defined at QDC, operations resource managers frequently complain that they are not part of the start-up project planning and feasibility activities. The project plan is given to them after all crucial data have been set in concrete, with little flexibility and regard for existing commitments. The dialogue between the director of manufacturing and the vice-president of operations during a staff meeting three months ago was typical of this situation.

Dir. Mfg.: Walt, every time we transfer a new product from engineering into manufacturing there are major producibility and testability problems. There must be a better way of doing it. In the end it's always us who look bad when the product has to be returned to engineering for redesign. However, these past problems will appear minor to what we will see when the MK 2000 gets into production. My people tell me that they know nothing about the new design and apparently nobody looked over the shoulder of engineering to assure that the new design is producible at the desired volume and cost.

Dir. Eng.: Chuck, don't put the blame on engineering. We run many projects and can't ask for your endorsement on every move we are making. It is our policy, though, to invite your people to every major design review. So, if you want to get involved early in a product design cycle, these meetings are your chance to participate. We completed our critical design review on the MK 2000 last week, I don't remember seeing a soul from your shop.

Dir. Mfg.: You are right, George. That's why I am concerned. We don't even have a project leader assigned for the MK 2000.

Dir. Eng.: Hear, hear! Who do you want to blame for this, Chuck?

Dir. Mfg.: We have not seen any work authorization or budget release on the MK 2000. In fact, if I didn't come to these staff meetings, I wouldn't even know where the project stands. I thought it was Don Patterson's (PDO) responsibility to make sure that everyone communicates and the project gets integrated across functional lines. Last week I took it upon myself to find out who controls what. I finally spoke to Don, who referred me to his new whiz kid Mark . . . what's his name? Mark complained to me that he received little cooperation from Manufacturing Engineering regarding the project transition. Apparently, he tried to sign up some of my people right on the spot. I couldn't believe my ears! I always supported an across-functions coordination position. But we need someone who understands the organization and can work with our senior managers.

Dir. Eng.: Okay, Chuck, you made your point. Let me find out why the process doesn't work for Manufacturing. All I can say is that the MK 2000 is moving along nicely within engineering. This is one of our biggest development programs and our engineers, and myself included, really take pride in being part of the MK team.

VP Oper.: Folks, I am really worried. The interface problems seem to be as bad as or worse than last year, when we didn't have the central coordination function. Don's PDP office was supposed to eliminate all of that. . . .

LEARNING ACTIVITIES BASED ON MK 2000 CASE

Discussion Questions

1. What are the major challenges in building a multifunctional project team, such as the MK 2000 product development?
2. What type of problems should you anticipate in interfacing a project between engineering and manufacturing?
3. What are the drivers and barriers toward unifying a multifunctional team?
4. Why did the engineering group apparently have a better project team?
5. What was the role of the technical program manager? How would you rate his performance?
6. What skills are required by the technical program manager to perform his role effectively?
7. Where and how could senior management have facilitated the team-building process?
8. Can you suggest any changes in the organizational structure or management process of the Quantum Data Corporation that might improve the effectiveness of the product development process?
9. Analyze the new product development process and suggest procedural changes for improving multifunctional team effectiveness.
10. Evaluate the project definition and feasibility phase and suggest improvements.

Major Assignments

Each assignment can be performed as an individual or group exercise. They all require careful prior reading of the case and an analysis of the major issues and problems involved.

1. Write a report to the vice-president of operations evaluating the effectiveness of the new product development team at Quantum Data. Include an assessment of the challenges, problems, and your recommendations for improving the team performance.
2. Write a policy statement and an operating procedure for organizing a multifunctional product development team.
3. Outline a long-range personnel development plan for building high-performing multifunctional project teams.

Role Play/Simulation

Set-Up. Divide the group of students or participants into various functions. At the minimum, these functions should include engineering, manufacturing, and the product development program office. These functions could be expanded to include any other work group found on the organization chart of the Quantum Data Corporation. Each group should elect its leader. One group should represent the Vice-President of Operations. *The time limits* shown are for a concentrated simulation exercise. A factor of two should be applied for a more detailed simulation, or a factor of ten for a real-world problem-solving session. The exercises are listed below:

Phase 1: Self Study (30 minutes). Each group (**a**) analyzes its strengths, weaknesses, limitations, and the challenges of working toward a successful new product; (**b**) lists what it needs from other groups, in terms of cooperation, help, data, criteria, etc., to be successful, and (**c**) lists the major strengths and contributions that the other groups offer to the total multifunctional team effort. (**d**) Document each of the categories on flip charts.

Phase 2: Information-Sharing (20 minutes). Each group presents its analysis from Phase 1 to the other groups. The others can ask questions of clarification and add items to the lists of other groups. No problem-solving is allowed at this phase.

Phase 3: Problem Definition (20 minutes). Each group evaluates the feedback received during Phase 2 and defines (**a**) five problems that could be solved within the group, and (**b**) five problems that need cross-functional involvement for resolution.

Phase 4: Organizing for Problem-Solving (10 minutes). Form cross-functional problem-solving groups, consolidate, and priority-rank-order all problems.

Phase 5: Problem Solving (30 minutes). Cross-functional groups solve the multifunctional problems. The remaining teams solve their own internal team problems. Develop and list specific recommendations.

Phase 6: Recommendations (15 minutes). Present recommendations to senior management. Engage in discussion, feedback, and fine-tuning of solutions. Final endorsement or rejection of each recommendation (count *yes*/*no* vote ratio) by **(a)** all team members/participants, and **(b)** senior management. (In a real-world situation, Phases 4, 5, and 6 would be reiterated until a satisfactory acceptance level has been reached.)

APPENDIX 2.3

CASE STUDY: THERMODYNE, INCORPORATED*

INTRODUCTION

The Thermodyne, Inc. case study discusses an engineering prototype development program, Pegasus, from its proposal phase through start-up and into its main phase, execution. Many problems develop which finally lead to a work stoppage and request for status assessment.

While the Pegasus program was funded by the U.S. government, the problems, methods, and difficulties discussed in the case are not necessarily unique to government programs, but are typical for any technical program, independent of its sponsor and size.

The Thermodyne, Inc. case can be used as independent reading or for group discussions to gain insight into the problems which are encountered during the acquisition, organization, and execution of an engineering program. The case provides the basis for discussing the total spectrum of program management tools, methods, and issues, including (1) skill requirements, (2) bid proposal process, (3) project staffing, (4) planning and control, (5) performance measuring, (6) direction and leadership, (7) role of support departments, (8) dealing with risks, (9) program tracking, (10) program organization, (11) upper management leadership and control, and (12) customer interface. In addition the case provides the basis for exploring alternative ways of organizing and managing the program to minimize or avoid the problems identified in the case, and for developing a program recovery plan.

*The Thermodyne case was developed under the direction of J. Sterling Livingston, Sterling Institute, Washington, D.C. It was previously published in Hans J. Thamhain, *Engineering Program Management*, New York: Wiley, 1986, and is reprinted by permission.

CASE DESCRIPTION

On February 13 Charles Lanzell, manager of the Pegasus program of Thermodyne, Inc., was worried and deeply discouraged as he reviewed the January budget report (Table A2.1) for his Pegasus project. Lanzell had conceived Thermodyne's role in this program, had prepared the proposal, and had managed the project since having received the contract in February of last year. Now that the originally scheduled completion date was approaching, it was apparent that the program could not be completed satisfactorily for at least another four to five months. Contract costs were overrun, and a request for additional funding, submitted the previous November, had not yet been approved. Even this would not be adequate to see the project through to completion. Interim spending authority, granted by the board of directors pending action on the funding request, would be exhausted in another month at the present rate of expenditure. An even more serious problem was the possibility of losing essential prospective follow-on business because of late delivery.

History of Thermodyne, Inc.

Thermodyne, Inc., was founded five years ago by several scientists and engineers from the propulsion division of Cascade Industries, Inc. After having first operated

TABLE A2.1 Thermodyne, Inc., Monthly Budget Report, Pegasus Project for the month ending January 31 (Thousands of Dollars)

	This Month (Jan.)			Cumulative for Project		
Item	*Budget	Actual	Variance	*Budget	Actual	Variance
Direct labor						
Laboratory	—	—	—	$ 10	$ 8	$ (2)
Engineering and design	$15.4	$18.5	$3.1	155	182	27
Manufacturing	6.0	5.8	(0.2)	40	30	(10)
Test	7.4	5.7	(1.7)	60	40	(20)
Total	28.8	30.0	1.2	265	260	(5)
Overhead	28.8	29.8	1.0	236	232	(4)
Material and direct charges	2.0	6.1	4.1	42	26	(16)
General and administrative	10.2	11.2	1.0	85	81	(4)
Grand total (000s)	69.8	77.1	7.3	628	599	(29)

*Original contract budget	$501,000
Budget authorized pending approval of funding request	200,000
Total budget	$701,000

as a consulting group on thermodynamics problems, the company undertook several development projects. Although not highly profitable, these projects had been technically successful and had helped establish Thermodyne's reputation for competent development work in the thermodynamics field.

One especially successful project was a lightweight, highly efficient portable power unit. Called the Sirocco, it was developed for use in the Phobos space program. It was believed that Thermodyne's success in this project, and its general capability in aerodynamics and thermodynamics, were instrumental in obtaining a contract for the boundary-layer control system on the Aurora, an experimental transport aircraft under development for the U.S. Air Force. This was, by Thermodyne's standard, a large contract, and one that showed good future volume and profit potential for the production phase.

The Aurora contract was received in July, 19 months ago. It necessitated a heavy recruiting program for technical personnel and a considerable refinement of the organization. From June to October of that year, employment rose from 350 to 600. In the same period, manufacturing was established as a separate organizational function, and technical operations were reorganized. Figure A2.2 shows the general organization of the company at that time.

Charles Lanzell had joined Thermodyne two years earlier. He had been employed by Cascade Industries, and although not one of the original founders of Thermodyne, he was well known and respected as an unusually able and promising physicist. Lanzell became senior engineer of the thermodynamics group on the Sirocco project, and it was believed that his work had contributed appreciably to the success of the Sirocco unit.

The Pegasus Proposal

Lanzell came across the Pegasus job almost accidentally. While making a presentation on the Sirocco unit to an Army evaluation group in January of last year, Lanzell learned of the Army's interest in an alternate system of emergency and auxiliary power for the Pegasus, a new high-capacity troop- and cargo-carrying helicopter.

The prime systems contractor for the Pegasus program was the Vertech Corporation, but Western Kinetics, Inc., was responsible for the propulsion and power subsystems. Under the approach taken by Western Kinetics, primary lift and propulsion power was furnished by gas turbine engines, with emergency lifting power supplied by rotor tip ramjet units. After discussing the system with Army engineers, Lanzell concluded that the tactical applications of the Pegasus offered a good potential for high-efficiency rocket-type thrust units. Although these would involve more complicated fuel handling, tankage, and hardware, and more difficult integration, these disadvantages would be offset by substantially higher performance for short intervals.

Lanzell discussed his thoughts with Army engineers, who expressed enthusiastic interest, and urged him to submit a proposal for developing the system. Since very little time was available for proposal preparation, Lanzell immediately started work

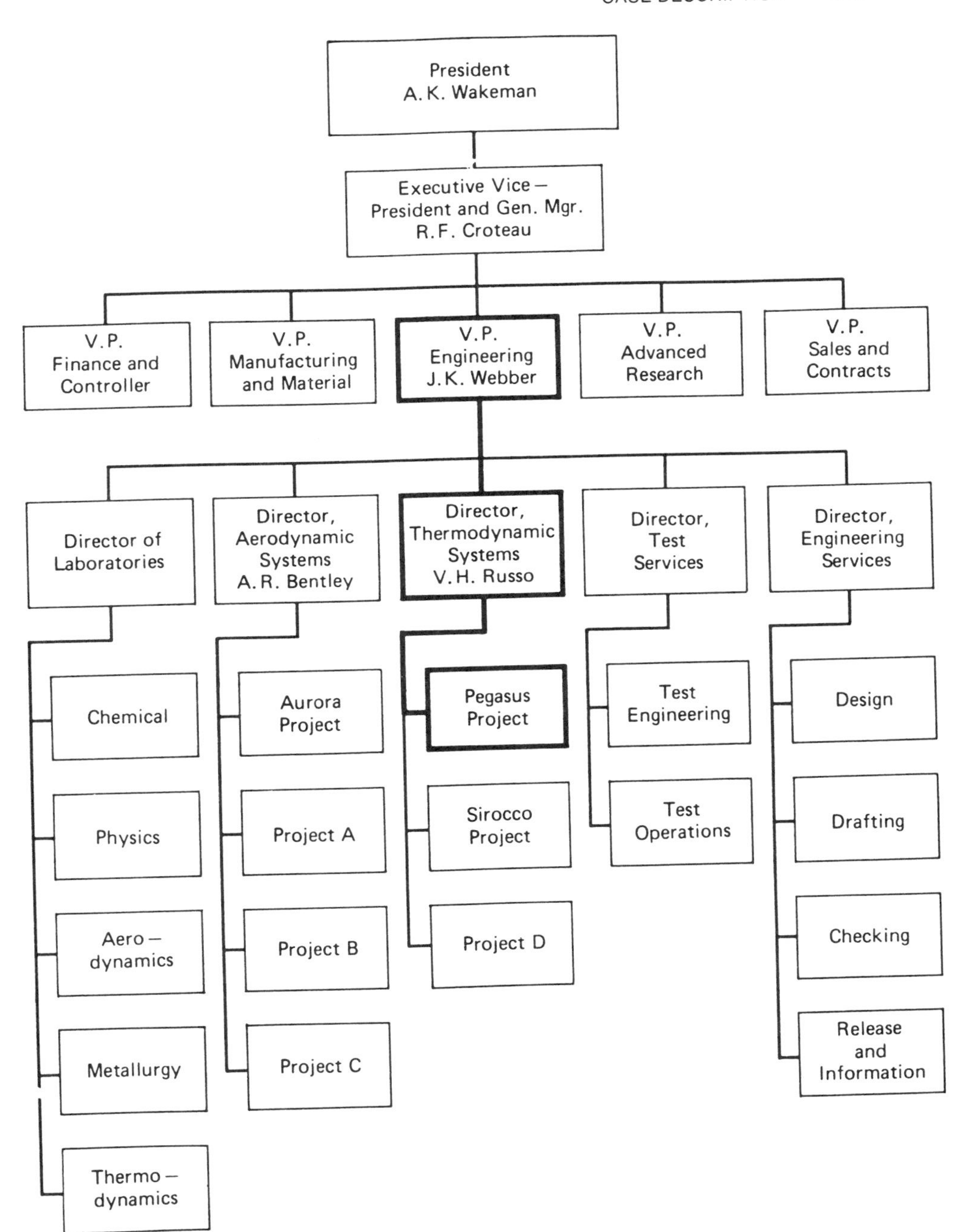

FIGURE A2.2 Thermodyne, Inc., organizational chart.

on it. The following day, a Saturday, Lanzell telephoned J. K. Webber, Vice-President of Engineering, to discuss the Pegasus proposal. His immediate senior, Victor Russo, Director of Thermodynamic Systems, was vacationing in Florida. Webber, who was about to depart on a business trip, expressed enthusiasm and instructed Lanzell to "drop whatever else you're doing. Take charge of this thing and get out a proposal."

Lanzell accordingly worked around the clock Saturday and Sunday. The technical approach offered no problem. To obtain estimates, he relied heavily on his Sirocco experience and on rate projection material, which he located in J. K. Webber's office. The time estimates had to be compatible with the overall program schedule, which called for auxiliary power unit testing and evaluation in March of the same year. Figure A2.3 shows the worksheet used by Lanzell in developing these time estimates. Table A2.2 shows the development of direct labor costs, based on hourly estimates priced at the rates shown in J. K. Webber's rate projection data. For current pricing use, this also showed an overhead rate of 80% of direct labor and a general and administrative expense rate of 15%.

In summary, Lanzell's estimate was as shown in Table A2.3.

Although proposals normally were reviewed by the vice-president of finance and the vice-president of sales and contracts, neither could be reached by telephone when the proposal was finished late Sunday evening. Consequently Lanzell mailed his finished proposal, satisfied that the proposed program offered an excellent new business opportunity for the company.

A week after the proposal was submitted, Victor Russo returned from vacation, and Lanzell briefed him on the proposed Pegasus work. Russo was impressed with the new business potential presented by the proposed system, but was slightly dubious about the time and cost estimates.

Early in February (of last year) Thermodyne received a letter of intent authorizing work to proceed on the Pegasus project as proposed. Lanzell immediately assumed responsibility for the job, and Russo appointed him project manager. Russo would have preferred to select Stanley Fram, an experienced propulsion engineer, since Lanzell was relatively inexperienced and had never before held overall project responsibilities. Russo felt that Lanzell's appointment was necessary, however, since he had personally conceived and "sold" the idea, and since J. K. Webber had authorized Lanzell to "take charge." Furthermore, Russo was not sure that it would be fair, either to the project or to the personnel involved, to ask someone else to assume responsibility for carrying out the work proposed within the time and cost limits Lanzell had established.

The first serious problem was finding technical personnel to assign to the job. Russo's organization, Thermodynamics Systems, had been drained of qualified technical personnel to staff the Aurora project under Alvin Bentley, Director of Aerodynamics Systems. Russo now requested Bentley's cooperation in returning several experienced engineers. Unfortunately, the specific persons requested by Russo were deeply involved in critical tasks and could not be removed without creating serious technical problems. Bentley was able to make available two engineers, one of whom,

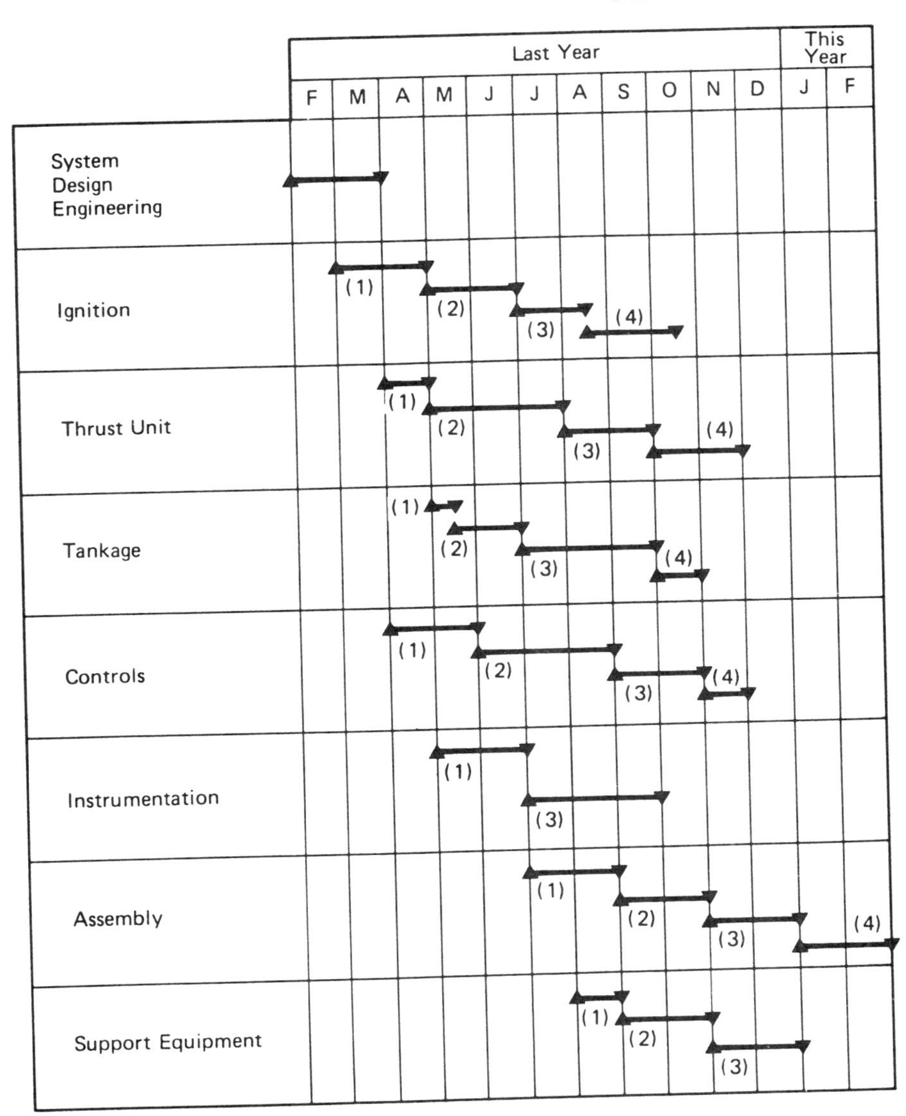

(1): Engineering
(2): Design and Drafting
(3): Procurement and Manufacturing
(4): Test

FIGURE A2.3 Thermodyne, Inc., time projection for the Pegasus project.

TABLE A2.2 Thermodyne, Inc., Direct Labor Estimate for Pegasus Project

Item	Feb	Mar	Apr	May	June	July	Aug	Sept	Oct	Nov	Dec	Jan	Feb	Total Hours	Total Dollars
Chemical lab:															
Hours	100	200	200												
$	600	1200	1200											500	3000
Metallurgy:		120	120	120											
Hours														360	2160
$		720	720	720											
Thermodynamic:		200	200	200	200										
Hours														800	4800
$		1200	1200	1200	1200										
Engineering:	450	1200	1200	1200	1200	1200	1200	900	900	600	600	500	500		
Hours														11,650	69,900
$	2700	7200	7200	7200	7200	7200	7200	5400	5400	3600	3600	3000	3000		
Design:				2000	2000	2000	2000	2000	2000						
Hours														12,000	60,000
$				10,000	10,000	10,000	10,000	10,000	10,000						
Shop:						1,000	1500	1500	1500	1000	1000				
Hours														7500	30,000
$						4000	6000	6000	6000	4000	4000				
Test:						1000	1000	1200	1500	1500	1000	1200	1600		
Hours														10,000	50,000
$						5000	5000	6000	7500	7500	5000	6000	8000		
Total hours	550	1720	1720	3520	3400	5200	5700	5600	5900	3100	26 00	1700	2000	42,810	219,860
Total $	3300	10,320	10,320	19,120	18,400	26,200	28,200	27,400	28,900	15,100	12,600	9000	11,000		
Cumulative $	3300	13,620	23,940	43,060	61,460	87,660	115,860	143,260	172,160	187,260	199,860	208,860	219,860		

TABLE A2.3 Lanzell's First Estimate

Direct labor	$220,000
Overhead at 80%	$176,000
Material and direct charges	$ 40,000
	$436,000
General and administrative expense at 15% (G&A)	$ 65,000
Total cost	$501,000
Fee	$ 30,000
Total price	$531,000

Bernard Hausmann, had an exceptional amount of aerodynamic experience. Hausmann had joined Thermodyne a year and a half earlier, too recently to have become critically involved in the Aurora project, and therefore could be made available to the Pegasus job. Since Hausmann was the most experienced engineer available, Lanzell decided to appoint him project engineer, and Russo agreed.

Monitoring Progress

Russo and Lanzell also discussed arrangements for the surveillance of progress in the Pegasus project:

Russo: Chuck, I haven't been very close to this job, and I think it's just going to spread me too thin if I try to get into it very deeply.

Lanzell: I agree, Vic. With all the planning I've done so far, it would be kind of tough for anyone to really get on board. I think I've got it pretty well under control.

Russo: I'm sure you have, and so is Webber. Said so to me the last time he was in town. Now, of course, I am going to need a monthly progress report.

Lanzell: Vic, we just don't have time for a lot of paper. These reports take a lot of work to prepare. Why don't we just have a chat every few weeks, and you can put down whatever you think is important.

Russo: Chuck, I have to have something in writing. Give me a memo—short, nothing elaborate. At a minimum, I have to have percent expended, percent completed, and a brief paragraph or two on technical progress and problems. I don't want to make a lot of work, but I have to answer for quite a few projects and can't keep them all at my fingertips without some kind of notes. Also, I have to have something to give J. K. (Webber) for monthly project review meetings.

Lanzell: Why don't I just carbon copy him on the memos to you?

Russo: O.K. with me. Incidentally, we'll probably be on the agenda for a presentation at several project review meetings, so keep it in mind. Now you've got the ball, so you're going to take care of the government reports and all that, right?

Lanzell: Right.

Following these arrangements, Lanzell used his original estimate to prepare a project budget, which projected monthly expenditures of engineering and laboratory labor, design labor, shop labor, test labor, and materials as well as other direct charges. He distributed this budget to affected managers. Within a few days several reactions occurred.

The assistant to the controller telephoned to say that the direct labor rates were inadequate, since they did not include a general increase in professional, technical, and scientific salaries and fringe benefits scheduled for April of last year.

J. K. Webber forwarded a memorandum he had received from the vice-president of manufacturing, questioning the adequacy of shop labor and material dollars and pointing out that he could not accept the budget without knowing exactly what was required. He also pointed out the immediate need for requisitioning long-lead-time materials.

Russo received a memo from the test director, who pointed out that his organization simply could not provide the hours required for Pegasus on a straight-time basis, and that half of the work, if done at all, would have to be at premium rates.

From these comments and from discussions with Hausmann, Lanzell concluded that contract funds might prove inadequate by approximately $50,000. He was confident, however, that overrun funds would be provided if necessary and felt that a request should be deferred until there had been considerable technical achievement and the adequacy of funding could be determined more precisely.

Because of staffing problems, Lanzell omitted the February progress report. The March report showed $21,000 expended. This was less than projected, but in Lanzell's opinion it was roughly equivalent to technical accomplishment. Progress was generally satisfactory, although metallurgical problems had proved more difficult than anticipated.

During May, $37,000 was expended, bringing total expenditures to about $75,000, or 15% of estimated costs. Technical performance was only 12% because of continuing metallurgical problems and release of designs in inefficient lots. Although the May schedule called for 2000 design hours, the preliminary design work had not progressed as rapidly as hoped and very little work was ready for design. At the insistence of the design manager, who had arranged for the necessary staff to be available, some work was released to design in a dubious state of readiness. This provided some work for the designers earmarked for Pegasus, although it was, in Lanzell's opinion, of questionable value.

At a project review meeting in mid-June, J. K. Webber was questioned on the status of Pegasus by the executive vice-president and general manager, and by the vice-president of manufacturing. Webber asked Lanzell to join him at the meeting and to present a brief report. Lanzell discussed the metallurgical and design problems but stated that he believed they had been overcome. At this meeting, Daniel McAlester, Vice-President of Manufacturing, questioned Lanzell:

McAlester: Lanzell, your schedule calls for shop work to start July 1—1000 shop hours next month. O.K., I've hired the men and I'll be waiting. What worries me is, are you going to be ready? What you've said so far hasn't eased my mind any.

Lanzell: I'd let you know if there were any problems.

McAlester: O.K. Now you understand, I hope, what we need—good prints in time to do our shop planning, get our tools and material together, and load the jobs. Incidentally, what with your metallurgy problems, I hope you have all your materials lined up in stores. No exotic stuff, now?

Lanzell: I can't tell you exactly, but we'll be using mostly stainless and nickel alloys.

McAlester: That's six-month stuff. We don't carry it in stores, and I know you haven't given purchasing one blessed requisition. Might as well have a layoff right now. We'll save everybody money in the long run.

Lanzell: Listen Dan, we won't need very much, and I'm almost sure I know where you can pick up whatever you need.

McAlester: I hope so. Now what about paper. I need good prints with two-week lead time. That means right now, today, I need pictures if I'm going to get my own shop paper together.

Lanzell: It's going to be pretty tight because we had to schedule this job close to meet the customer's needs. We're all going to have to give a little. With a little luck, we'll have usable prints ready the first of July. If that isn't enough time, we'll send our engineers right down to the machinists to make sure they understand what we need.

McAlester: I'll tell you one thing, you're not sending any wild-eyed junior achievement engineers into *my* shop.

Webber: Well now, look fellows, this isn't getting us anywhere. Dan, we'll see what we can do to help out—what more can we do?

After the meeting, Webber called Lanzell and stressed the importance of having some work ready for the shop by July 1. Lanzell assured him that some orders would be ready, although perhaps not worked out completely. Accordingly, design work was pushed and releases were readied on tankage and on two alternate ignition configurations. After considerable delay and difficulty in locating materials, the buyer was finally able to borrow some material from a nearby plant. Manufacturing was started in mid-July, although considerable difficulty was caused by the lack of tooling and shop operation sheets. Serious welding problems also developed due to faulty tankage dimensions, and extensive redesign work became necessary.

At the end of August expenditures were $226,000, or 45% of estimated costs. Lanzell and Hausmann estimated technical completion at 35%, but believed that the worst technical problems were past, and that performance would quickly overtake expenditures.

Early in October, Russo and Lanzell reported on the status of the Pegasus Project at the project review meeting. Lanzell said that as of the end of September, funds were 57% expended, and accomplishment perhaps 45 to 50%. One minor setback had occurred in September when both ignition configurations failed to meet their design specifications in preliminary tests. Some redesign would be necessary, and had already been started. It was now apparent that the overrun might be $50,000 to $70,000.

At this meeting Lanzell also reported on his visits to the Army evaluation center and to other contractors involved in the Pegasus Program. Army technical and contracting personnel had been very much impressed with his description of the design approach and progress and had been sympathetic with the design problems and the schedule pressure imposed on Thermodyne. It was Lanzell's impression that the contracting officer would be receptive to a request for overrun funds, if necessary. Lanzell had outlined some advanced approaches and hoped that these might be incorporated into the contract as scope of work changes, thus justifying additional fee for increased funding. In fact, considerable engineering time had already been applied to exploring these approaches.

Lanzell had also visited Vertech and Western Kinetics. Although Western Kinetics had received him cordially, Lanzell felt there had been some underlying reserve, if not hostility. He believed this meant that Western Kinetics was concerned that the Thermodyne approach indeed might prove superior to their own auxiliary lift system.

Following Lanzell's report, the problems of proposing overrun were discussed, and it was agreed to defer the discussion until the October results were known.

At this same meeting the controller reported that funding on the major Aurora effort was virtually exhausted. Unless some favorable action were taken on the pending proposals for Aurora continuation work, the company faced the immediate possibility of personnel reassignment or layoff and an increase in overhead. The vice-president of sales and contracts reported that he was confident the Air Force would accept the continuation work as proposed.

Additional problems were encountered during the remainder of October. The two test-unit thrust chambers exploded during static firings, and it became clear that the entire system needed very extensive redesign.

On October 27 the Air Force awarded a continuation contract on Aurora at approximately one-quarter of the level proposed. Its immediate effect was to raise overhead and G&A rates appreciably. It made available, however, several of the more experienced engineers for assignment to Pegasus.

At the end of October, Webber told Russo of his concern for the Pegasus project, "Vic, you'd better get close to that job. It's your responsibility, you know, and I'm afraid that Lanzell has been flying pretty high, wide, and handsome."

That same day Russo met with Lanzell, and after a thorough technical review, asked Lanzell for a complete status report and an estimate of completion time and

TABLE A2.4 Lanzell's Estimate of Costs to Complete (October)

Direct labor	$120,000
Overhead at 100%	$120,000
Material	29,000
	$269,000
G&A at 17%	$ 45,730
Required to complete	$314,730
Unexpended	$137,091
Additional funding required	$177,639

cost. Lanzell pointed out that meeting the schedule was mandatory, since the overall program schedule still called for delivery and evaluation of emergency lift subsystems in March, only five months away. They therefore would have to increase the effort in the remaining months through multiple-shift operations in the shop and in test, and extensive overtime in engineering and design. Lanzell's estimate of costs to complete was as shown in Table A2.4.

Russo felt that Lanzell's estimate might be unduly optimistic. Accordingly, he increased it to $200,000, prorating the increase among the various cost elements. Russo then advised Lanzell that he was assigning three additional Aurora senior engineers to Pegasus and that he expected them to be given broad responsibilities, commensurate with their experience. Lanzell objected to this, but Russo insisted.

At the November review meeting, Russo reported on the Pegasus project. Expenditures were running from $75,000 to $80,000 per month. Total expenditures were now $364,000, or 73% of contract funds. Due to setbacks arising from the test failures, technical completion was only 40%. Russo stressed the importance of completing the project on time and presented the revised estimate of the new funds required. He recommended, with the endorsement of J. K. Webber, that a funding request be submitted and negotiated with fee for increased work scope if possible, but without fee if necessary. The executive vice-president and general manager concurred and instructed the vice-president of sales and contracts to handle the matter. The vice-president of sales and contracts, however, was virtually certain that no scope increase could be obtained since he had heard that the Army Material Command (AMC) was extremely rigorous on granting overrun coverage. He was certainly not confident of rapid approval and warned the group that Thermodyne might have to provide its own financing to complete the work.

Russo replied that Lanzell had received more optimistic reports from the contracting officer, but that in any case the work should continue as rapidly as possible.

The funding request was submitted early in November.

November expenditures remained at about $77,000 for the month. (This figure was to remain constant during the following three months.) Overtime was heavy in all of the supporting departments. Overhead and G&A (general and administrative expenses) remained high because of the overall drop in volume caused by the reduced Aurora effort. Premium prices had to be paid for materials procurement and subcontracting because of short lead times. Operations in design, manufacturing, and test continued to be inefficient as a result of the piecemeal release of specifications, drawings, and hardware. Frequently, for example, design checkers assigned to check urgently needed design packages would wait most of the day without work, and then work long overtime hours in order to complete the check in time for reproduction and release to the shop on the following shift.

The vice-president of sales and contracts made frequent personal checks at a high level in AMC to expedite the additional funding. He was repeatedly advised that no definite answer could be given, but that cost reduction pressures from the Department of Defense were extremely heavy, and therefore the present time was not the most propitious for raising overrun questions.

Russo became more active in the project and was kept current on design and schedule problems. He contributed greatly to coordinating interface problems with manufacturing, purchasing, and test—areas that had been especially troublesome for Lanzell and Hausmann.

By the end of December, the original funding was exhausted. The contracting officer assured Lanzell that approval of new funding was almost certain, although AMC headquarters did not reflect this optimism. At their December 23 meeting the board of directors approved a recommendation made by the president and the executive vice-president and general manager to expend a maximum of $200,000 in corporate funds to cover additional work on the Pegasus project.

Substantial progress was made in December and January. The worst design problems were solved or at least alleviated. Test units of most major components were successful in meeting minimum test objectives. Nevertheless it became clear that it would be impossible to complete a fully integrated, tested, and instrumented system by the end of February. Russo estimated that this would require another $200,000 to $250,000 and four additional months.

Russo and Lanzell met on February 12 to explore the situation at length, discussing several alternative courses of action. Lanzell suggested setting up a rough system assembly not completed to specification, but capable of indicating the system's potential. He felt this could be accomplished by March 1, within the funding limitation established by the board of directors. On a recent trip to the evaluation center he had heard a rumor that Western Kinetics was also having schedule and cost difficulties on its emergency lift system. A preliminary systems evaluation might be advantageous to Thermodyne and lay the groundwork for more funding.

Another course was to request additional funding from the board. Russo was pessimistic about receiving executive endorsement for such a request, but felt that in the long run it might be advisable. Lanzell also suggested initiating an aggressive sales campaign among other propulsion and helicopter contractors, seeking additional markets, and perhaps financing, for the project.

After their discussion, Russo asked Lanzell to give him a full report of the situation with a recommended course of action. In the meantime Russo would formulate his own thoughts. After receiving Lanzell's report, Russo would submit his formal recommendations to J. K. Webber for consideration by company executives and officers.

ASSIGNMENTS BASED ON THE THERMODYNE CASE

Each assignment can be performed as an individual or group exercise. All assignments require a careful prior reading of the case and an analysis of the major issues and problems involved. One solution for each assignment is presented at the end of this case section. These solutions should not be considered as the only way to handle the assignment, but rather should stimulate the reader's thinking and show methods of approach. To ensure a learning benefit, readers should come to their own independent conclusions before consulting the solutions.

1. Analyze qualifications of Charlie Lanzell as a project manager, including his skills for planning, organizing, and leading the project.
2. Analyze the performance of the company's management with regard to direction, leadership, and assistance toward the Pegasus program.
3. Analyze the organizational structure of Thermodyne and its management systems with regard to its ability to handle projects such as Pegasus. Consider items such as Thermodyne's procedures for acquiring new business and contract management, financial controls, reviews and reports, communication channels to top management and the customer community, and integrated multifunctional project planning and implementation.
4. Assess the schedule and budget data with regard to project tracking and control.
5. Assuming you just have been hired by Mr. Webber as a consultant, what course of action would you suggest to Mr. Webber on February 15?
6. Assuming you are Mr. Lanzell back in January of last year, how would you handle the Pegasus program opportunity that you have identified with the U.S. Army?
7. Assuming you were just appointed program manager for Pegasus under the circumstances and developments given in the case, how would you organize and manage the program?
8. Assuming you are Mr. Russo back in January of last year, just returning from your trip, how would you react to the given situation? What options do you have?

CASE ANALYSIS AND SOLUTIONS

Each of the eight assignments is discussed in a separate, short essay to highlight the issues involved, to show potential approaches to overcoming the problems posted, and to stimulate further thoughts and discussions. The case situation provides plenty of room for various assumptions and different approaches for dealing with the complex project management scenarios that are presented.

Lanzell's Qualifications as Project Manager

The troubles of the Pegasus program cannot be blamed in their entirety on the program manager, but were compounded by upper management's apparent indifference to the program and their inability to manage the program as an integral part of Thermodyne's ongoing business. Furthermore, the organizational structure and management systems do not seem to be supportive to the multidisciplinary undertakings of Pegasus.

However, it is clear from the case that many of the problems and conflicts did develop to such a magnitude because of Lanzell's inexperience and inabilities as a program manager. For starters, Table A2.5 presents a listing of problems that got the program into trouble under Lanzell's leadership.

Before turning our attention to the weaknesses of Charlie Lanzell as project manager, his strengths should be pointed out. Lanzell deserves credit for taking the initiative in identifying, pursuing, and capturing the Pegasus project opportunity. He is an enthusiastic, motivated, and hard-working individual with some entrepreneurial ability, has a technical background, has been with Thermodyne for over three years, involved in ongoing projects, and is well-respected as a physicist.

However, as a project manager Lanzell scores very low. He lacks the qualifications needed to manage a million-dollar multifunctional engineering program successfully. Specifically, he lacks the know-how of basic management techniques in project planning, tracking, and controlling. In fact, he does not believe in planning and confuses established project management practices with red tape and paperwork. He appears unable to relate costs incurred to actual progress, nor is he able to predict potential problems and handle them while they are small. In spite of his technical background, it seems that Lanzell did not understand the total technology involved in the multifaceted Pegasus project. In addition, his attitudes toward planning and project management did not help. He is somewhat of a loner who does not delegate but tries to do everything by himself. He is often in conflict with others, cannot build a project team, nor can he get management to be involved in the project, supportive, and committed. He lacks the leadership skills needed to unify the project team and manage the project towards its integrated objectives.

Yet another problem area is Lanzell's insensitivity toward established cultures and value systems within Thermodyne. In spite of his three years with the company, he did not understand the organization and its interfaces well. This fact was especially evidenced during the argument with McAlester, when Lanzell suggested to send engineering personnel to the shop floor.

Perhaps because of his inexperience as project manager, Lanzell ignored many early warning signals and feedback from his boss and other senior professionals regarding potential problems and conflicts. Neither did he realize his limitations. He never asked for help. Because of his inexperience, Lanzell had a credibility problem

TABLE A2.5 Problems That Charlie Lanzell Encountered on the Pegasus Project

DURING BID PROPOSAL PHASE

1. Major oversights regarding work definition and costing.
2. No agreed-upon goals, objectives, or business reasons for pursuing Pegasus program.
3. Insufficient program planning.
4. No support from service departments, such as cost accounting and contracts.
5. No management involvement and commitment.
6. No formal bid proposal ever submitted.
7. Misinterpretation of contractual instrument. Lanzell thought he received a contract but really got a letter of intent. Thermodyne organization had no checks to pick this up.

DURING PROJECT START-UP PHASE

1. Insufficient program planning regarding task definition, timing, costs, and responsibility definition.
2. No formal planning tools, such as work breakdown structure, statement of work, work authorization, and cost tracking and projections.
3. Insufficient plan details and measurability.
4. No contingency planning.
5. Little involvement of resource personnel during project start-up. Lanzell tried to do almost everything alone.
6. Little agreement and commitment by functional support departments.
7. No management involvement and support.
8. No channels established for communicating among interdisciplinary groups, upper management, and the customer community. No strict system for program reporting and reviewing.
9. No charter for Lanzell, little authority.
10. Lanzell had low credibility as a (competent) project manager.
11. No project organization defined.
12. No clear project team definition.
13. No team spirit.
14. Lanzell was rushed into project kickoff.

DURING PROJECT EXECUTION

1. Lanzell could not build a unified project team.
2. Staffing problems persisted.
3. There was no involvement and support by upper management.
4. There was little ability to measure progress.
5. There was little ability to assess potential problems.
6. There was few or no project controls.
7. Lanzell was busy fighting "brush fires," could not see ahead or spot small problems early.
8. Lanzell was a poor listener, and ignored early warnings and feedback.
9. There were no formal project reviews and reports.
10. There was no customer communication.
11. Lanzell was insensitive to the established organizational culture and value system, such as interface with manufacturing.
12. There was no program plan revision after schedule slips became obvious.
13. There was no team spirit.
14. There was excessive interpersonal conflict.
15. There was little or no project leadership by Lanzell.
16. Finally major design problems surfaced. Preliminary estimates showed projected cost overruns of $250,000 (50%) and schedule slips of four months or more. The Pegasus project faced the reality of termination.

right from the start. The problem got bigger with increasing difficulties during the project life cycle, a situation that did not help in building quality images, motivating personnel, and getting management involved, supportive, and committed.

Even more important, on the human side, Lanzell lacks some fundamental interpersonal skills. He has little capacity to build teams and to handle conflict, he is a poor listener, is often abrasive, and has difficulties interacting with others.

Taken together, Lanzell lacks both (1) the administrative skills of planning, organizing, and tracking a project and (2) the leadership and interpersonal skills needed to unify, commit, and direct the effort toward its objectives. In addition, Lanzell did not have the technical skills needed to work with various technical groups, identify interdisciplinary technical problems, assess their impact, and participate in the search for solutions. A summary of Charlie Lanzell's qualifications as a project manager is provided in the profile of Table A2.6.

Thermodyne Management Performance

It might be tempting to put most of the blame for the poor Pegasus performance on the project manager. However, it was the Thermodyne management who let all these problems happen, without much direction and leadership, and without much help and support to Lanzell. Moreover it was upper management who, in a very unconventional way, violating established chain-of-command principles, appointed Lanzell to the post. Knowing that he was inexperienced as project manager and doubtful of his capabilities, they nevertheless let him struggle without any training, assistance, or support.

Table A2.7 identifies the major problems associated with Thermodyne's upper management, which contributed to the troubles and eventual failure of the Pegasus project.

In searching for reasons why upper management did not provide any direction and leadership for Lanzell and his Pegasus project and why Russo neglected his job as supervisor, we see that the Pegasus project was not related to any specific company goals and objectives. Possibly this could explain the apathy and disinterest of management toward Pegasus and the low team spirit, which apparently trickled down to the lower ranks of individual project contributors as well.

Besides the more glaring problems, there are some subtleties that did not help the troubled situation. Lanzell was never formally chartered to lead the Pegasus project. Thermodyne management seemed to be unclear as to what Lanzell's role as project manager really was, where his responsibilities started and ended, and what obligations the functional managers had to follow Lanzell's requests and directions. There were few or no functional managers to reschedule previously committed work. These managers may have perceived Lanzell as a person with little or no reward power, a person who was just a disruptive influence on their organizations.

Given this very poor foundation for management support, it would have taken a strong project leader with extraordinary leadership capabilities, credibility, and experience to build a priority image and to unify a Pegasus-focused project team.

TABLE A2.6 Personal Profile: Charlie Lanzell as Project Manager

TECHNICAL SKILLS

1. Good technical background, respected physicist.
2. Technical skills too narrow to manage multifunctional technology program.
3. Technical skills insufficient to anticipate engineering problems in their early stages or to participate in a search for solutions.
4. Does not know how to work with various technical groups.
5. Cannot build multidisciplinary technical teams.

PLANNING SKILLS

1. Does not believe in planning.
2. Few planning skills.
3. Cannot define work detail.
4. Does not know how to estimate.
5. Cannot assess risks.
6. Is overoptimistic.
7. No attention to details.
8. Cannot secure commitments.
9. Has no contingency plans.
10. Cannot work with company resources.
11. Does not know how to measure project performance.
12. Cannot set up communication channels.
13. Does not know how to set up and manage reviews and reports.
14. No measurability.
15. Does not communicate with customer or upper management.
16. No capability to control.
17. Does not know how to work with the organization and its interfaces.

LEADERSHIP AND INTERPERSONAL SKILLS

1. Little or no leadership ability.
2. Poor communications skills.
3. Poor listener.
4. Cannot motivate or build images.
5. Little team-building ability.
6. Little or no capacity to manage conflict.
7. Cannot delegate.
8. Is a loner.
9. Does not know how to work with upper management.

ENTREPRENEURIAL SKILLS

1. Good marketing and entrepreneurial instinct.
2. Used initiative in defining and pursuing new project business.
3. He is hard-working.
4. Initial credibility with customer.
5. Limited ability to manage a complete project business.
6. Little or no concern toward profits and overall business management.
7. Cannot integrate organizational resources.
8. Not a team player.

TABLE A2.7 Major Problems with Thermodyne's Upper Management

DURING PROPOSAL PHASE

1. Russo was unaware of the new business under development by Lanzell.
2. Russo did not have anyone acting in his place while vacationing. Therefore Lanzell had to get Webber, the vice-president of engineering, involved.
3. Webber made a very casual decision for Lanzell to proceed with the proposal to the Army, without really considering the consequences of (1) "drop whatever else you are doing," (2) "take charge," and (3) committing the company to Lanzell's bid proposal.
4. Not knowing Lanzell's qualifications and capabilities as proposal manager, Webber did not provide any assistance to Lanzell nor install any organizational checks and balances to avoid fundamental errors and oversights.
5. Apparently no bid proposal guideline or procedure existed.
6. No organizational checkpoints or sign-offs were used.
7. Sales and Contracts were not aware, or did not want to be aware, of Lanzell's proposal effort. They did not help or follow up.
8. The reason for the weekend proposal rush was never questioned.
9. The reason for pursuing the Pegasus program appears unclear. No business goals and objectives were articulated.
10. There was no effort by Thermodyne to check, validate, or correct the estimates sent to the customer.
11. There was no effort by Thermodyne to submit a formal proposal, even after the letter of intent was received.

DURING PROJECT START-UP PHASE

1. Russo appointed Lanzell as project manager under the assumptions (1) his boss wanted it and (2) we owe it to Lanzell, rather than assessing Lanzell's ability to perform.
2. Russo did not live up to his responsibilities of managing Lanzell and Pegasus.
3. Russo did not lead Lanzell through the project planning and organizing effort. He gave no guidance. He did not insist on proper planning.
4. There was no effort to validate original cost and time estimates.
5. There was no assistance on project staffing.
6. No charter was drawn up for Lanzell.
7. There was no visibility and business goals for Pegasus. There was no incentive for functional managers to support the project.
8. No management controls or checkpoints were set up.
9. Management appears indifferent toward Pegasus program.

DURING PROJECT EXECUTION

1. Russo did not manage the program. He felt uncommitted and was disinterested in the program.
2. No guidance, direction, and leadership were given to Lanzell.
3. Little effort to identify potential problems, or assist in existing problem situations, was made.
4. Upper management had no team spirit.
5. Management ignored early warning signals and feedback.
6. Webber trusted Russo too much.
7. Management did not establish any discipline or guidelines for running the Pegasus program. There was no reinforcement of any policies and procedures.
8. No formal project reviews and reports were made.
9. There was no customer interface at management level.
10. Little effort was made to diffuse interdepartmental conflicts or to seek conflict resolutions, as was evident in the Lanzell–McAlester confrontations.

Thermodyne's Organizational Structure and Management System

Thermodyne's current organization is very loosely structured, a situation that is quite common for smaller firms or technology-oriented companies that do not want to stifle innovation and creativity by rigid organizational systems. However, in the case of Thermodyne, not only was the company loosely structured in terms of authority and responsibility relations, but roles and charters of the various functional groups seem to be unclear regarding new project formation and support. For established projects, such as Sirocco and Aurora, a project management and project administrative team seems to form a project office, reporting to an autonomous business area, such as aerodynamic systems. The project office then contracts with the various functional departments, such as test and engineering, the services needed to support the project activities. In principle this operation is defined as a project matrix. It is unclear, however, how a new project can get organized properly without some integrated long-range resource planning throughout Thermodyne and some specific charters for each project office, including some means of tying project performance into the management appraisal and reward system.

Another organizational weakness is the apparent lack of any formal guidelines, policies, and procedures in the areas of bidding, organizing, and managing projects. Thermodyne does not seem to have any established norms of what are the ingredients of a project plan, who must review an estimate, who must sign off on a work package, or what type and frequency of reviews and reports are required.

Yet another critique is the lack of checks and balances, especially in the contractual and financial areas. The fact that (1) a budgetary proposal was sent to the customer without review or even a follow-up, (2) a letter of intent was received without contractual actions, and (3) the first time that concern about project performance surfaced to a business management level was when the total budget was spent points to a management system without checks and balances. In addition, the organization had few controls in place to catch problems early and to steer the project toward successful completion. Such controls could include (1) regular project reviews, (2) design reviews, (3) status reports, (4) action memos, (5) formal communication channels to customer and upper management, (6) financial status reports and analysis, (7) project performance measurement system and/or variance analysis, and (8) troubleshooters and special task forces. A summary of the shortcomings of the project management system at Thermodyne is given in Table A2.8.

Assessment of the Schedule and Budget Information

In analyzing Table A2.1 and Figure A2.3 some conclusions can be drawn with regard to the project reporting and tracking system that is in place for Pegasus. It is interesting to note that the data for financial tracking were available, but no attempt had apparently been made to use them. Here are some examples of conclusions that can be drawn from the monthly budget report shown in Table A2.1.

As indicated by the positive total variance for January, $7300 more was spent that month than originally planned. On the other hand, the cumulative variance for the

TABLE A2.8 Shortcomings of the Project Management System at Thermodyne

1. Organization too loosely structured for the size and complexity of projects managed and for the formality required in dealing with defense contracts.
2. No formal charter for Lanzell.
3. No apparent tie between project support and performance and the management appraisal and reward system.
4. No guidelines, policies, and procedures for project bidding, organizing, and managing.
5. No marketing and contract support for Pegasus.
6. No checks and balances, such as budget reviews, design reviews, sign-offs, contractual checks, financial and milestone checkpoints, and customer liaison.
7. Few management controls, such as regular project reviews, reports, action memos, design review, financial status and analysis, performance measurement system, and special task forces.

total project indicates an *underspending* of $29,000. This suggests the possibility of a late start and attempts to catch up now. The data by themselves do not indicate any performance problems, since they do not relate to the project completion or technical performance. Without this performance measure, no cost-to-complete estimate, variance analysis, or performance index can be determined.

The budget report of Table A2.1 provides some clues, however, about the flow of interdependent activities. Both the monthly report and especially the cumulative report show a substantial overrun (cumulative $27,000) for engineering and design activities, while the engineering-dependent activities in manufacturing and testing show a substantial underrun (cumulative $10,000 and $20,000). To the experienced project manager this indicates a potential problem at the engineering-manufacturing interface. Engineering seems to work very hard, maybe to fix a problem, while manufacturing and testing are possibly waiting for inputs. As we know from the case, this is indeed what happened. By the same logic, we may suspect that material ordering runs behind schedule; only 62% of the total material budget has been spent so far.

The bar chart schedule, shown in Figure A6.4, is a useful tool for the initial project planning, but also for overall project guidance. It defines the time phasing of the main activities as they move through the four project phases: (1) engineering, (2) design and drafting, (3) procurement and manufacturing, and (4) testing. Each one of the activities should relate to a level I task group on the work breakdown structure. Of course there is no evidence that such a work breakdown structure existed for Pegasus.

On the other hand there are some serious shortcomings to the schedule shown in Figure A6.4. First, the schedule is very coarse, allowing people to track activities only on a monthly basis. There is no evidence that the main activities were broken into subactivities with appropriate schedules. Second, apparently no attempt was made to update the Pegasus program. Obviously many schedule slips occurred, which had nominal effects on other activities. Third, the Pegasus program could have benefitted from a network master schedule, which shows the interdependence of the major activities. Fourth, nowhere in the case are the specific tasks and their results measurably defined. Therefore it is difficult, if not impossible, to actually determine the project performance as an integrated measure of technical accomplishment, schedule, and budget; that is, the current schedule and budget tracking system does not enable us to determine technical performance in terms of percent complete, and we have no basis for determining the earned value of the work complete, the cost to complete, the time to complete, or any of the other project performance measures discussed in Chapter 5.

How to Proceed with the Pegasus Project

Five basic options are open to Thermodyne in proceeding with the Pegasus project. As shown in Figure A2.4, we have some potential choices in handling the funding and the technical performance by either sticking to the original requirements or

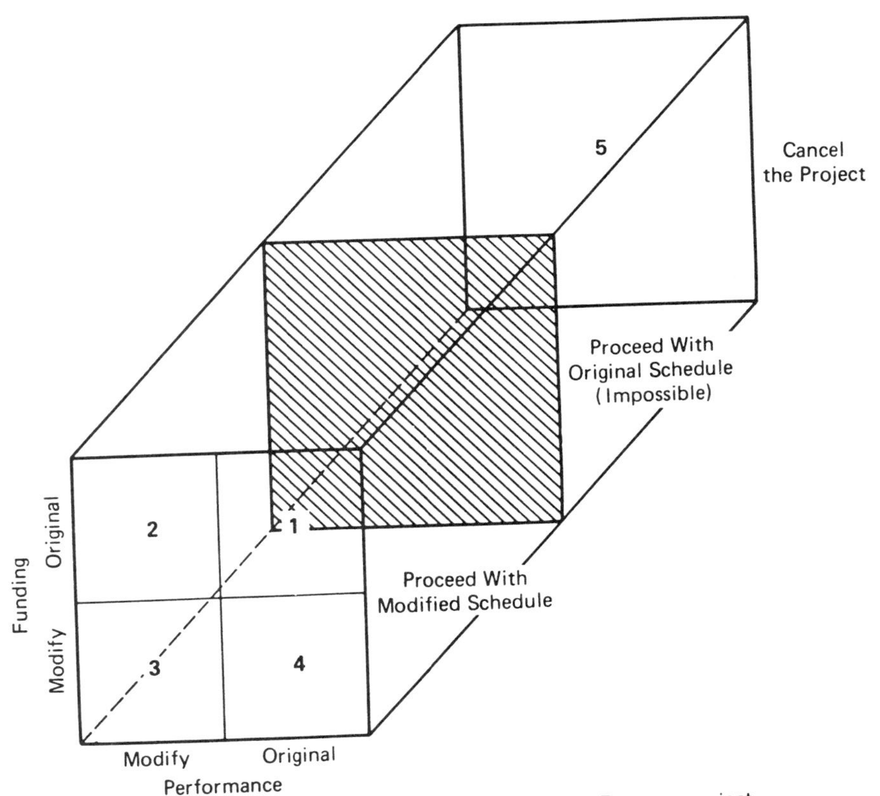

FIGURE A2.4 Five ways to proceed with the Pegasus project.

modifying the original plan. Since we are running out of time, no option exists in handling the schedule other than modifying it. Hence the five options are:

1. Plan to complete the project within originally agreed-upon performance and customer funding. Thermodyne covers for budget overrun.
2. Plan to complete the project with original customer funding, but with modified performance, to be agreed upon by customer.
3. Plan to complete the project with modified funding and modified performance, both to be negotiated with customer.
4. Plan to complete the project with modified funding, but within originally agreed-upon performance.
5. Cancel Pegasus.

Besides the five dichotomies stated above, there are a large number of potential in-between solutions to the problem that we are currently facing. It is clear that we need the customer's input to determine which of the options are feasible and to what degree any modification from the original plan is possible. We will exceed the planned project completion date, and the customer may not have any use for Pegasus beyond the February dateline.

On the other hand, it may be risky to approach the customer without a clear project status in hand and without a strategy for a preferred action plan. This suggests that we first perform a status assessment and formulate the preferred recovery approach before seeing the customer. The quandary is, however, that such a sequential approach has high risk factors: (1) the customer may learn through the grapevine about the current troubles, which could make the situation worse, (2) a status assessment without potential recovery options has a tendency to drift and to focus on who is guilty rather than on what should be done next, and (3) presenting a recovery plan to the customer without knowing its basic feasibility presents many risks and may exclude some other options.

A hybrid approach of initiating customer contact *while* performing a project status assessment might be most desirable. The sequence of events for such a project status assessment and recovery plan is discussed below.

Project Status Assessment. Three key questions must be answered at the outset: (1) who should perform the assessment, (2) how much time should be allocated for the task, and (3) what should be achieved.

1. Who Should Perform the Assessment? An effective project troubleshooter is someone who (1) is impartial to the Pegasus project, a person who is currently not associated with the project and its problems; (2) is an experienced and competent project leader, able to probe effectively into the current situation and to communicate with other project personnel; (3) has been given the authority and blessing by upper management to perform such as assessment; and (4) knows the technology involved and the organizational culture and value system. While the first three qualifications are absolutely essential for a meaningful assessment, the fourth is somewhat of a luxury that can be worked around by a competent project manager.

2. How Much Time Should Be Allowed for the Assessment? Project professionals often feel that this question depends on the size and complexity of the project to be assessed, and on the quality and depth requested. A different philosophy is presented here. First, for most projects in as much trouble as Pegasus, time is of essence. We do not have months or even weeks to determine where we are and whether it makes sense to negotiate for continuation or cancellation. Second, the longer the investigation lasts, the more of a risk there is that the status assessment turns into a witch-hunt, where project personnel are busy with finger-pointing, creating just-in-case files, and covering up project realities because they are afraid that these facts could be held against them at some point. An assessment that has deteriorated to such a level does not foster an environment for a candid assessment or for building a "can-do" team spirit. It is usually an environment that is full of mistrust, conflict, insecurity, and distortion of the facts that should be realistically assessed. Third, a quick status assessment, within a few days, is possible and is independent if the project approach is followed. The key to this discipline is an understanding of the scope of the assessment. We are not developing a detailed project recovery plan at this point, nor are we interested in all the minute details at the lowest project level. What we want to know in broad terms is: how complete is each of the major subsystems, where are the major problems and risk areas, what should be done about them, and how much more time and money are required to complete the project in its original or modified concept. Following a disciplined approach, such as the one discussed next, the time allowed for a project status assessment *should not exceed three days.*

3. What Results Should Be Achieved by the End of the Project Status Assessment? The results should form the basis for the management decision on the most appropriate course of action to be taken regarding the project recovery or its cancellation. The results further form the basis for the project recovery plan's development. Specifically, the status assessment should be subdivided into project subsystems, stating:

- Status of completion
- Problems faced: overall and by task element (technical, managerial, data requirements, resources, etc.)
- Specific help needed (modified specifications, more personnel, support from other departments, etc.)
- Alternative approaches suggested for project completion
- Time and money needed to complete each project alternative
- Risks.

How to Proceed with the Project Status Assessment. After the individual responsible for the status assessment has been selected, a brief kickoff meeting may be in order. This meeting should be attended by all project personnel who report directly to the project office, that is, all the task managers responsible for a work breakdown structure level I assignment. Also present should be key management personnel, such as Russo, Webber, McAlester, and, of course, the project manager, Lanzell. Someone like Webber may want to address the meeting and bring the forthcoming assessment into proper perspective. The results achieved on the project should be acknowledged and credited to the team, the management's willingness to support and rectify the problems should be stated, and the purpose of the assessment should be clear. It is important at this point to foster an environment of trust and motivation toward completing the project. The message should be: we need your help so that we can give you the proper support; we do not want to know who caused a problem, but rather what these problems are and how to overcome them. The meeting should be brief, maybe 10 minutes.

After kickoff the individual responsible for the status assessment meets with each level I task manager, or equivalent, on a one-on-one basis. These meetings should be scheduled well in advance. As a guideline, these meetings may be scheduled for 45 minutes for the first time and one hour for the next iteration. The agenda—such as the six points discussed in the previous section: (1) status, (2) problems, (3) help needed, (4) alternatives, (5) time and money to complete, and (6) risks—should be clearly communicated to all participants.

Besides determining the status and requirements for each project work package or subsystem, the troubleshooter must also try to determine the validity of the information by probing into the level of detail at which any subplan exists, the consistency of the subsystem results compared with the overall project specifications and requirements, and the capability of the task team to perform against the given assignment.

Usually the first round of meetings cross-validates much of the status information and identifies the major problem areas. During the second round of meetings, discrepancies of information from different task groups should be solidified and missing pieces filled in. It is often during the second round of assessment that the need for group meetings becomes obvious. These group meetings with various task leaders are often a useful vehicle to resolve interdisciplinary issues that involve several functional groups.

As an integral part of this interviewing, senior functional managers from the various supporting departments, such as engineering, manufacturing, contracts, and finance, should be consulted to round off the assessment and to validate the information obtained at the project level.

Meanwhile, someone is initiating contact with the customer community. These discussions might provide some insight as to what options are realistically open to Thermodyne regarding the schedule, funding, and technical requirements of Pegasus. These inputs will provide focus toward the alternatives that can be pursued.

The final step in the assessment process is to prepare a brief report and some overhead view graphs for a project status presentation to management. At the meeting, all personnel who participated in the assessment should be present. This will help validate the data, add credibility to the presentation, and build the team spirit needed for subsequently organizing the project toward successful completion.

How to Get the Project under Contract Again. All the previous activities are useless unless we can persuade the sponsor, in our case the U.S. Army, to continue the project with us. At this point we have all the components together for formally negotiating with the customer. We have:

1. Project status assessment completed, identifying the current problems, risks, desired alternatives, and time and resources needed to complete
2. Some idea of options that are acceptable to the customer
3. Management's decision regarding what is desirable and feasible
4. A basic recovery plan or approach.

When discussing the project status and options for recovery, it is crucial to build trust and credibility with the customer community. The ingredients of a can-do image are as follows: (1) we know exactly where we are on the project, (2) we understand the problems that caused the current situation, (3) we have workable solutions, and (4) we learned from the problems and will make sure that they cannot happen again. One of the best ways to build this credibility is to present a detailed project recovery plan, which follows the principles of good program planning discussed in Chapters 4 to 6. Thermodyne also must demonstrate that it indeed has the resources and capability to handle the Pegasus project, that it has fixed some fundamental problems, such as the lack of project tracking and review systems, and that Thermodyne management is fully committed to the project.

Prior to developing a detailed project recovery plan, a principal agreement must be worked out with the customer on the basic approach that is acceptable to both the customer and the contractor. Another reality of trying to recover from a major project setback is the approach often taken by the customer of canceling all current contractual arrangements and opening the project opportunity up again for competitive bidding. A summary of the steps involved in recovering from an adverse project situation is given in Table A2.9.

Handling the New Business Opportunity with Pegasus

Identifying, pursuing, and finally capturing new project business is hard work that often requires creative, innovative approaches outside the formal contractual framework established in a business community. One important ground rule is to work *with* the customer and to accommodate any special requests and concerns whenever possible. It is quite common for a government or industry customer to ask for a letter proposal, a preliminary proposal, or a budgetary estimate very early in a new project development in order to have a basis for comparison with competing ideas and for assessing initial feasibility. Response to such a request must usually be quick in order to maintain the customer's interest and to be relevant to the initial project work.

Lanzell certainly deserves credit for identifying the new business opportunity and for taking the initiative of pursuing it in a very timely manner. The critique centers

TABLE A2.9 How to Recover from an Adverse Project Situation

1. Organize for project status assessment
   - 1.1 Define management approach
   - 1.2 Select qualified individual to lead assessment
   - 1.3 Define detailed approach and gather background information
   - 1.4 Kickoff; message from management
2. Contact customer
   - 2.1 Open up communications.
   - 2.2 Explore basic options available
3. Perform status assessment
   - 3.1 Collect information from each work package leader (level I task manager) regarding: work status, problems, help needed, changes requested, time and money to complete, risks
   - 3.2 Collect information from managers of functional support departments
   - 3.3 Iterate meetings with work package leaders; resolve discrepancies in small group meetings
   - 3.4 Integrate data into overall project status assessment with focus on latest customer inputs
   - 3.5 Prepare for presentation to management
4. Present project status assessment to management and project team.
5. Make preliminary management decision
   - 5.1 Preferred options
   - 5.2 Customer approach
6. Define basic project recovery plan
7. Make formal status presentation to customer
8. Negotiate basic approach of recovery (or termination) that is acceptable to both customer and contractor
9. Develop detailed project recovery plan
   - 9.1 Show detailed project status
   - 9.2 Show understanding of current problems
   - 9.3 Show workable solutions
   - 9.4 Show that you learned from problems and fixed them
   - 9.5 Show detailed project plan for recovery
   - 9.6 Demonstrate available resources and capabilities
   - 9.7 Demonstrate management commitment
10. Submit and present project recovery plan (proposal) to customer and negotiate contract

on the process, most obviously on the way in which Lanzell rushed the proposal through, without any help and sign-offs, resulting in many oversights and inaccuracies. Although unnecessary, this would not have been such a terrible thing, given the realities of project marketing, if this initial budgetary estimate had been followed up with a detailed proposal and contract negotiation.

One possible approach that Lanzell could have taken in pursuing the new project opportunity is delineated below:

1. Identify new business opportunity during a presentation at customer site.
2. Discuss customer requirements. Suggest potential solutions. Get customer interest. Identify budget and schedule restraints.
3. Make oral proposal. Test relevance and basic feasibility with customer. Test principal interest. Does it make sense to pursue this opportunity any further?
4. Check actual interest of Thermodyne management. Talk to Russo or Webber. Get formal backing from management ("take charge" provides authority to proceed).
5. Develop a letter proposal stating (1) the technical baseline, including basic features and performance specifications, (2) the approach to executing the project, (3) the major milestone schedule, and (4) the budgetary estimate. Get as many functional support groups involved as possible. This will improve the quality of the letter proposal, check the initial feasibility of the project, and stimulate interest among the company personnel and management.
6. Ensure that cost accounting checks the cost estimate for proper process, rates, and format. Even better, let cost accounting prepare the final budgetary estimate. Advise the contracts department of his activities. Have someone in management endorse or sign off on the letter proposal.
7. Follow up with the customer. Fine-tune his initial proposal in customer discussion.
8. Try to obtain a Letter of Intent from the customer, which might provide some initial funding for a feasibility study, project start-up, or a more detailed proposal.
9. Follow up with a detailed proposal, which usually includes revised time and cost estimates. Or try to obtain a purchase order that delineates the contractual requirement feasible and acceptable to Lanzell's company.
10. The formal contractual document, such as the purchase order or contract, needs to be checked carefully and agreed upon in at least the three principal areas—technical, timing, and cost—by all senior personnel eventually responsible for the project implementation. Usually the final phase is a negotiation among the principal parties, resulting in a contract signed by at least one corporate officer in each contracting party.

Organizing the Pegasus Program

Charlie Lanzell wanted to become project manager of Pegasus. Therefore it is not surprising that he enthusiastically immersed himself in the job without too much concern about the tough situation he was facing. Right from the start, Lanzell had several strikes against him:

1. Russo was indifferent toward the project and unwilling to help in solving some of the start-up problems.

2. The project was not connected to a major business goal or part of a business plan. Therefore priority for support was low.
3. None of the personnel needed was involved during the proposal phase.
4. Lanzell was inexperienced and did not realize his limitations.
5. Thermodyne did not have a workable project support system.
6. Lanzell's credibility as project manager was low. He also had little organizational power, and no rewards or incentives for the functional departments he needed to support the Pegasus project.

Good practices for organizing engineering programs are described in Chapters 4 and 7. The key ingredients are management involvement; detailed planning, involving all key personnel; clear definition of project work packages (the work and results, schedules, resources, and responsibilities); and the interest and commitment of all project personnel and their management. The major steps of organizing an engineering program such as Pegasus are summarized below:

1. Break project into phases. For Pegasus this would have been:

- Project definition
- Project organization and start-up
- System design
- Engineering design and prototyping
- Test and integration
- Customer delivery and site testing.

2. Use the project definition phase to plan the project in sufficient detail. Get key personnel signed on and involved. Functional managers must get involved and commit the initially needed key personnel. At the end of the project definition a detailed *project plan* should be available with the following elements defined and agreed upon by all personnel and by the customer:

- Project baseline definition, including requirements, specifications, and results
- Project schedules and milestone charts
- Resource requirements
- Individual responsibility identification.

The evolving work breakdown structure is a good tool to provide focus on the various components to be defined.

3. Together with Thermodyne management, define the organization's and management's approach to executing the project. For example, should we assign personnel on a temporary basis (matrix) or projectize some or all of the activities. The

charter for the project manager, the project management guidelines, policies, and procedures, delineate the agreed-upon final ground rules.

4. Recruit key personnel for all level I work packages. In a matrix environment, personnel is assigned by the functional managers. Ideally, the key personnel are the same people who already came on board during the project definition phase.

5. The level I task managers then define their own work packages and sign on their own personnel, similar to the preceding step.

6. Establish project controls.

7. Organize and kickoff the project.

Options of Mr. Russo Last January

It is clear that Mr. Russo's choice for project manager was not Charlie Lanzell. He actually mentioned Stanley Fram as a candidate. However, in spite of his doubt about Lanzell's capabilities, he did not pursue any of the options that were available to him.

First, Russo made the assumption that his superior, Webber, wanted Lanzell to be project manager, an assumption that Russo never tested.

Second, Russo had several options available to him. The least appropriate option was for Russo to stay uninvolved and disinterested. Fundamentally Russo could have done one of the following:

1. Discuss the situation with his boss, Webber, and suggest his choice of project manager.
2. Inform Webber of his plans regarding Pegasus key personnel assignments, wait for reaction, and then do it.
3. Appoint a strong project manager and make Lanzell assistant project manager.
4. Appoint Lanzell as project manager and provide an administrative assistant and/or technical director, such as a project engineer. Although this choice would have been better than what Russo actually did, it seems a weak choice.
5. Appoint Lanzell as project manager and provide very close supervision and help in problem-solving.

APPENDIX 3

PROFESSIONAL SOCIETIES IN ENGINEERING AND TECHNOLOGY MANAGEMENT

Academy of Applied Science, 2 White Street, Concord, NH 03301; (603) 225-2072

Aerospace Electrical Society, Post Office Box 24883, Village Station, Los Angeles, CA 90024

American Association for the Advancement of Science, 1515 Massachusetts Avenue, NW, Washington, DC 20005; (202) 467-4400

American Association of Cost Engineers, 308 Monongahela Building, Morgantown, WVA 26505; (304) 296-8444

American Association of Industrial Management, 2500 Office Center, Maryland Road, Willow Grove, PA 19090; (215) 657-2100

American Chemical Society, 1155 16th Street, NW, Washington, DC 20036

American Institute for Decision Sciences, University Plaza, Atlanta, GA 30303; (404) 658-4000

American Institute of Aeronautics and Astronautics, 1290 Avenue of the Americas, New York, NY 10019

American Institute of Chemical Engineers, 345 East 47th Street, New York, NY 10017

American Institute of Industrial Engineers, 25 Technology Park/Atlanta, Norcross, GA 30092; (404) 449-0460

American Institute of Mining, 345 East 47th Street, New York, NY 10017; (212) 705-7695

American Institute of Physics, 335 East 45th Street, New York, NY 10017; (212) 661-9404

American Institute of Plant Engineers, 3975 Erie Avenue, Cincinnati, OH 45208; (513) 561-6000

American Management Associations, 135 West 50th Street, New York, NY 10020: (212) 586-8100

American Physical Society, 335 East 45th Street, New York, NY 10017; (212) 682-7341

American Society for Engineering Education, 11 Dupont Circle, Suite 200, Washington, DC 20036; (202) 293-7080

American Society for Engineering Management, 301 Harris Hall, University of Missouri–Rolla, Rolla, MO 65401; (314) 341-4560

American Society for Training and Development, 600 Maryland Avenue, SW, Suite 305, Washington, DC 20025; (202) 484-2390

American Society of Civil Engineers, 345 East 47th Street, New York, NY 10017

American Society of Heating, Refrigerating and Air Conditioning Engineers, 1791 Tullie Circle, Northeast, Atlanta, GA 30329; (404) 636-8400

American Society of Mechanical Engineers, 345 East 47th Street, New York, NY 10017

American Society of Metals, Metals Park, OH 44073; (216) 338-5151

American Society of Quality Control, 230 West Wells Street, Suite 7000, Milwaukee, WE 53203; (414) 272-8575.

American Society of Swedish Engineers, c/o Sune Ericson, ASEA Inc., 4 New King Street, White Plains, NY 10604; (914) 428-6000

Association for Computing Machinery, 1133 Avenue of the Americas, New York, NY 10036

Association for Science, Technology and Innovation, Post Office Box 57284, Washington, DC 20037

Association for Women in Science, 1346 Connecticut Avenue, NW, Room 1122, Washington, DC 20036; (202) 833-1998

Association of Consulting Chemists and Chemical Engineers, 50 East 41st Street, Suite 92, New York, NY 10017

Association of Management Consultants, 500 North Michigan Avenue, Chicago, IL 60611; (313) 266-1261

Chinese Institute of Engineers—United States of America, c/o Vincent Chu, 1310 Woodland Circle, Bethlehem, PA 18017; (215) 867-0909

Electro-Chemical Society, 10 South Main Street, Pennington, NJ 08534; (609) 737-1902

Industrial Relations and Research Association, 7226 Social Science Building, University of Wisconsin, Madison, WI 53706; (608) 262-2762

Industrial Research Institute, 100 Park Avenue, Suite 2209, New York, NY 10017; (212) 683-7626

Institute for the Advancement of Engineering, Box 26241, Los Angeles, CA 90026; (213) 413-4036

Institute of Electrical and Electronics Engineers, 345 East 47th Street, New York, NY 10017; (212) 705-7900

Institute of Industrial Engineers, 25 Technology Park/Atlanta, Norcross, GA 30092; (404) 449-0460

Institute of Management Sciences, 146 Westminster Street, Providence, NJ 02903; (401) 274-2525

Instrument Society of America, Post Office Box 12277, 67 Alexander Drive, Research Triangle Park, NC 27709; (919) 549-8411

Insulated Cable Engineers Association, Post Office Box P, South Yarmouth, MA 02664; (617) 394-4424

International Consultants Foundation, 5605 Lamar Road, Bethesda, MD 20816; (301) 320-4409

Internet—International Project Management Association, c/o CRB Switzerland, Zentralstrasse 153, CH-8003 Zurich, Switzerland

National Academy of Engineering, Office of Information, 2101 Constitution Avenue, Washington, DC 20418; (202) 334-2000

National Management Association, 2210 Arbor Boulevard, Dayton, OH 45439; (513) 294-0421

National Society of Professional Engineers, 2029 K Street, NW, Washington, DC 20006; (202) 463-2300

Operations Research Society of America, 428 East Preston Street, Baltimore, MD 21202; (301) 528-4146

Optical Society of America, 1816 Jefferson Place, NW, Washington, DC 20036; (202) 223-8130

Product Development and Management Association, c/o Thomas P. Hustad, Graduate School of Business, Indiana University, Indianapolis, IN 46202-5151

Project Management Institute, Post Office Box 43, Drexel Hill, PA 19026; (215) 622-1796

Robot Institute of America, Post Office Box 930, Dearborn, MI 48128; (313) 271-1500

Society for the Advancement of Management, 135 West 50th Street, New York, NY 10020; (212) 586-8100

Society of Automative Engineers, 400 Commonwealth Drive, Warrendale, PA 15096; (412) 776-4841

Society of Hispanic Professional Engineers, 670 Monterey Pass Road, Monterey, CA 91754; (213) 289-6231

Society of Research Administrators, 1505 Fourth Street, Suite 203, Santa Monica, CA 90401; (213) 393-3137

Technology Transfer Society, 9841 Airport Boulevard, Suite 800, Los Angeles, CA 90045; (213) 410-0295

Ukrainian Engineers' Society of America, 2 East 79th Street, New York, NY 10021; (212) 535-7676

INDEX